Wavelet Theory

Edited by Somayeh Mohammady

Published in London, United Kingdom

IntechOpen

Supporting open minds since 2005

Wavelet Theory
http://dx.doi.org/10.5772/intechopen.87895
Edited by Somayeh Mohammady

Contributors

Joan Nix, Bruce McNevin, Xi Zhang, Indrakshi Dey, Mohamed Tabaa, Safa Saadaoui, Mouhamad Chehaitly, Aamre Khalil, Fabrice Monteiro, Abbas Dandache, Mohammed S Mechee Mechee, Zahir Hussain, Zahrah Ismael Salman, Jüri Majak, Mart Ratas, Kristo Karjust, Boris Shvartsman, Caio Fleming Ferreira De Carvalho Cunha, Mariane Petraglia, André Carvalho, Antonio Lima, Maurice Omane-Adjepong, Imhotep Paul Alagidede, John Dramani Bosco, Karlton Wirsing, Kayupe Kikodio Patrick, Abdesselam Bassou, Nizar Albassam, Dr. Vidhya Lavanya Ramachandran, Sumesh EP, Mohd Fadzli Mohd Salleh, Jamaluddin Zakaria, Nicolae Tudoroiu, Mohamed Zaheeruddin, Roxana-Elena Tudoroiu, Sorin-Mihai Radu, Nikesh Bajaj, Paulo Silva, Shama Siddiqui

© The Editor(s) and the Author(s) 2021
The rights of the editor(s) and the author(s) have been asserted in accordance with the Copyright, Designs and Patents Act 1988. All rights to the book as a whole are reserved by INTECHOPEN LIMITED. The book as a whole (compilation) cannot be reproduced, distributed or used for commercial or non-commercial purposes without INTECHOPEN LIMITED's written permission. Enquiries concerning the use of the book should be directed to INTECHOPEN LIMITED rights and permissions department (permissions@intechopen.com).
Violations are liable to prosecution under the governing Copyright Law.

Individual chapters of this publication are distributed under the terms of the Creative Commons Attribution 3.0 Unported License which permits commercial use, distribution and reproduction of the individual chapters, provided the original author(s) and source publication are appropriately acknowledged. If so indicated, certain images may not be included under the Creative Commons license. In such cases users will need to obtain permission from the license holder to reproduce the material. More details and guidelines concerning content reuse and adaptation can be found at http://www.intechopen.com/copyright-policy.html.

Notice

Statements and opinions expressed in the chapters are these of the individual contributors and not necessarily those of the editors or publisher. No responsibility is accepted for the accuracy of information contained in the published chapters. The publisher assumes no responsibility for any damage or injury to persons or property arising out of the use of any materials, instructions, methods or ideas contained in the book.

First published in London, United Kingdom, 2021 by IntechOpen
IntechOpen is the global imprint of INTECHOPEN LIMITED, registered in England and Wales, registration number: 11086078, 5 Princes Gate Court, London, SW7 2QJ, United Kingdom
Printed in Croatia

British Library Cataloguing-in-Publication Data
A catalogue record for this book is available from the British Library

Additional hard and PDF copies can be obtained from orders@intechopen.com

Wavelet Theory
Edited by Somayeh Mohammady
p. cm.
Print ISBN 978-1-83881-947-7
Online ISBN 978-1-83881-948-4
eBook (PDF) ISBN 978-1-83881-955-2

For EU product safety concerns:
IN TECH d.o.o., Prolaz Marije Krucifikse Kozulić 3, 51000 Rijeka, Croatia, info@intechopen.com or visit our website at intechopen.com.

We are IntechOpen, the world's leading publisher of Open Access books

Built by scientists, for scientists

5,200+
Open access books available

128,000+
International authors and editors

150M+
Downloads

156
Countries delivered to

Our authors are among the
Top 1%
most cited scientists

12.2%
Contributors from top 500 universities

CLARIVATE ANALYTICS BOOK CITATION INDEX INDEXED

WEB OF SCIENCE™

Selection of our books indexed in the Book Citation Index
in Web of Science™ Core Collection (BKCI)

Interested in publishing with us?
Contact book.department@intechopen.com

Numbers displayed above are based on latest data collected.
For more information visit www.intechopen.com

Meet the editor

Dr. Somayeh Mohammady obtained a Ph.D. in Electrical and Electronic Engineering in 2012 from University Putra Malaysia (UPM). Prior to her current work as a certified lecturer, researcher, and publisher at the Technological University of Dublin (TU Dublin), she worked as a postdoc fellow and design engineer. Dr. Mohammady's research interests include digital signal processing (DSP), power efficiency improvement algorithms in wireless telecommunication systems, physical layer security improvement, bandpass filter design using optimization algorithms, the Internet of Things (IoT), and robotics and programmable logic control (PLC) automation systems. Her hobbies include illustrating, yoga, and swimming. She is skilled in communications, Arduino, and Allen-Bradley automation equipment. She is actively engaged in gender equality in Science, Technology, Engineering, and Mathematics (STEM) activities.

Contents

Preface

A wavelet is an important mathematical tool that appears in many fields of science and technology. It refers to analyzing data with special features and different scales depending on the application requirements.

The name "wavelet" first appeared in the early 1980s. It comes from the French word "ondelette," meaning "small wave." The original idea is rooted in many separate thoughts including Jean-Baptiste Joseph Fourier and his theory of approximation in which a complex function can be approximated as a weighted sum of simpler functions, Alfréd Haar and his theory of a sequence of rescaled "square-shaped" functions, and Dennis Gabor and his function for minimizing the deviation in the time and frequency domains.

The application of wavelet theory is rapidly growing in diverse fields and disciplines. As such, this book examines some of the most creative and popular applications including biomedical signal processing, image processing, communication signal processing, Internet of Things (IoT), acoustical signal processing, financial market data analysis, energy and power management, and COVID-19 pandemic measurements and calculations.

My personal interest in wavelet theory lies in its features, which are in contrast to Fourier transform, and its application in converting time domain signals to frequency domain signals and vice versa. The wavelet transform (WT) identifies what frequencies are present in a signal as well as when the signal experiences changes in the time domain, and thus the wavelet has information about where, what scale, and when the change occurred. This makes it very interesting for the study of high peaks in the time domain, which causes distortions in the frequency domain.

The mentioned research topics are also known as peak to average power ratio (PAPR) reduction or crest factor reduction (CRF) techniques. Any improvement of time domain peaks is directly related to the power efficiency of the amplification stage of telecommunication systems. Therefore, the study of the nonlinear behavior and distortion in high power amplifiers (HPA) is relative to the use of wavelets in this field. These topics are relevant to existing and future wireless telecommunication systems such as 5G and beyond.

I would like to thank all my family and friends for their encouragement and support. I also acknowledge that this publication is associated with CONNECT - the Science Foundation Ireland Research Centre for Future Networks and Communications.

Somayeh, Mohammady
School of Electrical and Electronic Engineering,
Technological University Dublin (TU Dublin),
Dublin, Republic of Ireland

Section 1

Wavelet Algorithm and Overview

Chapter 1

Time Frequency Analysis of Wavelet and Fourier Transform

Karlton Wirsing

Abstract

Signal processing has long been dominated by the Fourier transform. However, there is an alternate transform that has gained popularity recently and that is the wavelet transform. The wavelet transform has a long history starting in 1910 when Alfred Haar created it as an alternative to the Fourier transform. In 1940 Norman Ricker created the first continuous wavelet and proposed the term wavelet. Work in the field has proceeded in fits and starts across many different disciplines, until the 1990's when the discrete wavelet transform was developed by Ingrid Daubechies. While the Fourier transform creates a representation of the signal in the frequency domain, the wavelet transform creates a representation of the signal in both the time and frequency domain, thereby allowing efficient access of localized information about the signal.

Keywords: time-frequency analysis, Fourier transform, wavelet transform, signal processing, vanishing moment

1. Introduction

The Fourier transform has been the basis of digital signal processing since the development of the fast Fourier transform in 1965 by Cooley and Tukey in [1]. Its use for analysis goes back much farther with the development of the Fourier transform by Jean Baptiste Joseph Fourier in 1807 as a solution to thermodynamic equations. By using the Fourier transform, we can take any signal and obtain the amplitude of the sinusoids needed to recreate it. Then we can use this information to obtain the power spectrum of the signal, or we modify the amplitudes and take the inverse Fourier transform of the signal, which then filters the signal.

A fundamental limitation of the Fourier transform is that the all properties of a signal are global in scope. Information about local features of the signal, such as changes in frequency, becomes a global property of the signal in the frequency domain. There have been various methods proposed to address this limitation; the main two are the windowed Fourier transform and wavelets.

Gabor [2] created the windowed Fourier transform in 1946. It applies a window function of a short duration to the signal and the Fourier transform is applied to the resulting data. This method is frequently used; however, there are two limitations with this method. The first is that, since the filtering window is constant, it creates problems if the feature is larger or shorter than the window. The second is that the time resolution is the same for high frequencies as it is for low frequencies. Since as frequency increases, so does the rate of change of the signal, higher frequency

IntechOpen

signals can have more information in the same period of time as lower frequency signals, and so require a higher time resolution.

Wavelets overcome both these limitations in that the window is scaled in both time and frequency. The term wavelet was introduced by Ricker [3] in 1940 to describe the limited duration functions that he created to model seismic phenomena. The first wavelet was created earlier, in 1910, by Haar [4] as an alternative to the Fourier transform developed in 1807 by Fourier [5]. Work on the wavelet transform preceded slowly through the twentieth century until the 1980's when work on them increased dramatically with the development of the continuous wavelet transform. In the 1990's, the discrete wavelet transform and its inverse were developed, allowing filtering and compression of signals.

The wavelet transform has many more modes of operation and other options than the Fourier transform. This is one of the key problems with the use of wavelets; we can feel overwhelmed by all the options we have available with them. This chapter will go through some of these options and demonstrate their use.

2. Fourier transform

The Fourier Transform was first published in 1822 by Joseph Fourier [6]. It converts a mathematical function from the time domain to the frequency domain. This enables us to find new properties of the function that would otherwise be hidden. There are several different variations of the Fourier transform equation. In this chapter, we are using the traditional electrical engineering equation

$$\hat{f}(\omega) = \int_{-\infty}^{\infty} f(t)e^{-i\omega t}dt \tag{1}$$

to convert f(t) to the frequency domain.

The Fourier transform itself is for continuous functions. The Discrete Fourier transform was developed for astronomical observations. The goal was to calculate a trigonometric equation for the orbit of an object in the sky based on observations of its ascensions and declinations at various points in time. Most datasets consist of discrete points sampled in time. These can be converted to the frequency domain as well with the discrete Fourier transform. The computational complexity of this is $O(n^2)$.

An interesting note about the Fast Fourier Transform is that it actually predates the Fourier Transform. While the Fast Fourier Transform that we now use was published in a paper by Cooley and Tukey [1] in 1967, a functionally equivalent algorithm was found in an unpublished work by Carl Friedrich Gauss [7] that is presumed to date to 1805. A fascinating history of the Fast Fourier Transform is in [8]. Gauss was computing the discrete Fourier transform of 12 points and noted that the problem could be broken down into subproblems that could simplify the number of steps used [5].

The Fast Fourier Transform reduces the computational complexity of the Discrete Fourier Transform from $O(n^2)$ to $O(n \log_2 n)$. This enables efficient computation of time series. **Table 1** shows how the computational complexity increases for an $O(n^2)$ process versus an $O(n \log_2 n)$ process. The difference grows between the two processes until at 1 million data points, the discrete Fourier transform would require over 50,000 times the amount of time that the Fast Fourier transform would require.

The drawback with the Fourier transform is that all signal information is across the entire range of the transform. As stated in [9], "A local characteristic of the

n	$O(n^2)$	$O(n \log_2 n)$	Ratio
10	100	34	2.94
100	10,000	665	15.04
1000	1,000,000	9966	100.34
10,000	100,000,000	132,878	752.57
100,000	10,000,000,000	1,660,965	6020.60
1,000,000	1E+12	1,9931,569	50171.66

Table 1.
Computational complexity of $O(n^2)$ versus $O(n \log_2 n)$.

Figure 1.
Opening of Beethoven's 5th Symphony [11].

signal becomes a global characteristic of the transform". As illustrated by other authors [10], the best way that this can be explained is by a score of music as shown in **Figure 1**.

The score consists of many different notes, each with a finite duration, each happening at a precise time. A Fourier transform of this signal gives you the average amplitude of the individual frequencies over the entire piece, but obscures the duration and location of the notes. The Fourier power spectrum of music often approximates that of pink (1/f) noise [12]. That information is not lost, since the Fourier transform is reversible, but is encoded in the phase of the Fourier transform.

3. Windowed Fourier transform

In 1946, Gabor [2] proposed the windowed Fourier transform as a way to deal with this problem. In it, a window function of a short duration is applied to the signal and the Fourier transform is taken. This is repeated at different locations in the signal. An example of the use of Hamming window function is shown in **Figure 2**.

One limitation of the windowed Fourier transform is that the window length is constant. When a signal feature is much shorter than the window, information about it can be difficult to extract, since the any local property within the time span of the window becomes a global property of the Fourier transform of the window, as noted previously. Conversely, when a signal feature is larger than the windowing function, information about it spans multiple windows, and can also be difficult to extract.

Another limitation is that the time resolution for the windowed Fourier transform is the same for high frequency signals as it is for low frequency signals. The Heisenberg uncertainty principle states that the time resolution of the window is inversely proportional to the frequency resolution. Since a high frequency signal changes much faster than a low frequency signal, it would be ideal to have a transform with better time resolution for high frequency portions of the signal, and better frequency resolution for lower frequency portions of the signal.

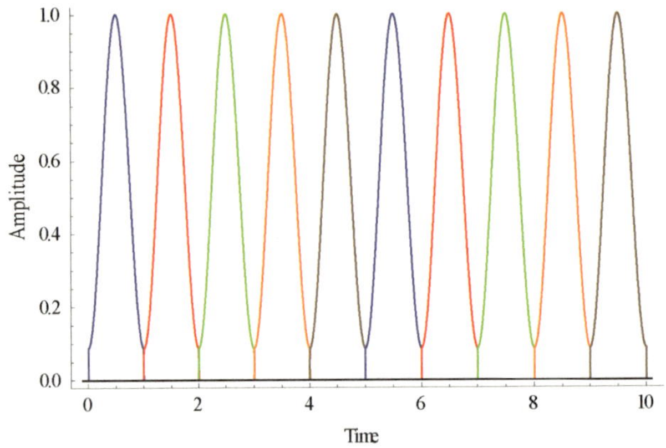

Figure 2.
Multiple Hamming window functions at successive locations in time.

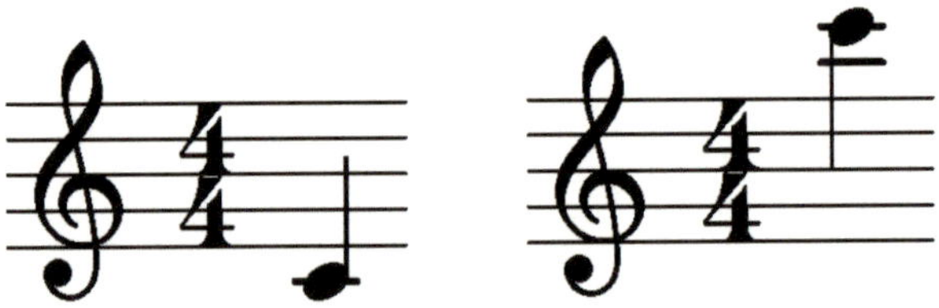

Figure 3.
Musical notation for middle C and C_6.

Going back to the music score, we can see this by looking at two different notes, Middle C, and one that is two octaves higher, called C_6, as shown in **Figure 3**.

The frequency for middle C is 261.63 Hertz, and the frequency for C_6 is 1046.50 [13]. The frequency for C_6 is quadruple the one for Middle C, which means that for every complete cycle of the middle C note, four complete cycles of C_6 have occurred, as shown in **Figure 4**. The windowed Fourier transform would have the limitation that both notes would be treated equally, when the time resolution for C_6 needs to be 4 times that of middle C for analysis purposes.

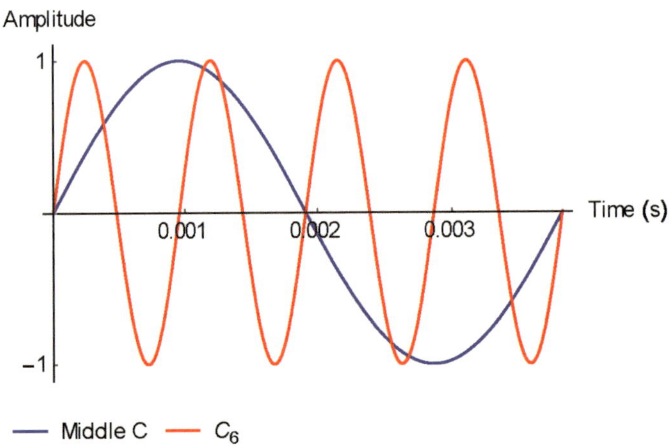

Figure 4.
Amplitude graph for Middle C and C_6.

4. Wavelet transform

The wavelet transform overcomes the limitation of the windowed Fourier transform by scaling the bandwidth of the filter inversely to the frequency. According to [14], while each box of the windowed Fourier transform has the same bandwidth, each level of the wavelet transform has the same Q as defined as

$$Q \equiv \frac{\Delta f}{f_0}. \quad (2)$$

This gives the transform the desired time resolution for the higher frequency portions of the signal and the desired frequency resolution for lower frequency portions.

5. Continuous wavelet

The continuous wavelet has a long history spanning from the 1940's to present. In 1940, Norman Ricker first proposed the term wavelet and various mathematical functions to model seismic waves as they traveled through the Earth's crust in [3]. He further refined this in a series of papers [15–17]. This was the first continuous wavelet. The functions in the time domain are given by

$$\psi_{f_m}(t) = \left(1 - 2\pi^2 f_m{}^2 t^2\right) e^{-\pi^2 f_m{}^2 t^2}, \quad (3)$$

called the three-loop equation, and

$$\psi_{f_m}(t) = t\, e^{-2\pi^2 f_m{}^2 t^2}, \quad (4)$$

called the two-loop equation [18]. Graphs for both of these are in **Figure 5**.

The next development for continuous wavelets was in the 1980's by Grossman and Morlet, and expanded on by Stephen Mallat and others [19]. The term continuous wavelet refers to the fact that it can be scaled to any time scale. Discrete wavelets can only use specific time scales, usually a power of 2.

Wavelet analysis centers around the use of a wavelet function, also called the mother function in literature, traditionally represented by the Greek letter upsilon (ψ). A key requirement is that it has finite energy, i.e.

$$\int |\psi(t)|^2 dt < \infty. \quad (5)$$

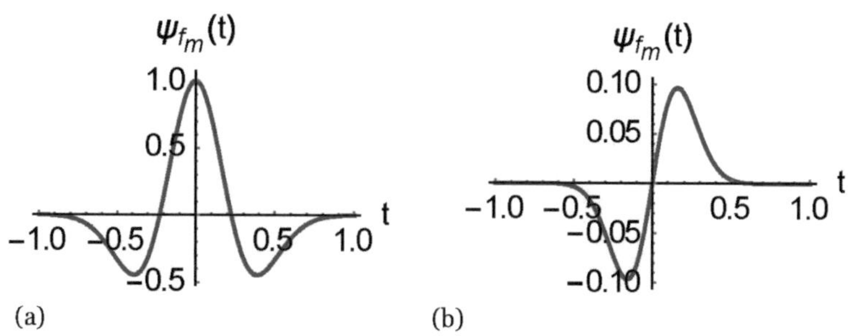

Figure 5.
Three loop (a) and two loop (b) Ricker wavelet equation with $f_m = 1$.

The energy of the wavelet function is usually one. Functions such as sine and cosine cannot be used as analyzing functions, because they violate this condition by having infinite energy. There is an implicit requirement that, while it has finite energy, it must have some energy, so the integration of the function must be greater than zero.

The second requirement is known as the admissibility condition, which states that the Fourier transform of the wavelet function cannot have a zero-frequency component, i.e.

$$C_g = \int \frac{|\hat{\psi}(f)|^2}{|f|} df < \infty \tag{6}$$

This can only be satisfied if $\hat{\psi}(f) = 0$, however this requirement is not absolute. The Gabor wavelet is a complex wavelet that violates the admissibility condition. The Morlet wavelet is a real valued wavelet that has a small but greater than zero value for the zero-frequency value for its Fourier transform.

A third condition is usually that the wavelet function must have zero mean, which means that it must oscillate, hence be a wavelet. Mathematically this is [20]

$$\int \psi(t)\, dt = 0. \tag{7}$$

Another condition is that the wavelet function has effective support. While the wavelet functions for the continuous wavelets are usually mathematical functions that extend to infinity, effective support means that the wavelet functions are effectively zero outside of a certain range. Since the continuous wavelet functions asymptotically approach 0 as x goes to either ∞ or -∞, the choice of the boundary of this range is a bit arbitrary and can vary from paper to paper.

6. Continuous wavelet transform

Morlet and Grossman formalized the continuous wavelet transform in 1984 in [21]. For the continuous wavelet transform, the wavelet function itself is shifted in time and is scaled to do the wavelet transforms [22] as the following equation illustrates:

$$\psi_{a,b}(t) = \frac{1}{\sqrt{|a|}} \psi\left(\frac{t-b}{a}\right), a, b \in R, a \neq 0. \tag{8}$$

The continuous wavelet transform is defined as the integration of the function to be analyzed with the complex conjugate of the wavelet function:

$$CWT\{f(t), a, b\} = \int_{-\infty}^{\infty} f(t)\psi^*_{a,b}(t)dt \tag{9}$$

In some papers such as [22], you will see the definition of the continuous wavelet transform without the complex conjugate definition. Since most wavelet functions are real valued and not complex, both definitions are equivalent, since the complex conjugate of a real number is equal to that number. The difference only comes up when the wavelet function is complex, such as the Gabor wavelet.

An alternate formula for the continuous wavelet transform is

$$W_n(s) = \sum_{n'=0}^{N-1} x_{n'} \psi^* \left[\frac{(n' - n)\delta t}{s} \right], \tag{10}$$

where $W_n(s)$ is the transformed sequence, $x_{n'}$ is the original sequence, and ψ^* is the complex conjugate of the analyzing wavelet function, n represents the time shift or dilation, and s represents the scale. Usually the time shift is calculated over the total number of data points of the function, and s goes over the scales that are being analyzed to give a two-dimensional picture of the data [23].

7. Discrete wavelets

The first discrete wavelet was created in 1910 by Alfred Haar as an alternative to the Fourier transform. This consists of two functions as shown in **Figure 6**, one a scaling function and a wavelet function. The scaling function is the unit step function and the wavelet function consists of offsets from that.

One of the drawbacks of the continuous wavelet transform is that it creates a lot of redundant data, since the coefficients between the scales are highly correlated. Ingrid Daubechies developed the theory of discrete wavelets in 1988, which generates compact data by eliminating the redundancy. Daubechies created an entire family of wavelet functions with the Haar wavelet forming the first level of the Daubechies wavelet.

The wavelet function for discrete wavelets is modified to

$$\psi_{j,k}(t) = \frac{1}{\sqrt{s_0^j}} \psi\left(\frac{t - k\tau_0 s_0^j}{s_0^j} \right), j, k \in Z, s_0 > 1, \tag{11}$$

where s_0 is the scale of the wavelet, usually 2 [20]. This condition as well as the condition that j and k are integers restricts the wavelet to only certain scales. The wavelet function has the properties of finite energy, oscillation, and the admissibility condition of the continuous wavelets, as well as the properties of compact support, vanishing moments, and orthogonality.

Compact support means that the wavelet function is defined by a series of coefficients over a finite region, and is zero at all other places. This contrasts with the continuous wavelets, which, as mentioned, are mathematical functions and have effective support in which the function continues to infinity, but is effectively zero outside of a finite range.

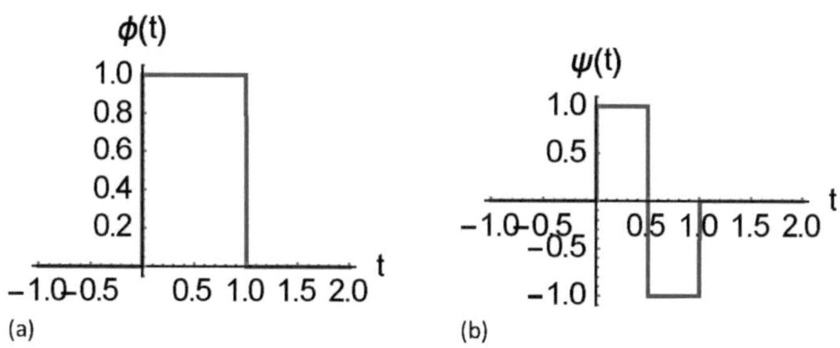

Figure 6.
Scaling (a) and wavelet (b) functions for Haar wavelet.

Vanishing moments are obtained when the following condition defined mathematically as

$$\int x^k \psi(x)dx = 0 \tag{12}$$

holds true for all integers $0 \leq k < N$, where N is the number of vanishing moments of the function [24]. This property is useful for analyzing functions that have an additive polynomial trend function given by

$$f(x) = g(x) + N(x). \tag{13}$$

Here, g(x) is the function to be analyzed and N(x) is the polynomial trend function (also termed a nuisance function in Economics).

The orthogonality condition removes the redundancy of the continuous wavelet transform. As stated earlier, the discrete wavelet transform can only be used at certain scales, most often a power of 2. Mathematically it is stated as

$$\int \psi_{j,k}(t)\psi^*_{m,n}(t)dt = \begin{cases} 1 & \text{if } j = m \text{ and } k = n \\ 0 & \text{otherwise} \end{cases}. \tag{14}$$

An orthogonal basis ensures that the signal is represented in the most compact way possible. However, by removing all the redundant information, this also removes information to handle shift variance. The exact same function sampled at two different places can yield very different results. In order to deal with this, some discrete wavelet transforms retain some of this redundant information.

Each wavelet of the discrete wavelet family consists of two functions, a wavelet function (ψ), as in the continuous wavelet families, and also a new function called a scaling function (ϕ). In literature, these are termed the mother and father functions respectively. The scaling function has its own admissibility condition, which ensures that it has the zero-frequency component that the wavelet function does not:

$$\int \phi(x)dx = 1. \tag{15}$$

This is necessary so that a discrete wavelet transform terminates in a finite number of steps and can completely regenerate the information in the signal [20]. Otherwise, the zero-frequency component could never be captured, since no amount of scaling value can cause the wavelet filter to have a zero-frequency component.

In addition, as specified in [25] the scaling equation is defined in terms of a finite set of coefficients p_k that are defined by the following equation

$$\phi(x) = \sum\nolimits_k p_k \phi(2x - k), \tag{16}$$

that adheres to the following conditions as specified in [25] as well:

$$P(z) = (1/2)\sum\nolimits_k p_k z^k, \tag{17}$$

$$P(1) = 1 \text{ (Averaging Property)}, \tag{18}$$

$$|P(z)|^2 = |P(-z)|^2 = 1 \text{ for } |z| = 1, \text{ and} \tag{19}$$

$$\left|P(e^{it})\right| > 0 \text{ for } |t| \leq \pi/2 \tag{20}$$

The wavelet function is defined by

$$\psi(x) = \sum_{k \in Z} (-1)^k \overline{p_{l-k}} \phi(2x - k), \tag{21}$$

where l is the length of the set of coefficients, so that the wavelet coefficients are basically the scaling coefficients in reverse order with alternating signs. These coefficients are used to implement the discrete wavelet transform as a filter bank of Finite Impulse Response (FIR) filters. Graph of the scaling and wavelet functions for Daubechies level 2 wavelet are shown in **Figure 7** and the frequency response is shown in **Figure 8**. As with the Haar wavelet, the wavelet function is a high pass filter and the scaling function is a low pass filter. Both are symmetric around $\pi/2$.

Different papers and software implementations have different coefficients for the Haar and Daubechies wavelet, depending on how they are normalized and whether the scale parameter from Eq. (8) is included is included in the filter. The coefficients for the Haar and the Daubechies level 2 wavelet are in **Tables 2** and **3** with b defined by the implementation. Mathematica uses 2 for b, which would normalize the sum of the coefficients to 1. PyWavelets uses $\sqrt{2}$ for b. In any implementation, the filter coefficients for the wavelet filter are the coefficients for the scaling filter in reverse order with every other coefficient multiplied by -1.

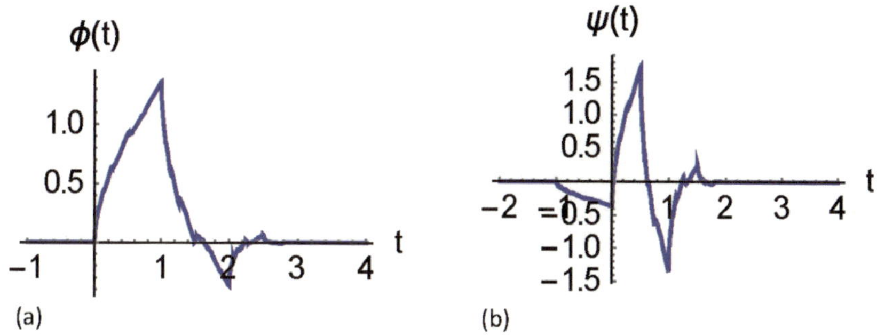

Figure 7.
Scaling (a) and Wavelet (b) functions for Daubechies level 2 wavelet.

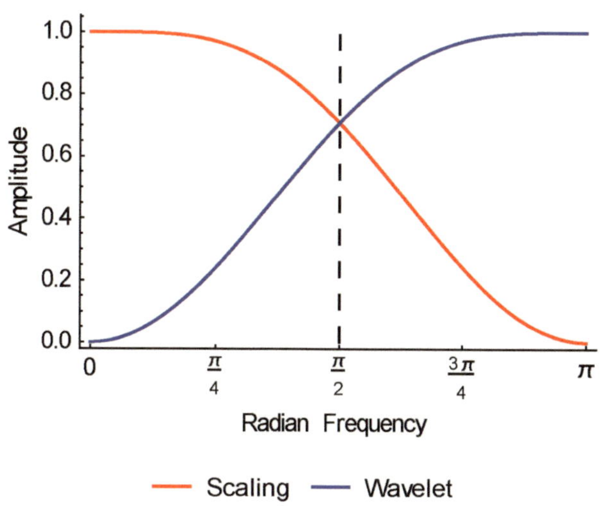

Figure 8.
Frequency response of scaling function (red) and wavelet function (blue) for Daubechies level 2 wavelet.

	Scaling coefficients		Wavelet coefficients
c_0	1/b	d_0	1/b
c_1	1/b	d_1	−1/b

Table 2.
Coefficients for Haar scaling and wavelet functions.

	Scaling coefficients		Wavelet coefficients
c_0	$\frac{1+\sqrt{3}}{4b}$	d_0	$\frac{1-\sqrt{3}}{4b}$
c_1	$\frac{3+\sqrt{3}}{4b}$	d_1	$\frac{-3+\sqrt{3}}{4b}$
c_2	$\frac{3-\sqrt{3}}{4b}$	d_2	$\frac{3+\sqrt{3}}{4b}$
c_3	$\frac{1-\sqrt{3}}{4b}$	d_3	$\frac{-1-\sqrt{3}}{4b}$

Table 3.
Coefficients for Daubechies level 2 scaling and wavelet functions.

8. Discrete wavelet transform

The class of discrete wavelet functions has many transforms available with the discrete wavelet transform in **Figure 9** the most common. Since this was the transform introduced with the Haar wavelet, it is sometimes referred to as the Haar transform [26] as well as the decimated wavelet transform [10]. Essentially, it works as a pyramid algorithm, where the number of coefficients of each lower level is roughly twice that of the preceding level, but each coefficient is influenced by half as much of the data set as the preceding level. Each level has two sets of coefficients, one is called coarse and the other is called details.

In **Figure 9**, g is the scaling filter defined by the set of scaling filter coefficients and h is the wavelet filter defined by the set of wavelet filter coefficients. At each level, the detail coefficients (W) are outputs, except for the final level, where the coarse coefficients (V) are given as outputs as well. Collectively, this set of coefficients contains enough information to reconstruct the signal perfectly.

One key part of the discrete wavelet transform is the down sampling operator, which is a function that removes every other position from a sequence. An example would be the sequence {a, b, c, d, e, f, g, h} would be {a, c, e, g} or {b, d, f, h}, after

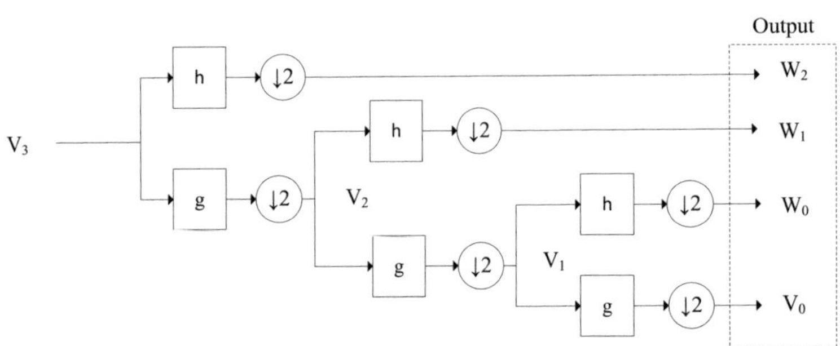

Figure 9.
Diagram of a three-level discrete wavelet transform.

the down sampling operator is applied, depending on whether the even or the odd positions are eliminated. Both are valid, however, by convention with the discrete wavelet transform, the even positions are eliminated, leaving only the odd positions. The down sampling operator is what makes the discrete wavelet transform a pyramid function and also reduces the set of coefficients to the minimum amount necessary to reconstruct the signal.

A problem with the decimation operator is aliasing. This is when different sequences map to the same sequence after the application of the operator. An example would be that the sequences {a, b, c, d, e, f, g} and {a, h, c, i, e, j, g} would both map to the sequence {a, c, d, g}. Therefore, just given the sequence {a, c, d, g}, it would be impossible to reconstruct the original. The filters of the discrete wavelets are designed to compensate for this, ensuring that the original sequence can be recovered. The combination of these filters with the down sampling operator is referred to as decimation.

The discrete wavelet transform also has an inverse transform. This process combines as described in **Figure 10** to form a perfect reconstruction of the signal, where $\tilde{g}$ is the inverse scaling filter coefficients and $\tilde{h}$ is the inverse wavelet filter coefficients. Just as the discrete wavelet transform had the decimation operator, the inverse transform has the upsampling operator. This takes a sequence and inserts 0 at every other position. For example, the sequence {a, c, d, g} would be {a, 0, c, 0, d, 0, g, 0} after the operator is applied.

Implementing the discrete wavelet transform as a finite impulse response filter and using decimation gives it a computational complexity of O(n). As **Table 4** shows, an O(n) process can be much faster than an O(n $\log_2$ n) process such as the fast Fourier Transform. At 1 million samples, an O(n) process requires almost 20 times less operations than an O(n $\log_2$ n) process (**Table 4**).

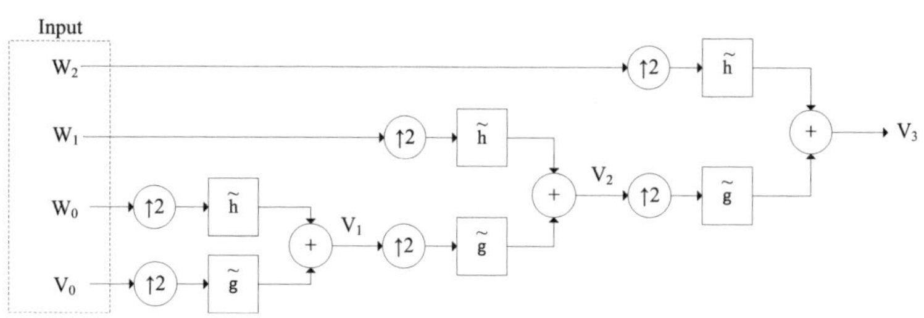

Figure 10.
Diagram of a three-level inverse discrete wavelet transform.

n	O(n $\log_2$ n)	O(n)	Ratio
10	34	10	3.40
100	665	100	6.65
1000	9966	1000	9.97
10,000	132,878	10,000	13.29
100,000	1,660,965	100,000	16.61
1,000,000	19,931,569	1,000,000	19.93

Table 4.
Computational complexity of O(n $\log_2$ n) versus O(n).

9. Stationary wavelet transform

Another wavelet transform for discrete wavelet functions is the stationary wavelet transform, also known as the undecimated discrete wavelet transform. Essentially the stationary wavelet transform is the discrete wavelet transform without the decimation operation for the data. Whereas the number of coefficients for each level is half that of the preceding level in the discrete wavelet transform, the number of coefficients is the same for each level in the stationary wavelet transform.

The procedure is diagrammed in **Figure 11**, where g_n is the set of the scaling filter coefficients and h_n is the set of the wavelet filter coefficients. The reason that the scaling filter and wavelet filter coefficients are different for each level is that instead of the decimation operator being applied to the wavelet data coefficients after each level, the upsampling operator is applied to the wavelet and scaling filter coefficients. The wavelet and scaling coefficients for each level are upsampled from the previous level, as shown in **Figure 12**.

Like the discrete wavelet transform, the stationary wavelet transform has an inverse transform, as shown in **Figure 13**. The difference between this and the inverse discrete wavelet transform is the absence of the upsampling operator. As with the stationary wavelet transform, the filter coefficients for the inverse stationary wavelet transform are changed instead of the data. In this case, the filters are down sampled. The retention of redundant data in the stationary wavelet

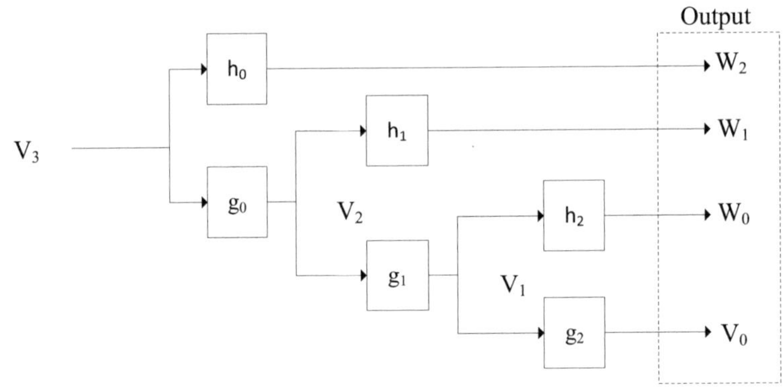

Figure 11.
Diagram of a three-level stationary wavelet transform.

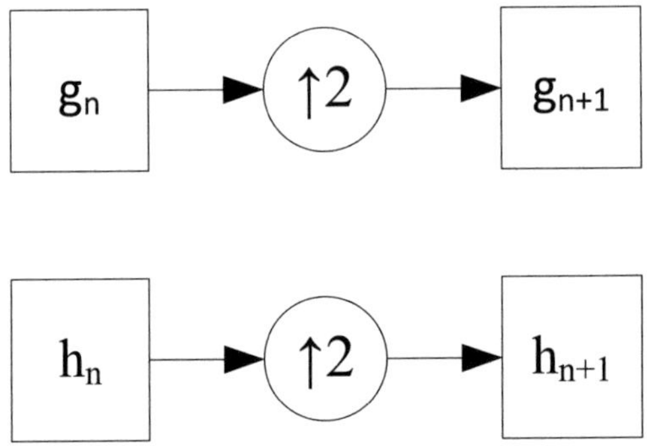

Figure 12.
Diagram of a filter upsampling for the stationary wavelet transform.

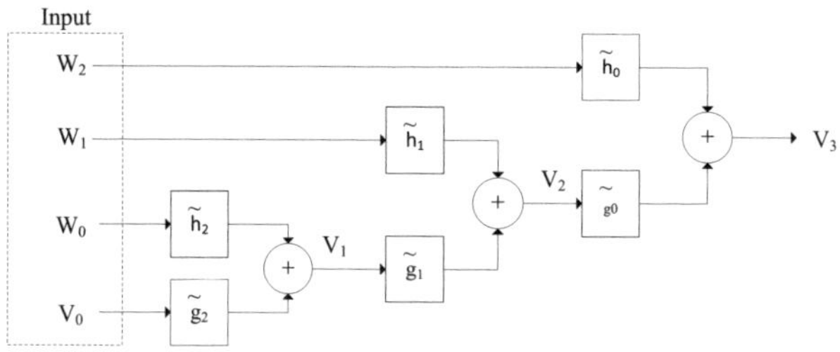

Figure 13.
Diagram of a three-level inverse stationary wavelet transform.

transform helps to make it translation invariant, which is useful for filtering applications (**Figure 13**).

Since the decimation step is not used, the stationary wavelet transform has a computational complexity of $O(n \log_2 n)$, the same as the Fast Fourier Transform. However, there is also memory complexity to consider. While the Fast Fourier Transform and the Discrete Wavelet Transform has an $O(n)$ memory complexity, the stationary wavelet transform has an $O(n \log_2 n)$ memory complexity. Therefore, the output will always be larger than the input.

10. Discrete wavelet packet transform

The two previous transforms applied the detail and the coarse filters to the data at each level. The output of the coarse filter is given as the input to the next level and the output of the detail filter at that level is included in the set of the outputs of the transform. In the final level, the output of both the detail and the coarse filters were included in the set of outputs of the transform; however, that is not the only possibility. The packet transform creates a binary tree where the detail and coarse filters are applied to each node, diagrammed in **Figure 14**. The output of the detail

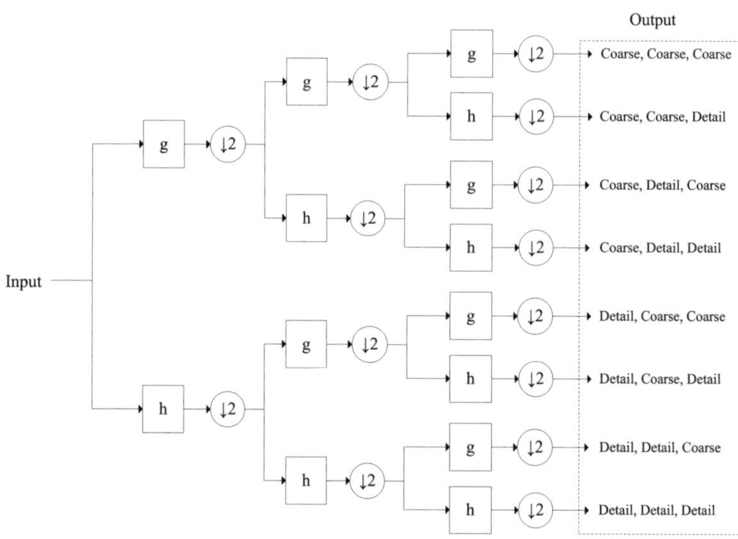

Figure 14.
Diagram of a three-level discrete wavelet packet transform.

filter becomes one child and the output of the coarse filter becomes the other. This process is repeated until the final level is reached, creating a set of output coefficients where each set is identified by the sequence of filters applied to it.

11. Stationary wavelet packet transform

The stationary wavelet packet transform is yet another transform for discrete wavelet functions. Basically, it combines the stationary wavelet transform with the wavelet packet transform, as diagrammed in **Figure 15**. Instead of the decimation operator, the filters themselves are upsampled for each level. The transform creates a binary tree, as with the discrete wavelet packet transform, where both filters for each level are applied at each node. As with the wavelet packet transform, the output from the detail filter becomes one child and the output from the scaling filter becomes the other, and the process is repeated until the final level is reached. Each set of output coefficients are also identified by the sequence of filters applied to it, with the difference that since there is no decimation applied between levels, the number of each set of output coefficients is the same as the input data. This leads to the total number of output coefficients to be 2 times the number of levels multiplied by the length of the input data. Both the discrete wavelet packet transform and the stationary wavelet packet transform have inverse transforms.

The wavelet packet transform introduces many more possibilities for use, some of which are discussed here. Depending on the application, you can do different combinations of the scaling and wavelet filters. Computational complexity depends on the filter combinations selected. If it is taken to the maximum level with the maximum filter combinations, then the discrete wavelet packet transform has a complexity of $O(n \log_2 n)$ and the stationary wavelet packet transform has a complexity of $O(n^2)$.

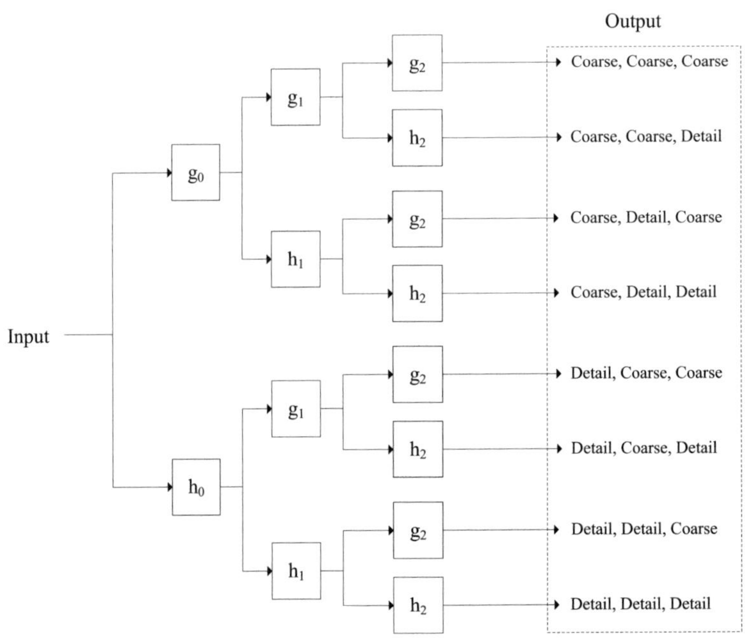

Figure 15.
Diagram of a three-level stationary wavelet packet transform.

12. Conclusions

The Fast Fourier Transform has been listed as one of the top algorithms of the 20th century [27]. Its development has been instrumental to digital signal processing. However, recently a new algorithm, the wavelet transform, has started to have a significant impact on digital signal processing. The wavelet transform improves on the Fourier Transform in that it can analyze a signal by time and frequency simultaneously, thereby easily recovering localized signal information. This is key to many applications, including fractal and multifractal analysis, compression, and filtering.

The wavelet transform introduces many possibilities for use and this chapter has only touched the surface of it. Different wavelets can be used and the transform itself can be customized to fit the application as shown with the wavelet packet transform. Future research will be to determine the proper combination of features for various applications. In addition, there are other possibilities, such as the lifting wavelet transform, which wasn't covered in this chapter. Only orthogonal wavelets that use the same set of wavelets for the forward and inverse transform were covered in this chapter. Biorthogonal wavelets that use different wavelets for the forward and inverse transforms are also available.

The key to wavelet compression and filtering is the sparse signal representation generated by the wavelet transform. The wavelet transform can reduce a signal to minimal set of coefficients. Coefficients that are near zero can be rounded to zero, reducing the size of the signal. In addition, fractional parts of the coefficients can be rounded, also reducing signal size. One of the first uses of this was to compress fingerprints for the FBI [28]. As stated in [29], in the 1990's the FBI had 25 million cards, each containing 10 fingerprints. Digitized, each card contained 10 megabytes of information, for a total of 250 terabytes. Using the two-dimensional discrete wavelet packet transform gives a compression ratio of 20 to 1, enabling the archive to be stored on approximately 12.5 terabytes, while still being able to search and match unknown fingerprints against the ones in the archive. The recently developed JPEG format at the time was based on using the discrete cosine transform on blocks of the image, which left unacceptable artifacts in the image.

JPEG 2000 was developed using the two-dimensional wavelet transform to be the successor to JPEG, although it hasn't caught on. JPEG 2000 allows both lossy and lossless compression. It also doesn't have the lossy artifact generation that the JPEG format has as mentioned previously. Both lossy and lossless compression use the discrete wavelet transform, the difference is that the lossless one uses a wavelet transform that is reversible, while the lossy one uses a wavelet transform that introduces quantization noise that making it irreversible.

Compressed sensing deals with the fact that we that we can obtain a vast amount of information and a lot of it can be discarded and still retain what is relevant. As stated in [24], "singularities and irregular structures often carry the most important information in signals." This is due to the fact that they represent changes to one or more of the properties of the signal. An example of this would be the edges in an image. Compressed sensing removes the redundant, unnecessary information from a signal and analyzes the remaining part of the signal. This is an ideal application for the wavelet transform.

The discrete wavelet transform has been used for Iris recognition for biometric identification in patent US 20020150281A1 [30]. After taking a picture of the eye, the iris is extracted from the image and then converted to polar coordinates. Using the discrete wavelet transform, the high frequency components are extracted, which are the detail coefficients as referenced in this paper. These form the characteristic vector that is used to identify an iris from the previously recorded data.

The wavelet transform can provide an efficient way to filter white noise from a signal. The procedure consists of applying one of the discrete wavelet transforms to the data and then executing a threshold algorithm that modifies the detail coefficients. After the coefficients are modified, then the inverse transform is applied; the resulting output is a representation of the signal with the noise component significantly reduced.

There are numerous packages available for experimenting with the wavelet transform. The discrete and stationary wavelet transforms are available in Mathematica, Maple, Matlab, R, and PyWavelets to name a few, with the wavelet packet transform available in Mathematica, Matlab, and PyWavelets.

The wavelet transform provides many possibilities for signal analysis depending on the application. A few potential applications were touched on here. The reader is encouraged to develop their own uses and applications for the wavelet transform.

Notes

A lot of this has been previously published under my Master's thesis [31].

Author details

Karlton Wirsing
Virginia Polytechnic and State University, Manassas, Virginia, United States

*Address all correspondence to: kwirsing@vt.edu

IntechOpen

© 2020 The Author(s). Licensee IntechOpen. This chapter is distributed under the terms of the Creative Commons Attribution License (http://creativecommons.org/licenses/by/3.0), which permits unrestricted use, distribution, and reproduction in any medium, provided the original work is properly cited.

References

[1] Cooley JW, Tukey JW. An Algorithm for the Machine Calculation of Complex Fourier Series. *Mathematics of Computation*. 1965;**19**(90):297-301

[2] Gabor D. Theory of communication. Part 3: Frequency compression and expansion. *Electrical Engineers - Part III: Radio and Communication Engineering, Journal of the Institution of*. 1946;**93**(26): 445-457

[3] Ricker NH. The form and nature of seismic waves and the structure of seismograms. *Geophysics*. October 1, 1940;**5, 1940**(4):348-366

[4] Haar A. Zur Theorie der orthogonalen Funktionensysteme. *Mathematische Annalen*. 1910;**69**(3): 331-371

[5] Briggs WL, Henson VE. *The DFT : an owner's manual for the discrete Fourier transform*. Philadelphia: Society for Industrial and Applied Mathematics; 1995 xv, 434 p

[6] J. B. J. Fourier, *Théorie analytique de la chaleur*. 1822.

[7] C. F. Gauss, "Nachlass, Theoria Interpolationis Methodo Nova Tractata," in *Carl Friedrich Gauss Werke, Band 3, Königlichen Gesellschaft der Wissenschaften* Göttingen, 1866, pp. 265-330.

[8] Heideman MT, Johnson DH, Burrus CS. Gauss and the History of the Fast Fourier Transform. *Archive for History of Exact Sciences*. 1985;**34**(3): 265-277

[9] Hubbard BB. *The world according to wavelets : the story of a mathematical technique in the making*. Wellesley, Mass: A.K. Peters; 1998

[10] D. L. Fugal, *Conceptual wavelets in digital signal processing : an in-depth, practical approach for the non-mathematician*, 1st ed. San Diego, Calif.: Space & Signals Technical Pub., 2009, pp. xxii, 302, 50 p.

[11] L. v. Beethoven, "Symphonie No. 5, Op. 67," E. Pauer, Ed., ed: B. Schott's Söhne, 1808.

[12] R. F. Voss and J. Clarke, ""1/f noise" in music: Music from 1/f noise," *The Journal of the Acoustical Society of America,* vol. 63, no. 1, pp. 258-263, 1978.

[13] B. H. Suits. (1998, September 2, 2013). Physics of Music - Notes. Available: http://www.phy.mtu.edu/~suits/notefreqs.html

[14] J. Y. Stein, Digital signal processing : a computer science *perspective* (Wiley series in telecommunications and signal processing). New York: Wiley, 2000, pp. xx, 859 p.

[15] N. Ricker, "Further developments in the wavelet theory of seismogram structure," *Bulletin of the Seismological Society of America,* vol. 33, no. 3, pp. 197-228, July 1, 1943 1943.

[16] Ricker N. Wavelet functions and their polynomials. *Geophysics*. July 1, 1944;**9, 1944**(3):314-323

[17] Ricker N. The form and laws of propagation of seismic wavelets. *Geophysics*. January 1, 1953;**18, 1953**(1): 10-40

[18] Hosken J. Ricker wavelets in their various guises. *First Break*. January 1988; **6**:24-33

[19] Graps A. An introduction to wavelets. *Computational Science & Engineering, IEEE*. 1995;**2**(2):50-61

[20] C. Valens, "A Really Friendly Guide to Wavelets," Available: http://www.rob

ots.ox.ac.uk/~parg/mlrg/papers/arfgtw.pdf

[21] Grossmann A, Morlet J. Decomposition of Hardy Functions into Square Integrable Wavelets of Constant Shape. *SIAM Journal on Mathematical Analysis*. 1984;**15**(4):723-736

[22] Y. Yu, J. Zhou, Y. Wang, F. Li, and C. Ge, "On the computation of wavelet series transform," in *Signal Processing Proceedings, 1998. ICSP '98. 1998 Fourth International Conference on*, 1998, pp. 313-316 vol.1.

[23] C. Torrence and G. P. Compo, "A Practical Guide to Wavelet Analysis," *Bulletin of the American Meteorological Society,* vol. 79, no. 1, pp. 61-78, 1998/01/01 1998.

[24] Mallat S, Hwang WL. Singularity detection and processing with wavelets. *IEEE Trans. Inf. Theor.* 1992;**38**(2): 617-643

[25] A. Boggess and F. J. Narcowich, *A first course in wavelets with Fourier analysis*, 2nd ed. Hoboken, N.J.: John Wiley & Sons, 2009, pp. xv, 315 p.

[26] Strang G, Nguyen T. *Wavelets and filter banks*. Wellesley. MA: Wellesley-Cambridge Press; 1996. p. xxi 490 p

[27] Rockmore DN. The FFT: an algorithm the whole family can use. *Computing in Science & Engineering*. 2000;**2**(1):60-64

[28] Brislawn CM. The FBI Fingerprint Image Compression Specification. In: Topiwala PN, editor. *Wavelet Image and Video Compression*. Boston, MA: Springer US; 1998. pp. 271-288

[29] "Wavelets, Fingerprints, and Image Processing," in *Discovering Wavelets*, 1999, pp. 1-21.

[30] S.-W. Cho, "Method of recognizing human iris using daubechies wavelet transform" USA Patent US 2002/0150281 A1, 2001.

[31] K. Wirsing, "Application of Wavelets to Filtering and Analysis of Self-Similar Signals," Master of Science, Electrical and Computer Engineering, Virginia Polytechnic Institute and State University, 2014.

Chapter 2

Wavelet Theory: Applications of the Wavelet

Mohammed S. Mechee, Zahir M. Hussain and Zahrah Ismael Salman

Abstract

In this Chapter, continuous Haar wavelet functions base and spline base have been discussed. Haar wavelet approximations are used for solving of differential equations (DEs). The numerical solutions of ordinary differential equations (ODEs) and fractional differential equations (FrDEs) using Haar wavelet base and spline base have been discussed. Also, Haar wavelet base and collocation techniques are used to approximate the solution of Lane-Emden equation of fractional-order showing that the applicability and efficacy of Haar wavelet method. The numerical results have clearly shown the advantage and the efficiency of the techniques in terms of accuracy and computational time. Wavelet transform studied as a mathematical approach and the applications of wavelet transform in signal processing field have been discussed. The frequency content extracted by wavelet transform (WT) has been effectively used in revealing important features of 1D and 2D signals. This property proved very useful in speech and image recognition. Wavelet transform has been used for signal and image compression.

Keywords: Haar wavelet, continuous wavelet function, wavelet transform, B-cubic spline base

1. Introduction

Wavelets are special mathematical functions which have advantages over traditional Fourier methods in analyzing physical situations where the signal contains discontinuities and sharp spikes. The fields of applied mathematics such as quantum physics, seismic geology and electrical engineering have used and developed independently wavelets during last twenty years ago which leads to new wavelet applications such as image compression, radar, and earthquake prediction. Haar wavelet was initiated and independently developed by some authors. Wavelets can be summarized as a family of functions constructed from transformation and dilation of a single function called mother wavelet. From various types of continuous and discrete wavelets, Haar wavelet is the discrete type of wavelet which was first proposed and the first orthonormal wavelet basis is the Haar basis. Differential equations (DEs) are most important tools in mathematical models for physical phenomena. Many basis used to approximate the solutions of DEs. Haar wavelet is simple basis used to approximate the solution of DEs. [1] established a simple numerical method based on Haar wavelet operational matrix of integration for solving two dimensional elliptic partial differential equations (PDEs) of the form $\nabla^2 u(x,y) + ku(x,y) = f(x,y)$, [2]

IntechOpen

used Haar wavelet operational matrix for the numerical solutions of FrDEs, [3] used Haar wavelet-quasi linearization technique for FrDEs, [4] used Haar wavelet method for solving FrPDEs numerically, [5] applied Haar wavelet transform to solve integral equations (IEs) and DEs, [6] solved 2D and 3D Poisson equations and biharmonic equations by the Haar wavelet method while [7] presented a numerical method for inversion of Laplace transform using the method of Haar wavelet operational matrix. The implementations of FrDEs which are used as mathematical models in many physically significant fields and applied science. Recently, the approximated solutions of the FrDEs have been studied using Haar wavelet method which shows to be more suitable to approximate the solutions of them. Nowadays, Haar wavelets are most widely and simplest due to their simplicity, the Haar wavelets are effective tools for approximating solutions of DEs. When this type of problem arises, mainly approximated solutions come to be available. From the many approximated methods, Haar wavelet approach is one to find the solutions of DEs. If the approximated solution gives less error than other methods, then, the method be an efficient method. However, one of interesting applications of wavelets bases is the approximation of DEs. Also, Haar wavelet technique is used to approximate the solutions of DEs of fractional-order. Wavelet transform is a mathematical approach widely used for signal processing applications. It can decompose special patterns hidden in mass of data. Wavelet transform has the advantage to simultaneously display functions and manifest their local characteristics in time-frequency domain. Wavelet transforms have had tremendous impact on the fields of signal processing, signal coding, estimation, pattern recognition, applied sciences, process systems engineering, econometrics, and medicine. Wavelet transforms are mainly divided into two groups; continuous wavelet transform (CWT) and discrete wavelet transform (DWT). The discretization of a voice transform generated by a representation of the Blaschke group on the Hardy space of the unit disk leads to the construction of analytic rational orthogonal wavelets. In this chapter, we introduce the concept of continuous wavelet functions together to the approximations solutions of DEs with ordinary- or fractional-order using Haar wavelet functions. Also, a comparison between Haar wavelet base with cubic spline base has been introduced. Wavelet transform as a mathematical approach has been discussed together to the applications of wavelet transform in signal.

1.1 Objectives of chapter

This chapter aims at achieving the following objectives:

1. To introduce continuous wavelet functions.

2. To use Haar wavelet approximations in solving of differential equations (DEs).

3. To imply Haar wavelet functions to approximate the solutions of DEs of fractional-order.

4. To compare Haar wavelet base with cubic spline base.

5. To study wavelet transform as a mathematical approach.

6. To discuss the discrete wavelet transform (DWT).

7. To study the applications of wavelet transform in signal processing field.

1.2 Scope of study

This chapter entailed the studying of continuous wavelet functions and Haar wavelet approximations. Wavelet transform introduced as a mathematical approach with some of applications of wavelet transform which is widely used in signal processing field. The approximation of DEs using Haar wavelet base was implemented with comparing to B-cubic spline base.

2. Preliminary

In this section, we introduce the definitions of two types of continues Haar wavelet functions and linear, quadratic and cubic spline functions base.

2.1 Continues Haar wavelet functions

Haar functions have been introduced by Hungarian mathematician. The orthogonal set of Haar functions is defined as a square waves with magnitude of ± 1 in some interval and zero elsewhere. The first curve is that $h_0(x) = 1$ during the whole interval $0 \leq x \leq 1$. The second curve $h_1(x)$ is the fundamental square wave, or mother wavelet which also spans the whole interval $[0, 1]$. All the other subsequent curve are generated from $h_1(x)$ with two operation translation and dilation. Haar wavelet functions defined as follows on $(0, X]$ [8].

$$h_0(x) = \frac{1}{\sqrt{M}} \qquad 0 \leq x < X, \tag{1}$$

$$h_1(x) = \frac{1}{\sqrt{M}} \begin{cases} 1 & 0 \leq x < \frac{X}{1} \\ -1 & \frac{X}{2} \leq x < X \\ 0 & o.w. \end{cases} \tag{2}$$

$$h_i(x) = \frac{1}{\sqrt{M}} \begin{cases} \sqrt{2^j}, & \frac{k-1}{2^j} X \leq x < \frac{k - \frac{1}{2}}{2^j} X \\ -\sqrt{2^j}, & \frac{k - \frac{1}{2}}{2^j} X \leq x < \frac{k}{2^j} X \\ 0, & o.w. \end{cases} \tag{3}$$

for $i = 1, 2, 3, \ldots, m-1$, $M = 2^j$ and $i = 2^j + k - 1$. We say that $h_1(x)$ is mother function and $h_i(x) = 2^{\frac{j}{2}} h_1(2^j x - k)$ for $i = 2, 3, \ldots, m-1$. In general, we have the following: $h_0(x) = h_1\left(2^j x - \frac{k}{2^j}\right)$, where $n = 2^j + k, j \geq 0, 0 < k \leq 2^j$. Note that:

$$(h_p(x), h_q(x)) = \int_0^x h_p(x) h_q(x) dx = \frac{X}{m} \delta_{pq}.$$

To approximate $f(x)$ using Haar functions consider $f(x) = \sum_{i=0}^{m-1} a_i h_i(x)$..
Then,

$$a_j = \frac{\int_0^x h_i(x) h_j(x) dx}{\int_o^x h_j^2(x) dx} = \frac{m}{X} \int_0^x f(x) h_j dx$$

for

$$j = 0, 1, 2, \dots m - 1.$$

All Haar wavelets are orthogonal to each other:

$$\int_0^1 h_i(x)h_j dx = 2^{-j}\delta_{ij} = \begin{cases} 2^{-j}, & i = j = 2^{j+k} \\ 0, & i \neq j \end{cases}$$

2.2 Spline functions

The spline is used to refer to a wide class of smooth functions that are used in applications requiring data interpolation [9, 10]. The data may be either one-dimensional or multi-dimensional. Spline functions for interpolation are normally determined as the minimizers of suitable measures of roughness (for example integral squared curvature) subject to the interpolation constraints. Smoothing splines may be viewed as generalizations of interpolation splines where the functions are determined to minimize a weighted combination of the average squared approximation error over observed data and the roughness measure. For a number of meaningful definitions of the roughness measure, the spline functions are found to be finite dimensional in nature, which is the primary reason for their utility in computations and representation. For the rest of this section, the focus is entirely on one-dimensional, polynomial splines and the use of the term spline in this restricted sense. The base $\Phi(x) = \{\Phi_1(x), \Phi_2(x), \dots, \Phi_n(x)\}$ is called spline base of order n if the basis functions satisfy $\Phi_i(x) \in C^{n-1}(-\infty, \infty)$ for $i = 1, 2, \dots, n$. First of all, we partition $[0, 1]$ by choosing a positive integer n and defining $h = \frac{1}{n+1}$. This produces the equally-spaced nodes $x_i = ih$, for each $i = 0, 1, \dots, n + 1$. We then define the basis functions $\{\phi(x)\}_{i=0}^{n+1}$ on the interval $[0, 1]$.

2.2.1 Linear spline

The simplest spline is a piecewise polynomial function, with each polynomial having a single variable. The spline S takes values from an interval $[a, b]$ and maps them to $\Re$ where $S : [a, b] \to \Re$ Since S is piecewise defined, choose k subintervals to partition $[a, b]$. The simplest choice of spline functions basis involves piecewise-linear polynomials. The first step is to form a partition of $[0, 1]$ by choosing points $x_0, x_1, \dots, x_{n+1}$. Letting $h_i = x_{i+1} - x_i$, for each $i = 0, 1, \dots, n$. We have defined the basis functions $\Phi_1(x), \Phi_2(x), \dots, \Phi_n(x)$. Linear spline is linear polynomial $S(x)$ which satisfies $S(x) \in C(-\infty, \infty)$. To construct linear spline base in which it can satisfy the boundary conditions $\phi_i(0) = \phi_i(1)$ for $i = 1, 2, \dots, n$. we have constructed the following component linear spline functions:

$$\Phi_i(x) = \begin{cases} 0, & 0 \le x \le x_{i-1}, \\ \frac{1}{h_{i-1}}(x - x_{i-1}), & x_{i-1} < x \le x_i, \\ \frac{1}{h_i}(x_{i+1} - x), & x_i < x \le x_{i+1}, \\ 0, & x_{i+1} < x \le 1. \end{cases} \tag{4}$$

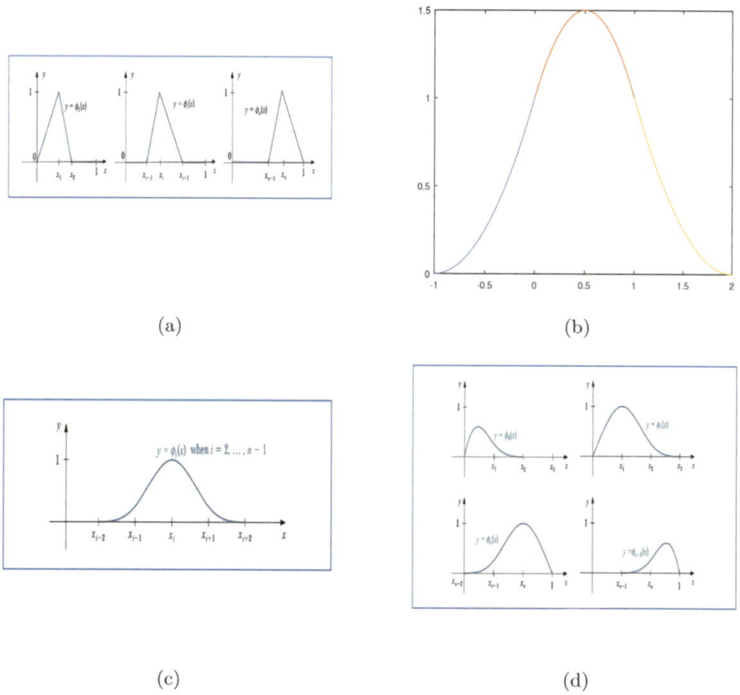

Figure 1.
(a) Linear spline and (b) quadratic spline (c) cubic spline function and (d) compound cubic spline function.

for each $i = 1, 2, \ldots, n$. (See **Figure 1(a), Table 1(a)**). We can prove that the functions are orthogonal because $\Phi_i(x)$ and $\Phi_i'(x)$ are nonzero only on (x_{i-1}, x_{i+1}) such that $\Phi_i(x)\Phi_j(x) = 0$ and $\Phi_i'(x)\Phi_j'(x) = 0$ if $i \neq j, j-1, j+1$, consequently, $\Phi_i(x) \in C(-\infty, \infty)$ for i = 1,2,3, ... n.

x_i	$\Phi_i(x_i)$	$\Phi_i'(x_i)$	$\Phi_i''(x_i)$
x_{i-1}	0		
x_i	1		
x_{i+1}	0		
(a)			
x_{i-1}	$\frac{1}{2}$	2	
x_i	1	0	
x_{i+1}	$\frac{1}{2}$	-2	
(b)			
x_{i-2}	0	0	0
x_{i-1}	$\frac{1}{4}$	$\frac{3}{4}$	$-\frac{3}{2}$
x_i	1	0	$-\frac{3}{4}$
x_{i+1}	$\frac{1}{4}$	$-\frac{3}{4}$	$-\frac{3}{2}$
x_{i+2}	0	0	0
(c)			

Table 1.
Values at node points (a) linear spline, (b) quadratic B-spline, and (c) cubic B-spline.

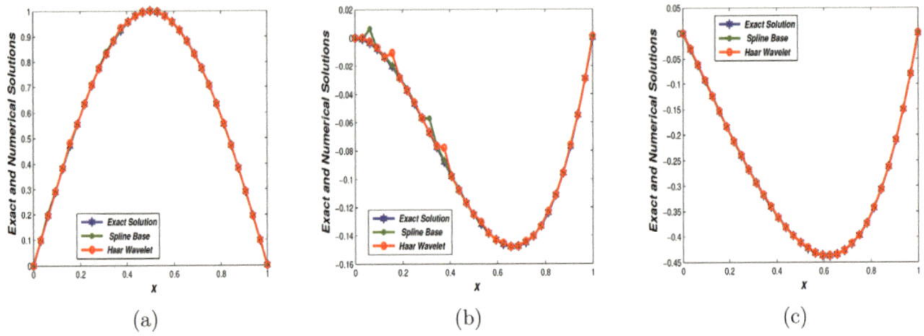

Figure 2.
Comparison of Haar wavelet and Spline Base with exact solutions for examples (a) 3.5 (b) 3.6 (c) 3.7.

2.2.2 Quadratic B-spline

Quadratic B-spline is quadratic B-Spline polynomial $S(x)$ which satisfy $S(x) \in C^2(-\infty, \infty)$. To construct quadratic spline base in which satisfy the boundary conditions $\phi_i(0) = \phi_i(1)$ for $i = 1, 2, \dots, n$ we have constructed the following component quadratic spline functions (**Figure 2**):

$$\phi_i(x) = \frac{1}{h^2}\begin{cases} (x_{i+2}-x)^2 - 3(x_{i+1}-x)^2 + 3(x_i - x)^2; & [x_{i-1}; x_i]; \\ (x_{i+2}-x)^2 - 3(x_{i+1}-x)^2; & [x_i; x_{i+1}]; \\ (x_{i+2}-x)^2; & [x_{i+1}; x_{i+2}]; \\ 0; & o.w. \end{cases} \tag{5}$$

See **Figure 1(b)** and **Table 1(b)**.

2.2.3 Cubic B-spline

Many researchers used B-cubic spline base which defined as follows:

$$S(x) = \frac{1}{4}\begin{cases} 0, & x < -2 \\ (2+x)^3, & -2 \le x \le -1 \\ (2+x)^3 - 4(1+x)^3, & -1 < x \le 0 \\ (2-x)^3 - 4(1-x)^3, & 0 < x \le 1 \\ (2-x)^3, & 1 < x \le 2 \\ 0, x > 2. & \end{cases} \tag{6}$$

Consequently, $S(x) \in C_0^2(-\infty, \infty)$.

To construct cubic spline base in which satisfy the boundary conditions $\phi_i(0) = \phi_i(1)$ for $i = 1, 2, \dots, n$ we have constructed the following component cubic spline functions:

$$\phi_i(x) = \begin{cases} S\left(\frac{x}{h}\right) - 4S\left(\frac{x+h}{h}\right), & i=0 \\ S\left(\frac{x-h}{h}\right) - S\left(\frac{x+h}{h}\right), & i=1 \\ S\left(\frac{x-ih}{h}\right), & 2 \leq i \leq n \\ S\left(\frac{x-nh}{h}\right) - S\left(\frac{x(n+2)h}{h}\right), & i=n \\ S\left(\frac{x-(n+1)h}{h}\right) - 4S\left(\frac{x-(n+2)h}{h}\right), & i=n+1. \end{cases} \quad (7)$$

See **Figure 1(c)**, **(d)** and **Table 1(c)**.

3. Approximation of differential equations (DEs)

Mathematics has several tools to describe the problems in real life, engineering and science. ODEs and PDEs are significant tools in applied mathematics. They played significant rule in describing the mathematical models in applications of engineering, science and economics. High-order DE arises in some fields of engineering and science such as nonlinear optics and quantum mechanics. The approximated solutions of DEs should be studied when the ODEs and PDEs have no analytical solutions or it is very difficult to find the analytical solutions. The numerical or approximated solutions of DEs are very important in scientific computation, as they are widely used to model real life problems. In this section, we have studied the approximation solutions of DEs using spline and Haar wavelet bases (**Table 2**).

3.1 Approximation of ordinary differential equations (ODEs)

In this section, we have studied approximation solutions of ODEs using spline and Haar wavelet bases.

$n\backslash t$	0	0.25	0.5	0.75	1
5	0	1.3345e-3	0.0015	5.0673e-3	3.6339e-3
	0	1.311e-3	0.0005	5.0683e-3	3.6229e-3
10	0	1.3232e-5	2.6342e-5	1.5634e-6	4.1443e-5
	0	1.3211e-5	2.1212e-5	1.2341e-6	4.0101e-5
50	0	2.3416e-7	1.6611e-7	5.1126e-7	2.1233e-7
	0	2.1414e-7	1.2211e-7	5.2233e-7	2.1266e-7
100	0	4.9383e-8	3.4453e-8	5.0347e-8	6.4332e-7
	0	4.9121e-8	3.4564e-8	5.0111e-8	6.4222e-7

Table 2.
Absolute errors of example 3.1 using numerical collection method with (a) polynomial basis (b) Haar wavelet basis.

3.1.1 Approximation of DEs using spline functions

In this section, we have introduced the linear, quadratic and cubic B-spline base and their applications in solving ODEs. The operational matrices of the fractional-order integration of the B-spline base has been studied.

3.1.2 Rayleigh-Ritz Metod

Rayleigh-Ritz metod is variational technique for solving boundary value problems (BVPs) which is the first reformulated as a problem of choosing, from set of all sufficiently differentiable functions satisfying the boundary conditions, the function to minimize a certain integral. To describe the Rayleigh-Ritz method, we consider the approximation of the solution to a linear of two-boundary value problem from beam-stress analysis. This BVP is described by the following DE:

$$-\frac{d}{dx}\left(p(x)\frac{dy}{dx}\right)+q(x)y(x)=f(x), \quad 0\leq x\leq 1. \tag{8}$$

with boundary conditions

$$y(0)=y(1)=0.$$

The DE describes the deflection $y(x)$ of a beam of length 1 with variable cross section represented by $q(x)$. The deflection is due to the added stresses $p(x)$ and $f(x)$. We have the following functional that is equivalent to Eq. (8).

$$I(u(x))=\int_0^1\left(p(x)(u'(x))^2+q(x)(u(x))^2-2f(x)u(x)\right)dx. \tag{9}$$

An approximation

$$u(x)=\sum_{i=0}^{n}\Phi_i(x). \tag{10}$$

to the solution $y(x)$ of Eq. (9) can be obtained by finding the constants $c_1, c_2, c_3, \ldots c_n$ to minimize the integral Eq. (9): When considering $I(c_1, c_2, c_3, \ldots c_n)$ as a function of $c_1, c_2, c_3, \ldots c_n$ to have

$$\frac{\partial I}{\partial c_j}=0$$

for i = 1,2,3, ... ,n.
Lastly, we have obtained the linear system of equations $Ac=b$, where,

$$a_{ij}=\int_0^1\left(p(x)\Phi_i'(x)\Phi_j'(x)+q(x)\Phi_i(x)\Phi_j(x)\right)dx$$

and

$$b_i=\int_0^1 f(x)\Phi_i(x)dx,$$

for $i,j = 1, 2, \ldots n$ [10]. To impalement the Ritz method we consider the following problems.

Example 3.1 [8]

Consider

$$y''(t) + y(t) = t^2 - t + 2, \qquad 0 \le t \le 1, \tag{11}$$

subject to the initial condition is

$$y(0) = y(1) = 0,$$

with the exact solution $y(t) = t^2 - t$

Let n = 5, then,

$$a_{ij} = \int_0^1 \left(-h_i'(x)h_j'(x) + h_i(x)h_j(x)\right)dx = \frac{m}{X}\delta_{ij},$$

and

$$b_i = \int_0^1 (x^2 - x + 2)h_i(x)dx,$$

for $i,j = 1, 2, \ldots n$. However,

$$c = \left[-\frac{1}{3}, 0, \frac{1}{4\sqrt{2}}, -\frac{1}{4\sqrt{2}}, \frac{.0234375}{2}\right].$$

Example 3.2 [8].

Consider

$$y''(t) + \pi^2 y(t) = 0, \qquad 0 \le t \le 1, \tag{12}$$

subject to the initial condition is

$$y(0) = y(1) = 0,$$

with the exact solution $y(t) = \sin \pi t$

Let n = 10, then,

$$a_{ij} = \int_0^1 \left(-h_i'(x)h_j'(x) + \pi^2 h_i(x)h_j(x)\right)dx = \pi^2\frac{m}{X}\delta_{ij},$$

and

$$b_i = 0,$$

for $i,j = 1, 2, \ldots n$. However, c = [1,1,2,3,1,1,1,1,1,1].

3.1.3 Analysis of collection method

Let the differential operator L defined on the interval $I = [a, b]$. Define the collocation points $x_i = a + ih$ for $i = 0, 1, \ldots, n;$ where $h = \frac{b-a}{n}$ and n is the number of partitions on I. Discretize the functions

$$\Phi(x) = \{\Phi_1(x), \Phi_2(x), \Phi_3(x), \dots, \Phi_n(x)\}.$$

Suppose

$$y(x) = \sum_{i=1}^{n} c_i \Phi_i(x).$$

Put the approximation of $y(x)$ at the point x_j in the DE, we get the function coefficient matrix $\Phi_{i,j} = \Phi_i(x_j)$ and $\Phi'_{i,j} = \Phi'_i(x_j)$. The matrix of coefficients has the dimension $n \times n$. Any function $y(x)$ which is square integrable in the interval $(0, 1)$ can be expressed as an infinite sum of Haar wavelet. The above series terminates at finite terms if $y(x)$ is piecewise constant or can be approximated as piecewise constant during each subinterval.

3.2 The quadratic B-Spline Base

Consider the quadratic B-spline Base

$$S(x) = \{S_1(x), S_2(x), S_3(x), \dots, S_n(x)\}.$$

Suppose $y(x) = \sum_{i=1}^{n} c_i S_i(x)$. The general ODE of first-order has the following form

$$a_0(t)y'(t) + a_1(t)y(t) = f(t), \qquad 0 \le t \le 1, \tag{13}$$

subject to the initial condition is $y(0) = \alpha$..
Example 3.3 [8].

$$y'(t) + y(t) = \sin(t) + \cos(t), \qquad 0 \le t \le 1, \tag{14}$$

subject to the initial condition is

$$y(0) = 0,$$

The coefficients are $a_0(t) = a_1(t) = 1$ and $f(t) = \sin(t) + \cos(t)$..
Consider the quadratic B-spline base, then, the matrix of coefficients has the following formula:

$$A_{ij} = S'_i(t_j) + S_i(t_j),$$

and

$$b_i = \sin(t_i) + \cos(t_i)$$

for $i, j = 1, 2, \dots, n$. By solving the system of coefficients $Ac = b$ we will obtain the coefficients of approximation.

3.3 Approximation of DEs using Haar wavelet Functions Base

We introduce the Haar wavelet technique for solving general linear first-order ODEs [11].

3.3.1 First-order linear ODEs

Consider the following general linear first-order ODE:

$$y'(t) + f(t)y(t) = g(t); 0 \le t \le a; f(t) \ne 0, \tag{15}$$

$$y(0) = \beta. \tag{16}$$

Substituting $t = ax$ in Eq. (15) which reduces to

$$y'(x) + af(x)y(x) = ag(x); 0 \le t \le 1; f(t) \ne 0, \tag{17}$$

$$y(0) = \beta. \tag{18}$$

We assume that

$$y'(x) = \sum_{i=1}^{k} c_i h_i(x), \tag{19}$$

where $c_i's$ for i = 1,2, ... ,k are Haar coefficients to be determined. Integrating Eq. (19) with respect to x, we get the following

$$y(x) = \beta + \sum_{i=1}^{k} c_i P_{1,i}(x). \tag{20}$$

Substituting Eqs. (19) and (20) in Eq. (17), we get the following system of equation:

$$\sum_{i=1}^{k} c_i h_i(x) + af(x) \sum_{i=1}^{k} c_i h_i(x) = ag(x). \tag{21}$$

Put $x = t_j$ for $j = 1, 2, \dots, n$. in Eq. (21), we get linear system in which the matrix of coefficients has the following formula:

$$A_{ij} = (1 + af(t_i))h_i(t_j)$$

and

$$b_i = ag(t_j),$$

for $i, j = 1, 2, \dots, n$. By solving the linear system of coefficients $Ac = b$, we obtain the coefficients of approximated solution.

3.4 Approximation of fractional differential equations (FrDEs)

In this section, we have studied approximation of DEs using spline and Haar wavelet bases.

3.4.1 Operational matrix of the fractional order integration of the B-Spline Base

In this section, we have evaluated the operational matrices of the fractional-order integration of the linear, quadratic and cubic B-spline Base.

3.4.2 Linear spline

This subsection examines the cubic linear spline operational matrix $FS\alpha$ of integration of the fractional order as follows:

$$J_{xi}^{\alpha}(x) = \frac{1}{\Gamma(\alpha+2)} \begin{cases} 0, & 0 \le x \le x_{i-1}, \\ \frac{1}{h_{i-1}} x_{i-1}^{\alpha+1}, & x_{i-1} < x \le x_i, \\ \frac{1}{h_{i-1}} x_{i-1}^{\alpha+1} + \frac{1}{h_i}\left(\alpha h_i x_i^{\alpha} - x_i^{\alpha+1}\right), & x_i < x \le x_{i+1}, \\ 0, & x_{i+1} < x \le 1. \end{cases} \tag{22}$$

where $x_{i-1} = x - x_{i-1}, x_i = x - x_i$.

3.4.3 Quadratic B-spline

This subsection introduced the quadratic B-spline operational matrix $FS\alpha$ of integration of the fractional order as follows:

$$J_x^{\alpha}(x) = JJ * \frac{1}{\Gamma(\alpha+3)},$$

where

$$JJ = \begin{cases} 2(x-1)^{\alpha+2}; & x \in [1,2]; \\ 2(x-1)^{\alpha+2} - 6(x-2)^{\alpha+2}; & x \in [2,3]; \\ 2(x-1)^{\alpha+2} - 6(x-2)^{\alpha+2} + 6(x-3)^{\alpha+2}; & x \in [3,4]; \\ 0; & o.w. \end{cases} \tag{23}$$

3.4.4 Cubic B-spline

This subsection introduced the cubic B-spline operational matrix FSα of integration of the fractional order as follows:

$$J_x^{\alpha}(x) = JJ * \frac{1}{\Gamma(\alpha+4)},$$

where

$$JJ = \begin{cases} 0, & x < -2 \\ \frac{3}{2} x^{\alpha+3}, & -2 \le x \le -1 \\ \frac{3}{2} x^{\alpha+3} - 6x_1^{\alpha+3}, & -1 < x \le 0 \\ \frac{3}{2} x^{\alpha+3} - 6x_1^{\alpha+3} + 9x_2^{\alpha+3}; & 0 < x \le 1 \\ \frac{3}{2} x^{\alpha+3} - 6x_1^{\alpha+3} + 9x_2^{\alpha+3} - 6x_3^{\alpha+3}; & 1 < x \le 2 \\ 0, & x > 2, \end{cases} \tag{24}$$

where $x_i = x - i$; $i = 1,2,3$.

3.5 Numerical Solutions of fractional differential equations

3.5.1 Numerical solutions of fractional differential equations using Haar base

We will introduce the Haar wavelet technique for solving FrDEs.
Example 3.4 [8].
Consider the general fractional-order linear DE

$$y^{\alpha}(t) = A(t) + B(t)y(t) = C(t); 0 \le t \le a; n - 1 < \alpha < n, \tag{25}$$

subject to initial conditions $y_j(0) = a_j$ for $j = 0, 1, \dots, n - 1$. where $A(t), B(t)$ and $C(t)$ are given functions, are arbitrary constants and α is a parameter describing the order of the fractional derivative. The general response expression contains a parameter describing the order of the fractional derivative that can be varied to obtain various responses.

Substituting $t = ax$ in Eq. (25) which reduces to

$$y^{\alpha}(x) = aA(x) + aB(x)y(x) = C(t); \quad 0 \le x \le 1; \qquad n - 1 < \alpha < n, \tag{26}$$

$$y_j(0) = a_j; g = 0, 1, \dots, n - 1. \tag{27}$$

We assume that

$$y^{\alpha}(x) = \sum_{i=1}^{k} c_i h_i(x). \tag{28}$$

If $\alpha = \frac{1}{2}$, integrating Eq. (28) once, we get

$$y(x) = a_0 + \sum_{i=1}^{k} c_i FH_{\frac{1}{2},i}(x). \tag{29}$$

Substituting Eqs. (28) and (29) in Eq. (26), we get

$$\sum_{i=1}^{k} c_i h_i(x) - aA(x) - aB(x)\left(a_0 + \sum_{i=1}^{k} c_i FH_{\frac{1}{2},i}(x)\right) = C(x), \tag{30}$$

If $\alpha = \frac{3}{2}$, integrating Eq. (26) twice, we get

$$y^{\frac{1}{2}}(x) = a_1 + \sum_{i=1}^{k} c_i FH_{\frac{1}{2},i}(x), \tag{31}$$

and

$$y(x) = a_0 + a_1 x + \sum_{i=1}^{k} c_i FH_{\frac{3}{2},i}(x). \tag{32}$$

Substituting Eqs. (28) and (37) in Eq. (26), we get

$$\sum_{i=1}^{k} c_i h_i(x) - aA(x) - aB(x)\left(a_0 + \sum_{i=1}^{k} c_i FH_{\frac{3}{2},i}(x)\right) = C(x). \tag{33}$$

Put $x = t_j$ for $j = 1, 2, \dots, n.$ in Eq. (30) in case $\alpha = \frac{1}{2}$, or in Eq. (33) in case $\alpha = \frac{3}{2}$, we get the linear system in which the matrix of coefficients has the following formula:

$$A_{ij} = h_i(x)(t_j) + aB(t_j)FH_{\alpha,i}(t_j)$$

and

$$b_i = C(t_j) + aA(t_j) - aa_0B(t_j),$$

for $i, j = 1, 2, \dots, n$. By solving the linear system of coefficients $Ac = b$ we obtain the coefficients of approximated solution $y(t)$ of Eq. (26).

3.5.2 Numerical solutions of fractional differential equations using B-spline base

We will introduce the B-spline technique for solving FrDE (26).
Consider the quadratic B-spline base

$$S(x) = \{S_1(x), S_2(x), S_3(x), \dots, S_n(x)\},$$

Suppose

$$y(x) = \sum_{i=1}^{n} c_i S_i(x).$$

We assume that

$$y^{\alpha}(x) = \sum_{i=1}^{k} c_i S_i(x). \tag{34}$$

If $\alpha = \frac{1}{2}$, integrating Eq. (34) once, we get

$$y(x) = a_0 + \sum_{i=1}^{k} c_i FS_{\frac{1}{2},i}(x). \tag{35}$$

Substituting Eqs. (34) and (35) in Eq. (26), we get

$$\sum_{i=1}^{k} c_i S_i(x) - aA(x) - aB(x)\left(a_0 + \sum_{i=1}^{k} c_i FS_{\frac{1}{2},i}(x)\right) = C(x). \tag{36}$$

If $\alpha = \frac{3}{2}$, integrating Eq. (34) once, we get

$$y(x) = a_0 + a_1 x + \sum_{i=1}^{k} c_i FH_{\frac{3}{2},i}(x). \tag{37}$$

Substituting Eqs. (28) and (29) in Eq. (26), we get

$$y^{\frac{1}{2}}(x) = a_1 + \sum_{i=1}^{k} c_i FH_{\frac{1}{2},i}(x), \tag{38}$$

and

$$y(x) = a_0 + a_1 x + \sum_{i=1}^{k} c_i FH_{\frac{3}{2},i}(x), \tag{39}$$

$$\sum_{i=1}^{k} c_i S_i(x) - aA(x) - aB(x)\left(a_0 + \sum_{i=1}^{k} c_i FS_{\frac{3}{2},i}(x)\right) = C(x). \tag{40}$$

Put $x = t_j$ for $j = 1, 2, \ldots, n$. in Eq. (36) in case $\alpha = \frac{1}{2}$, or Eq. (40) in case $\alpha = \frac{3}{2}$,, we get the linear system in which the matrix of coefficients has the following formula:

$$A_{ij} = \left(h_i(x)\left(t_j\right)\right) + aB\left(t_j\right)FH_{\alpha}, i\left(t_j\right)\big)$$

and

$$b_i = C\left(t_j\right) + aA\left(t_j\right) - aa_0 B\left(t_j\right),$$

for $i, j = 1, 2, \ldots, n$. By solving the linear system of coefficients, we obtain the coefficients of approximated solution $y(t)$ of Eq. (26).

3.5.3 Numerical solution of fractional Lane differential equation

We generalize the definition of Lane-Emden equations up to fractional order as following:

$$D^{\alpha}y(t) + \frac{k}{t^{\alpha-\beta}}D^{\beta}y(t) + f(t,y) = g(t); 0 < t \leq 1, k > 0, \tag{41}$$

with the initial condition $y(0) = A; y'(0) = B$ where $1 < \alpha \leq 2, 0 < \beta \leq 1$ and A, B are constants and $f(t,y)$ is a continuous real-valued function and $g(t,y) \in [0,1]$. The theory of singular boundary value problems has become an important area of investigation in the past three decades. One of the equations describing this type is the Lane-Emden equation. Lane-Emden type equations, first published by [12], and further explored in detail by [13], represents such phenomena and having significant applications, is a second-order ODE with an arbitrary index, known as the polytropic index, involved in one of its terms. The Lane-Emden equation describes a variety of phenomena in physics and astrophysics. [14] imposed the Lane-Emden DE of fractional order and the approximate solution is obtained by employing the method of power series and a numerical solution is established by the least squares method for these Eqs. [14] approximate the solution of DE by employing the method of power series and the numerical solution is established by collection method.

3.5.3.1 Analysis of numerical method of fractional Lane differential equation

[15] studied the solution of DEs based on Haar operational matrix, [16] studied the solution of DEs using Haar wavelet collocation method, [17] studied the numerical solution of DEs by using Haar wavelets, [18] used Haar wavelet approach to ODEs, [19] solved the fractional Riccati DEs using Haar wavelet while [14] studied the fractional DEs of Lane-Emden type numerically by method of collocation. [20] introduced an operational Haar wavelet method for solving fractional Volterra integral equations, [21] solved fractional integral equations by the Haar wavelet method, [22] used Haar wavelet-quasi linearization technique for fractional nonlinear DEs, [21] solved the fractional integral equations by the Haar wavelet method, [4] used Haar wavelet method for solving fractional PDEs numerically. In Eq. (41), consider $\alpha > \beta, f(t,y) = \frac{1}{t^{\alpha+2}}y(t)$ and $g(t) = 0$,.

However, $D^{\alpha}W(t) = ah(t) = \sum_{i=0}^{m} c_i h_i(t)$ and

$$\begin{aligned} D^{\beta}W(t) &= \left(I^{\alpha-\beta}D^{\alpha}\right)W(t) + W^{\beta}(0) \\ &= ap^{\alpha-\beta}h(t) + W^{\beta}(0) \\ W(t) &= (I^{\alpha}D^{\alpha})W(t) + W(0) \\ &= ap^{\alpha}h(t) + A. \end{aligned}$$

Hence,

$$ah(t) + \frac{k}{t^{\alpha-\beta}} ap^{\alpha-\beta}h(t) + W^{\beta}(0) + ap^{\alpha}h(t) + A = Ch(t).$$

If we consider $\alpha = \frac{3}{2}$ and $\beta = \frac{1}{2}$ we solve the system of equations to obtain the coefficients $(c_0, c_1, c_2, \dots, c_m)$.

3.6 Comparison study using numerical collection method

Collocation method for solving DEs is one of the most powerful approximated methods. This method has its basis upon approximate the solution of FrDEs by a series of complete sequence of functions, a sequence of linearly independent functions which has no non-zero function perpendicular to this sequence of functions. In general, $y(t)$ is approximated by [14].

$$y(t) = \sum_{i=1}^{n} a_i \Theta_i(t), \tag{42}$$

where a_i for $i = 1, 2, \dots, n$ are an arbitrary constants to be evaluated and Θ_i for $i = 1, 2, \dots, n$ are given set of functions. Therefore, the problem in Eq. (41) of evaluating $y(t)$ is approximated by (42) then, is reduced to the problem of evaluating the coefficients for $i = 0, 1, 2, \dots, n$.

Let $\{t_1, t_2, \dots, t_n\}$ is a partition to interval $[0, 1]$ and $t_j = jh$ and $h = \frac{1}{n}$ and $j = 0, 1, 2, \dots, n$. See the comparison of absolute errors of the problem using numerical collection method with polynomial basis and Haar wavelet basis.

Example 3.5 [8]

Consider

$$w''(t) + \pi^2 w(t) = 0, \tag{43}$$

with the boundary conditions

$$w(0) = w(1) = 0.$$

The exact solution is $w(t) = sin\pi t$.

Example 3.6 [8]

Consider

$$t^2 w''(t) - 6w(t) = 4t^2, \tag{44}$$

with the boundary conditions

$$w(0) = w(1) = 0.$$

The exact solution is $w(t) = t^2(t-1)$..
Example 3.7 [8]
Consider

$$w''(t) = w(t) + 4te^t, \tag{45}$$

with the boundary conditions

$$w(0) = w(1) = 0.$$

The exact solution is $w(t) = t(t-1)e^t$.

4. Wavelet transform (WT)

Fourier transform (FT) of a time signal $x(t)$ reveals the frequency content of the signal by decomposing the signal using complex sinusoids as follows:

$$X(f) = F\{x(t)\} = \int_{-\infty}^{\infty} x(t)e^{-j2\pi ft}dt.$$

However, FT cannot reveal the time information associated with a specific frequency. This drawback enhanced research in the time-frequency domain [23]. One of the most important time-frequency distributions (TFD's) is the wavelet transform (WT), which is a time-frequency representation of signals. While not all TFD's are invertible, a big advantage of WT over many other TFD's is invertibility. WT proved to be successful in revealing spectral features of signals. Instead of sustainable waves like sinusoidal waves as in the case of Fourier transform, WT is based on decomposing signals using decaying waves (small waves, or wavelets), all are shifted and dilated versions of a specific wavelet called mother wavelet. The continuous wavelet transform (CWT) of a signal $x(t)$ using a mother wavelet $\psi(t)$ is given by:

$$W_x^{\psi}(t,s) = \frac{1}{\sqrt{s}}\int_{-\infty}^{\infty} x(\lambda).\psi^*\left(\frac{\lambda - t}{s}\right)d\lambda,$$

where λ is a representation of time inside the convolution integral, ψ^* is the complex conjugate of the wavelet ψ, and $s \in \mathfrak{R}^+ = \mathfrak{R} - \{0\}$ is called the "scale", which we expect to be inversely related to the radian frequency $\omega = 2\pi f$ for the above structure to be comparable to the structure of the sinusoidal waves $\sin\omega(t)$ used in the Fourier transform; the actual scale-frequency relationship is given by:

$$s \approx \frac{K}{f}$$

where $K = f_m.f_s; f_m = arg(max\{\psi(f)\}); \psi(f) = F\{\psi(t)\}; f_s$ is the sampling frequency used to discretize $\psi(t)$ while computing $\psi(f)$ via DFT. It is apparent that, for a fixed scale s, the wavelet transform $W_x^{\psi}(t,s)$ is given by the convolution between the signal and the time-reversed wavelet as follows:

$$W_x^{\psi}(t,s) = x(t) \odot h(t)$$

where

$$h(t) = \left[\frac{1}{\sqrt{s}}\psi^*\left(\frac{-t}{s}\right)\right]$$

and $\odot$ refers to the 1D convolution process:

$$x(t) \odot h(t) = \int_{-\infty}^{\infty} x(\lambda)(h)(t-\lambda)d\lambda$$

This fact gives another equivalent expression for $W_x^{\psi}(t,s)$ using Fourier transforms of the signal and the wavelet as follows:

$$W_x^{\psi}(t,s) = F^{-1}\{X(f).H(f)\} = \int_{-\infty}^{\infty} X(f).H(f)e^{+j2\pi ft}df$$

where $H(f) = F\{h(t)\}$. Hence, $W_x^{\psi}(t,s)$ can be implemented via filtering the signal $x(t)$ by a filter whose impulse response is $h(t)$. This will be the basis for implementing the discrete 1D and 2D wavelet transforms as explained below.

Generally speaking, Fourier transform $X(f) = F\{x(t)\}$ decomposes the signal $x(t)$ using the same sinusoidal wave $e^{-j2\pi ft}$ at different values of frequency f, while the wavelet transform $W_x^{\psi}(t,s)$ decomposes the signal $x(t)$ using the same mother wavelet $\psi(t)$ at different values of scale s (hence, frequency, f) and time t; where both time and frequency information are revealed.

The WT is invertible, giving it a great advantage in applications:

$$x(t) = \frac{1}{c_{\psi}} \int_{-\infty}^{\infty} \int_{-\infty}^{\infty} W_x^{\psi}(\lambda,s).\frac{1}{s^2}\psi\left(\frac{t-\lambda}{s}\right)d\lambda.ds$$

where $c_{\psi} = \int_{-\infty}^{\infty} \frac{|\psi(f)|}{|f|}df$, which implies that $\psi(0) = 0 \rightarrow \int_{-\infty}^{\infty}\psi(t)dt = 0$, hence, $\psi(t)$ must be oscillating. Also, to satisfy Parseval's Theorem we should have $\int_{-\infty}^{\infty}\psi(t)dt = 1$. The above continuous wavelet transform can be discretized to give the discrete wavelet transform (DWT), which can be implemented (as 1D DWT) by passing the signal $x(t)$ through a low-pass filter followed by down-sampling with a factor of 2 (giving approximation coefficients), and a high-pass filter then down-sampling by a factor of 2 (giving detail coefficients). These filters differ according to the analyzing wavelet [24]. The 2D DWT (for images) can be designed based on 1D DWT via tensor products, and it results into a decomposition of approximation coefficients at level k into four components: low-pass component that contains the approximation coefficients at level k + 1, and three high-pass components that contain the detail coefficients in three directions (horizontal, vertical, and diagonal). Note that approximation at level k = 0 is equivalent to the original 2D signal [24].

4.1 Some applications of the wavelet transform

The frequency content extracted by wavelet transform (WT) has been effectively used in revealing important features of 1D and 2D signals. This property proved very useful in speech and image recognition [25]. Also, the orthogonality of WT paved the way for using WT in orthogonal frequency division multiplexing (OFDM), a pivotal technique for 4th and 5th generations of digital communication [26]. In addition to that, WT proved to put high focus on the low-frequency part of the signal, in which most of the information resides, hence, WT has been used for signal and image compression [27]. The compression process can be performed using hard-thresholding of the WT as follows:

$$t(x) = \begin{cases} x & |x| \geq T \\ 0 & |x| < T \end{cases} \tag{46}$$

where $t(x)$ is the new WT coefficient value to replace the original coefficient x, and T is the threshold. Better compression results (in terms of signal size) can be obtained by increasing the threshold, however, larger deviation from the original signal (i.e., larger error) is obtained. Hence, choice of the threshold involves a trade-off between size and error. Original signal can be obtained from the compressed one via inverting the thresholded WT.

4.2 Noise removal using WT

An important application of the Wavelet Transform is noise removal from signals and images. As most of the information content of real-life signals is in the low-frequency regions, removal of high frequency regions in the WT of signals can help in removing the majority of noise. This can be done via thresholding WT coefficients or by removing the details coefficients of WT and considering only the approximation coefficients of WT. This property of separating low-frequency content from high-frequency content in the WT is mainly due to the filtering involved in the structure of WT as explained above. Noise removal using WT is more efficient for 1D signals corrupted by 1D noise process, where the 2D structure of WT in joint time-frequency domain can spread the 1D noise effect into a 2D plane, hence the noise power is greatly reduced. For noise removal, a soft-threshold can be used to cut out high-frequency coefficients as follows:

$$t(x) = \begin{cases} sign(x)(|x| - T) & |x| \geq T \\ 0 & |x| < T \end{cases} \tag{47}$$

where $t(x)$ is the new WT coefficient value to replace the original coefficient x, T is the threshold, and $sign(x)$ is the signum function defined as follows:

$$sign(x) = \begin{cases} +1 & x > 0 \\ 0 & x = 0 \\ -1 & x < 0 \end{cases} \tag{48}$$

Figure 3 shows the use of WT to denoise an image, while **Figure 4** shows the denoising of 1D signal using WT, where WT is performed on MATLAB via the wavelet Daubechies 3,

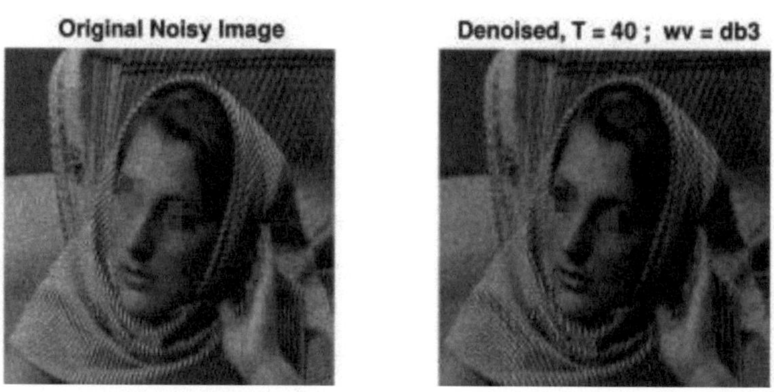

Figure 3.
Image denoising using WT with soft threshold.

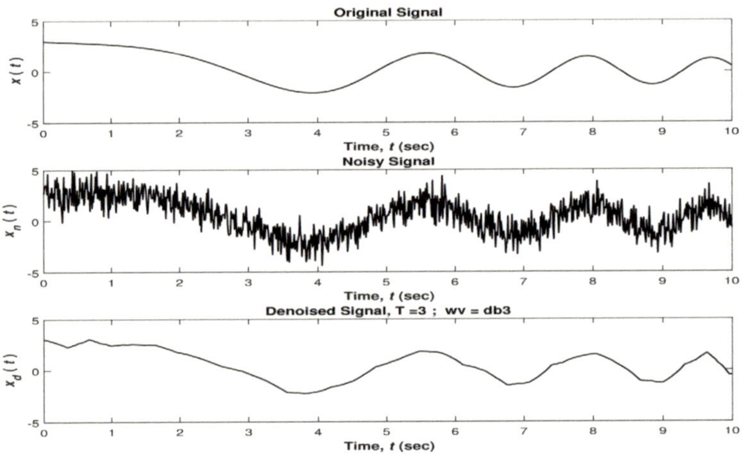

Figure 4.
1D signal denoising using WT with a hard threshold.

5. Discussion and conclusion

The numerical solutions of differential equations using Haar wavelet technique have been studied. Haar wavelet technique is used to approximate the solutions of DEs. The results which obtained form numerical solutions of ordinary differential equations as well as fractional differential equations by Haar collection method are compared with spline base. The numerical results have clearly shown the advantage and the efficiency of the techniques in terms of accuracy and computational time. Special initial value problem of Lane-Emden equation has been solved to show the applicability and efficacy of the Haar wavelet method. Wavelet transform as a mathematical approach has been studied and the applications of wavelet transform in signal processing field have been introduced. The wavelet transform has been effectively used to reveal on the features signals and the compression of signal and image.

Author details

Mohammed S. Mechee[1*], Zahir M. Hussain[1] and Zahrah Ismael Salman[2]

1 University of Kufa, Najaf, Iraq

2 Department of Mathematics, Faculty of Basic Education, Misan University, Amarah, Iraq

*Address all correspondence to: mohammeds.abed@uokufa.edu.iq

IntechOpen

© 2020 The Author(s). Licensee IntechOpen. This chapter is distributed under the terms of the Creative Commons Attribution License (http://creativecommons.org/licenses/by/3.0), which permits unrestricted use, distribution, and reproduction in any medium, provided the original work is properly cited.

References

[1] N. A. Ghani, "Numerical solution of eliiptic partial differential equations by haar wavelet operational matrix method," *Thesis, university Malaya*, 2012.

[2] F. A. Shah and R. Abbas, "Haar wavelet operational matrix method for the numerical solution of fractional order differential equations," *Nonlinear Engineering*, vol. 4, no. 4, pp. 203–213, 2015.

[3] U. Saeed and M. ur Rehman, "Haar wavelet–quasilinearization technique for fractional nonlinear differential equations," *Applied Mathematics and Computation,* vol. 220, pp. 630–648, 2013.

[4] L. Wang, Y. Ma, and Z. Meng, "Haar wavelet method for solving fractional partial differential equations numerically," *Applied Mathematics and Computation,* vol. 227, pp. 66–76, 2014.

[5] Ü. Lepik, "Application of the haar wavelet transform to solving integral and differential equations," in *Proceedings of the Estonian Academy of Sciences, Physics, Mathematics,* vol. 56, no. 1, 2007.

[6] Z. Shi and Y. Cao, "Application of haar wavelet method to eigenvalue problems of high order differential equations," *Applied Mathematical Modelling,* vol. 36, no. 9, pp. 4020–4026, 2012.

[7] S. M. Aznam, "A study of the hyperbolic heat conduction problem and laplace inversion using generalized haar wavelet operational matrix method," *Thesis, university Malaya,* 2012.

[8] O. I. A.-S. . M. S. Mechee, "A study of haar wallet for solving differential equations with some applications," *Thesis, university of Kufa,* 2018.

[9] M. S. Mechee, "Numerical and approxmited solutions of partial differential equations," *Thesis, university of Bagdad,* 1991.

[10] D. Faires and Burden, *Numerical Methods.* Thomson Learning, 2003.

[11] M. S. Mechee, O. I. Al-Shaher, and G. A. Al-Juaifri, "Haar wavelet technique for solving fractional differential equations with an application," in *AIP Conference Proceedings*, vol. 2086, no. 1. AIP Publishing LLC, 2019, p. 030025.

[12] J. Homer Lane, "On the theoretical temperature of the sun under the hypothesis of a gaseous mass maintaining its volume by internal heat and depending on the laws of gases known to terrestrial experiment," *Am. J. Sci., 2d Ser,* vol. 50, pp. 57–74, 1870.

[13] R. Emden, "Gaskugeln, bg teubner, leipzig and berlin," *Google Scholar,* p. 448, 1907.

[14] M. Mechee and N. Senu, "Numerical study of fractional differential equations of lane-emden type by method of collocation," *Applied Mathematics,* vol. 3, no. 08, pp. 851–856, 2012.

[15] N. Berwal, D. Panchal, and C. Parihar, "Solution of differential equations based on haar operational matrix," *Palestine journal of mathematics,* vol. 3, no. 2, pp. 281–288, 2014.

[16] B. Sahoo, "A study on solution of differential equations using haar wavelet collocation method," Ph.D. dissertation, 2012.

[17] Z. Shi, L.-Y. Deng, and Q.-J. Chen, "Numerical solution of differential equations by using haar wavelets," in *Wavelet Analysis and Pattern Recognition, 2007. ICWAPR'07. International*

Conference on, vol. 3. IEEE, 2007, pp. 1039–1044.

[18] R. Chen, "Haar wavelet approach to ordinary differential equation," Ph.D. dissertation, California State Polytechnic University, Pomona, 2016.

[19] Y.-l. Li and L. Hu, "Solving fractional riccati differential equations using haar wavelet," in *Information and Computing (ICIC), 2010 Third International Conference on,* vol. 1. IEEE, 2010, pp. 314–317.

[20] H. Saeedi, N. Mollahasani, M. M. Moghadam, and G. N. Chuev, "An operational haar wavelet method for solving fractional volterra integral equations," *International Journal of Applied Mathematics and Computer Science,* vol. 21, no. 3, pp. 535–547, 2011.

[21] Ü. Lepik, "Solving fractional integral equations by the haar wavelet method," *Applied Mathematics and Computation,* vol. 214, no. 2, pp. 468–478, 2009.

[22] U. Saeed and M. Rehman, "Haar wavelet picard method for fractional nonlinear partial differential equations," *Applied Mathematics and Computation,* vol. 264, pp. 310–322, 2015.

[23] Z. M. Hussain and B. Boashash, "Design of time-frequency distributions for amplitude and if estimation of multicomponent signals," in *Proceedings of the Sixth International Symposium on Signal Processing and its Applications (Cat. No. 01EX467),* vol. 1. IEEE, 2001, pp. 339–342.

[24] Y. Meyer and D. H. Salinger, "Wavelets and operators," 1995.

[25] M. P. Sampat, Z. Wang, S. Gupta, A. C. Bovik, and M. K. Markey, "Complex wavelet structural similarity: A new image similarity index," *IEEE transactions on image processing,* vol. 18, no. 11, pp. 2385–2401, 2009.

[26] K. Abdullah, A. Z. Sadik, and Z. M. Hussain, "On the dwt-and wpt-ofdm versus fft-ofdm," in *2009 5th IEEE GCC Conference & Exhibition.* IEEE, 2009, pp. 1–5.

[27] N. Al-Hinai, K. Neville, A. Z. Sadik, and Z. M. Hussain, "Compressed image transmission over fft-ofdm: A comparative study," in *2007 Australasian Telecommunication Networks and Applications Conference.* IEEE, 2007, pp. 465–469.

Section 2

Wavelet Theory and Communication Systems

Chapter 3

Wavelet Theory and Application in Communication and Signal Processing

Nizar Al Bassam, Vidhyalavanya Ramachandran and Sumesh Eratt Parameswaran

Abstract

Wavelet analysis is the recent development in applied mathematics. For several applications, Fourier analysis fails to provide tangible results due to non-stationary behavior of signals. In such situation, wavelet transforms can be used as a potential alternative. The book chapter starts with the description about importance of frequency domain representation with the concept of Fourier series and Fourier transform for periodic, aperiodic signals in continuous and discrete domain followed by shortcoming of Fourier transform. Further, Short Time Fourier Transform (STFT) will be discussed to induce the concept of time frequency analysis. Explanation of Continuous Wavelet Transform (CWT) and Discrete Wavelet Transform (DWT) will be provided with the help of theoretical approach involving mathematical equations. Decomposition of 1D and 2D signals will be discussed suitable examples, leading to application concept. Wavelet based communication systems are becoming popular due to growing multimedia applications. Wavelet based Orthogonal Frequency Division Multiplexing (OFDM) technique and its merit also presented. Biomedical signal processing is an emerging field where wavelet provides considerable improvement in performance ranging from extraction of abnormal areas and improved feature extraction scheme for further processing. Advancement in multimedia systems together with the developments in wireless technologies demands effective data compression schemes. Wavelet transform along with EZW, SPIHT algorithms are discussed. The chapter will be a useful guide to undergraduate and post graduate who would like to conduct a research study that include wavelet transform and its usage.

Keywords: 1-D and 2-D signals, continuous wavelet transform (CWT), discrete wavelet transform (DWT), orthogonal frequency division multiplexing (OFDM), image compression, cough detection

1. Introduction

We are familiar with real world signals such as speech signal, temperature of a patient in every hour etc. Generally, signals are visualized as a time domain graph. In literature, it is possible to express same information in many different languages; in a similar fashion signals can be represented in frequency domain to convey the message [1]. These signals can be processed to achieve desired outputs or to carry out certain actions according to the application.

IntechOpen

2. Concept of frequency and time domain analysis

2.1 Frequency domain

Figure 1 represents a sinusoidal signal expressed in time and frequency domain. The representation in frequency domain shows there is only one frequency component in the waveform. This representation is simple to extract behavior of signal compared to that in time domain in majority of real-world applications.

In real world, not the signals are processed in frequency domain to arrive at desired output, since they simplify the analysis mathematically [2]. In applications like control engineering, differential equations are used to represent systems. Frequency domain analysis converts the differential equations to algebraic equations which are relatively easy to solve. Applications related to speech, image and video also gets simplified with the use of frequency domain approach since the sensory organs interprets the signal in frequency domain.

2.2 Fourier analysis - the tool used to convert to frequency domain

Any signal which satisfy Dirichlet conditions (signals with finite number of discontinuity, finite maximum or minimum magnitude) can be converted to frequency domain by Fourier analysis. For analog signals Fourier series or Fourier transform if signal is periodic or aperiodic, respectively. The corresponding counterparts for discrete signals are Discrete Fourier series and Discrete Fourier transform [1]. Fourier analysis can be visualized as inner product of a kernel function with the signal. For example, to find the coefficient corresponding to 100 Hz, find the inner product of signal with a kernel which is a unit sine wave of 100 Hz.

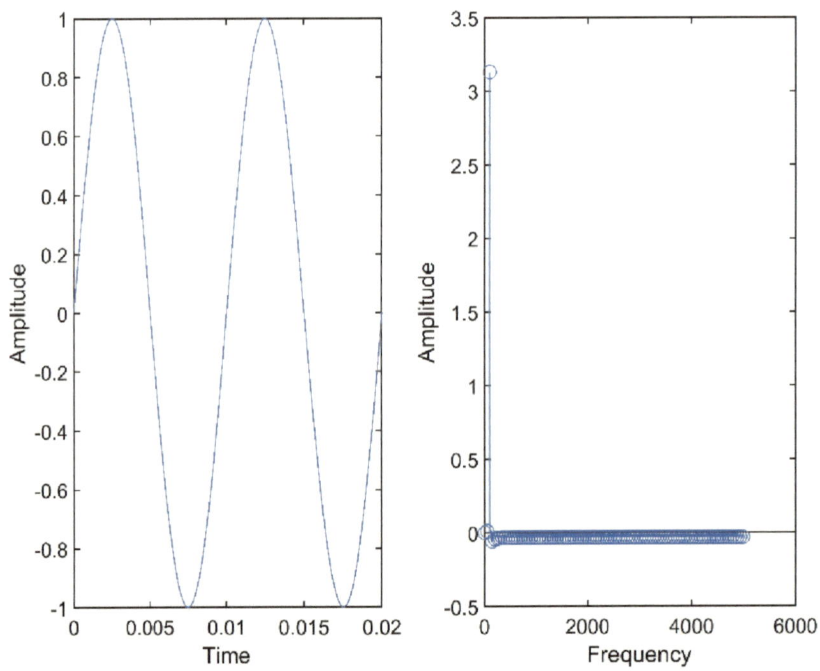

Figure 1.
Representation of a 100 Hz sinusoidal signal in time and frequency domain.

2.2.1 Time frequency analysis - short time Fourier transform

Fourier transforms (FT) assumes periodicity if waveform is known for a finite duration of time and it works only for stationary signals. For non-stationary signals it fails to provide complete behavior of signal. As an example, consider the waveform given in **Figure 2**. The signal has constantly changing frequency. But Fourier

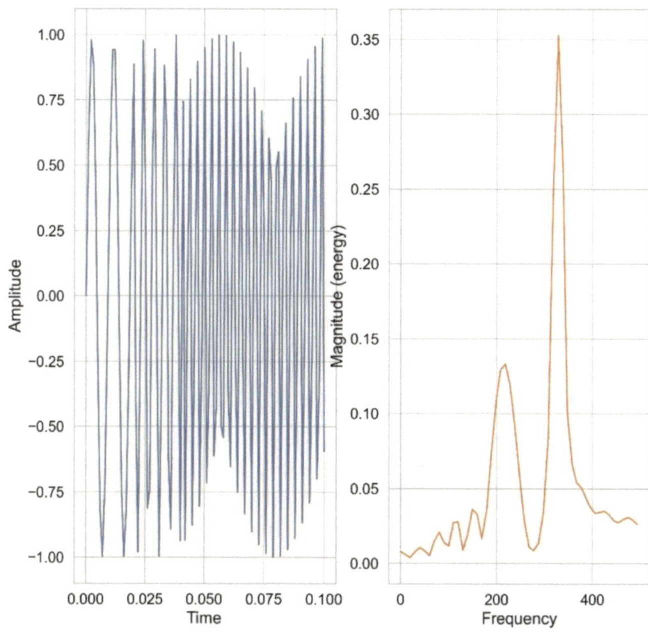

Figure 2.
Non-stationary sinusoidal signal of varying frequency in time and frequency domain.

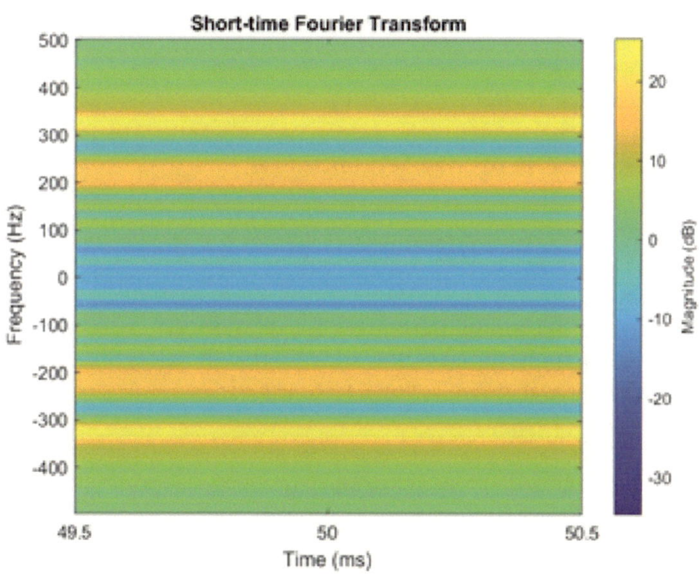

Figure 3.
STFT of chirp signal.

analysis represents it as a combination of 1 Hz and 2 Hz present all the time. If the application demands to initiate some operations when the frequency starts to change, Fourier analysis cannot be used for the application. In order to tackle these difficulties Short Time Fourier Transform (STFT) is used.

In STFT signal is divided into small sub intervals or windows and FT is calculated for each sub interval separately. The problem of STFT is the choice of time windows; low frequencies require large time windows and high frequencies need a short time windows to provide required resolution in time and frequency. This resolution issue is similar to the Heisenberg uncertainty principle relating position and speed which are analogous to time and frequency respectively [3]. **Figure 3** shows STFT of the chirp signal given in **Figure 2**. It is evident that the time localization of frequency is not clearly defined in STFT output. Frequency values corresponding to the 200 Hz and 300 Hz are present across the time interval 49.5 to 50 ms. But **Figure 2** indicated only 300 Hz was present between time 49.5 to 50 ms.

3. Wavelets

As discussed in Section 1.2 the kernel of FT is not time limited (exists for all time values) and this is the bottleneck for analyzing a non-stationary signal. Wavelets are waveforms which are time limited or exists only for a given time period only. Wavelets are useful for examining aperiodic, noisy signal in both time and frequency domain simultaneously. The word “wavelet” means a “small wave”. There are variety of wavelets available which are selected according to the application. The short duration wavelet is superimposed to the signal under consideration for a short duration of time and decompose them to useful form. This process is called wavelet transform. The method of transforming the decomposed signal to original wave is called inverse wavelet transform.

There are the two ways the wavelets are manipulated. The transform of the entire signal is done by translating the wavelet to the next instance of time called as translation. If the signal is of different frequency the mother wavelet is expanded or contracted. This method is called as dilation.

3.1 Continuous wavelet transform (CWT)

Mathematical expression for wavelet function is

$$\psi_{a,b}(t) = \frac{1}{\sqrt{|a|}}\psi\left(\frac{t-b}{a}\right) \tag{1}$$

where a represents the scaling parameter for dilation and ‘b’ represents the moving parameter for translation for the entire signal location [4].

Continuous wavelet transform follows two properties.

$$\int_{-\infty}^{\infty} \psi(t)dt = 0 \tag{2}$$

$$\int_{-\infty}^{\infty} |\psi(t)|^2 = 0 \tag{3}$$

For the CWT the mathematical expression C(a, b) obtained by integrating the input function with the wavelet is

$$C(a,b) = \int_t f(t)\frac{1}{\sqrt{|a|}}\psi\left(\frac{t-b}{a}\right)dt \tag{4}$$

Figures 4 and **5** represent the translation and dilation of wavelets respectively.

3.2 Discrete Wavelet Transform (DWT)

Eq. 1 represents the CWT, on substituting $a = 2^{-j}$ and $b = k2^{-j}$ the equation will become as $D(a,b) = \int_t f(t)2^{-j}\psi(2^j t - k)dt$, where j refer to the scale.

Haar scaling and wavelet functions are given as

$$\phi(t) = \begin{cases} 1, & 0 \le t < 1 \\ 0 & otherwise \end{cases} \tag{5}$$

$$\psi(t) = \begin{cases} 1, & 0 \le t < 1/2 \\ -1 & 1/2 \le t < 1 \end{cases} \tag{6}$$

The scaling coefficients are [h(0), h(1)]= [1, 1] and wavelet coefficients are [g(0), g(1)] = [1, −1]. The scaling and the wavelet functions shown in **Figure 6** are orthogonal with each other. The decomposition of coefficients done by scaling function is done by the low pass filter and wavelet function is by the high pass filter [5].

For eg. Let us consider a signal having function f(t) as

$$f(t) = \begin{cases} 2, & 0 < t \le 1 \\ 4, & 2 < t \le 3 \\ 3, & 4 < t \le 5 \end{cases} \tag{7}$$

If the signal is decomposed by Haar scaling, the function will be

$$f(t) = 2\phi(t) + 4\phi(t-1) + 4\phi(t-2) + 3\phi(t-3) + 3\phi(t-4) \tag{8}$$

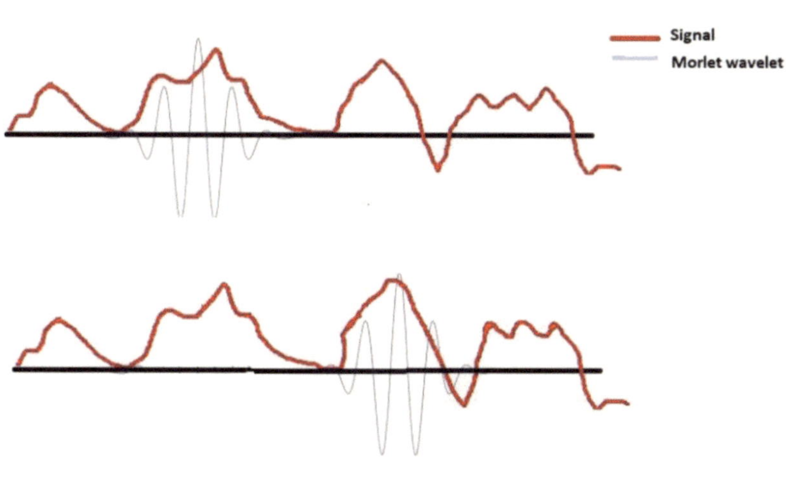

Figure 4.
Signal with the translated Morlet wavelet [5].

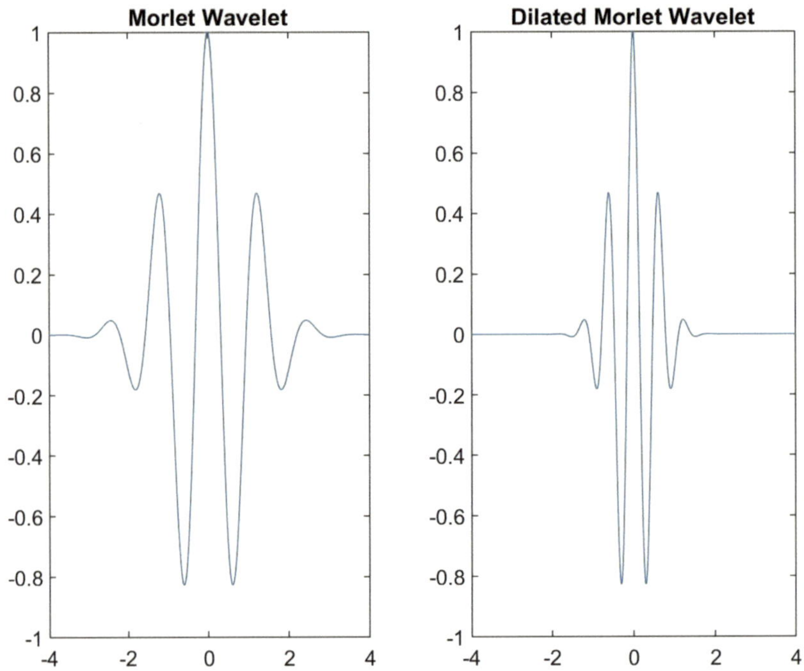

Figure 5.
Dilated Morlet wavelet [5].

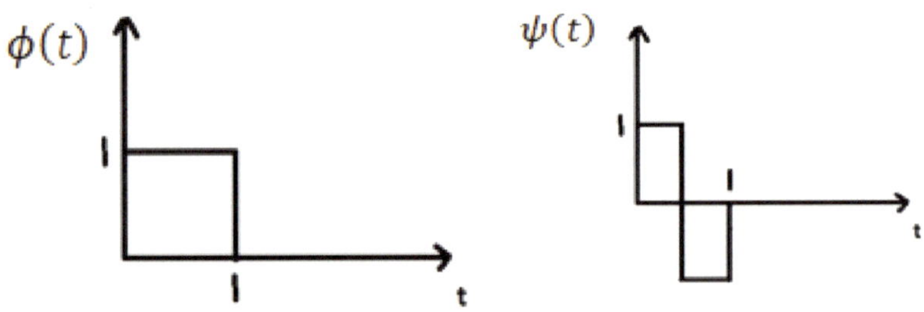

Figure 6.
Haar scaling ϕ (t) and wavelet function ψ(t) [5].

3.3 Properties of the scaling and wavelet functions

1. The scaling and wavelet functions are orthogonal to each other

2. The scaling function and translates are orthogonal to each other.

3. The area of the scaling function is equal to one.

4. The scaling function and translates of wavelet function are orthonormal to each other.

All the above conditions are satisfied by all the scaling and wavelet coefficients. Nonlinear equations are derived using the numerical values of the coefficients [5].

4. Analysis and synthesis of signals

As discussed in Section 2, h(n) and g(n) are the dilation coefficients.
The refinement relation can be written as

$$\phi(t) = \sum_{n=0}^{N-1} \mathrm{h(n)}\sqrt{2}\phi(2t - n) \tag{9}$$

$$\psi(t) = \sum_{n=0}^{N-1} \mathrm{g(n)}\sqrt{2}\phi(2t - n) \tag{10}$$

where h(n) and g(n) are the scaling and wavelet coefficients, N is the support of wavelet [4].

4.1 Decomposition of signal

The analysis of the signal or the decomposition of the signal can be represented in a block diagram as decomposing input into low frequency and high frequency samples.

Let us consider a signal having function f(t) as

$$f(t) = \begin{cases} 2, & 0 < t \le 1 \\ 4, & 2 < t \le 3 \\ 3, & 4 < t \le 5 \end{cases} \tag{11}$$

If the signal is decomposed by Haar scaling, the function will be

$$f(t) = 2\phi(t) + 4\phi(t-1) + 4\phi(t-2) + 3\phi(t-3) + 3\phi(t-4) \tag{12}$$

The decomposed output of the low pass filter shown in **Figure 7** can be visualized as the average of the immediate signals.

Figure 7 represents the decomposition and reconstruction of one level of the signal. Let us consider the vector space of the input function f(t) is V_2. After decomposition the samples are decomposed as f1(t) of low frequency samples spanning the vector space of V_1 and high frequency samples g1(t) of high frequency spanning the vector space of W_1.

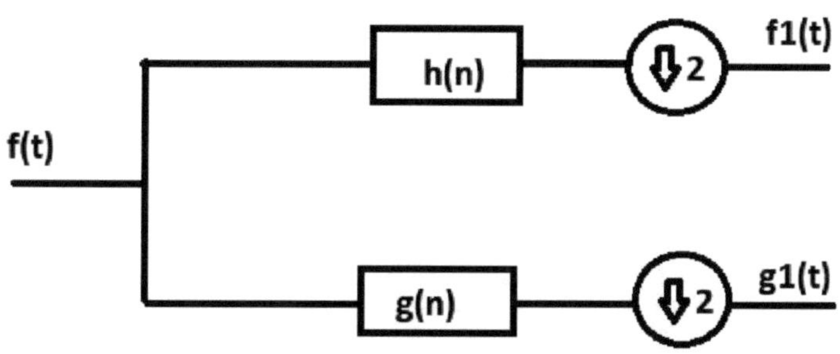

Figure 7.
Block diagram of one level of decomposition of signal.

If the signal f1(t) is again decomposed (second level of decomposition), then they are still split up as low frequency samples say f2(t) spanning the vector space V_0 and the high frequency samples say g2(t) spanning the vector space W_0. Like this, the signal can be decomposed for any level there by splitting the signal with the cut off frequency as per the coefficients of the filter h(n) and g(n).

For V_2 space the vector space can be written as

$$V_2 = V_1 \oplus W_1$$
$$V_2 = V_0 \oplus W_0 \oplus W_1 \tag{13}$$

Frequency band split and the vector space is shown in the **Figure 8**.

Thus, decomposition is the process in which the signal of high resolution is converted to signal of two orthogonal signal of low resolution [5].

4.2 Reconstruction of signal

The synthesis or the reconstruction of the signal can be represented in a block diagram (**Figure 9**) combining input low frequency and high frequency samples.

For a perfectly matched filter the signal f(t) given as input in the analysis filter and the output of the reconstruction filter r(t) will be equal.

Any signal resolution in the present level and previous level are orthogonal. This property was useful for the wavelets to be used for Multiresolution Analysis (MRA) [4].

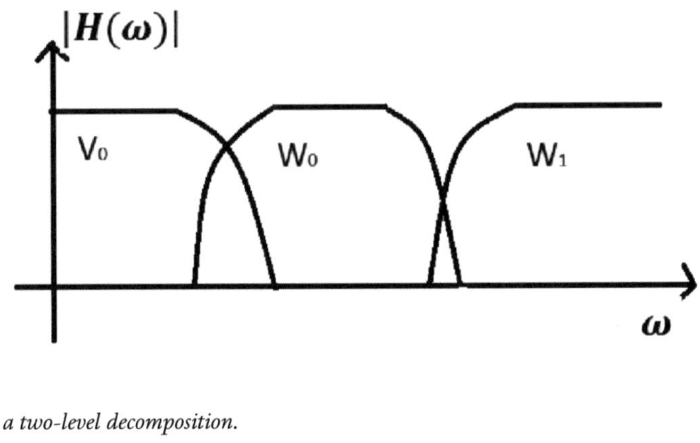

Figure 8.
Frequency split for a two-level decomposition.

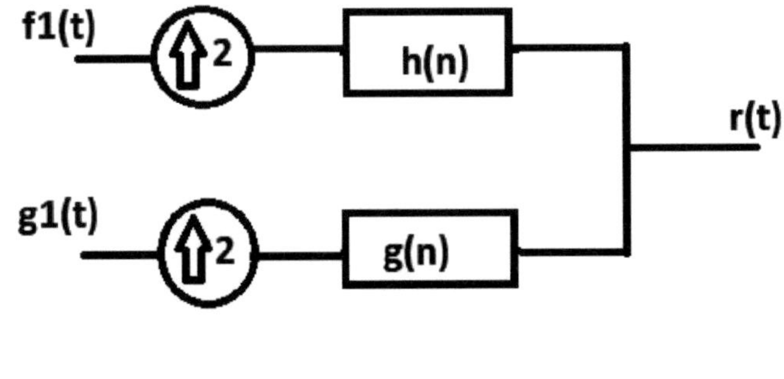

Figure 9.
Block diagram of one level of reconstruction of signal.

4.3 Decomposition of 2-D signal

Two-dimensional signal is an image that has row and column pixel or samples. Wavelet decomposition of the image can be done by scanning the pixel values in a regular fashion and given as input to the decomposition block as explained in Section 3.1. Two types of scanning methods are used for reading the row-column pixel values thereby converting the 2-D signal to 1-D. They are raster scanning and Morton scanning.

Figure 10 shows the scanning of pixel values in the horizontal direction. Once the samples are decomposed by sending it to the block in **Figure** 7, then the pixels are then scanned vertically and again decomposed. This completes one level of decomposition of 2-D signal. After horizontal decompose, the image will be separated as Low pass and high pass pixels in horizontal fashion. Once the vertical decompose is done followed by horizontal the image is divided into two rows and two columns named as LL, LH, HL, HH as shown in **Figure 11**.

For a second level of decomposition the scanning process is done to the LL band of the first level and LL band is again decomposed as LL1, LH1, HL1, HH1. Thus, the high-resolution signal is transformed to Low frequency and high frequency bands. For higher levels, the process is repeated for the LL bands leading to the coarser resolutions [5].

4.4 Reconstruction of 2-D signal

The decomposed signal after processing can be reconstructed by up sample the data by two and then convoluted with the filter of low and high frequency coefficients. This need to be done for all the sub bands and the values are summed up to move from the coarser scale to high resolution scale. This process is repeated to reach the final image is reached [5].

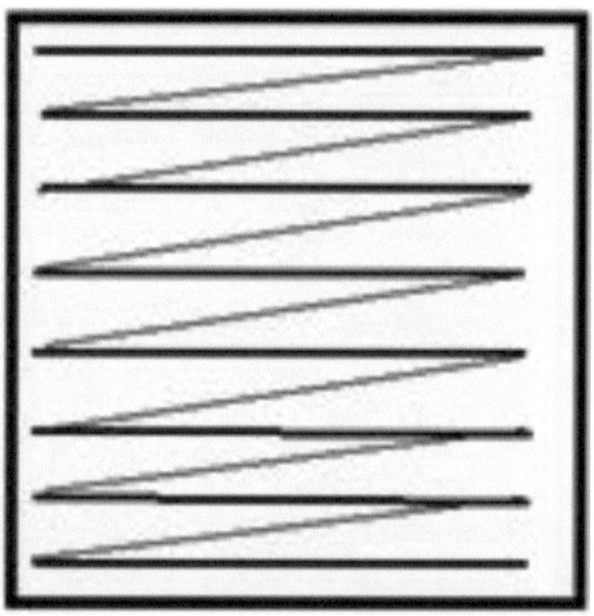

Figure 10.
Scanning of pixels.

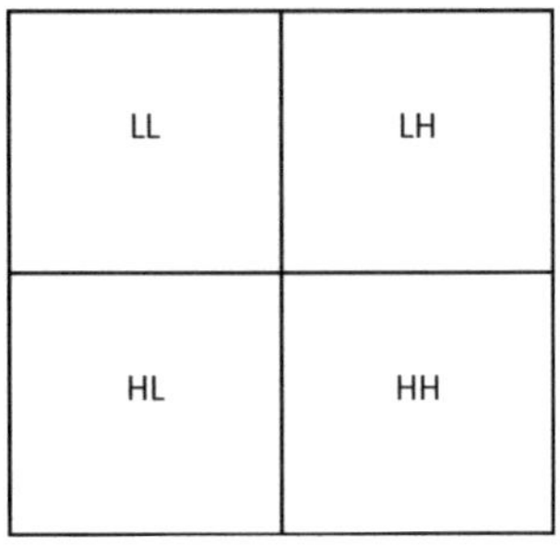

Figure 11.
Bands of 2-D after one level of decomposition.

5. Applications

5.1 Wavelets in communication system

Wavelets are suited for communication systems due to its potential to handle non stationary behavior and to segregate information into uncorrelated segments. This section provides a brief about various methodologies to extract benefits of wavelets into communication systems.

5.1.1 OFDM systems using wavelets

Orthogonal Frequency Division Multiplexing (OFDM) uses multicarrier technique by dividing the spectrum into many subcarriers and each subcarrier gets modulated by low data rate. If M symbols are required to be transmitted, M orthogonal subcarriers are used to modulate each symbols.

Wavelet packet transforms can be effectively used in OFDM systems **Figure 4**.1 shows a wavelet packet based system.

In **Figure 12**, M band IDWT is used to split the signal to smaller sub bands while the M band DWT is used to combine the sub bands. The wavelet packet transform structure is given in **Figure 13**. Here sub band division is done in both approximation and detail coefficients compared with a discrete wavelet transform.

Wavelet based OFDM systems provides improved Bit Error Rate (BER) and peak-to-average-power ratio (PAPR) performance compared to conventional OFDM system using FFT. Due to large number of sub band carriers, OFDM has high PAPR and makes it sensitive to nonlinear effects. Wavelet based system controls the PAPR ratio and results in improved performance compared to FFT based

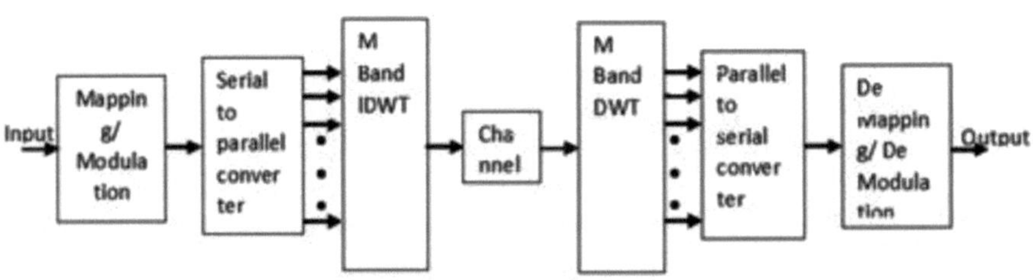

Figure 12.
Wavelet based OFDM system [6].

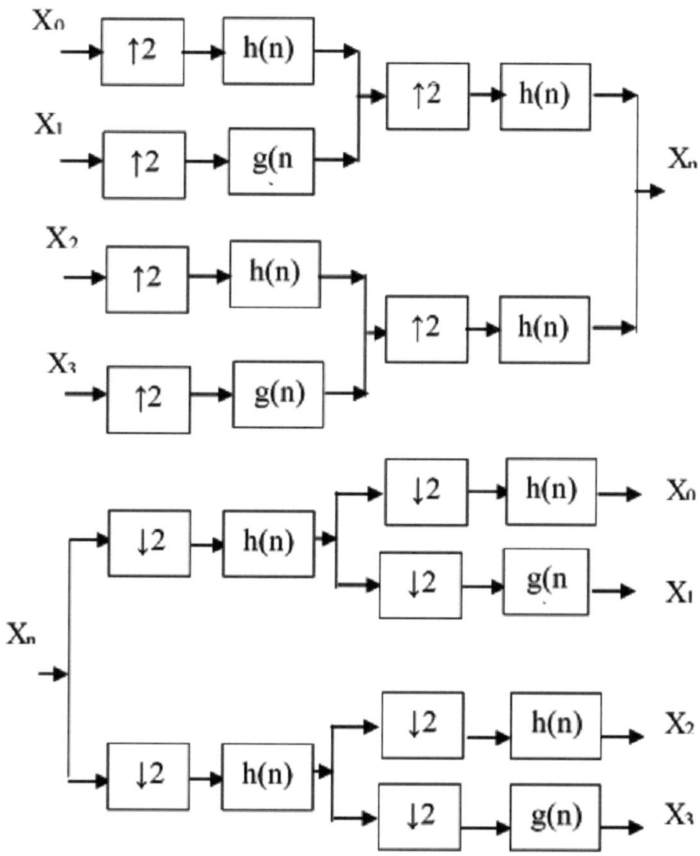

Figure 13.
Wavelet packet based combining or splitting into sub bands in OFDM system.

systems. Applications of Wavelets in 4G, 5G and beyond was reported in literature [6–8]. Selection of wavelet and level plays an important role and is promising field for future research.

5.2 Wavelets in biomedical applications

Bio medical signal processing is a promising field for future research. Wavelet transform can be used effectively in combination with artificial intelligence to provide solution to many problems. Bio medical signals are generally one dimensional time series data (Electro Cardiogram- ECG, electroencephalogram -EEG) or an image (X ray, ultrasound scan, MRI). Accordingly a 1D or 2D wavelet transform can be used to process the signal. Wavelet transform helps to divide the signal to uncorrelated sub bands due to orthogonality property. The transform coefficients or a part of it (say certain level coefficients) are used as feature for classifying the signal is a common methodology that can be adopted for a variety of applications. Recent advancement in neural networks like CNN with wavelet coefficients as input features opens up stage for a wide variety of research solutions. Another promising category of application is in signal preprocessing to remove unwanted information in biomedical signals [9] with thresholding techniques. A complete list of applications is beyond the scope of this book chapter, a few applications of wavelets are described briefly in coming sections.

5.2.1 Wavelets in EEG signal processing

Wavelet transform with suitable mother wavelet and appropriate level of transform has to be selected to extract maximum performance for a given application. **Figure 14** gives a general block level representation for an EEG signal classification system to detect abnormality type.

EEG signal after required preprocessing steps to remove noise is passed through a DWT step to divide it to different sub bands. The required sub bands are further passed through a thresholding step (optional) to remove certain features that do not contribute to abnormalities. The DWT coefficients are fed as input to a neural network classifier to classify the signal to different categories based on the abnormality. The challenges in these signal processing is selection of wavelets, level of transform, threshold calculations, selection of neural network of appropriate level and availability of data to train the network to achieve a desired accuracy [10]. In [11] EEG classification system 5 level decomposition with Db4 wavelet is used to extract features for a neural network classifier is described to arrive at success rate of 94% to identify healthy, epilepsy syndrome and seizure cases with 250 and 50 training and testing samples of EEG respectively. Similar kind of systems can be used to classify ECG signals as well. In [12] Haar wavelet transform was used to identify characteristics of ECG wave. QRS complex detection sensitivity above 99% was reported. Here wavelet packet transform for 20 levels are done and D20 coefficient was used to detect R wave of ECG signal.

5.2.2 Wavelets in cancer detection

Wavelets can be effectively used for abnormality detection in biomedical images. In [13] wavelet decomposition coefficients are used to extract features by calculating 2 level Haar wavelet transform and extract mean, standard deviation and energy of the transform coefficients as features for extraction of abnormal areas in image.

Figure 15 shows the block schematics of the system described in [13].

5.2.3 Research scope for wavelets in COVID 19

Cough detection is one of the important application in monitoring of public places and to monitor people in isolation or quarantine. Ability of wavelet to model nonstationary signal is a potential strength here since cough signals (audio) are non stationary time series signal. Currently Arduino based cough detection systems are being developed as low cost cough detection system. In these system Mel Frequency Cepstral Coefficient (MFCC) are used to extract feature from cough signals to train neural network model like Convolutional Neural Networks (CNN) [14]. Wavelets can be used instead of MFCC for feature extraction to improve system accuracy. This require further research to arrive at optimized set of wavelets and appropriate level of transform.

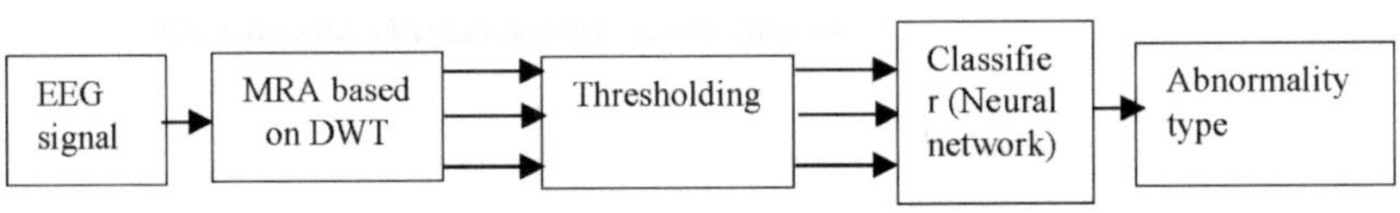

Figure 14.
Wavelet based EEG signal classification system.

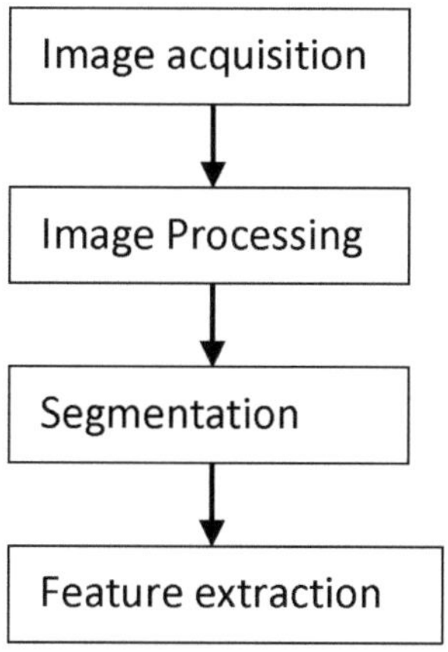

Figure 15.
Block diagram of breast cancer diagnosis system [13].

5.3 Image compression

In today's world handling of huge amount of data is complex. Conversion of large data files into smaller one is important for storing and transmission. This conversion process is called as data compression. This is important in this digital era. The reverse process is called as decompression. Several algorithms are proposed for efficient conversion process. Lossy and Lossless compression techniques are used in different applications. JPEG image compression technique is the standard technique used. It uses Discrete Cosine Transform (DCT) for the frequency domain conversion followed by Huffman coding [5].

5.3.1 EZW (embedded zero tree) coder

EZW algorithm was proposed by Shapiro. After applying Wavelet transform on an image the pixels will be converted to four bands of pixels as explained in Section 3.3. then the wavelet coefficients are passed to the EZW encoder. It is a progressive encoding which is embedded encoding with high accuracy. EZW is a lossless compressor which can be applied for any level of decomposition [15].

EZW takes care of coding both time and space. After decomposition, the pixels are represented as trees. One coefficient in the lower sub band will have four children. Each of the children will have four more grandchildren in the next higher sub band. **Figure 16** represents the relationship between the coefficients of lower sub bands with the higher sub bands.

Each tree is coded with a symbol as P, N, T, Z. For every decomposition levels, the pixels are categorized to two passes as Dominant pass and Subordinate pass. Dominant pass will find the pixel values which are above the threshold value and subordinate pass quantize all the pixel values in the present pass and the previous dominant pass. The threshold value is calculated using the equation

$$T_0 = 2^{\lfloor \log_2(Max(|C(x,y)|)) \rfloor} \tag{14}$$

where (x, y) are the coordinates of the image and C (x, y) represent the pixel coefficient. The main advantage of the EZW coder is that the output can be coded for the desired size. The eliminated pixels are least significant bits and the low frequency pixel are not eliminated entirely.

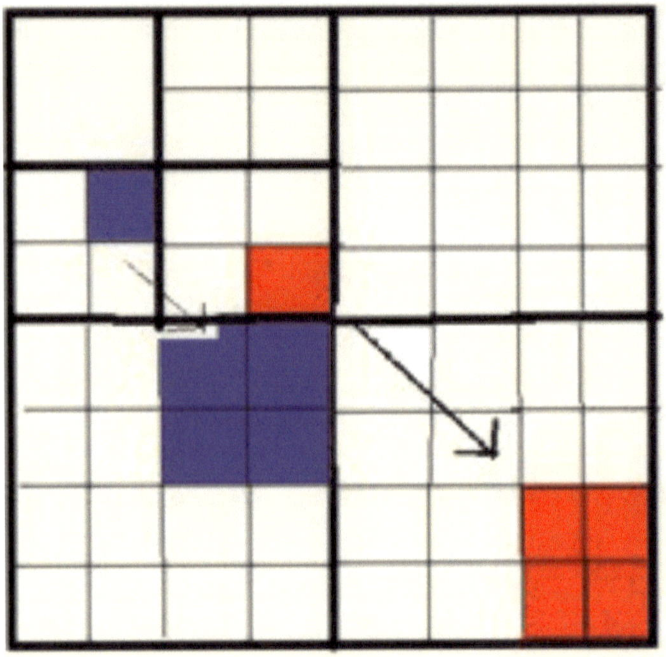

Figure 16.
Wavelet coefficients and their relationships with sub bands.

5.3.2 SPIHT Coder:

SPIHT algorithm was designed by Said and Pearlman. This algorithm has a spatial relationship with the children and grandchildren of the pixels. As in EZW algorithm SPHIT algorithm has two passes, sorting and refinement pass. In the sorting pass, the List of Insignificant Pixel (LIP) is updated comparing with the threshold value. Then the List of Insignificant Set (LIS) is updated. The sorting of LIS is to have a single zero for the set of coefficients and zero trees. In the refinement pass, the List of significant Pixel (LSP) is used to refine the coefficients with the current threshold. For the next level the threshold will be halved [16].

5.3.3 Performance analysis.

The performance of the compressor algorithms is evaluated by PSNR (Peak Signal to Noise Ratio) and MSE (Mean Square Error)

$$PSNR(db) = 10\,log_{10}\left(\frac{255^2}{MSE}\right) \tag{15}$$

$$MSE = \frac{1}{mn}\sum_{i=0}^{m-1}\sum_{j=0}^{n-1}\left[(I(i,j) - K(i,j))^2\right]$$

where m and n are the number of rows and columns.

where $I(i,j)$ is the coefficients of the original image and $K(i,j)$ is the coefficients of the decompressed image. When the original image and the decompressed image are same, they subtracts each other and the value of MSE is equal to zero. So lesser the value of MSE the better the compression by the coder. The performance of SPHIT algorithm and EZW algorithm outperforms the JPEG standard compression algorithm.

6. Conclusion

Frequency domain approach is a powerful tool in signal processing since many of the natural system like human auditory system works based on frequency content of audio signals. Fourier transform is a conventional tool to convert the signal from time to frequency domain. However, FT fails for non-stationary signals. In such a situation wavelet transforms are primarily useful for processing non stationary signals. Applications demanding time and frequency information simultaneously wavelets are considered as a potential tool to provide tangible results. Ability of wavelet transform to split signals into orthogonal bands makes them suitable in communication systems. Improved BER and PAPR performance is highlight of wavelet based system. Biomedical signals like speech, scan images etc. can be processed effectively with wavelet transform in conjunction with techniques like neural network classifier can be thought as an efficient method to solve a multitude of problems. Compression and preprocessing of signal are other categories of wavelet applications. Artificial Intelligence techniques are considered as a hot research topic in recent technology. A combination of wavelet based feature extraction and AI techniques can be applied to problems demanding processing of big data.

Acknowledgements

The research leading to these results has received funding from the Ministry of Higher Education, Research and Innovation in the Sultanate of Oman under Covid19 Funding Program. TRC Block Funding Agreement No [TRC/CRP/MEC/COVID-19/20/09].

Author details

Nizar Al Bassam*, Vidhyalavanya Ramachandran and Sumesh Eratt Parameswaran
Middle East College, Muscat, Oman

*Address all correspondence to: nazarhooby@yahoo.co.uk

IntechOpen

© 2021 The Author(s). Licensee IntechOpen. This chapter is distributed under the terms of the Creative Commons Attribution License (http://creativecommons.org/licenses/by/3.0), which permits unrestricted use, distribution, and reproduction in any medium, provided the original work is properly cited.

References

[1] Oppenheim A, Schafer R. Discrete-Time Signal Processing. 3rd ed. Essex: Pearson New International Edition; 2013. ISBN-10:1–292–02572-7

[2] Cadence PCB Solutions. Time Domain Analysis vs Frequency Domain Analysis: A Guide and Comparison [Internet]. 2020. Available from: https://resources.pcb.cadence.com/blog/2020-time-domain-analysis-vs-frequency-domain-analysis-a-guide-and-comparison [Accessed: 2020-08-20]

[3] Robi Polikar. Wavelet tutorial [internet]. Available from: http://users.rowan.edu/~polikar/WTpart1.htmll [Accessed: 2020-08-28]

[4] Strang G, Nguyen T. Wavelets and Filter Banks. 2nd ed. Wellesly – Cambridge Press; 1996. ISBN- 10: 0961408871

[5] Soman KP. Insights into Wavelets : From Theory to Practice. Gardners Books; 2004. ISBN-10 : 8120326504

[6] M. F. Raji, J. P. Li, A. U. Hao, E. Raji, M. Happy. Performance Evaluation of Orthogonal Wavelet Division Multiplex for 5G and Beyond. In Proceedings of the IEEE International Computer Conference on Wavelet Active Media Technology and Information Processing, Chengdu, China: IEEE; 2019, pp. 1–5, doi: 10.1109/ICCWAMTIP47768.2019.9067577

[7] Wenwen L, Zheng D, Lin Q, Chengzhuo S. Wavelet Transform Based Modulation Classification for 5G and UAV Communication in Multipath Fading Channel . Physical Communications. Elsevier, 34, 2019, pp 272–282.

[8] R. Kaur, A. Singh. PAPR and Spectrum Analysis of 4G and 5G Techniques. In Proceedings of 2019 International Conference on Wireless Communications Signal Processing and Networking (WiSPNET), Chennai, India, 2019, pp. 18–21, doi: 10.1109/WiSPNET45539.2019.9032858.

[9] Giulia B, Eleonora S, Monica U, Roberto T, Luigi R, Danilo P. Wavelet Denoising as a Post-Processing Enhancement Method for Non-Invasive Foetal Electrocardiography. Computer Methods and Programs in Biomedicine. Elsevier, 195, 2020, 105558.

[10] I. Omerhodzic, S. Avdakovic, A. Nuhanovic, K. Dizdarevic Energy Distribution of EEG Signals: EEG Signal Wavelet-Neural Network Classifier, https://arxiv.org/ftp/arxiv/papers/1307/1307.7897.pdf accessed on 14/10/2020]

[11] Prochazka A, Kukal J, Vysata O. Wavelet transform use for feature extraction and EEG signal segments classification. In Proc. 2008 IEEE Communications, Control and Signal Processing, 3rd International Symposium, pp. 719–722.

[12] Hsieh JC, Tzeng WC, Yang YC, Shieh S M. Detecting ECG characteristic points by novel hybrid wavelet transforms: an evaluation of clinical SCP-ECG database. Computers in Cardiology, 2005, Lyon, 2005, pp. 751–754, doi: 10.1109/CIC.2005.1588213.

[13] S. Benmazou, Merouani HF. Wavelet based feature extraction method for breast cancer diagnosis, *2018 4th International Conference on Advanced Technologies for Signal and Image Processing (ATSIP)*, Sousse, 2018, pp. 1–5, doi: 10.1109/ATSIP.2018.8364477.

[14] Igor D. S, Miranda, Andreas H. D, Thomas RN. A Comparative Study of Features for Acoustic Cough Detection Using Deep Architectures [Internet]. Available from: https://dsp.sun.ac.za/

~trn/reports/miranda+diacon+niesler_embc19.pdf

[15] Mischel M, Yves M, Oppenheim G, Jean Michel P. The EZW Algorithm [Internet]. 2007. Available from: [Accessed 2020-09-21] https://onlinelibrary.wiley.com/doi/pdf/10.1002/9780470612491.app1

[16] Mukherjee D, Mitra S. Mint: Vector SPIHT for Embedded Wavelet Video and Image Coding, IEEE Trans. Circuits and Syst. Video Tech. March 2003.

Chapter 4

Wavelet Based Multicarrier Modulation (MCM) Systems: PAPR Analysis

Jamaluddin Zakaria and Mohd Fadzli Mohd Salleh

Abstract

Orthogonal frequency division multiplexing (OFDM) is a prominent system in transmitting multicarrier modulation (MCM) signals over selective fading channel. The system offers to attain a higher degree of bandwidth efficiency, higher data transmission, and robust to narrowband frequency interference. However, it incurs a high peak-to-average power ratio (PAPR) where the signals work in the nonlinear region of the high-power amplifier (HPA) results in poor performance. Besides, an attractive dynamic wavelet analysis and its derivatives such as wavelet packet transform (WPT) demonstrates almost the same criteria as the OFDM in MCM system. Wavelet surpasses Fourier based analysis by inherent flexibility in terms of windows function for non-stationary signal. In wavelet-based MCM systems (wavelet OFDM (WOFDM) and Wavelet packet OFDM (WP-OFDM)), the constructed orthogonal modulation signals behaves similar to the fast Fourier transform (FFT) does in the conventional OFDM (C-OFDM) system. With no cyclic prefix (CP) need to be applied, these orthogonal signals hold higher bandwidth efficiency. Hence, this chapter presents a comprehensive study on the manipulation of specified parameters using WP-OFDM, WOFDM and C-OFDM signals together with various wavelets under the additive white Gaussian noise (AWGN) channel.

Keywords: multicarrier modulation (MCM), orthogonal frequency division multiplexing (OFDM), peak-to-average power ratio (PAPR), wavelet transform, wavelet packet transform (WPT)

1. Introduction

Orthogonal Frequency Division Multiplexing (OFDM) technique provides a number of advantages: In OFDM since the subcarriers are overlapped, accomplishes a higher degree of spectral efficiency that results in higher transmission data rates. Considering the use of the efficient FFT technique, the process is considered computationally lower. Besides, in the Single-Carrier Modulation (SCM) the ISI problem which commonly occurs the use of the cyclic prefix (CP) greatly eliminates the problem [1]. The division of a channel into several narrowband flat fading (subchannels) results in the subchannels being more resilient towards frequency selective fading. The loss of any subcarrier(s) due to channel frequency selectivity, proper channel coding schemes can recover the lost data [1]. Thus, this technique

IntechOpen

offers robust protection against channel impairments without the need to implement an equalizer as in the SCM, and this greatly reduces the overall system complexity. However, the high Peak-to-Average Power Ratio (PAPR) has been the major drawback in the OFDM system. This situation happens when the peak OFDM signals surpass the specified threshold and as a result the high-power amplifier (HPA) operates in a nonlinear region. This produces spectral regrowth of the OFDM signals and broken the orthogonality among the subcarriers. Thus, the effect on bit error rate (BER) performance at the receiver is poor.

To deliver massive high-speed data over a wireless channel, Multi-carrier-modulation (MCM) scheme has been widely used transmission technique. Despite its advantages, the MCM scheme is prone to high PAPR signal transmission, which has been single out as the main difficulty. In the MCM scheme, the conventional way to obtain orthogonal subcarrier signals is by using a Fourier transform. The emergence of wavelet transforms has paved the way for new promising techniques to obtain orthogonal subcarrier signals in future MCM systems. Wavelet transforms have been testified practical for the MCM system due to the orthogonal overlapping symbols property that they possess in time and frequency domains, respectively.

In order to mitigate PAPR, there have been many techniques proposed in literature either to reduce the peak power with fixed average power or alter the distribution so that the average power produced has smaller peak power [2–6]. Due to this, there are two categories of PAPR reduction techniques which are called as signal distortion technique and signal scrambling technique. A prominent technique known as Partial Transmit Sequence (PTS) has been first introduced in [7]. This technique categorized as signal scrambling offers big potential for further exploration as explored in the works [8–11].

This chapter presents the analysis of various wavelet families in their applicability towards MCM systems and their PAPR profiles. Details analysis is presented for obtaining the BER results for various Wavelets.

2. Background

2.1 Wavelet transform

In this section the basic concept of wavelet and wavelet packet transform (WPT) are presented. The WPT is constructed based on the continuous wavelet transform (CWT) and wavelet transform (WT) theory. For ease of reading, all the following equations in these subsections are mostly taken from [12–15].

2.1.1 Discrete wavelet transform (DWT)

The computation cost for wavelet coefficients in the CWT is high since they are highly redundant data, which is not desirable for real application. Therefore, discrete wavelets offer as the alternative for practical applications. In discrete wavelets, the scalable and translatable wavelets are discrete. The process of discrete scaling and translation of the mother wavelet can be expressed as

$$\psi_{\alpha,\beta} = \sqrt{a_0^{\alpha}}\psi\left(a_0^{\alpha}t - \beta_{b_0}\right) \tag{1}$$

where a_0 represents the fixed step of dilation and b_0 indicates the translation factor. The integer α and β signify the indices scale and translation, respectively. The scaling in time domain correlates with an inverse scaling in frequency domain, therefore the product of $(\Delta t_{a,b})\left(\Delta f_{a,b}\right)$ is independent of the dilation parameter a. If any time resolution gain is obtained, this inversely effect the cost of frequency resolution and vice versa. Therefore, this detains the Heisenberg uncertainty principle for the dilated and translated wavelet $\psi_{CWT_{a,b}}(t)$ and the mother wavelet $\psi_{CWT}(t)$. The most natural choice for dilation step is 2 that results in octave bands or dyadic scales. The wavelet is compressed in frequency domain by a factor of 2 for each successive value of scale index. This produces the stretched in time domain by the same factor. The translation factor is set to the value of "1" to get the dyadic sampling fashion. The time-shift and scaling function are set as [16];

$$\varphi_\beta = \varphi(t-\beta), \beta \epsilon Z, \varphi \epsilon L^2 \tag{2}$$

where Z is the set of all integer numbers, and $L^2(\mathbb{R})$ is the vector space of square integrated function. The parameter v_0 is a space spanned by scaling function, which is defined as

$$v_0 = \frac{\overline{Span\{\varphi_\beta(t)\}}}{\beta}, \beta \in Z \tag{3}$$

In this subspace, if $x(t) \in v_0$, it can be expressed as

$$\boldsymbol{x}(\boldsymbol{t}) = \sum_{\beta=-\infty}^{+\infty} \boldsymbol{\alpha_\beta \varphi_\beta(t)} \tag{4}$$

One can increase the size of the subspace by changing the time scale of the scaling functions. The two-dimensional parameterization (time and scale) of scaling function $\varphi(t)$ from v_0 to v_α can be expressed as

$$\boldsymbol{\varphi_{\alpha,\beta}} = \boldsymbol{2^{\alpha/2}\varphi(2^\alpha t - \beta)} \tag{5}$$

Then, the new function for the expanded subspace v_α is given as

$$v_\alpha = \frac{\overline{span\{\varphi_\beta(2^\alpha t)\}}}{\beta} = \frac{\overline{span\{\varphi_{\alpha,\beta}(t)\}}}{\beta} \tag{6}$$

In the extended subspace, whenever $x(t) \in v_a$, then it can be expressed as

$$\boldsymbol{x}(\boldsymbol{t}) = \sum_{\beta=-\infty}^{+\infty} \boldsymbol{\alpha_\beta \varphi(2^\alpha t + \beta)} \tag{7}$$

From (Eq. (7)), the span v_α is larger than v_0, for $\alpha > 0$ and $\varphi_{\alpha,\beta}(t)$ able to represent the finer detail (due to its finer scale). For $\alpha < 0$ this condition is true that represents for the coarse scale. Wavelet obeys to multi-resolution concept's requirement, where every signal is decomposed into finer detail gradually as expressed as [17, 18].

$$\dots C v_{-2}\ C v_{-1}\ C v_0\ C v_1\ C v_2\ C \dots . \tag{8}$$

where the terms $v_{+\infty} = L^2$, and $v_{-\infty} = \{0\}$ indicate that within the same vector space of L^2, there exist both high resolution and low-resolution coefficients.

Consequently, if $x(t) \in v_a$, then $x(2t) \in v_{a+1}$. Additionally, the $\phi(t)$ term is expressed as the weighted sum of the time-shifted scaling function

$$\varphi(t) = \sum_{n=-\infty}^{+\infty} h(n)\sqrt{2}\varphi(2t - n), n \in Z \tag{9}$$

where the term $h(n)$ represents the scaling function coefficients (sequence of real or imaginary numbers). The $v_{\alpha+1}$ is the expanded space of v_α and w_α represents the corresponding orthogonal complement. Therefore, a new set of spaces is produced. Suppose that $w_{\alpha+1}$ be the subspace spanned by the wavelet, the enlargement of v_1 and v_2 space are written as (Eq. (10)) below and as illustrated as in **Figure 1** [19].

$$\begin{aligned} v_1 &= v_0 \bigoplus w_0 \\ v_2 &= v_1 \bigoplus w_1 = (v_0 \bigoplus w_0) \bigoplus w_1 \\ &\vdots \\ v_{\alpha+1} &= v_\alpha \bigoplus w_\alpha = v_0 \bigoplus_{l=0}^{\alpha} w_l, \forall \alpha \in Z \end{aligned} \tag{10}$$

The definition of the wavelet function $\psi(t)$ is the same as the scaling space v_0. Let the space spanned by the wavelet function $\psi_\beta(t)$ be w_0, and the expanded space spanned by $\psi_{\alpha,\beta}(t)$ be w_α that is obtained after using (Eq. (3)) to (Eq. (6)). The w_α term is orthogonal to v_α and thus the orthogonality between $\varphi(t)$ and $\psi(t)$ is given as [19];

$$\langle \varphi_{\alpha,\beta}(t), \psi_{\alpha,\beta}(t) \rangle = \int \varphi_{\alpha,\beta}(t)\psi_{\alpha,\beta}(t)dt = 0 \tag{11}$$

Due to these wavelets are in space spanned by the next finer scaling function, the wavelet function $\psi(t)$ can be expressed by the sum of the weighted time-shifted wavelet function given as

$$\psi(t) = \sum_{n=-\infty}^{+\infty} g(n)\sqrt{2}\varphi(2t - n), n \in Z \tag{12}$$

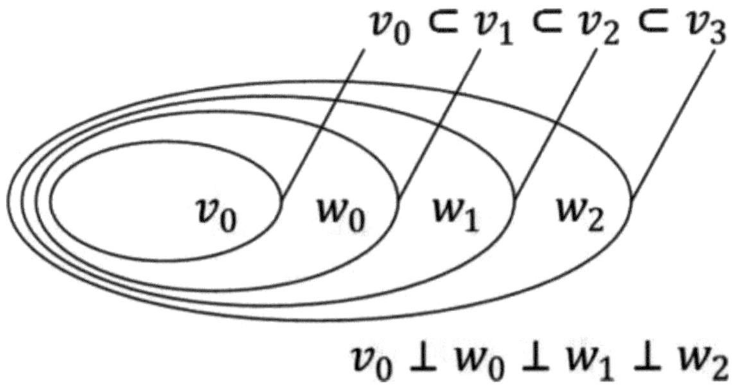

Figure 1.
Wavelet vector spaces and scaling function.

where $g(n)$ is called the wavelet function coefficient. The relationship between wavelet filter $g(n)$ and scaling filter $h(n)$ can be expressed as [19];

$$g(n) = (-1)^n h(1-n) \tag{13}$$

Both coefficients are restricted by the orthogonality condition. If $h(n)$ has a finite even length N, then the (Eq. (13)) can be rewritten as

$$g(n) = (-1)^n h(N-1-n) \tag{14}$$

The wavelet function coefficients $g(n)$ is normally required by the orthonormal perfect reconstruction (PR) process. In the communication system point of view, this PR process offers advantage to the receiver whereby the received signals can be reconstructed perfectly. For example, Haar wavelet below is analyzed with the wavelet function $\psi(t)$ can be expressed as

$$\psi(t) = \begin{cases} 1 & 0 \leq t < 0.5 \\ -1 & 0.5 \leq t < 1 \\ 0 & otherwise \end{cases} \tag{15}$$

and its scaling function is

$$\varphi(t) = \begin{cases} 1 & 0 \leq t \leq 1 \\ 0 & otherwise \end{cases} \tag{16}$$

Furthermore, the basic version of Haar wavelet for wavelet and scaling function is shown in **Figure 2**.

The Haar filter coefficients are obtained by applying (Eq. (9)) and (Eq. (12)).

$$g(n) = \left(\frac{1}{\sqrt{2}}, \frac{-1}{\sqrt{2}}\right) \tag{17}$$

$$h(n) = \left(\frac{1}{\sqrt{2}}, \frac{1}{\sqrt{2}}\right) \tag{18}$$

Furthermore, the signal $x(t) \in L^2(\mathbb{R})$ has its discrete wavelet expansion given as [14].

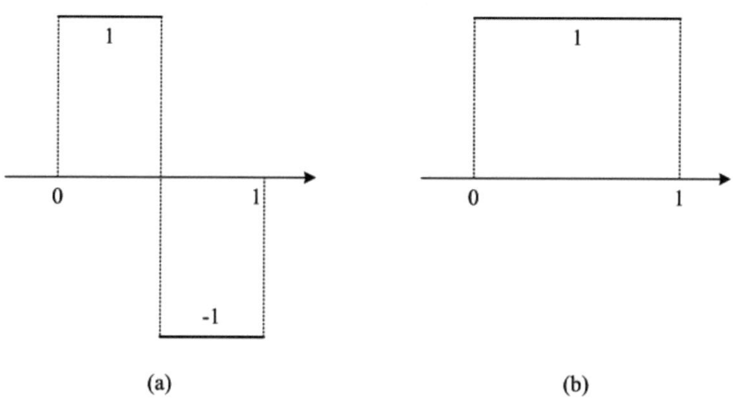

Figure 2.
Haar Wavelet transform (a) mother wavelet function, (b) scaling function.

$$x(t) = \sum_{\beta=-\infty}^{+\infty} c_{\alpha_0}(\beta)\varphi_{\alpha_0,\beta}(t) + \sum_{\beta=-\infty}^{+\infty}\sum_{\alpha=\alpha_0}^{+\infty} d_\alpha(\beta)\psi_{\alpha,\beta}(t) \tag{19}$$

where $\alpha, \beta, \in Z$ which Z is real integer. The α_0 is an arbitrary integer, and $L^2(\mathbb{R})$ is the vector space of the square integrated function. The frequency (or scale) and time localizations are provided by the parameters α and β respectively. The approximation coefficient and the detail coefficient have been deduced as $c_\alpha(\beta)$ and $d_\alpha(\beta)$ respectively.

In the wavelet expansion, by manipulating (Eq. (9)) and (Eq. (19)), the higher scale (i.e. $\alpha + 1$) can also be obtained that results the approximation coefficient as

$$c_\alpha(\beta) = \langle x(t), \varphi_{\alpha,\beta}(t)\rangle = \int x(t)2^{\alpha/2}\varphi(2^\alpha t - \beta)dt = \sum_m h(m - 2\beta)c_{\alpha+1}(m) \tag{20}$$

while the detail coefficient is expressed as

$$d_\alpha(\beta) = \langle x(t), \psi_{\alpha,\beta}(t)\rangle = \int x(t)2^{\alpha/2}\psi(2^\alpha t - \beta)dt = \sum_m g(m - 2\beta)c_{\alpha+1}(m) \tag{21}$$

Both the terms of $c_\alpha(\beta)$ and $d_\alpha(\beta)$ in (Eq. (20) and (21)) are computed by taking the weighted sum of DWT coefficients of higher scale $(\alpha + 1)$. In order to obtain the scaling of the DWT coefficients $(c_\alpha(\beta))$ at scale α, the scaling function coefficient $h(n)$ is convolved with the scaling DWT coefficients $(c_{\alpha+1}(\beta))$ at scale $\alpha + 1$, followed by subsampling with a factor of 2. Similarly, to obtain the wavelet DWT coefficients $(d_\alpha(\beta))$ at scale α, the wavelet function coefficient $g(n)$ is convolved with the scaling DWT coefficients $(c_{\alpha+1}(\beta))$ at scale $\alpha + 1$, followed by subsampling with a factor of 2. Hence, as shown in **Figure 3**, that both of these expressions can be illustrated as 2-channel filter banks analysis [20].

The input signal to the 2-channel filter bank is split into two parts. The first portion of the signal goes to filter H and second goes to filter G. Subsequently, both the filtered signals are subsampled by 2. Each filtered signal contains half of the number of original samples and spans half of the frequency band. However, the number of samples at the output of the filter bank is the same as the original signal since there are two filters used. The decomposition process starts at the largest scale of $c(\beta)$. If there are three level of decompositions involved, this implies the term $c_3(\beta)$ exist and produces the terms $c_0(\beta)$, $d_0(\beta)$, $d_1(\beta)$ and $d_2(\beta)$ at the decomposition branches, as illustrated in **Figure 4**.

On the other hand, the reconstruction of the DWT coefficients process is expressed by (Eq. (19)). If (Eq. (9)) (for scaling refinement) and (Eq. (12))

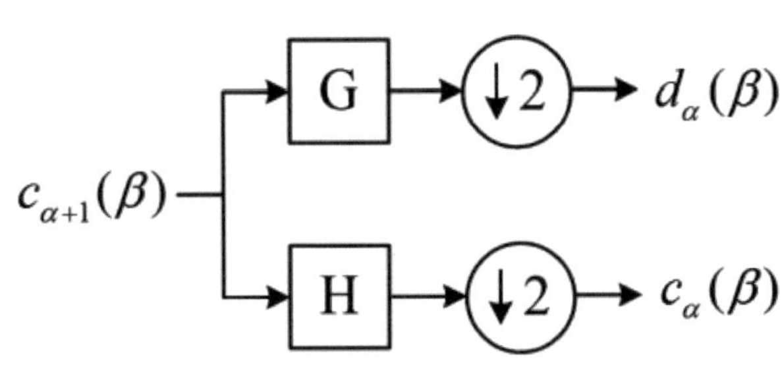

Figure 3.
DWT decomposition (single level).

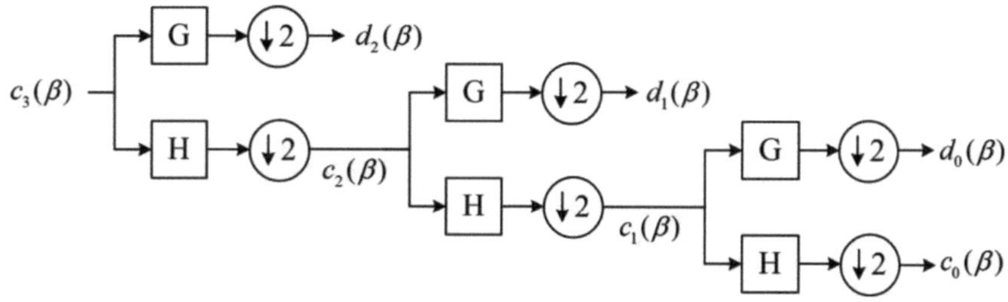

Figure 4.
DWT decomposition (three level).

(wavelet function) are substituted into (Eq. (19)) (reconstruction function), thus produces

$$\begin{aligned} x(t) &= \sum_{\beta=-\infty}^{+\infty} c_{\alpha_0}(\beta)\varphi_{\alpha_0,}\beta(t) + \sum_{\beta=-\infty}^{+\infty}\sum_{\alpha=\alpha_0}^{+\infty} d_\alpha(\beta)\psi_{\alpha,}\beta(t) \\ &= \sum_{\beta=-\infty}^{+\infty} c_{\alpha_0}(\beta) \sum_{n=-\infty}^{+\infty} h(n)\left(\sqrt{2}\right)^{\alpha+1}\varphi\left(2^{\alpha+1}-2\beta-n\right) \\ &+ \sum_{\beta=-\infty}^{+\infty}\sum_{\alpha=-\infty}^{+\infty} d_\alpha(\beta) \sum_{n=-\infty}^{+\infty} g(n)\left(\sqrt{2}\right)^{\alpha+1}\varphi\left(2^{\alpha+1}-2\beta-n\right) \end{aligned} \tag{22}$$

By multiplying both sides of (Eq. (22)) with $\varphi\left(2^{\alpha+1}-\beta\right)$ and taking the integral produces the lower scale of DWT coefficients [18], the scaling DWT coefficients of higher scale is given as

$$c_{\alpha+1}(\beta) = \sum_m c_\alpha(m)h(\beta-2m) + \sum_m d_\alpha(m)g(\beta-2m) \tag{23}$$

This implies that the scaling DWT coefficients at a certain value $(\alpha+1)$ can be computed by taking the weighted sum of wavelet DWT coefficients that are multiplied with the scaling DWT coefficients at scale α. **Figure 5** illustrates this process which is known as a 2-channel synthesis filter bank. The scaling DWT coefficients $(c_\alpha(\beta))$ and wavelet DWT coefficients $(d_\alpha(\beta))$ at scale α are first up-sampled by factor 2. Then, the scaling DWT coefficients $(c_\alpha(\beta))$ are filtered with a LPF $\hat{H}$, and the wavelet DWT coefficients $(d_\alpha(\beta))$ are filtered with a HPF $\hat{G}$ respectively. Finally, the two filtered signals are added together to form the scaling DWT coefficients at scale $\alpha+1$ i.e. $(c_{\alpha+1}(\beta))$.

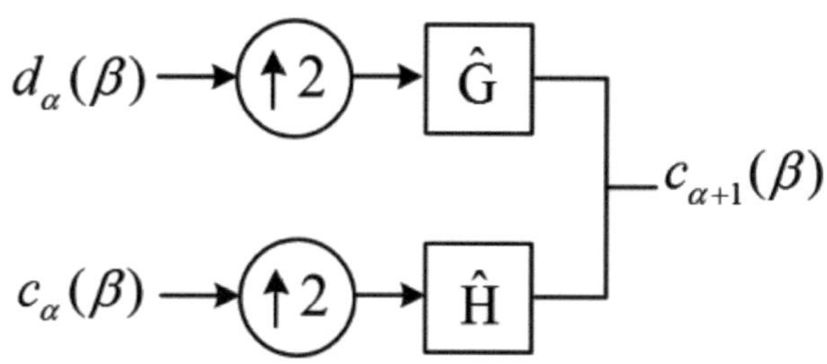

Figure 5.
DWT reconstruction (single level).

In short, the DWT decomposes signals into coefficients. The IDWT reconstructs the original signals from coefficients which can be implemented efficiently by iterating the 2-channel synthesis filter bank.

2.1.2 Wavelet packet transform (WPT)

In DWT decomposition, the direction of decomposition is heading towards the low pass branches, i.e. the sequence of iteration for the 2-channel filter bank is always taking the low pass filters. At the end of decomposition, the low frequencies portion contains fewer numbers of coefficients, hence occupying a narrow bandwidth. The high frequencies portion contains larger number of coefficients, hence occupying a wide bandwidth.

On the other hand, wavelet packet transform (WPT) executes the iteration of 2-channel filter bank on both sides, i.e. on the low pass and high pass filter branches for decomposition. Therefore, the WPT has evenly space frequency resolution and similar bandwidth size since both the high frequency and low frequencies components are decomposed. In WPT, the filter bank structure is expanded into a full binary tree. A set of WPT coefficients is labeled by ζ and the level that corresponds to the depth a node in the tree structure is indicated by l and parameter p indicates the position at current node. Every parent node is split by the WPT in two orthogonal subspaces W_l^P which is located at the next recursive level, and is given as [19];

$$W_l^p = W_{l+1}^{2p} \bigoplus W_{1+1}^{2p+1}, W_l^p = \overline{Span\{2^{l/2}\zeta_l^p(2^l t - \beta)\}} \tag{24}$$

In WPT, the scaling WPT coefficients are denoted as ζ_{l+1}^{2p} and wavelet WPT coefficients are labeled as ζ^{2p+1}, given as in the following expressions, and are depicted as in **Figure 6**.

$$\zeta_{l+1}^{2p}(\beta) = \sum_m h(m - 2\beta)\zeta_l^p(m) \tag{25}$$

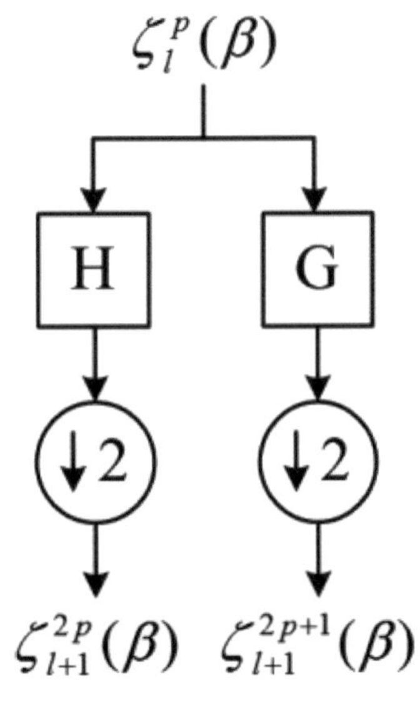

Figure 6.
WPT decomposition at single level.

$$\zeta_{l+1}^{2p+1}(\beta) = \sum_m g(m - 2\beta)\zeta_l^p(m) \tag{26}$$

In WPT, the number of iterations by the 2-channel filter bank increases exponentially as the number of levels increased. Therefore, WPT has higher computational complexity than the regular DWT. The WPT requires $O(Nlog(N))$ operation (by using fast filter bank algorithm), while Fast Fourier Transform (FFT) requires only $O(N)$ operations to complete DWT [16]. The reconstruction (inverse WPT) is executed by taking the reverse direction of the tree in **Figure 6**. The wavelet packet coefficients $\zeta_l^p(\beta)$ at any level l can be expressed as

$$\zeta_l^p(\beta) = \sum \zeta_{l+1}^{2p}(m)h(\beta - 2m) + \sum \zeta_{l+1}^{2p+1}(m)g(\beta - 2m) \tag{27}$$

2.2 Multicarrier modulation (MCM) system

Multicarrier modulation (MCM) scheme is a technique that transforms the high-speed serial signals into multiple low-speed parallel signals with N overlapping subcarriers. This special multicarrier modulation scheme was introduced by Chang [21], and is known as the orthogonal frequency division multiplexing (OFDM). The technique is widely used in various applications such as in European Digital Audio Broadcasting (DAB), IEEE 802.11 (WiFi) and IEEE 802.16 (WiMAX). OFDM has high spectral efficiency and consecutive subcarriers experience no crosstalk if the orthogonality is preserved.

In this study two wavelet-based MCM systems are used i.e. the wavelet-based OFDM (WOFDM) and wavelet packet-based OFDM (WP-OFDM) systems. As seen above. The primary difference between these two MCM systems is the way the wavelet tree being expanded. Therefore, in wavelet-based OFDM (WOFDM), the decomposition process expands the branches in dyadic way. In wavelet packet-based OFDM (WP-OFDM), the decomposition process expands the nodes as a full binary tree. Hence, wavelet packet process possesses richer signal analysis than wavelets process and for the detail analysis, wavelet packet process is capable to focus on any of the tree nodes. This main difference of the two MCM systems is illustrated in **Figure 7**. Notice that the wavelet decomposition produces different range of bandwidth divisions. The wavelet bandwidth is in form of dyadic division, while wavelet packet bandwidth is uniform. Therefore, the use of wavelet packet

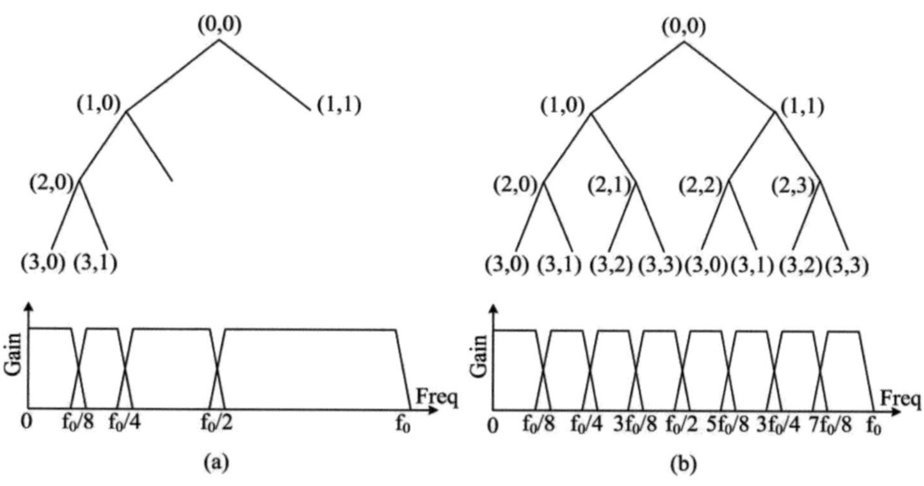

Figure 7.
Decomposition and bandwidth division for (a) DWT and (b) WPT.

transform in MCM system is preferable since its major characteristic resembles the conventional OFDM [22].

In wavelet packet-based OFDM (WP-OFDM) scheme that wavelet packet transform is utilized to change a series of parallel signals into a single composite signal. Both OFDM and WP-OFDM possess high spectral efficiency since their subcarriers are orthogonal that overlap between each other. The only difference between the two schemes is in term of the shape of the subcarriers produced. In ordinary OFDM the Fourier bases are used i.e. the sine or cosine terms. However, in WP-OFDM scheme the wavelet packet provides flexibility for modification of the filter banks property to suit the characteristic of system transmission [14]. The general multicarrier modulation system is shown in **Figure 8**.

WP-OFDM is implemented by the using the inverse orthogonal transform at the transmitter which is known as the inverse discrete wavelet packet transform (ID-WPT) as illustrated in **Figure 9** (left-hand side). The forward orthogonal transform is implemented at the receiver called as discrete wavelet packet transform (DWPT) as depicted in **Figure 9** (right-hand side). The implementation of WP-OFDM that utilizes the wavelet packet transform has been derived from MRA concept [23]. It is commenced by introducing a pair of filters called as quadrature mirror filters

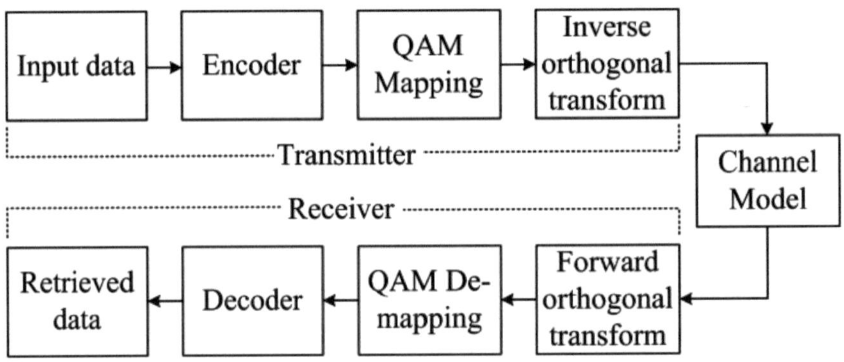

Figure 8.
General schemes for multicarrier modulation.

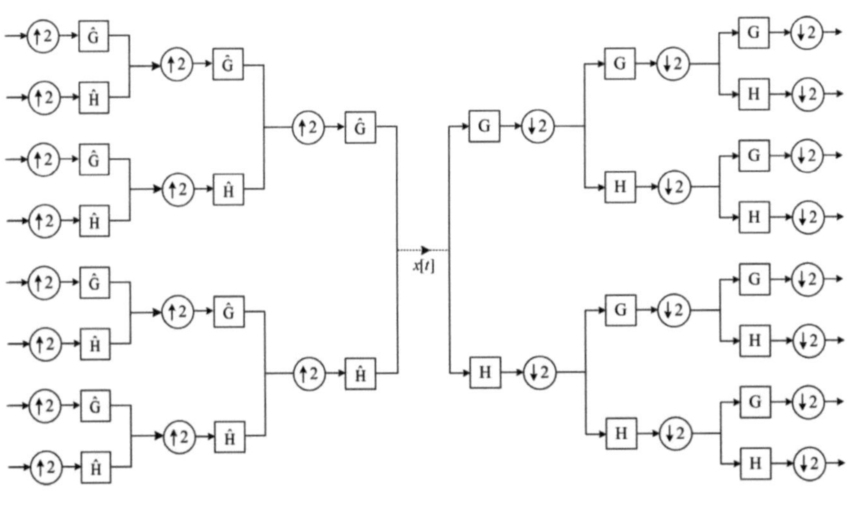

Figure 9.
IDWPT and DWPT in MCM scheme.

(QMF) that contain half-band of the low and high-pass filters, i.e. $h[n]$ and $g[n]$ respectively of length L. The relationship of the filters is described as the following;

$$g[L-1-n] = (-1)^n h[n] \tag{28}$$

The complex conjugate time reversed variant is given by [24];

$$h'[n] = h^*[-n] \; and \; g'[n] = g^*[-n] \tag{29}$$

The pair of $h'[n]$ and $g'[n]$ is the synthesis filter-pair which is used to produce wavelet packet carriers for modulation at the end of the transmitter, while pair of $h[n]$ and $g[n]$ is the analysis filter-pair for demodulation at the end of the receiver. The wavelet packet coefficients Υ_l^p are obtained from QMF filters which are derived via MRA as [24];

$$\Upsilon_{l+1}^{2p}(t) = \sqrt{2}\sum_m h[n]\Upsilon_l^p(2t-n) \tag{30}$$

$$\Upsilon_{l+1}^{2p+1}(t) = \sqrt{2}\sum_m g[n]\Upsilon_l^p(2t-n) \tag{31}$$

where p is subcarrier index at any tree depth l.

3. PAPR profile of wavelet-based multicarrier modulation signals

This section presents a comprehensive study on the PAPR profile of multicarrier modulation (MCM) signals. The performance of the transmitted signal is measured by the ratio of peak power signal to its corresponding average power signal within similar MCM frame, known as the peak-to-average power ratio (PAPR). It is desired to have a minimum PAPR as possible in order to reduce the complexity of high power amplifier (HPA) and at the same time, the average transmitting power can be boosted up efficiently as maximum as possible in a linear region of a HPA. Besides, it is disadvantageous of having high PAPR as the signals may be distorted in the nonlinear region of the HPA and results in poor reception and bit error rate (BER) performance. In order to cope with high PAPR, this chapter provides a study that investigates the wavelet-based OFDM (WOFDM), wavelet packet-based OFDM (WP-OFDM) and conventional OFDM systems performances. This investigation is carried out by replacing different orthogonal base modulations, which is normally used in Fourier based MCM (as the conventional OFDM system).

3.1 Multicarrier modulation system models

This section presents the general multicarrier modulation system model structures for implementation. The condition for determining the initial data value and maximum potential number of symbols to be carried by system subcarriers are also discussed.

3.1.1 System models descriptions

The three evaluated multicarrier modulation (MCM) system models are represented by a single general MCM model as illustrated in **Figure 10**. The information bits are generated based on the uniform random distribution binary

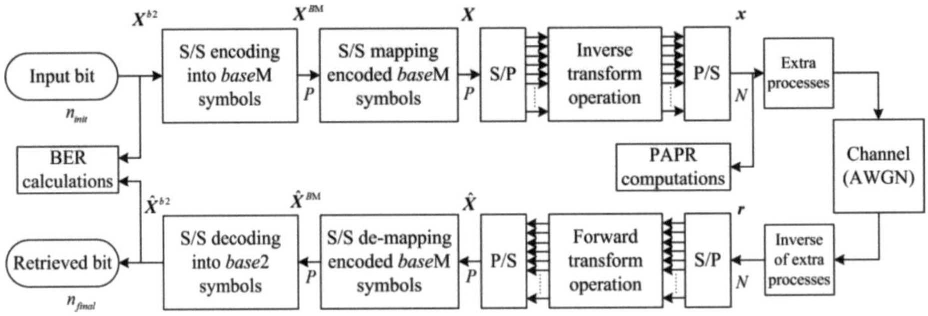

Figure 10.
Model for general MCM scheme with data sequence details.

number. The data are arranged (in every frame) in a horizontal matrix $1 \times n_{init}$, and are converted into *base*16 number format. Subsequently they are encoded by Reed-Solomon (RS) codes, and converted into *baseM* number format, where *M* corresponds to the total mapping points in a particular QAM constellation. In this work, the Reed-Solomon of RS(n, k) is used, where n is the encoded data, and k is original data. In particular, the RS(15,11) is used throughout the work for protecting the original data. This channel coding scheme is compatible with hexadecimal number for encoding and decoding processes. In addition, RS encoded symbols are converted to the *baseM* symbols to achieve the same configuration that adapts with modulation constellation mapping.

Table 1 shows four possible of *baseM* number format types associated with the number of bits per symbol, N_{bps} as well as the corresponding constellation mapping modulation types. Then, the data frames are transformed into time-domain MCM signals by using a particular inverse transform prior transmission and they are retrieved back by the corresponding forward transform at the receiver. The particular inverse and forward transforms applied are labeled in block diagrams as shown in **Figures 11** and **12**.

In **Figure 11**, there are two types of wavelet-based MCM models to be considered i.e. the wavelet-based OFDM (WOFDM) and wavelet packet-based OFDM (WP-OFDM) systems. At the transmitter, either the inverse discrete wavelet transform (IDWT) or inverse discrete wavelet packet transform (IDWPT) is used. At the receiver, either the discrete wavelet transform (DWT) or discrete wavelet packet transform (DWPT) is used. These modulation techniques offer higher spectral efficiency since there is no for the system to use the cyclic pre-fix (CP) codes as in the conventional OFDM.

Figure 12 shows the conventional OFDM (C-OFDM) which is included for comparison system model. This model utilizes the inverse fast Fourier transform (IFFT) and fast Fourier transform (FFT). Additional blocks are required for appending and re-moving the CP codes where 25 percent of the OFDM frames tail are copied and appended to OFDM frames head [25, 26].

BaseM	No. of bits per symbol, N_{bps}	Suitable mapping type
*Base*2	1	BPSK
*Base*4	2	QAM
*Base*16	4	16QAM
*Base*64	6	64QAM

Table 1.
BaseM *and its appropriate constellation mapping.*

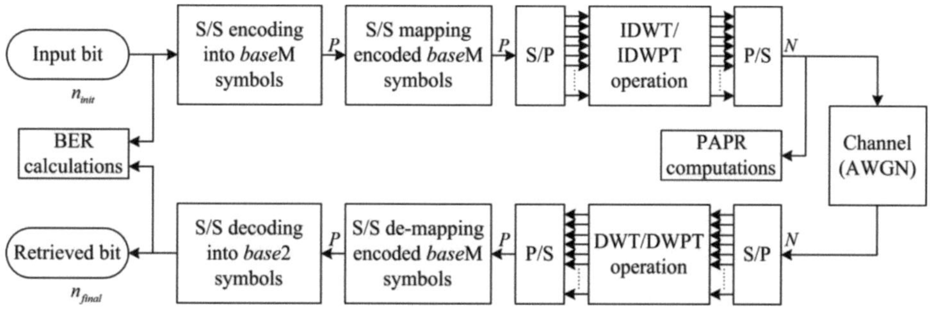

Figure 11.
Model for wavelet- and wavelet packet-based OFDM schemes.

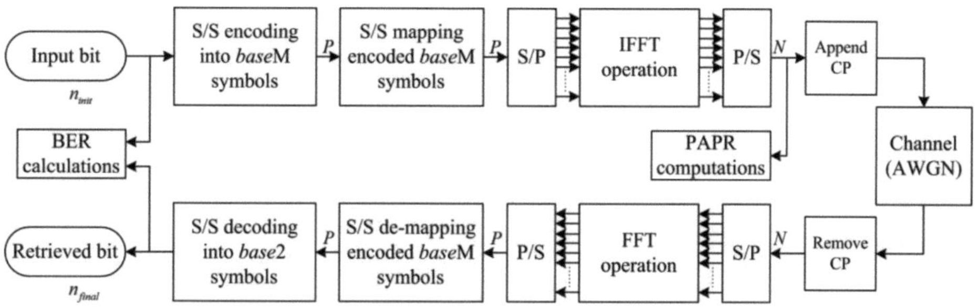

Figure 12.
Model for conventional OFDM scheme.

Each frame must contain P symbols and is always less than or equal to the total number of subcarriers N, i.e. $P \leq N$. The specified number of base for every modulation type is fixed as in **Table 1**. The number of initial binary information n_{init} increases as the number of bits per symbol N_{bps} increased with constant value of N. In this work, the RS(15, 11) is used, and the encoded data ($n = 15$), and the original data ($k = 11$) respectively. This implies that each time a sequence of symbols to be encoded, the number of original symbols taken is eleven and this produces total fifteen encoded symbols afterwards. Therefore, during the encoding process, the raw binary data (*base*2) need to be converted to *base*16 symbols to suites the requirement of RS(15, 11) coding scheme where each encoded symbol should have a value between 0 to 15.

3.1.2 Determination transmission parameter

This section describes how the transmission parameters values of P is obtained by manipulating the base number of the symbols. **Figure 10** above shows the block diagram of the of S/S encodes where raw input data bits are converted into *baseM* symbols. **Figure 13** shows further details of this process.

Figure 13 denotes three conversion processes for the initial input bit n_{init} which are indicated as the α, β and γ processes. The α process converts every four bits ($v = 4$) of binary source data to a *base*16 symbol. For example, if $n_{init,base2} = \{1,0,0,1,0,0,1,1,0,0,1,1,0,0,1,0,1,0,0,0,0,0,0,0,0,1,1,1,0,0,0,1,0,1,1,1,0,0,1,0,1,0,0,1\}$ then, the output from process α is $n_\alpha = \{9,3,3,2,8,0,7,1,7,2,9\}$. This means the number of overall symbols is reduced to one-fourth.

Next, β process converts every eleven *base*16 symbols ($k = 11$) into fifteen *base*16 encoded symbols ($n = 15$). For instance, when $n_\alpha = \{9,3,3,2,8,0,7,1,7,2,9\}$ then, the output of β process is $n_\beta = \{9,3,3,2,8,0,7,1,7,2,9,15,2,7,11\}$ which increases

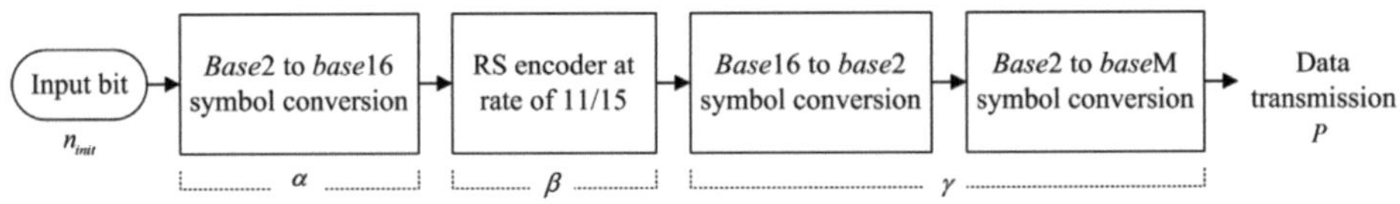

Figure 13.
S/S encoding into baseM *symbols block diagram.*

Mapping type	Base number	Output of γ process	Number of symbols (P) at output γ process
BPSK	*Base2*	$n_{\gamma,base2}$ = {1,0,0,1,0,0,1,1,0,0,1,1,0,0,1,0,1,0,0,0, 0,0,0,0,0,1,1,1,0,0,0,1,0,1,1,1,0,0,1,0, 1,0,0,1,1,1,1,0,0,1,0,0,1,1,1,1,0,1,1}	60
QAM	*Base4*	$n_{\gamma,base4}$ = {2,1,0,3,0,3,0,2,2,0,0,0,1,3, 0,1,1,3,0,2,1,3,3,0,2,1,3,2,3}	30
16QAM	*Base16*	$n_{\gamma,base16}$ = {9, 3, 3, 2, 8, 0, 7, 1, 7, 2, 9, 15, 2, 7, 11}	15
64QAM	*Base64*	$n_{\gamma,base64}$ = {36, 51, 10, 0, 28, 23, 10, 31, 9, 59}	10

Table 2.
Output of γ *process based on mapping type selection.*

the number of symbols along the encoding processes with the additional redundancy required for channel error protection. This implies the number of overall symbols now has been increased by 15/11. After that, γ process converts every single *base*16 symbol into four *base*2 symbols ($w = 4$). Then this follows by converting every $(N_{bps} = \log_2 M)$ number of *base*2 symbols back to a single *baseM* symbol. Suppose that $M = 16$, continuing the above example, when $n_\beta = \{9, 3, 3, 2, 8, 0, 7, 1, 7, 2, 9, 15, 2, 7, 11\}$ then, the output of γ process is $n_{\gamma,base16} = \{9, 3, 3, 2, 8, 0, 7, 1, 7, 2, 9, 15, 2, 7, 11\}$. The rest of other mappings are as listed in **Table 2**. The number of transmission symbols P, thus can be expressed as

$$P = n_{init} \times \left(\frac{1}{v}\right) \times \left(\frac{n}{k}\right) \times w \times \left(\frac{1}{N_{bps}}\right) \tag{32}$$

where $P \leq N$.

Using (Eq. (32)), the number of transmission symbols P and initial input bit n_{init} can be obtained after defining number of subcarriers (N). Thus, number of bits per symbol, N_{bps} can be obtained and the quantitative relationships between these parameters are shown in **Table 3**.

Figure 14 shows the partitions between the occupied slot positions of the encoded data (P) and the remaining slot positions (R) for three different values of

Parameters	Value											
No. of subcarrier, N	64				128				256			
No. of bits per symbol N_{bps}	1	2	4	6	1	2	4	6	1	2	4	6
No. of initial binary information, n_{init}	44	88	176	264	88	176	352	528	176	352	748	1100
No of symbols per frame P	60	60	60	60	120	120	120	120	250	250	250	250

Table 3.
The relationship between n_{init}, P, N *and* N_{bps}.

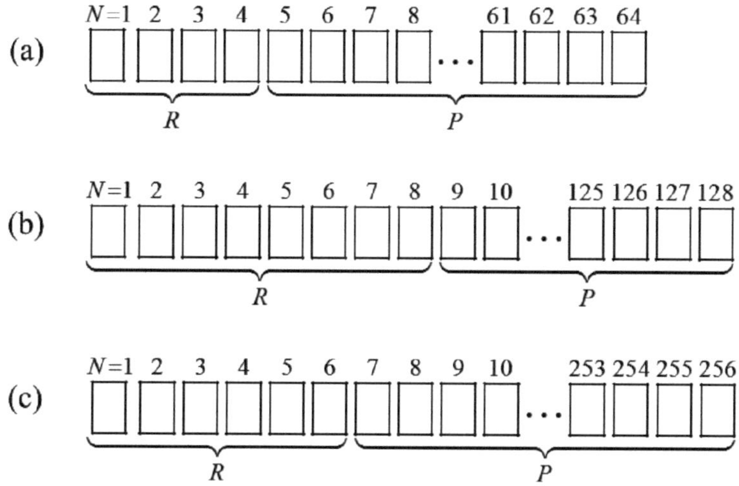

Figure 14.
MCM partition with respect to N. Partition of the occupied slot: the encoded data (P) and the remaining slot positions (R) for total subcarriers {N = 64; 128; 256}.

total subcarriers $\{N = 64; 128; 256\}$. It can be observed that the subcarriers are not fully occupied for the whole slot positions with frames, hence the remaining positions, R is filled up with zeros. There are four zero frames ($N = 64$), eight zero frames ($N = 128$) and six zero frames ($N = 256$) respectively.

4. PAPR profile: results and analysis

This section presents the results and discussions on the PAPR profile performances based on several important parameters i.e. modulation types, number of subcarriers, the orthogonal bases (Fourier/Wavelets) and filter length. The BER performance is also included to investigate the efficiency of the system models. The common parameters used in the experiments are as list in **Table 4** below.

The effect of modulation constellation mapping on PAPR is analyzed in the following paragraphs. The list of parameters involved are shown as in **Table 5**. **Figure 15** shows that both the conventional C-OFDM and WP-OFDM systems are having almost the same PAPR profiles, regardless of the modulation mapping types used. The reason for the PAPR profiles of the wavelet based OFDM (WOFDM) outperform the PAPR profiles of the WP-OFDM, is that the WOFDM system contains a smaller number of signal analysis than the WP-OFDM system. The PAPR profile for WOFDM system is superior since the decomposition and reconstruction signals are only involved the low pass branches. Thus, there is lower probability for the peak to be above the average signals leading to slightly superior PAPR profile.

Parameter	Descriptions
System Model	WOFDM, WP-OFDM, C-OFDM
Encoder type	RS(15,11)
Channel model	AWGN
CP for conventional MCM	25% of total subcarriers

Table 4.
Common parameters for experiments.

Parameter	Descriptions
Mapping type	QAM, 16QAM, 64QAM
Number of subcarriers	128
Orthogonal bases	Fourier, wavelet (Haar)

Table 5.
Parameters used for studying the effect of different type of mapping modulation.

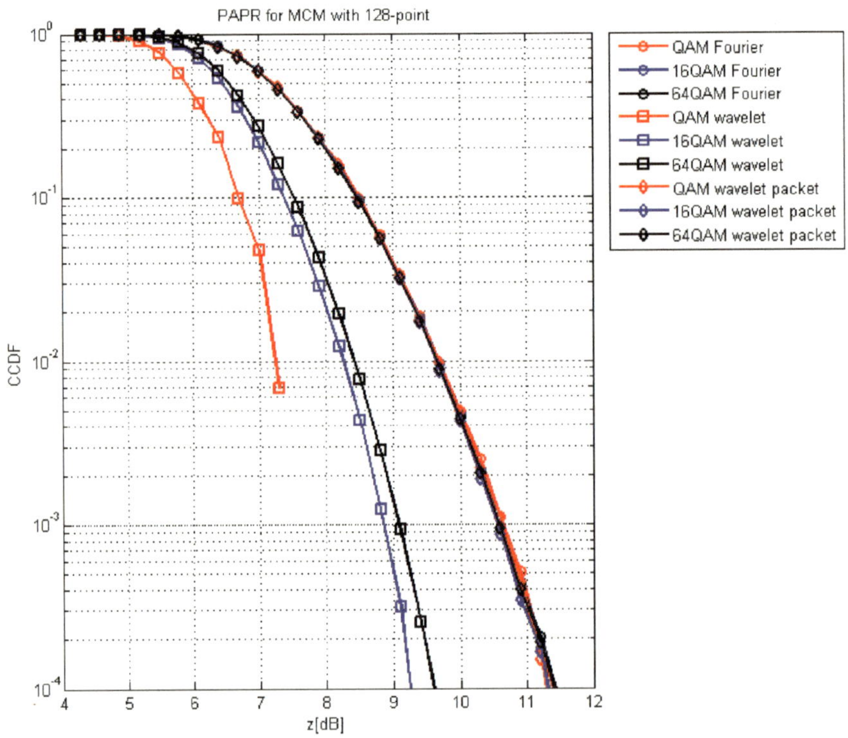

Figure 15.
CCDF of the PAPR with variation of mapping type.

However, it can clearly be seen in **Figure 16**, the BER performances are indeed worse for all three MCM systems as the type of mapping changes from QAM towards 16QAM and 64QAM. The BER performance is highly related with the type of the signal mapping used. Theoretically, the error probability at the receiver increases as the number of constellation points increased. In order to reduce the error probability, in general higher E_b/N_0 is required. **Table 6** shows for probability of bit error at 10^{-6} the corresponding of E_b/N_0 (dB) values for all modulation mapping types. The channel impairment used for evaluating the performance is using the AWGN channel. From **Figure 16**, the 64QAM modulation mapping type that can hold higher bit information, where each symbol represents 6 bits, even though it requires much higher transmitting power.

The following paragraphs analyze the effect on varying the number of subcarriers on the PAPR profiles. **Table 7** lists all parameters required for this experiment. It is found that, when the number of subcarriers N decreases i.e. from N = 256 until N = 64, the PAPR profile (CCDF) of any modulation scheme is gradually improves as shown in **Figure 17**. Explicitly, the PAPR profile of WOFDM

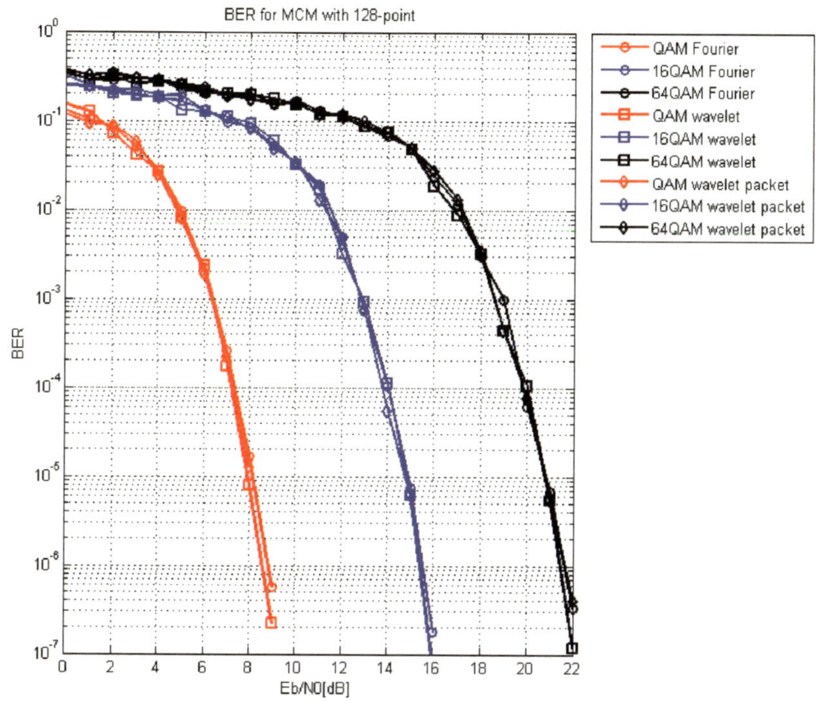

Figure 16.
Corresponding BER performances due to variation of mapping type.

Mapping type	E_b/N_0 (dB) at BER level of 10^{-6}
QAM	9.0
16QAM	15.5
64QAM	21.5

Table 6.
Common value of E_b/N_o *for the corresponding mapping type.*

Parameter	Descriptions
Mapping type	64QAM
Number of subcarriers	64, 128, 256
Orthogonal bases	Fourier, wavelet (Haar)

Table 7.
Parameters used for studying the effect of different No. of subcarriers.

model outperforms the PAPR profile of C-OFDM and WP-OFDM models by 1.5 dB at the CCDF level of 10^{-5} for fixed $N = 64$. The PAPR profiles for C-OFDM and WP-OFDM systems are similar. The PAPR profile for WOFDM system is superior since the decomposition and reconstruction signals are only involved the low pass branches. Thus, there is lower probability for the peak to be above the average signals leading to slightly superior PAPR profile.

Figure 18 shows that there is no significant different in term of BER performances, although different numbers of subcarriers are used for modulation. At BER of 10^{-6}, it is found that the difference between the lowest and highest value of

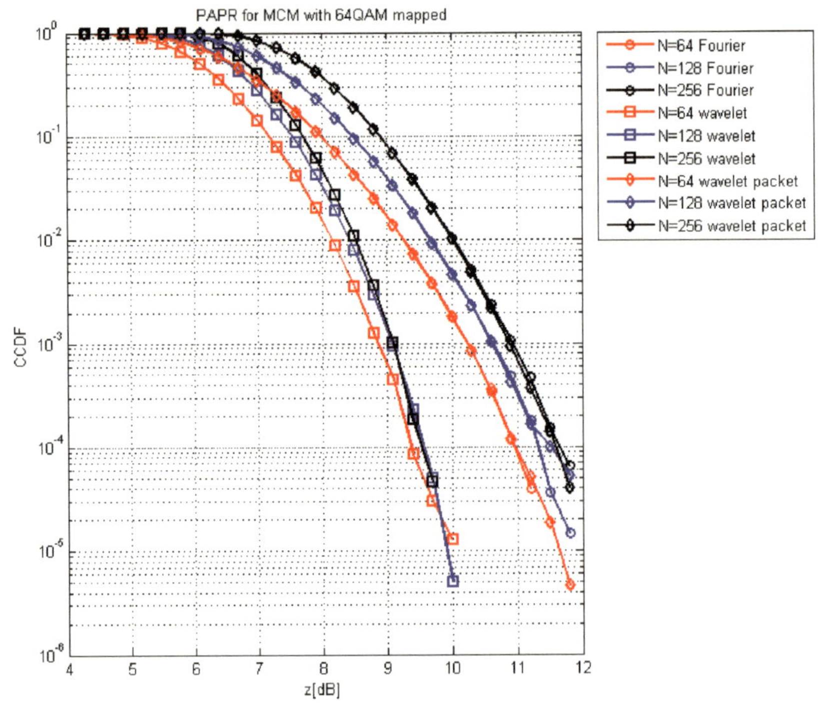

Figure 17.
CCDF of the PAPR with variation of number of subcarriers.

E_b/N_0 is less than 1 dB. Thus, it can be deduced that, the number of subcarriers gives less impact to the BER performance. The E_b/N_0 is quite high (i.e. nearly 22 dB for all profiles). The increase in the number of subcarriers worsen the PAPR profile. Therefore, for practical application, the number of subcarriers $N = 128$ is selected since it is a moderate choice as compare to the other number of subcarriers.

The following paragraphs analyze the influence of different orthogonal bases, wavelet types and their filter lengths on the PAPR profile. Several wavelet families applied includes the Daubechies, Symlet, Coiflet and Meyer wavelets with various lengths of coefficients. The parameters are briefly listed as in **Table 8**. This analysis is mainly focuses on the wavelet OFDM and wavelet packet-based OFDM systems. However, the C-OFDM scheme is also included as a performance reference. Additional information regarding the characteristic of the wavelet families are included in **Table 9**.

Figure 19 shows the PAPR profiles for the three OFDM systems, where Daubechies wavelet with different filter lengths are used (Fourier based OFDM profile is just for reference only). In analyzing the effect of wavelet filter length, various filter lengths are used in the experiment. From **Figure 19**, looking at the WOFDM profiles (red color), as the filter length increases, the PAPR profiles become worse. In other words, the Daubechies wavelet (in WOFDM) with higher filter length produces inferior PAPR profiles. This is due to the fact that with higher filter length, the wavelet has richer signal analysis. Thus, there is higher probability for the peak to be above the average signals leading to slightly inferior PAPR profile.

However, for WP-OFDM profiles (blue color), there is no significant difference in the PAPR performance. Since the signal analysis in WP-OFDM is in full binary tree analysis rather than dyadic (lower-half band) analysis in WOFDM system. There is already high amount of data involved in decomposition and reconstruction which makes the effect of wavelet filter length insignificant.

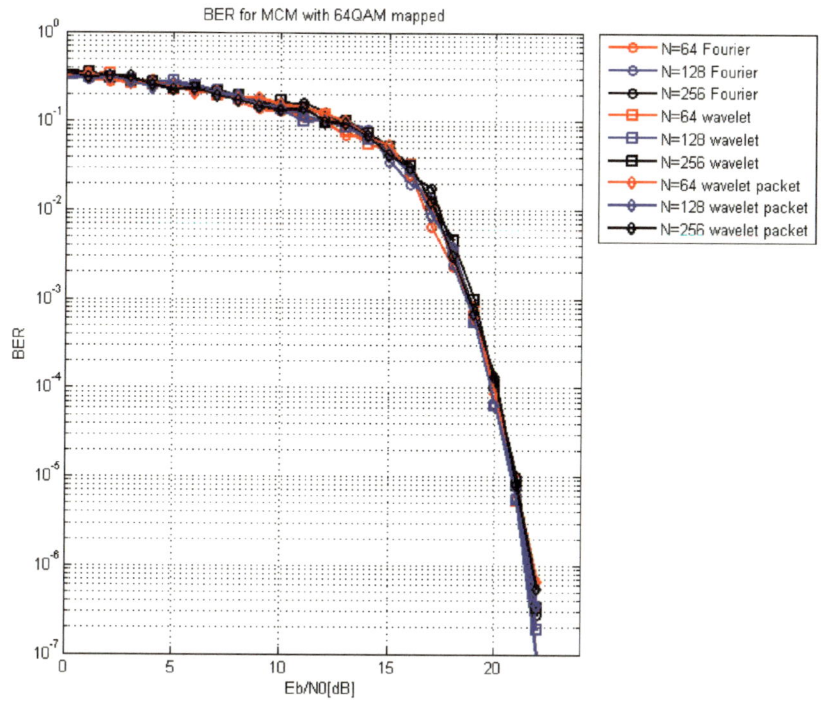

Figure 18.
Corresponding BER performance due to variation of number of subcarriers.

Parameter	Descriptions
Mapping type	64QAM
Number of subcarriers	128
Orthogonal bases	Fourier, wavelet (*db1, db2, db3, db5, db10, db20, sym2, sym3, sym5, sym10, coif1, coif3, coif5, dmey*)

Table 8.
Parameters used for studying the effect of different bases and filter length.

Full name	Abbreviated name	Vanishing order	Length, L
Haar	Haar	1	2
Daubechies	*dbN*	N	$2N$
Symlets	*symN*	N	$2N$
Coiflet	*coifN*	N	$6N$
Discrete Meyer	*dmey*	—	62

Table 9.
Wavelet family characteristic [23].

In **Figures 20** and **21**, different wavelet types (Daubechies, Symlet, Coiflet and Discrete Meyer wavelets) are used but the filter length is fixed $L = 6$ (short category). For long category the filter lengths are mixed, i.e. $L = \{18(coif\,3), 20\,(db10, sym10), 62(dmey)\}$ respectively. From these figures, there can be observed

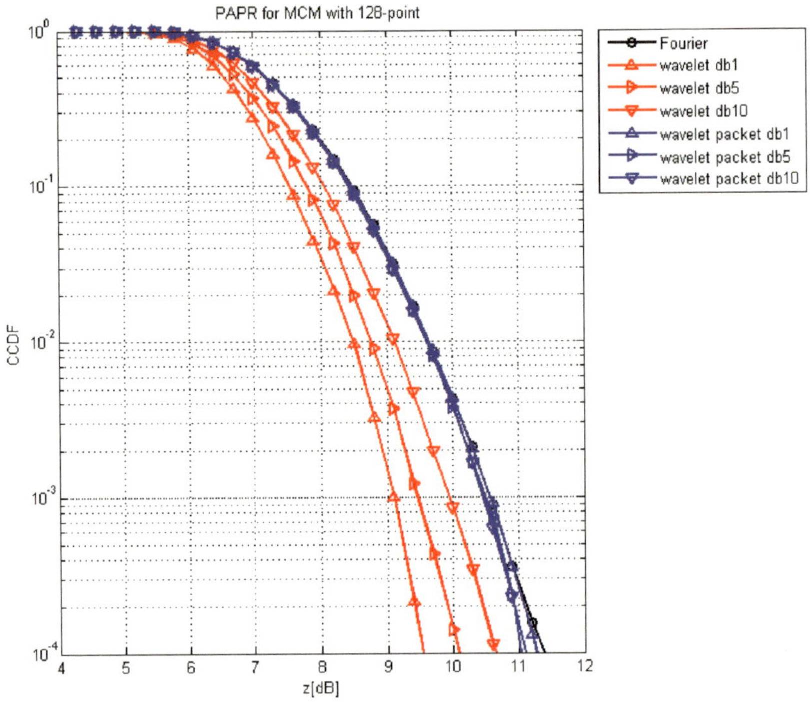

Figure 19.
CCDF of the PAPR with Daubechies wavelet with different filter lengths.

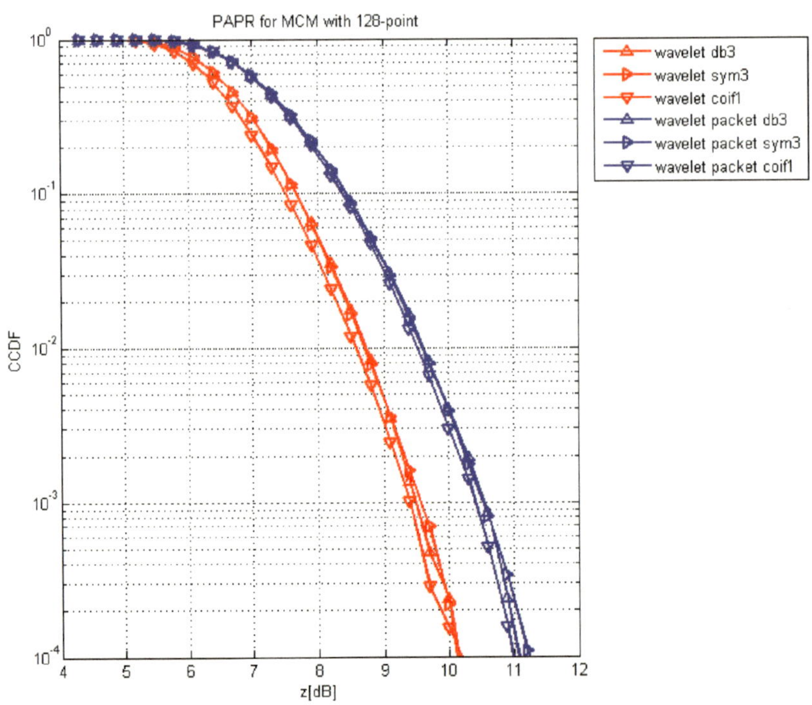

Figure 20.
*CCDF of the PAPR with different orthogonal bases modulation and **short** filter lengths.*

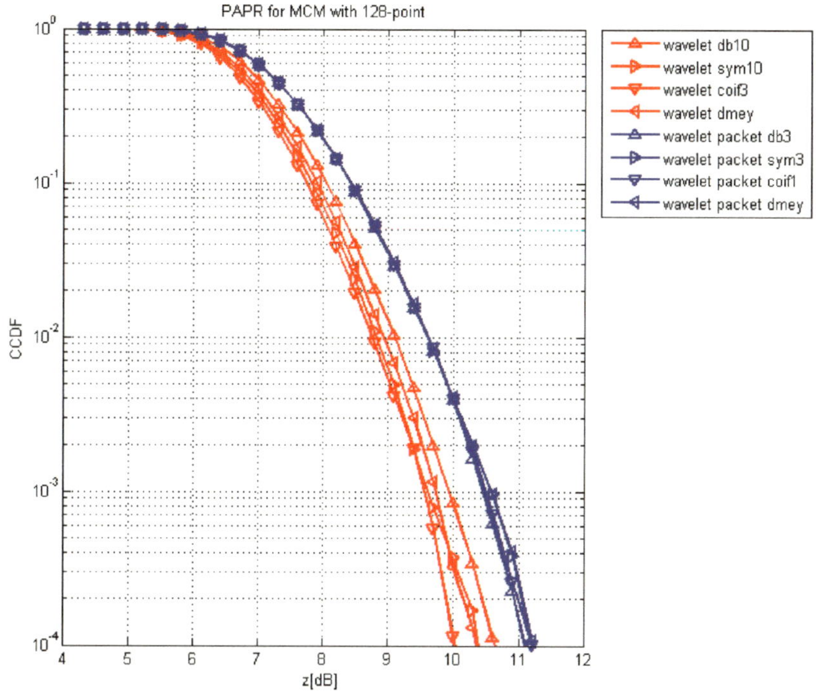

Figure 21.
*CCDF of the PAPR with different orthogonal bases modulation and **long** filter lengths.*

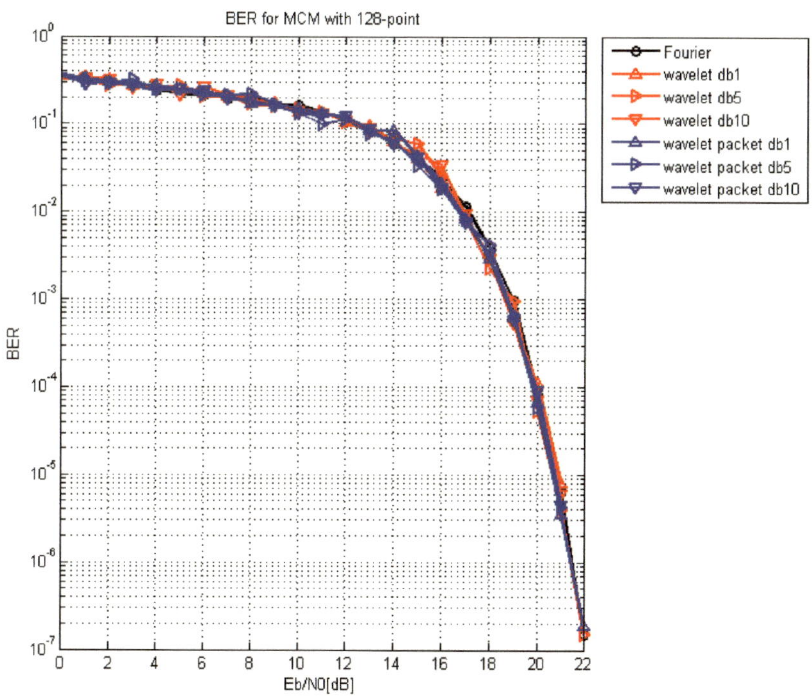

Figure 22.
Corresponding BER performances for Daubechies wavelet with different filter lengths.

that no explicit difference found from PAPR profiles of WP-OFDM signals either by changing the wavelet's type or length.

The BER performances are shown in **Figure 22**, where the experiment is carried out on the Daubechies wavelet with different filter lengths. It can be observed that no significant difference between BER performances. For example, at BER 10^{-6} all profiles having the same value of E_b/N_0.

5. Conclusion

The phenomenon of high PAPR in MCM system cannot be avoided since the signals consist of multiple low-rate parallel signals, which can be seen as the composite subcarriers in time domain representation. It is expected by using different orthogonal base for modulation, the PAPR profile can be reduced. Hence, discrete wavelet transform (DWT) and discrete wavelet packet transform (DWPT) are used for this purpose instead of fast Fourier transform (FFT). In comparison to the C-OFDM system, WOFDM and WP-OFDM systems do not need any cyclic prefix (CP) codes for their MCM frame in order to avoid intercarrier interference (ICI) and inter symbol interference (ISI).

Although, WOFDM system provides superior PAPR performance than other systems, data are lost at higher frequencies branches since signals decomposition are in dyadic (lower half-band) fashion. On the other hand, WP-OFDM system decomposes the signals in both lower and upper-band frequencies, that enrich signals analysis. The results obtained in Section 4 proves the characteristics. In addition, applying various wavelet bases do not offer much improvement in PAPR profile.

Acknowledgements

The authors would like to acknowledge the USM RU grant (Grant No. 1001/PELECT/814100), for funding this research work.

Author details

Jamaluddin Zakaria and Mohd Fadzli Mohd Salleh*
School of Electrical and Electronic Engineering, Universiti Sains Malaysia, Seri Ampangan, 14300 Nibong Tebal, Pulau Pinang, Malaysia

*Address all correspondence to: fadzlisalleh@usm.my

IntechOpen

© 2020 The Author(s). Licensee IntechOpen. This chapter is distributed under the terms of the Creative Commons Attribution License (http://creativecommons.org/licenses/by/3.0), which permits unrestricted use, distribution, and reproduction in any medium, provided the original work is properly cited. CC BY

References

[1] Van Nee R, Prasad R. OFDM for Wireless Multimedia Communications. Boston, London: Artech House; 2000.

[2] Zakaria J, Salleh M. F. M. Wavelet-based OFDM analysis: BER performance and PAPR profile for various wavelets. In: Proceedings of the IEEE Symposium on Industrial Electronics and Applications (ISIEA '12); 23–26 September 2012; Bandung. Indonesia: IEEE; 2012. P. 29–33.

[3] Chafii M, Palicot J, Gribonval R, Burr A. G. Power spectral density limitations of the wavelet-OFDM system. In: Proceedings of the *24th European Signal Processing Conference (EUSIPCO 16);* 28 August-2 September 2016; Budapest, Hungary: 2016. P. 1428–1432, doi: 10.1109/EUSIPCO.2016.7760484.

[4] Hsu C. Y., Liao, H. C. Generalised precoding method for PAPR reduction with low complexity in OFDM systems. IET Communications. 2018; 12 (7): 796–808. DOI: 10.1049/iet-com.2017.0824

[5] Kaur J, Sharma V. A-STBC incorporated power-efficient Radio over Fibre system, Optics Communications. 2019; 441: 84–89.

[6] Sarowa S, Kumar N, Agrawal, Balwinder S S. Evolution of PAPR Reduction Techniques: A Wavelet Based OFDM Approach, Wireless Personal Communications. 2020; https://doi.org/10.1007/s11277-020-07643-1

[7] Muller S, Huber J. A Comparison of Peak Power Reduction Schemes for OFDM. In: Proceedings of the IEEE Global Telecommunications Conference (GLOBECOM '97), 3–8 Nov. 1997; Phoenix, Arizona: 1, P. 1–5.

[8] Bouhlel A, Sakly A, Mansouri M. N. Partial Transmit Sequence technique based on Particle Swarm Optimization for WOFDM PAPR reduction. In: Proceeding of the 2nd International Conference on Advanced Technologies for Signal and Image Processing (ATSIP '16), March 21–23, 2016; Monastir, Tunisia: P. 710–714, doi: 10.1109/ATSIP.2016.7523183.

[9] Zakaria J, Salleh, M. F. M. PAPR reduction scheme: wavelet packet-based PTS with embedded side information data scheme. IET Communications. 2017; 11 (1), 127–135

[10] Yoon E, Hwang D, Jang C, Kim J, Yun U. Blind Selected Mapping with Side Information Estimation Based on the Received Pilot Signal, Wireless Communications and Mobile Computing. 2018; (5):1–9, DOI: 10.1155/2018/8523680

[11] Ahmed M S, Boussakta S, Al-Dweik A, Sharif B, Tsimenidis C C. Efficient Design of Selective Mapping and Partial Transmit Sequence Using T-OFDM. IEEE Transactions on Vehicular Technology. 2020; 69(3), 2636–2648, doi: 10.1109/TVT.2019.2928361.

[12] Daubechies I. Ten Lectures on Wavelets. Society for Industrial and Applied Mathematics. 1992. doi.org/10.1137/1.9781611970104

[13] Li A, Shieh W, and Tucker R. Wavelet Packet Transform-Based OFDM for Optical Communications, Journal of Lightwave Technology. 2010; 28(24): 3519–3528.

[14] Torun B. Peak-to-Average Power Ratio Reduction Techniques for Wavelet Packet Modulation. PhD thesis, Microwave Technology and Systems for Radar (MTSR), Department of Telecommunications, Faculty of Electrical Engineering, Mathematics and Computer Science, Delft University of Technology. 2010.

[15] Gargour C, Gabrea M, Ramachandran V, Lina J M. A Short Introduction to Wavelets and Their Applications, IEEE Circuits and Systems Magazine. 2009; 9(2): 57–68.

[16] Burrus S C, Gopinath R A, Guo H. Introduction to Wavelets and Wavelet Transforms: A Primer. Prentice Hall. 1997.

[17] Akansu A N, Haddad R A. Multiresolution Signal Decomposition: Transforms, Subbands, and Wavelets. Academic Press, Inc., Orlando, FL, USA. 1992.

[18] Mallat S G. Multiresolution Approximations and Wavelet Orthonormal Bases of $L^2(R)$, Transactions of the American Mathematical Society. 1989; 315(1): 69–87.

[19] Li A, Shieh W, Tucker R. Wavelet Packet Transform-Based OFDM for Optical Communications, Journal of Lightwave Technology. 2010; 28(24): 3519–3528.

[20] Goswami J C, Chan A K. Fundamentals of Wavelets: Theory, Algorithms, and Applications, Second Edition, John Wiley & Sons, Inc; 2011.

[21] Chang R W. Synthesis of Band-Limited Orthogonal Signals for Multichannel Data Transmission, Bell Systems Technical Journal. 45: 1775–1796. see also U.S. Patent 3,488,445, Jan. 6, 1970.

[22] Erdol N, Bao F, Chen Z. Wavelet Modulation: A Prototype for Digital Communication Systems. In: Proceeding of the IEEE Conference Record Southcon '95, 7–9 March, Fort Lauderdale, FL: 1995. P. 168–171.

[23] Jamin A, Mähönen P. Wavelet Packet Modulation for Wireless Communications: Research Articles, Wireless Communications and Mobile Computing. 2005; 5(2): 123–137.

[24] Vetterli M, Kovacevic J. Wavelets and Subband Coding, Prentice Hall Signal Processing Series, Prentice Hall PTR. 1995.

[25] Jr R M W. IEEE 802.16 Broadband Wireless Access Working Group. Technical report. SciCom, Inc. 2001.

[26] Zhang L. A Study of IEEE 802.16a OFDM-PHY Baseband. Master's thesis, Department of Electrical Engineering, Linköping Institute of Technology. 2005

Section 3

Signal Processing and Wavelet Theory

Chapter 5

Wavelets for EEG Analysis

Nikesh Bajaj

Abstract

This chapter introduces the applications of wavelet for Electroencephalogram (EEG) signal analysis. First, the overview of EEG signal is discussed to the recording of raw EEG and widely used frequency bands in EEG studies. The chapter then progresses to discuss the common artefacts that contaminate EEG signal while recording. With a short overview of wavelet analysis techniques, namely; Continues Wavelet Transform (CWT), Discrete Wavelet Transform (DWT), and Wavelet Packet Decomposition (WPD), the chapter demonstrates the richness of CWT over conventional time-frequency analysis technique e.g. Short-Time Fourier Transform. Lastly, artefact removal algorithms based on Independent Component Analysis (ICA) and wavelet are discussed and a comparative analysis is demonstrated. The techniques covered in this chapter show that wavelet analysis is well-suited for EEG signals for describing time-localised event. Due to similar nature, wavelet analysis is also suitable for other biomedical signals such as Electrocardiogram and Electromyogram.

Keywords: EEG, artefacts, wavelet analysis, CWT, DWT, WPD, artefact removal algorithms, time-frequency analysis

1. Introduction

Biomedical signals are electrical activities recorded by sensors from a part of the body, such as the brain, heart, muscles, etc. They can be recorded as images e.g. functional Magnetic resonance Image (fMRI) from brain or a temporal signal e.g. Electrocardiogram (ECG), Electroencephalogram (EEG), Electromyogram (EMG), Galvanic Skin Response (GSR), etc. These signals contain useful information to analyse and understand the underlying physiological response of the body, thus they are also referred to as physiological signals. Biomedical signals are extensively used in healthcare to diagnose deceases and monitor health. With recent advancements and ease of using the devices to record the biomedical signals have open a window to use it to analyse and understand the day-to-day activities, emotions, and, experiences [1–3]. While recording the physiological activities through sensors, the signals are usually contaminated by noise and various artefacts [4]. Corrupted signals mislead the analysis and understanding of the underlying physiology [5]. The characteristics of wavelet to identify the time-localised events makes it suitable for the biomedical signals to clean, process, feature extraction, and analyse for various applications. Recent studies have shown the promising results of using wavelet in biomedical signals [6].

In this chapter, first, we introduce one kind of biomedical signal - EEG. We will explain the conventional features used in EEG studies. We will introduce the artefacts that commonly contaminate EEG signals, which makes it harder to use. The

IntechOpen

chapter then will move towards a short description of Wavelet analysis techniques, namely Continues Wavelet Transform (CWT), Discrete Wavelet Transform (DWT), and Wavelet Packet Decomposition (WPD). We would, then, compare CWT and STFT for EEG signal. Then, we will discuss artefact removal algorithms, with more details on Wavelet-based algorithms. The chapter will show the comparative analysis of artefact removal algorithm. The approach and analysis shown in this chapter for EEG signals can easily be applied to other biological signals.

2. Electroencephogram - EEG

The brain processes any information by means of neurons that use electrical and chemical signals to communicate by releasing and receiving neurotransmitters. The neural activity in the human brain is an electrical change. The brain generates electrical signals throughout the day for various activities. Studying these electrical signals is vital to understanding the neurophysiological behaviour of the brain [4]. A number of techniques are used to study brain activities. Functional magnetic resonance imaging (fMRI), Functional Near-Infrared Spectroscopy (firs), and Electroencephalography (EEG) recordings widely used techniques. The fMRI measures brain activity by scanning the blood flow. The fNIRS measures brain activity by measuring hemodynamic response in the brain through detecting the temporal changes in infrared light source. The EEG measures the electrical activity of the brain by electrodes placed on the scalp. Comparing to the other two, EEG measures brain activity directly, with high temporal resolution and most accessible and portable for the research. The fMRI has a high spatial resolution but very expensive, therefore it is mostly limited to medical diagnosis and treatments.

2.1 The EEG measure

The EEG signal is measured by placing multiple electrodes on the scalp that measure the current flow from neurons. A setup for EEG recording is shown in **Figure 1**. Each neuron (brain cell), when activated, it produced an electrical and magnetic field around the scalp. Since there are 100 billion neurons in the brain, when an electrode is placed on the scalp, it measures the accumulative activity of many neurons together. The complex structure of the brain attenuates the electrical signals, therefore an electrode can record the brain activity, only when a large number of neurons generate enough potential. The EEG devices amplify the recorded signal to store and process it [4].

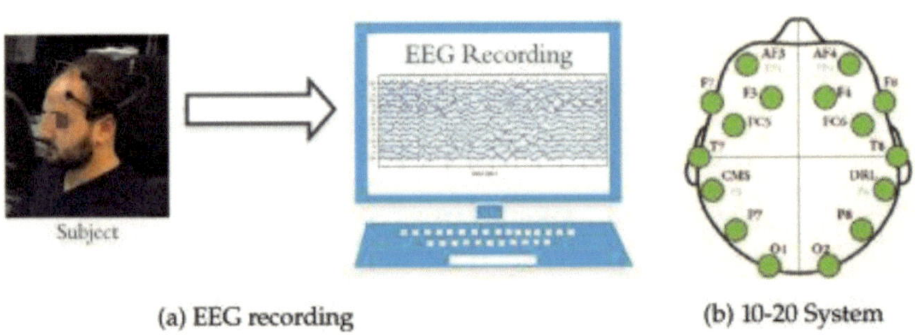

Figure 1.
EEG recording setup: (a) a wireless device Emotiv Epoch mounted on a subject, transmitting EEG signal to a computer. (b) Electrode positions as 10–20 system, source: https://www.emotiv.com/.

The placement of electrodes has been standardised with the specific anatomical landmarks with a distance between electrodes as 10% or 20% of total length. This placement is called the 10–20 system, as shown in **Figure 1b**. The number of electrodes used for EEG recording varies, depending on the device. One of the low spatial resolutions can be of a 14-channel EEG device and high spatial resolution with 128 or 256 channels. The name of the electrode position is labelled as character followed by a number to identify the part of the brain. The characters are *Fop* for pre-frontal, *F* for frontal, *P* for parietal, *T* for temporal, *O* for occipital, and *C* for central lobe of the brain. A few in between two landmarks are named with two characters, such as *AF*, between *Fp* and *F* and *FC*, between *F* and *C* [4]. An example of 14-channel is shown in **Figure 1b**.

The raw recording of EEG signal in the time-domain is complex to interpret. Similar to many other signals, frequency domain analysis has been widely used. The decades of work on EEG studies have identified five major frequency bands for EEG signals and established the correlation between behaviour and neural activity of a certain part of the brain. The frequency bands widely used are; Delta (0.1 – 4 Hz or 0.5 – 4 Hz), Theta (4 – 8 Hz), Alpha (8 – 14 Hz), Beta (14 – 30 Hz), Gamma (30 – 63 Hz). A raw EEG signal from a channel and corresponding signal in different bands are shown in **Figure 2**. It can be observed that low frequency, Delta activity, is the dominating wave in raw EEG and high-frequency Gamma is almost noise like with a little amplitude [4].

Due to multichannel signals, it is usually viewed as topographical brain activity (heatmap over an image of head) under different frequency bands. An example of 5 seconds EEG recording with a 14-channel device is shown in **Figure 3**. The first second of all the channels are used to compute the energy distribution over brain regions. In **Figure 3**, the top left shows the raw EEG signal and corresponding brain activity, which shows a high activity in the frontal lobe of the brain. However, under different frequency bands, the different part of the brain shows higher activity.

The frequency bands; Delta, Theta, Alpha, Beta, and Gamma, are also called brain rhythms. Brain rhythms have been investigated over decades and a few characteristic behaviour of these brain rhythms have been established [4].

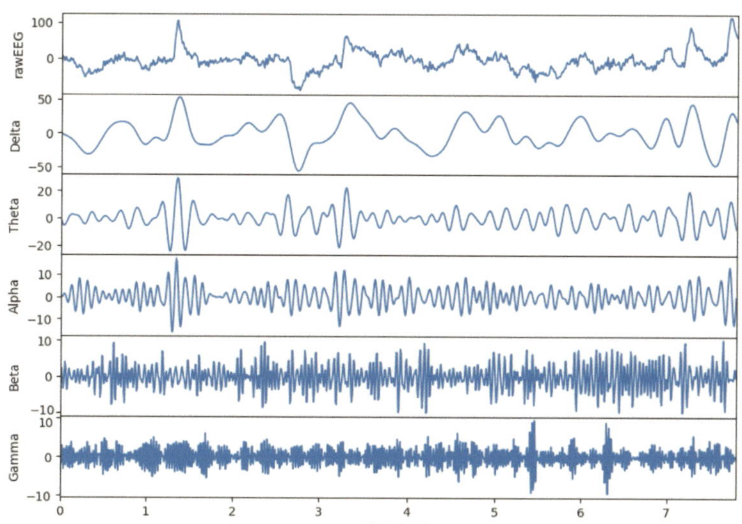

Figure 2.
The signal channel raw EEG signal and corresponding frequency bands: Delta (0.1 – 4 Hz), theta (4 – 8 Hz), alpha (8 – 14 Hz), Beta (14 – 30 Hz), gamma (30 – 63 Hz).

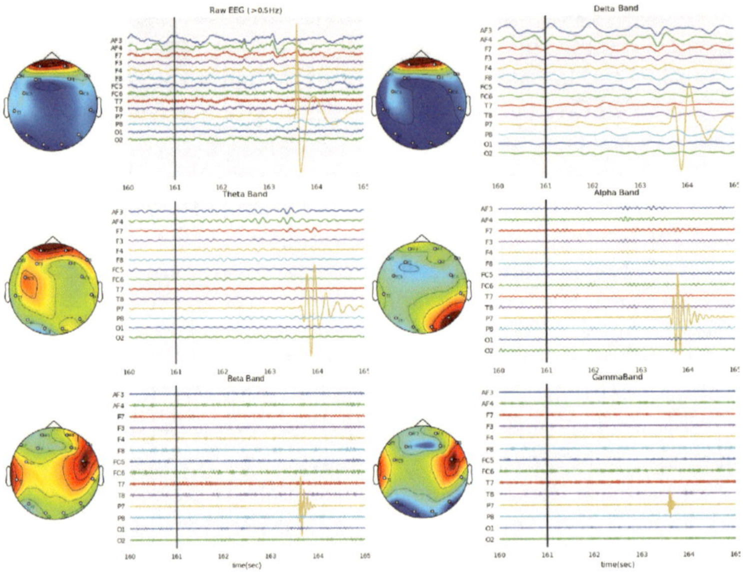

Figure 3.
Topographical view of brain activity: Energy distribution of EEG recording over different brain regions under five frequency bands and raw signal.

- **Delta**: Delta waves were first introduced by Walter in 1936, it ranges from 0.1 (or 0.5) to 4 Hz in frequency. Delta waves are usually observed in deep sleep. Since delta wave is the low-frequency wave, it is easily confused by the movement artefact, due to similar nature. Delta waves have also been linked to continuous attention tasks.

- **Theta**: Theta waves were introduced by Dovey and Wolter, ranges from 4 to 8 Hz in frequency. Theta waves are linked to drowsiness and deep meditation state.

- **Alpha**: Alpha waves, perhaps are the most widely investigated waves in EEG studies. Alpha waves were introduced by Berger in 1929. They lie in a range from 8 to 14 Hz. Alpha waves usually appear on the occipital lobe of the brain. Alpha waves are the most common indication of a relaxing state of mind and are also linked to closing eyes. Any sign of anxiety or attention reduces the alpha waves.

- **Beta**: Beta waves lie in the range of 14–30 Hz of frequency. Beta waves have been associated with active thinking, anxious, high alert, and focus of the brain.

- **Gamma**: Gamma waves are the higher frequency waves, ranges from 30 to onwards. Gamma wave is considered to play a complex role in brain functionality, such as combining information from two different sensory inputs. It is also used to confirm certain brain diseases.

2.2 Artefacts in EEG

While recording, EEG signals are frequently contaminated with various artefacts. The most common types of artefacts are motion, muscular, ocular, and cardiac artefacts [4], which are shown in **Figure 4**. The motion artefacts are caused by

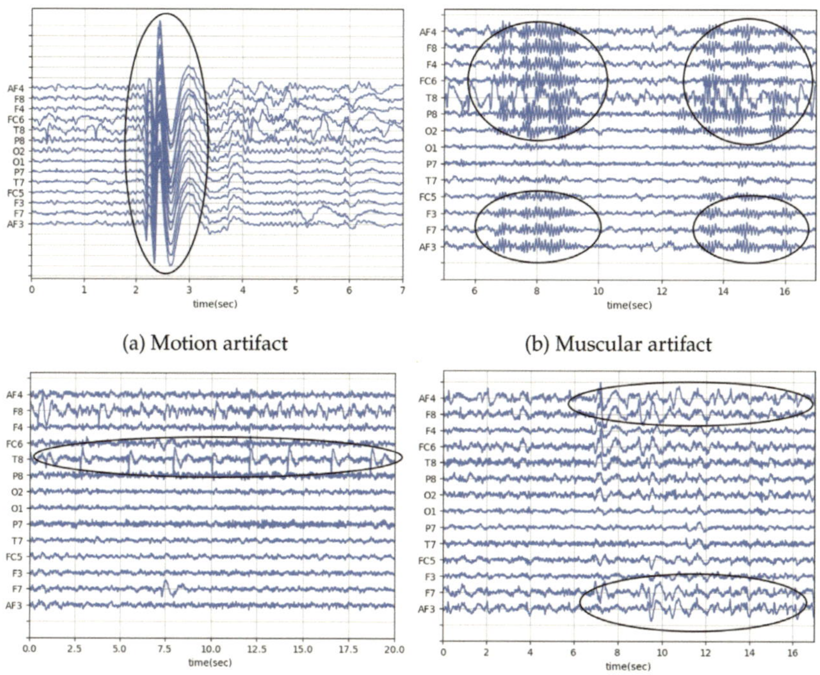

(a) Motion artifact

(b) Muscular artifact

(c) Cardiac artifact

(d) Ocular artifact

Figure 4.
Common type of artefacts in EEG. Corresponding artefacts are circled in the figure.

the physical movement of the person's body. As shown in **Figure 4a**, motion artefacts produce a sudden high valued spike in all the channels of EEG recording. The muscular artefacts, shown in **Figure 4b** are caused by any muscular contraction such as grinding the teeth. It produces high-frequency bursts in EEG recording as circled in the **Figure 4b**. The cardiac artefacts, shown in **Figure 4c**, are caused by the electrical activities of the heart. They appear as a weak form of QRS wave of heart and most likely to be appeared in the channels near to ears (temporal lobe), though it can be sometimes present in channels from the frontal lobe [7]. The ocular artefacts are slow oscillating waves appear on the frontal lobe, caused by the eye movements or closed eyes, as circled in the **Figure 4d**. The higher magnitude of the artefacts corrupts the EEG recording and leads to misinterpretations of the results and analysis [5]. Even though there are many algorithms to remove the artefacts, but there is always a possibility of losing the cerebral information while removing the artefacts.

3. Wavelet analysis

Most of the real-life signals are non-stationary in behaviour, which means their properties change over time. To localise the events of interest, time-frequency analysis is widely used. The conventional way of time-frequency analysis is the Short-Time Fourier Transform (STFT), where Fourier transform of the signal is taken over short-windows, resulting spectrogram plot. STFT has limitations on resolutions, due to Heisenberg's uncertainty principle, e.g. improvement in time resolution results in poorer frequency resolution and vice-versa. The alternative to STFT is wavelet transform, which exploits the property of low-frequency signals

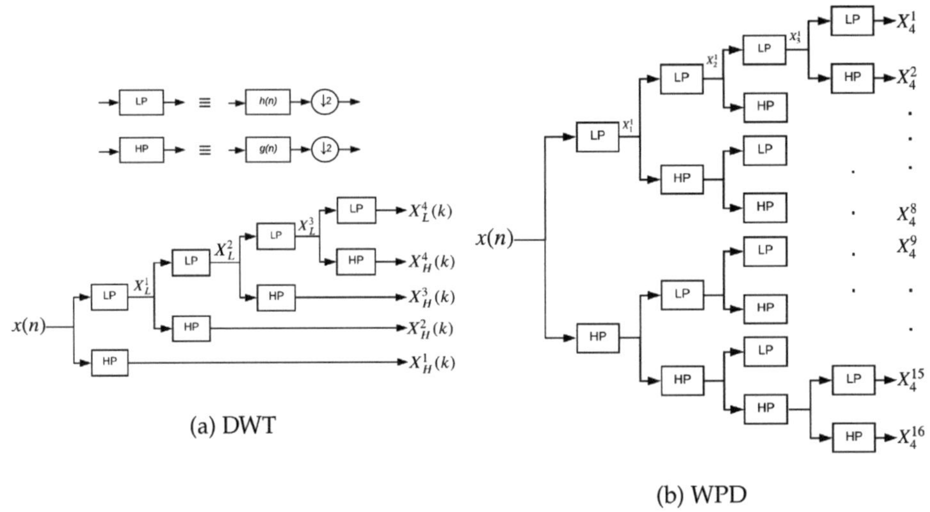

Figure 5.
4-level decomposition tree for (a) discrete wavelet transform (DWT), (b) wavelet packet decomposition (WPD).

being widespread over time and high-frequency bursts occurring on short intervals. Wavelet transform uses the variable size of windows with a wavelet function.

Wavelet analysis is usually applied in two ways, Continuous Wavelet Transform (CWT) and Discrete Wavelet Transform (DWT). CWT uses a wavelet function $\psi(t)$ and produces a scalogram, similar to a spectrogram for time-frequency analysis. However, DWT decomposes a signal into two (1) average or lowpass signal, using scaling function and (2) difference or highpass signal using wavelet function. The conventional DWT recursively decomposes lowpass signal with the same scaling and wavelet functions to the desired level of decomposition. A decomposition tree for DWT is shown in **Figure 5a**. For some applications, it is useful to decompose highpass signal at each level too, this is called Wavelet Packet Decomposition (WPD). A tree for WPD is shown in **Figure 5b**.

As shown in **Figure 5**, block LP is a lowpass filter $h(n)$ and block HP is a highpass filter $g(n)$, both followed by downsampler $(\downarrow 2)$. The coefficients of lowpass filter are corresponds to scalling function $\phi(n)$ and coefficients of highpass filter are corresponds to wavelet funciton $\psi(n)$. A N-level DWT decomposes a signal $x(n)$ into set of signals: $[X_L^N, X_H^N, X_H^{N-1}, \dots, X_H^1]$, each with different dimensions. However, a N-level WPD decomposes a signal $x(n)$ into set of packets: $[X_N^1, X_N^2, \dots, X_N^{2^N}]$, each with same dimensions.

4. Time-frequency analysis of EEG using CWT

As discussed, a conventional way to time-frequency analysis is STFT, however, using CWT with different wavelet functions can enrich the analysis with more details. In this section, we will show, how a continuous wavelet function $(\psi(t))$, can be applied to a discrete EEG signal $x(n)$, and compare the spectrogram with scalogram of different wavelet functions.

A spectrogram is obtained using STFT, which is Fourier Transform computed for a short windows. STFT $X_{stft}(\tau, \omega)$ of signal $x(t)$ as given as Eq. (1), where $w(t)$ is a window function. On the other hand, CWT $X_{cwt}(a, b)$ of a signal $x(t)$ is given by

Eq. (2), where $\psi^*_{a,b}(t)$ is a complex conjugate of scaled and shifted version of mother wavelet $\psi(t)$, a is scaling parameter and b is shifting parameter.

$$X_{stft}(\tau,\omega) = \int_{-\infty}^{\infty} x(t)w(\tau - t)e^{-j\omega t}dt \tag{1}$$

$$X_{cwt}(a,b) = \int_{-\infty}^{\infty} x(t)\psi^*_{a,b}(t)dt. \tag{2}$$

CWT operation from Eq. (2) can be seen as convolution of input signal $x(t)$ with scaled version of wavelet function $\psi(t)$.

$$X_{cwt}(a) = x(t)\circledast\psi^*_a(t) \tag{3}$$

$$X_{cwt}(f) = X(f)\psi_a * (f) \tag{4}$$

where $X(f)$ is Fourier transform of $x(t)$, and the same for others. For computations with discrete signal $x(n)$, both equations; 3 and 4 can be used with discrete operations, e.g. convolution and multiplication and discrete wavelet function $\psi(n)$, while for Fourier Transform, Fast Fourier Transform (FFT) is used. For computational efficiency, however, Eq. (4) is widely used, by multiplying FFT of $x(n)$ and FFT of scaled and discrete version $\psi(n)$. Even though, for discrete signal $x(n)$, discrete wavelet function $\psi(n)$ is used, however, the conventional definitions of wavelet functions for CWT are defined in continues time-domain. A set time-domain and frequency-domain equations for six complex wavelet functions are defined below. **Figure 6** shows all the six wavelet functions, with their real and imaginary part. All six functions are similar, in terms of smoothness and being derived from exponential and sinusoidal functions, however, they have different parameters to control the oscillation and frequency band to be captured.

Gaussian Wavelet: A time-domain wavelet it derived from a Gaussian function centered at t_0 and modulated by a complex exponential function with frequency f_0 [8].

$$\psi(t) = e^{-a(t-t_0)^2} \cdot e^{-2\pi j f_0(t-t_0)} \tag{5}$$

$$\psi(f) = \sqrt{\pi/a}\left(e^{-2\pi j f t_0} \cdot e^{-\pi^2\left((f-f_0)^2\right)/a}\right) \tag{6}$$

where $a = \left(\frac{f_0}{Q}\right)^2$.

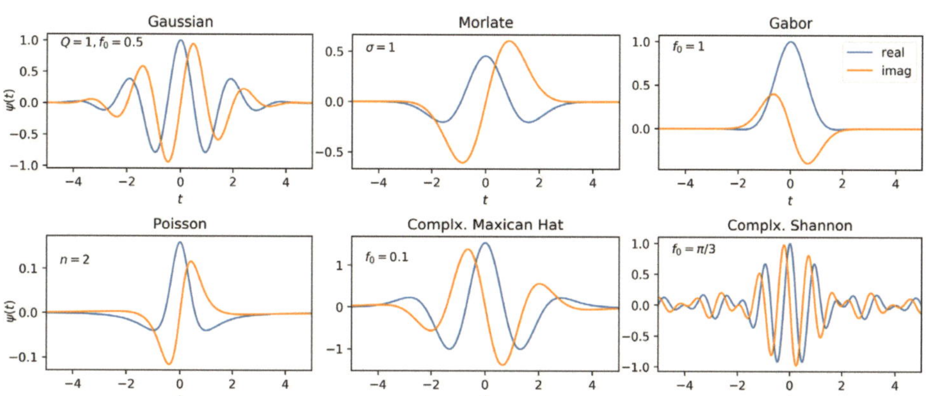

Figure 6.
Continues wavelet functions.

Gabor Wavelet: Gabor wavelet is perhaps the most widely used function for various applications. It is essentially the same as Gaussian wavelet function, with simplified equations, as follow [8, 9];

$$\psi(t) = e^{-(t-t_0)^2/a^2} e^{-jf_0(t-t_0)} \tag{7}$$

$$\psi(f) = e^{-((f-f_0)a)^2} e^{-jt_0(f-f_0)} \tag{8}$$

where a is oscillation rate and f_0 is center frequency, t_0 is centred time.

Morlet Wavelet: Morlet is considered very similar to Gabor wavelet and Gabor filters. The oscillation of Morlet wavelet is controlled by σ. A higher value of σ results in higher oscillation [10].

$$\psi(t) = C_\sigma \pi^{-0.25} e^{-0.5t^2} \left(e^{j\sigma t} - K_\sigma\right) \tag{9}$$

$$\psi(w) = C_\sigma \pi^{-0.25} \left(e^{-0.5(\sigma-w)^2} - K_\sigma e^{-0.5w^2}\right) \tag{10}$$

where $C_\sigma = \left(1 + e^{-\sigma^2} - 2e^{-\frac{3}{4}\sigma^2}\right)^{-0.5}$, $K_\sigma = e^{-0.5\sigma^2}$, and $w = 2\pi f$.

Poisson Wavelet: Poisson wavelet is defined by positive integers (n), unlike other, and associated with Poisson probability distribution [11, 12].

$$\psi(t) = \frac{1}{2\pi}(1-jt)^{-(n+1)} \tag{11}$$

$$\psi(w) = \frac{1}{\Gamma n + 1} w^n e^{-w} u(w) \tag{12}$$

where $w = 2\pi f$ and $u(w)$ is a unit step function, e.g. $u(w) = 1$ if $w >= 0$, 0 else.

Complex Mexican hat wavelet: Complex Mexican hat wavelet is derived from the conventional Mexican hat wavelet. It is a low-oscillation wavelet which is modulated by a complex exponential function with frequency f_0 [13].

$$\psi(t) = \frac{2}{\sqrt{3}} \pi^{-\frac{1}{4}} \left(\sqrt{\pi}(1-t^2)e^{-\frac{1}{2}t^2} - \left(\sqrt{2}jt + \sqrt{\pi} erf\left[\frac{j}{\sqrt{2}}t\right](1-t^2)e^{-\frac{1}{2}t^2}\right)\right) e^{-2\pi j f_0 t} \tag{13}$$

$$\psi(w) = 2\sqrt{\frac{2}{3}} \pi^{-1/4} (w-w_0)^2 e^{-\frac{1}{2}(w-w_0)^2} \quad \text{if } w \geq 0, \quad 0 \text{ else} \tag{14}$$

where $w = 2\pi f$ and $w_0 = 2\pi f_0$.

Complex Shannon wavelet: Complex Shannon wavelet is the most simplified wavelet function, exploiting Sinc function by modulating with sinusoidal, which results in an ideal bandpass filter. Real Shannon wavelet is modulated by only a cos function [14].

$$\psi(t) = Sinc(t/2) \cdot e^{-2j\pi f_0 t} \tag{15}$$

$$\psi(w) = \prod\left(\frac{w - w_0}{\pi}\right) \tag{16}$$

where $\prod(x) = 1$ if $x \leq 0.5$, 0 else and $w = 2\pi f$ and $w_0 = 2\pi f_0$.

An example of using the above six wavelet functions for a small single-channel EEG segment is shown in **Figure 7**, along with spectrogram. It can be observed, spectrogram highlights a few events in signal (sharp peaks and lowpass wave),

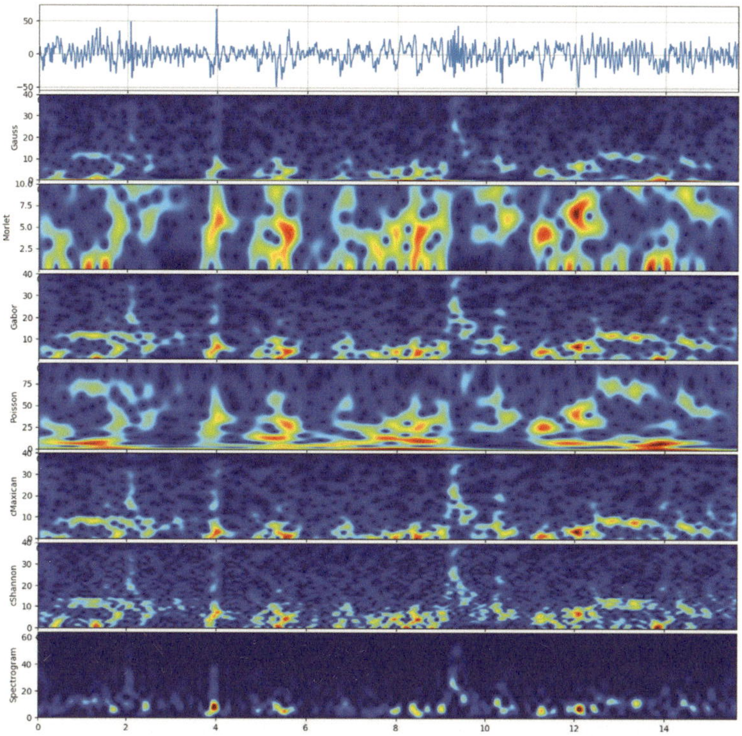

Figure 7.
Scalogram and spectrogram of a segment of signal channel EEG signal with six wavelet functions and STFT. Figure obtained using spkit python library - https://spkit.github.iohttps://spkit.github.io

however, using CWT with different Wavelet functions, much richer information can be observed. Since, we observed that in the formulation of wavelet functions that they are similar to the underlying principle, we could also observe the similarities across different scalograms. Specifically, spectrogram using Complex Shannon and Complex Mexican hat wavelet are much similar. Interestingly, Morlate and Poisson wavelet functions are able to produce a better resolution towards lower frequencies.

5. Artefact removal algorithms using DWT and WPD

Artefacts in EEG recording is a primary obstacle that all researchers have to deal with. There are decades of research work in literature to remove these artefacts [15, 16]. A range of methods have been proposed to remove the artefacts, starts with a statistical with interpolation method [17] and regression method [18]. The most commonly used approaches are based on Blind Source Separation (BSS) using Independent Component Analysis (ICA) [19, 20]. ICA based approach have been widely explored with statistical measures [21–24], and variant of ICA as FastICA, InfoMax, and Extended InfoMax [25–27]. Wavelet-based approaches are well suited for time-localised short events, as opposed to ICA. This property has been exploited to remove artefacts from single-channel EEG. In contrast to a single channel, wavelet has also been used for multi-channel EEG [28] and in combination with ICA [29–34], in which identified artifactual component is cleaned with wavelet rather than removed. The ICA-based approaches can only be applied to multi-channel EEG and need an expert to select artifactual component, which has been automated with heuristics [21, 35, 36]. In contrast, most wavelet-based algorithms remove artefacts from each channel individually.

The key idea of wavelet-based artefact removal algorithms is to apply DWT on single-channel EEG signal $x(n)$ and remove (set to zero) wavelet coefficients that fall above some statistical threshold and reconstruct signal back using inverse-DWT (IDWT) $\hat{x}(n)$ [37–39]. With linear property of electrical activities, recorded EEG signal is considered as $x(n) = s(n) + v(n)$, where $s(n)$ is source signal of brain activity and $v(n)$ is artifactual components. The two most widely used threshold formulations are used with wavelet.

Global Threshold: Also known as the optimal threshold for removing white-gaussian noise from any signal [40] using DWT. Global Threshold (T_G) is defined as;

$$T_G = \hat{\sigma}\sqrt{2logN} \quad \text{for}\, \hat{\sigma} = \frac{median(|w|)}{0.6745} \tag{17}$$

where N is the length of signal and for wavelet coefficients w, $\hat{\sigma}$ is the estimate of noise variance. To denoise a signal, wavelet coefficients with magnitude below the threshold T_G are set to zero and reconstruct the signal back. However, for recorded EEG signal $x(n)$, source signal $s(n)$ is considered to be zero mean and normally distributed, i.g. $s(n) \sim \mathcal{N}(0, \sigma)$ [4]. For which any wavelet coefficients with a magnitude above the threshold T_G is considered to be artifactual and removed (set to zero).

Standard Deviation (STD) Threshold: As name suggests, STD threshold is based on Standard Deviation (STD) of wavelet coefficients [39].

$$T_{STD} = 1.5 \times STD(w) \tag{18}$$

ATAR algorithm: A recent study has shown that approaches based on above thresholds are very aggressive, since, statistically, a few wavelet coefficients of any signal will always fall above these thresholds [14]. In contrast, an Automatic and Tunable Artefact Removal (ATAR) Algorithm based on WPD was proposed [41], which provides three different wavelet filtering modes and a tunable parameter. As shown a block diagram of ATAR algorithm in **Figure 8**, a single channel EEG signal $x(n)$ is first split into smaller windows $x_w(n)$, apply L-level WPD to get wavelet coefficients $w = X_L(k) = WPD(x_w(n))$, then wavelet coefficients are filtered using wavelet filtering $\hat{w} = \lambda(w)$ to reconstruct signal $\hat{x}(n)$ from corrected windows $\hat{x}_w(n)$.

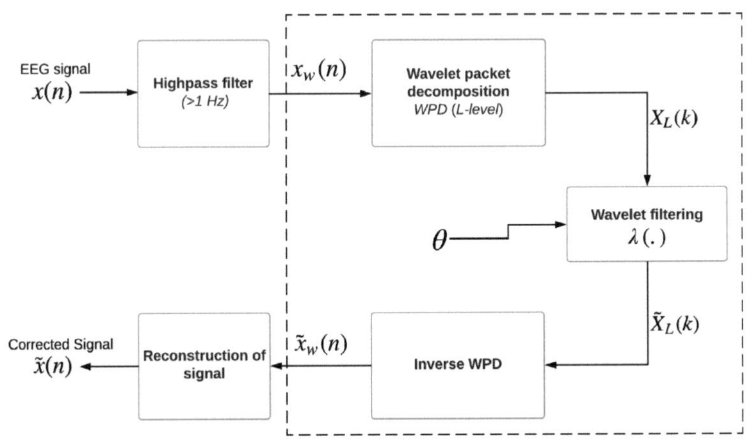

Figure 8.
A block diagram of ATAR algorithm [41].

The three filtering modes in ATAR algorithm are namely; Elimination $\lambda_e(\cdot)$, Linear attenuation $\lambda_a(\cdot)$, and soft thresholding $\lambda_s(\cdot)$, which are defined below;

$$\lambda_e(w) = \begin{cases} w & \text{if } |w| \le \theta_\alpha \\ 0 & \text{else} \end{cases} \tag{19}$$

$$\lambda_a(w) = \begin{cases} w & \text{if } |w| \le \theta_\alpha \\ sgn\,(w)\theta_\alpha\left(1 - \dfrac{|w| - \theta_\alpha}{\theta_\beta - \theta_\alpha}\right) & \text{if } \theta_\alpha < |w| \le \theta_\beta \\ 0 & \text{else} \end{cases} \tag{20}$$

$$\lambda_s(w) = \begin{cases} w & \text{if } |w| < \theta_\gamma \\ \dfrac{1 - e^{-\alpha w}}{1 + e^{-\alpha w}}\theta_\alpha & \text{otherwise} \end{cases} \quad \text{where } \alpha = -\frac{1}{\theta_\gamma}\log\frac{\theta_\alpha - \theta_\gamma}{\theta_\alpha + \theta_\gamma} \tag{21}$$

where w is a wavelet coefficient, $sgn\,(\cdot)$ is the signum function, and $\theta_\alpha > \theta_\gamma$. A default setting for θ_γ and θ_β is; $\theta_\gamma = 0.8\theta_\alpha$ and $\theta_\beta = 2\theta_\alpha$. The characteristics of wavelet filtering mode are shown in **Figure 9**. From **Figure 9**, it can be seen that Elimination mode of filtering is the same as conventional filtering, however, Linear attenuation and soft-thresholding modes do not remove the wavelet coefficient, rather suppress them softly. Another distinction ATAR algorithm has over others is the threshold selection. The threshold θ_α is computed from Interquartile Range (IQR) of wavelet coefficients using Eq. (22), which is robust against outliers, as oppose to *STD*.

$$\theta_\alpha = \begin{cases} f_\beta(r) & \text{if } f_\beta(r) \ge k_1 \\ k_1 & \text{else} \end{cases} \quad \text{where } f_\beta(r) = k_2\, exp\left(-\beta\frac{100}{k_2}\frac{r}{2}\right) \tag{22}$$

where r is Interquartile Range (IQR) of coefficients i.e. $r = IQR(w)$ and k_1 and k_2 are lower and upper bounds on filtering.

Figure 10 shows a visual comparative analysis of wavelet-based approaches (i.e. Global threshold, STD threshold, and ATAR algorithm) and ICA based approaches (FastICA, InfoMax, and Extended-InfoMax) to remove the artefacts. It is visually apparent that wavelet-based approaches are better than ICA-based approaches.

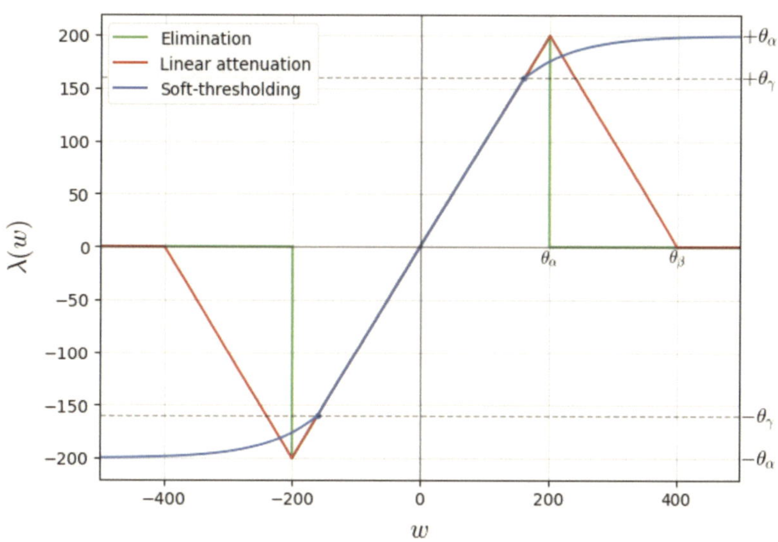

Figure 9.
Wavelet filtering modes for ATAR algorithm. For $\theta_\alpha = 200, \theta_\gamma = 0.8\theta_\alpha = 160, \theta_\beta = 2\theta_\alpha = 400$ [41].

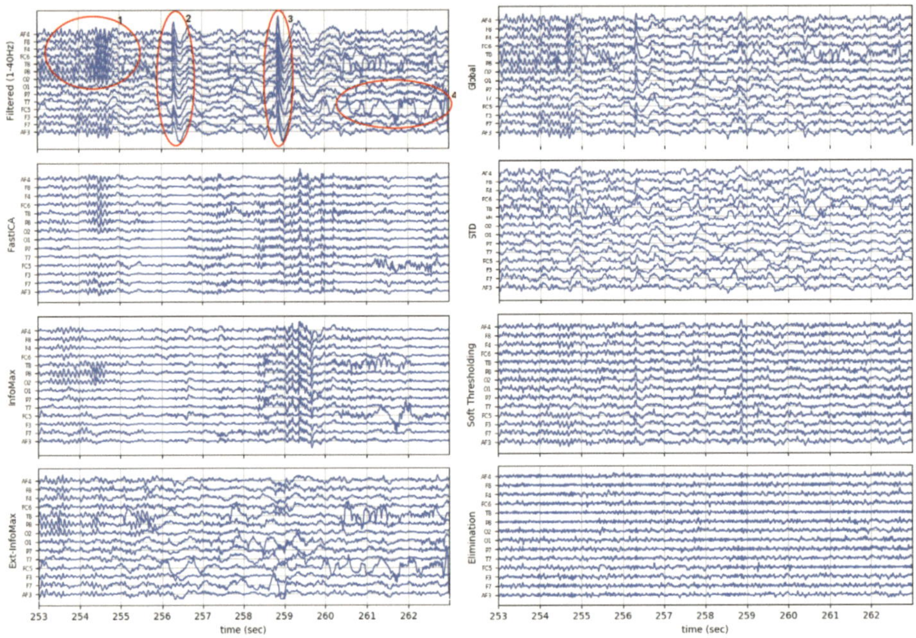

Figure 10.
Comparison of artefact removal approaches from [41].

Among wavelet-based approaches, using ATAR gives much control over Global and STD based threshold selection. Other quantitative analyses of the above-mentioned approaches are discussed in the article [41], which also demonstrate the effect of tuning parameter and filtering modes on different predictive tasks of EEG signal. The formulation of relationship, algorithmic implementation details, and comparative results are given in article [41].

6. Conclusions

This chapter presents the overview of Wavelet for EEG analysis. The first chapter introduces EEG signal, commonly used features for predictive analysis, and artefacts that often contaminate EEG signal. Then chapter discusses the Wavelet analysis approaches, namely CWT, DWT, and WPD. The richness of CWT over STFT for time-frequency analysis using various wavelet functions is demonstrated. Finally, the artefact removal algorithms based on wavelet and ICA are discussed. The comparative analysis present in the chapter shows that the wavelet-based approach outperforms ICA based approach. Specifically, a recent algorithm (ATAR) allows controlling the removal or suppression of assumed artifactual components in the signal, which can be tuned to improve the performance of any predictive tasks. The techniques presented in this chapter show how wavelet can be used for EEG studies to extract rich information and removing the artefacts. The comparative analysis shows wavelet based approaches are well suited for EEG signal processing. Further, similar approaches can be used with other biomedical signals such as electrocardiogram (ECG or EKG), Electromyography (EMG) etc.

Author details

Nikesh Bajaj[1,2]

1 University of East London, London, United Kingdom

2 Queen Mary University of London, London, United Kingdom

*Address all correspondence to: n.bajaj@uel.ac.uk; n.bajaj@qmul.ac.uk

IntechOpen

© 2020 The Author(s). Licensee IntechOpen. This chapter is distributed under the terms of the Creative Commons Attribution License (http://creativecommons.org/licenses/by/3.0), which permits unrestricted use, distribution, and reproduction in any medium, provided the original work is properly cited.

References

[1] D. Petit, J.-F. Gagnon, M. L. Fantini, L. Ferini-Strambi, J. Montplaisir, Sleep and quantitative eeg in neurodegenerative disorders, Journal of psychosomatic research 56 (2004) 487–496

[2] M. L. Perlis, M. T. Smith, P. J. Andrews, H. Orff, D. E. Giles, Beta/gamma eeg activity in patients with primary and secondary insomnia and good sleeper controls, Sleep 24 (2001) 110–117.

[3] A. R. Clarke, R. J. Barry, R. McCarthy, M. Selikowitz, Eeg analysis in attention-deficit/hyperactivity disorder: a comparative study of two subtypes, Psychiatry research 81 (1998) 19–29.

[4] S. Sanei, J. A. Chambers, EEG signal processing, John Wiley & Sons, 2013.

[5] D. Hagemann, E. Naumann, The effects of ocular artifacts on (lateralized) broadband power in the eeg, Clinical Neurophysiology 112 (2001) 215–231.

[6] H. Olkkonen, Discrete Wavelet Transforms: Biomedical Applications, BoD–Books on Demand, 2011.

[7] G. Dirlich, T. Dietl, L. Vogl, F. Strian, Topography and morphology of heart action-related eeg potentials, Electroencephalography and Clinical Neurophysiology/Evoked Potentials Section 108 (1998) 299–305.

[8] T. S. Lee, Image representation using 2d gabor wavelets, IEEE Transactions on pattern analysis and machine intelligence 18 (1996) 959–971.

[9] Gabor wavelet, https://en.wikipedia.org/wiki/Gabor wavelet, 2020.

[10] Morlate wavelet, https://en.wikipedia.org/wiki/Morlet wavelet, 2020.

[11] K. A. Kosanovich, A. R. Moser, M. J. Piovoso, A new family of wavelets: the poisson wavelet transform, Computers & chemical engineering 21 (1997) 601–620.

[12] Poisson wavelet, https://en.wikipedia.org/wiki/Poisson wavelet, 2020.

[13] Complex mexican hat wavelet, https://en.wikipedia.org/wiki/Complex Mexican hat wavelet, 2020.

[14] Shannon wavelet, https://en.wikipedia.org/wiki/Shannon wavelet, 2020.

[15] J. A. Urigüen, B. Garcia-Zapirain, EEG artifact removal—state-of-the-art and guidelines, Journal of neural engineering 12 (2015) 031001.

[16] M. Fatourechi, A. Bashashati, R. K. Ward, G. E. Birch, Emg and eog artifacts in brain computer interface systems: A survey, Clinical neurophysiology 118 (2007) 480–494.

[17] M. Junghöfer, T. Elbert, D. M. Tucker, B. Rockstroh, Statistical control of artifacts in dense array EEG/MEG studies, Psychophysiology 37 (2000) 523–532.

[18] J. Woestenburg, M. Verbaten, J. Slangen, The removal of the eye-movement artifact from the eeg by regression analysis in the frequency domain, Biological psychology 16 (1983) 127–147.

[19] T.-P. Jung, S. Makeig, C. Humphries, T.-W. Lee, M. J. Mckeown, V. Iragui, T. J. Sejnowski, Removing electroencephalographic artifacts by blind source separation, Psychophysiology 37 (2000) 163–178.

[20] L. Shoker, S. Sanei, W. Wang, J. A. Chambers, Removal of eye blinking artifact from the electro-encephalogram, incorporating a new constrained blind source separation

algorithm, Medical and Biological Engineering and Computing 43 (2005) 290–295.

[21] H. Nolan, R. Whelan, R. Reilly, FASTER: fully automated statistical thresholding for EEG artifact rejection, Journal of neuroscience methods 192 (2010) 152–162.

[22] R. R. Vázquez, H. Velez-Perez, R. Ranta, V. L. Dorr, D. Maquin, L. Maillard, Blind source separation, wavelet denoising and discriminant analysis for eeg artefacts and noise cancelling, Biomedical Signal Processing and Control 7 (2012) 389–400.

[23] A. Mognon, J. Jovicich, L. Bruzzone, M. Buiatti, Adjust: An automatic eeg artifact detector based on the joint use of spatial and temporal features, Psychophysiology 48 (2011) 229–240.

[24] A. Belouchrani, K. Abed-Meraim, J.-F. Cardoso, E. Moulines, A blind source separation technique using second-order statistics, IEEE Transactions on signal processing 45 (1997) 434–444.

[25] D. Langlois, S. Chartier, D. Gosselin, An introduction to independent component analysis: Infomax and fastica algorithms, Tutorials in Quantitative Methods for Psychology 6 (2010) 31–38.

[26] A. J. Bell, T. J. Sejnowski, An information-maximization approach to blind separation and blind deconvolution, Neural computation 7 (1995) 1129–1159.

[27] T.-W. Lee, M. Girolami, T. J. Sejnowski, Independent component analysis using an extended infomax algorithm for mixed subgaussian and supergaussian sources, Neural computation 11 (1999) 417–441.

[28] G. Inuso, F. La Foresta, N. Mammone, F. C. Morabito, Brain activity investigation by eeg processing: wavelet analysis, kurtosis and renyi's entropy for artifact detection, in: 2007 International Conference on Information Acquisition, IEEE, pp. 195–200.

[29] N. P. Castellanos, V. A. Makarov, Recovering eeg brain signals: artifact suppression with wavelet enhanced independent component analysis, Journal of neuroscience methods 158 (2006) 300–312.

[30] M. T. Akhtar, W. Mitsuhashi, C. J. James, Employing spatially constrained ica and wavelet denoising, for automatic removal of artifacts from multichannel eeg data, Signal Processing 92 (2012) 401–416.

[31] H. Ghandeharion, A. Erfanian, A fully automatic ocular artifact suppression from eeg data using higher order statistics: Improved performance by wavelet analysis, Medical engineering & physics 32 (2010) 720–729.

[32] N. K. Al-Qazzaz, S. H. B. M. Ali, S. A. Ahmad, M. S. Islam, J. Escudero, Discrimination of stroke-related mild cognitive impairment and vascular dementia using eeg signal analysis, Medical & biological engineering & computing 56 (2018) 137–157.

[33] R. Mahajan, B. I. Morshed, Unsupervised eye blink artifact denoising of eeg data with modified multiscale sample entropy, kurtosis, and wavelet-ica, IEEE journal of Biomedical and Health Informatics 19 (2014) 158–165.

[34] N. Al-Qazzaz, S. Hamid Bin Mohd Ali, S. Ahmad, M. Islam, J. Escudero, Automatic artifact removal in eeg of normal and demented individuals using ica–wt during working memory tasks, Sensors 17 (2017) 1326.

[35] W. Kong, Z. Zhou, S. Hu, J. Zhang, F. Babiloni, G. Dai, Automatic and

direct identification of blink components from scalp eeg, Sensors 13 (2013) 10783–10801.

[36] R. N. Vigário, Extraction of ocular artefacts from EEG using independent component analysis, Electroencephalography and clinical neurophysiology 103 (1997) 395–404.

[37] S. V. Ramanan, N. Kalpakam, J. Sahambi, A novel wavelet based technique for detection and de-noising of ocular artifact in normal and epileptic electroencephalogram (2004).

[38] V. Krishnaveni, S. Jayaraman, S. Aravind, V. Hariharasudhan, K. Ramadoss, Automatic identification and removal of ocular artifacts from eeg using wavelet transform, Measurement science review 6 (2006) 45–57.

[39] P. S. Kumar, R. Arumuganathan, K. Sivakumar, C. Vimal, Removal of ocular artifacts in the eeg through wavelet transform without using an eog reference channel, Int. J. Open Problems Compt. Math 1 (2008) 188–200.

[40] D. L. Donoho, J. M. Johnstone, Ideal spatial adaptation by wavelet shrinkage, biometrika 81 (1994) 425–455.

[41] N. Bajaj, J. R. Carrión, F. Bellotti, R. Berta, A. De Gloria, Automatic and tunable algorithm for eeg artifact removal using wavelet decomposition with applications in predictive modeling during auditory tasks, Biomedical Signal Processing and Control 55 (2020) 101624.

Chapter 6

Ultra-High Performance and Low-Cost Architecture of Discrete Wavelet Transforms

Mouhamad Chehaitly, Mohamed Tabaa, Fabrice Monteiro, Safa Saadaoui and Abbas Dandache

Abstract

This work targets the challenging issue to produce high throughput and low-cost configurable architecture of Discrete wavelet transforms (DWT). More specifically, it proposes a new hardware architecture of the first and second generation of DWT using a modified multi-resolution tree. This approach is based on serializations and interleaving of data between different stages. The designed architecture is massively parallelized and sharing hardware between low-pass and high-pass filters in the wavelet transformation algorithm. Consequently, to process data in high speed and decrease hardware usage. The different steps of the post/pre-synthesis configurable algorithm are detailed in this paper. A modulization in VHDL at RTL level and implementation of the designed architecture on FPGA technology in a NexysVideo board (Artix 7 FPGA) are done in this work, where the performance, the configurability and the generic of our architecture are highly enhanced. The implementation results indicate that our proposed architectures provide a very high-speed data processing with low needed resources. As an example, with the parameters depth order equal 2, filter order equal 2, order quantization equal 5 and a parallel degree $\boldsymbol{P} = 16$, we reach a bit rate around 3160 Mega samples per second with low used of logic elements (≈400) and logic registers (≈700).

Keywords: Mallat binary tree algorithm, DWPT, IDWPT, lifting scheme wavelet, FIR filter, parallel-pipeline architecture, VHDL-RTL modeling, FPGA

1. Introduction

We notice in the last year a wide usage of wavelet transform theory in different domain like telecommunications, image and video processing, data compression, optical fiber, encryption and others. But these domains are evolved extremely which require a new wavelet transform architecture with low cost target technology that can provide a high-speed data processing and low power consummation. In parallel FPGA technology is massively blossomed to come very popular and to be a target technology of many applications, in particular of Discrete Wavelet Packet Transform (DWPT).

Although there are tons of research elsewhere, the talking of efficient hardware implementation of wavelet transform is still a complex mission and depend directly on the target application. Where in each application, there is a compromise between the different constraints: processing speed, implementation cost, and power consumption.

IntechOpen

1.1 Related works

Since 1980, the crucial date of the born of "Wavelet Transform (WT)" with its founder J. Morlet, we found many works describe the hardware implementations of wavelet transforms. We note that the first work was done by Vishwanath Denk and Parhi [1], the authors propose an orthonormal DWT architecture combine a digit-serial processing technique with a lattice structure of quadratic mirror filter (QMF). After that, Vishwanath [2] and Motra [3], describe an efficient hardware implementation for DWT and Inverse DWT (IDWT). In 2001, Hatem et al. [4] worked in the reducing of the number of multipliers in the filters structure in a mixed parallel/sequential DWT architecture.

Wu and Hu [5] describe an implementation of DWPT/IDWPT in a strategy to minimize the number of multipliers and adders in symmetric filters using Embedded Instruction Codes (EIC). In other way to improve the data processing of DWT, Jing and Bin [6] implement the architecture on FPGA based on advanced distributed arithmetic (IDA), while Wu and Wang [7] used a multi-stage pipeline structure, although Palero et al. [8] work on the implementation of two-dimensional DWT architecture. Also, Hu and Jong [9] present two-dimensional DWT based on lifting scheme architecture that ensure a high throughput data processing.

Based on lifting scheme architecture, Fatemi and Bolouki [10] describe a pipeline and programmable DWPT architecture. Other important work, to optimize the hardware complexity of DWT based on coextensive distributive computation developed by Sowmya and Mathew [11]. Paya et al. [12] used a classical recursive pyramid algorithm (RPA) and polyphase decomposition to develop a new architecture for IDWPT based on the lifting scheme. Acharya [13] developed a systolic architecture for both DWPT/IDWPT with a fixed number of requirement pages. Farahani and Eshghi [14] described a new DWPT implementation based on a word-serial pipeline architecture and on parallel FIR filters banks. Sarah et al. [15] presented a convolution block suitable for DWT decomposition. Radhakrishnana and Themozhib [16] developed a new DWT architecture by using XOR-MUX adders and Truncations multipliers instead of the conventional adders and multipliers. Taha et al. [17] developed a parallel execution to perform Lifting Wavelet transform implementation with real time, while Shaaban Ibraheem et al. [18] presented a high throughput parallel DWT hardware architecture based on pipelined parallel processing of direct memory access (DMA).

Also, we have to notice that we found recently some orientation to software approach to compute DWPT/IDWPT on parallel processes to increase the data processing speed with optimization of the distributed computation. But the problem is still the required computing resources (concurrent network processors or processor cores) while the energy consumption is one of the critical criteria in most application domains, for that we do not include it in our bibliography.

1.2 Wavelet theory

In the previous work, we present a detailed review of the wavelet theory. Where we focus here on:

- the discrete wavelet packed transform which know as first generation of Discrete Wavelet Transform based on Mallat algorithm [19]

- the lifting scheme approach which know as first generation of Discrete Wavelet Transform based on

1.2.1 Review of DWPT and IDWPT

From the definition of wavelet theory, the DWPT and IDWPT of a signal $x[n]$ is set of approximation coefficients and detailed coefficients based on Mallat algorithm (or Mallat tree) and using FIR bank filter and inversely.

Based on Mallat the DWPT transform can be presented like decomposition, as shown in **Figure 1**.

Where the input signal is presented by the coefficients $D_0^0[k]$ in level zero with data sampling D_{in}. This amount of data (input signal) will be decomposed into two part:

i. High frequency signal presented by approximation coefficient $D_1^1[k]$ with half data sampling of original signal $(D_{in}/2)$ by using low-pass filter $h(n)$ and down sampling by a factor of two.

ii. Low frequency signal presented by detailed coefficient $D_1^0[k]$ with half data sampling of original signal $(D_{in}/2)$ by using high-pass filter $g(n)$ and down sampling by a factor of two.

Then the data path will be following the same processing in the next level with the same filters' characteristics. The depth in Mallat tree algorithm is equal to the number level and describe the needed filters equal to 2^{level} in each level. In general, the corresponding approximation and detailed coefficients in different levels in **Figure 1** are calculated as follows:

$$D_l^{2i}(k) = \sum_n h(n) D_{l-1}^i(2k - n) \quad (1)$$

$$D_l^{2i+1}(k) = \sum_n g(n) D_{l-1}^i(2k - n) \quad (2)$$

Where l presents the level and $i = 0, 1, \dots, (2^{(l-1)} - 1)$.

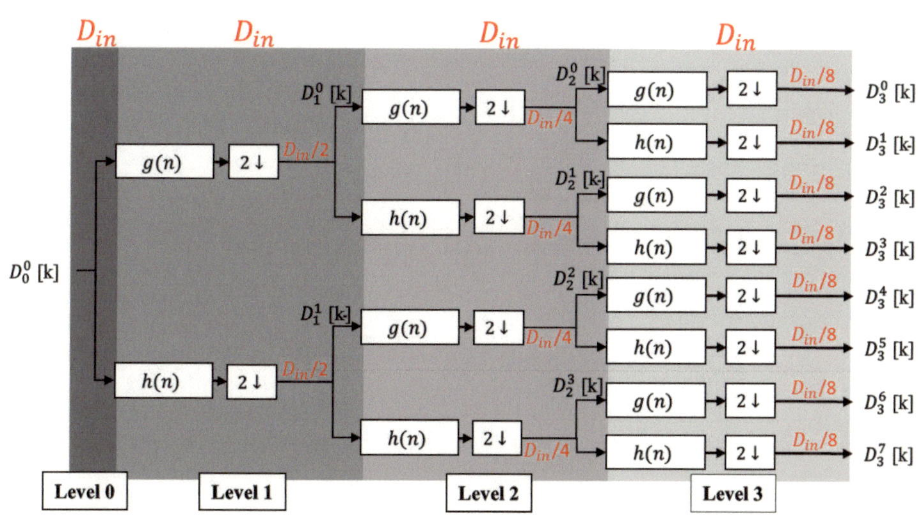

Figure 1.
DWPT three level transform based on Mallat algorithm.

As proposed Mallat in [19], the corresponding transfer functions of $\boldsymbol{h}(\boldsymbol{n})$ and $\boldsymbol{g}(\boldsymbol{n})$ are derived in the following equations:

$$H(Z) = H_0 + H_1 z^{-1} + H_2 z^{-2} + \dots + H_{L-1} z^{-(L-1)} \tag{3}$$

$$G(Z) = G_0 + G_1 z^{-1} + G_2 z^{-2} + \dots + G_{L-1} z^{-(L-1)} \tag{4}$$

where z^{-1} indicates the delay for one ***samplingperiod*** and $\boldsymbol{L}$ is the order of filters depends on the used mother wavelet.

In inverse way, the reconstruction of signal or IDWPT without loss of information is possible based on two important properties of wavelets: admissibility and regularity. Similar to decomposition way, the reconstruction operation is following an iterative method and the corresponding coefficients in different levels are calculated as follows:

$$D_l^i(k) = \sum_n \overline{h}(n) D_{l+1}^{2i}(2k - n) + \sum_n \overline{g}(n) D_{l+1}^{2i}(2k - n) \tag{5}$$

For example, the reconstruction of signal in three level based always on Mallat algorithm is presented in **Figure 2**.

Where $\overline{h}(n)$ and $\overline{g}(n)$ are the conjugated low-pass and high-pass of $\boldsymbol{h}(\boldsymbol{n})$ and $\boldsymbol{g}(\boldsymbol{n})$. Mallat used the quadratic mirror filter (QMF) of corresponding transfer functions $H, G, \overline{H}$ and $\overline{G}$ to ensure the perfect reconstruction of the original signal.

1.2.2 Review of lifting scheme discrete wavelet transform

Based on the wavelet theory, we can consider that the lifting wavelet theory is the second generation of DWPT. The strategy in this generation is to reduce the impact of the high pass and the low pass filters by replacing it into a sequence of smaller filters: update filters and predict filters. Therefore, the convolution

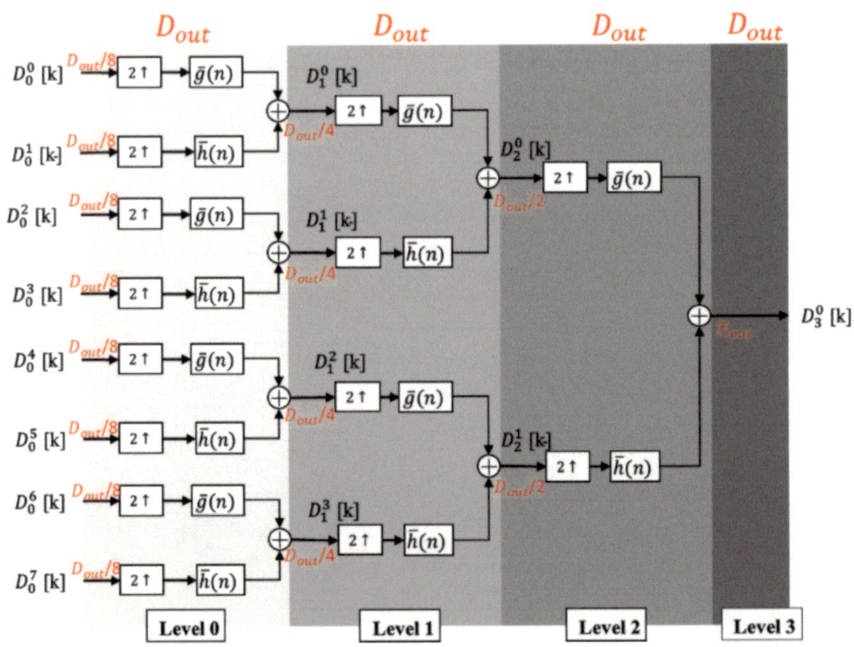

Figure 2.
IDWPT three level transform based on Mallat algorithm.

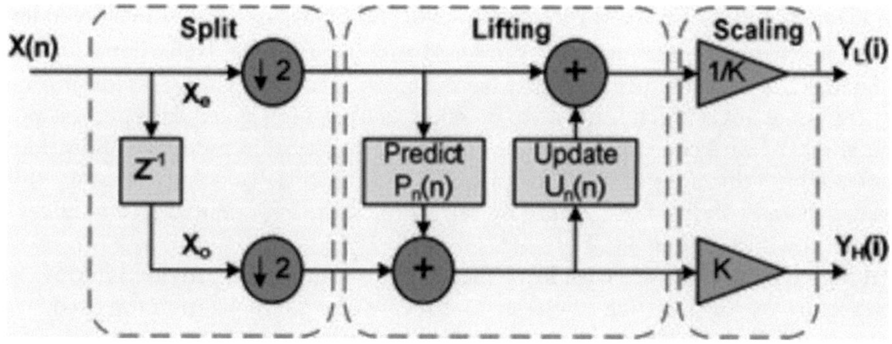

Figure 3.
Kernel of the lifting wavelet transform.

computations are reduced by comparison to the first generation which naturally reduce the design complexity by maintaining the same quality and speed.

By definition, the lifting wavelet transform is dividing to three steps: Split, Lifting, and Scaling, as shown in **Figure 3**.

In the split steps, the input signal ***X(n)*** will be divided into two sub sequences odd and even. The obtained sub-signal will be modified in lifting steps, by using alternating prediction and updating filters. And finally, a scaling operation is used to obtain an approximated and detailed signal.

1.3 Contributions and work organization

In this work, our goal is to develop a high performance, low cost implementation and configurable new hardware architecture of discrete wavelet transform based on Mallat algorithm [19]: first generation (based on Discrete Packet Wavelet Transform - DWPT) and second generation (lifting scheme Discrete Wavelet Transform) by exploitation of this suitable FPGAs environment. In order to provide the low hardware cost and the high processing speed by design, we develop a new generic parallel-pipeline architecture avoiding the complexity of the traditional architectures with the massif need for hardware resource by: i) intelligent sharing of hardware computing resources (multipliers and adders) among the different filters and stages, ii) design a linear architecture to limited impact of filter and wavelet order. To improve the high performance (data processing speed and hardware cost) of our proposal, we will perform different simulation function of selected wavelet family, transformation depth, filtering order and coefficient quantization. In VHDL at the RTL level, we modeled our architectures and we synthesized it using Altera Quartus Prime Lite, targeting an Intel/Altera Cyclone-V FPGA.

This work is organized as follows: in Section 2, we introduce our linear non-parallel and P-parallel architecture of first generation for both the DWPT and IDWPT along with simulation results. In Section 3, our linear non-parallel architecture for second generation based on lifting scheme is described. Finally, conclusion is given in Section 4.

2. Hardware implementation of first generation

2.1 DWPT

As shown in **Figure 1**, we notice that in a given stage ***k***, each filter proceeds the same amount of data and half data rate by comparison of filter in the adjacent level

$k - 1$. The number of needed filters (low-pass and high-pass filters) in a given level k is 2^k. Furthermore, the amount of procced data in each level is the same.

So, the tree architecture of Mallat have a big regularity of the behavior of filters in different levels. Which leads us to develop an ultra high speed data processing with low hardware consumption (this constraint is critical in modern application that need high throughput with low power consumption). To achieve that we think to develop an evolving architecture by retransform the exponential tree to linear one, as shown in **Figure 4**.

A high throughput rate with lower hardware resources are provided in this architecture by linearization of classic Mallat tree and parallelization the used transposed FIR filter. To achieve our goal by minimizing the hardware consumption, we proposed a shared computational resource (multipliers and adders) between the low-pass and high pass filters as shown in **Figure 5**.

In this structure, we propose a modified transposed FIR filter corresponding to H/G blocks in **Figure 4**, this model is look like the serial FIR filter in the theory of FEC coding. The H/G blocks can process in parallel P inputs sampling (signals) and consequently P outputs sampling (signals) in each clock cycle and consequently the P-parallel DWPT (**Figure 4**) are able to transform P sampling in each clock cycle.

Furthermore, this architecture is suitable for all wavelet family where we need just to change the coefficients of high-pass and low-pass for each family. Where the data handling (filter coefficients or signal sampling) of the low-pass and high-pass filter between different stages is dedicated to specific block in our architecture; we called it "buffers block". The main role of "buffers block" is to interleaving data from stage to the next stage and to manage data between low-pass and high-pass filter in the same stage. Their structure is detailed in **Figure 6**.

To procced the same amount of data in the original Mallat binary tree b (of course multiplied by the degree of parallelize P), the buffer blocks should be working with this mechanism:

a. The parameter k describe the stage, change from 1 to max depth of wavelet transform. The parameter P present the degree of parallelism and must respect the dyadic rule, $P = 2^k$, x $\mathbb{N}^+$.

b. The structure of buffer block is based on the concept of manipulation of data speed transfer in register level, where we built up inside P sub-blocks, each block has two registers/buffers level speed: "Fast Buffer" and "Slow Buffer". On each clock cycle, the Fast Buffer take data from the output of previous stage and achieve P-shift. While, Slow Buffer sub-blocks are powered to take data from the Fast Buffer registers of the same stage and then achieve P-shift on two-clock cycles.

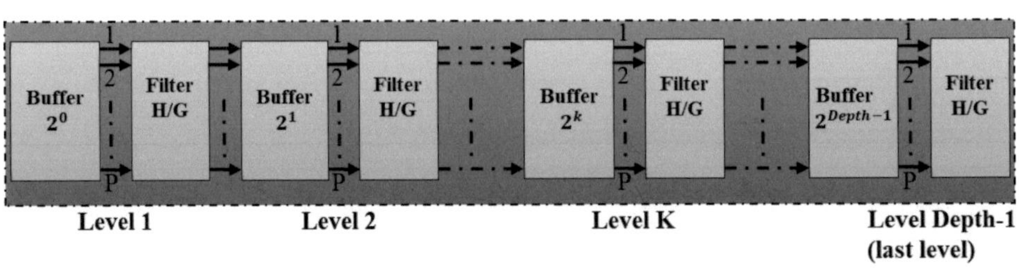

Figure 4.
Datapath diagram of linear and P-parallel proposed DWPT architecture.

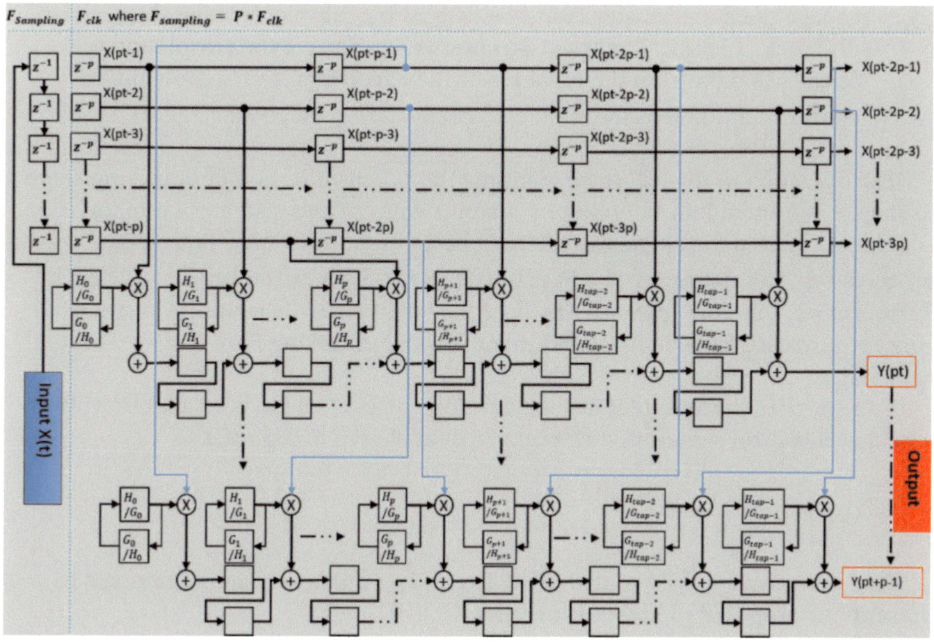

Figure 5.
P-parallel transposed FIR filter structure.

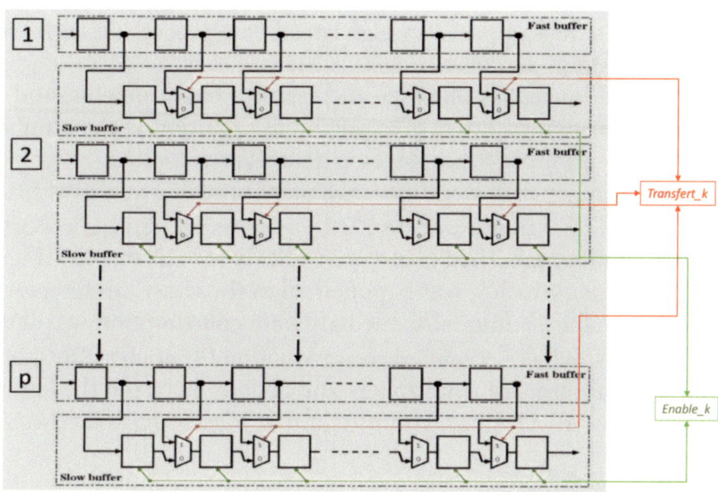

Figure 6.
*General view of buffer block structure (in stage **k**) of parallel DWPT architecture.*

c. The size Buffer blocks (number of fast registers and slow registers) depends on two parameters: the stage presented by $\boldsymbol{k}$ and the parallel degree presented by $\boldsymbol{P}$.

d. To manage the data path between "Slow Buffer" and "Fast Buffer", we specified two control signal: "$enable_k$" and "$transfer_k$", where the $enable_k$ signal (in green) is dedicated to control the shift rate between the different registers in Slow Buffers sub-blocks and "$transfer_k$" signal (in red) is to manage the data transfer from the fast buffer sub-block to the slow buffer sub-block. Technically these two control signals give the permission to transfer all data from the registers of the Fast Buffer to the Slow Buffer registers simultaneously in each $\mathbf{2}^{k}$ clock cycle (in a given stage $\boldsymbol{k}$).

The operation in the "**d**" stage combine the synchronization of data from stage $\boldsymbol{k}$ to stage $\boldsymbol{k} - \mathbf{1}$ and down sampling by factor **2** without using an extra memories or DSP block. Where the playing in the time between buffers, give us the possibility to procced only half data from the Fast Buffer to the Slow Buffer on every $\mathbf{2}^{k}$ cycle. Furthermore, the slow speed of the Slow Buffer ensures the twice (to respect the concept proposed by Mallat algorithm) presented of leaving Slow Buffer data to the next stage.

To centralize the architecture, we developed a control block or control unit to manage all control signals in different stage, as shown in **Figure** 7.

2.2 IDWPT

As a reverse way of P-parallel DWPT transform, this section is dedicated to present our proposed model of P-parallel IDWPT.

As we mention in the section of P-parallel DWPT transform, the reconstruction process has also a big regularity, where in **Figure 2**, we notice that each filter proceeds the same amount of data and half data rate by comparison to filter in the adjacent level. The number of needed filters (low-pass and high-pass filters) in a given level $\boldsymbol{K}$ is $\mathbf{2}^{k}$. Furthermore, the amount of proceed data in each level is the same. This leads us to develop an ultra high speed data processing with low cost resources consummation.

We introduce the concept of linearize and serialize in our pipeline and P-parallel architecture to eliminate the impact of exponential evolution of the number of used filters. So, as shown in **Figure 8**, we develop a novel architecture.

In this architecture, in each stage we implement only one modified filter instead of using $\boldsymbol{P} * \mathbf{2}^{k}/\mathbf{2}$ low pass filters and $\boldsymbol{P} * \mathbf{2}^{k}/\mathbf{2}$ high pass filters. It is important to mention that the number modified transposed FIR filter bank increased linearly as a function of depth order which it was exponential in the classic architecture.

To achieve our goal by minimizing the hardware consumption, we develop a Blocks Filter $\overline{\boldsymbol{H}}$ / $\overline{\boldsymbol{G}}$ which is a modified reconstruction P-parallel FIR filters by shared computational resource (multipliers and adders) between the low-pass and high pass filters and with a similar structure of that present in **Figure 4**. The only

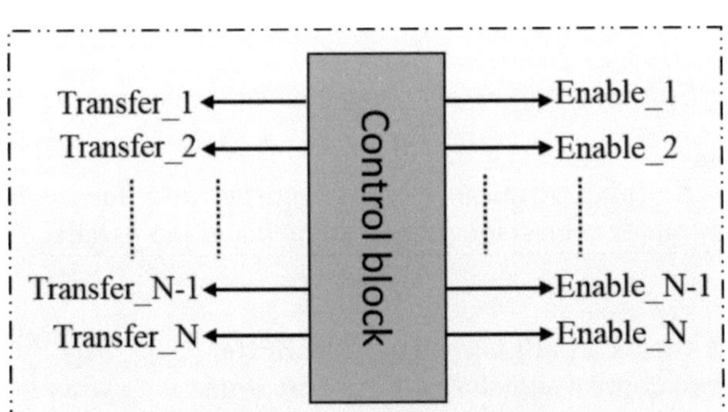

Figure 7.
Control block.

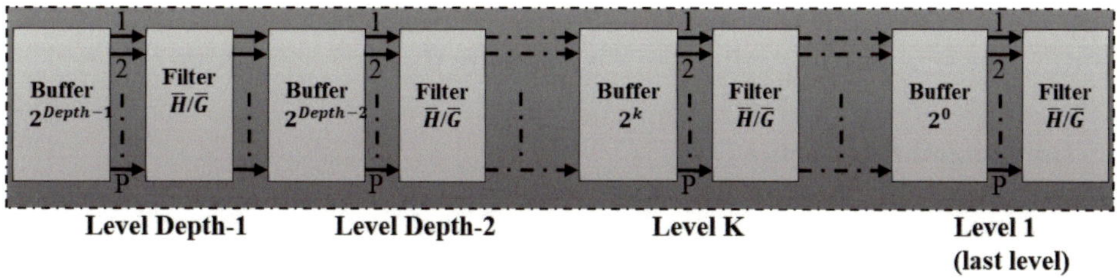

Figure 8.
Data path diagram of linear and P-parallel proposed IDWPT architecture.

difference is the coefficients filters values. Consequently, the P-parallel FIR filter is able to filter ***P*** sampling in each clock cycle.

The data manage and interleaving between filters from the first to the end stages is dedicated to the buffer block. Their structure is detailed in **Figure 9**.

To ensure the data management between reconstruction high-pass and low-pass filter, we play on the timing of buffer register: slow buffer and fast buffer. To procced the same amount of data in the original Mallat binary tree b (of course multiplied by the degree of parallelize ***P***), the buffer blocks should be working with this same mechanism as that used in the previous section.

The "fast buffer" achieve P-shift on each clock cycle while the "slow buffer" achieve P-shift on two-clock cycles. To manage the data follow path between "Slow Buffer" and "Fast Buffer", we specified two control signals: "$\boldsymbol{enable_k}$" and "$\boldsymbol{transfer_k}$". The "$\boldsymbol{enable_k}$" signal (in green) is dedicated to control the shift rate between the different registers in Slow Buffers sub-blocks and the "$\boldsymbol{transfer_k}$" signal (in red) is to manage the data transfer from the fast buffer sub-block to the

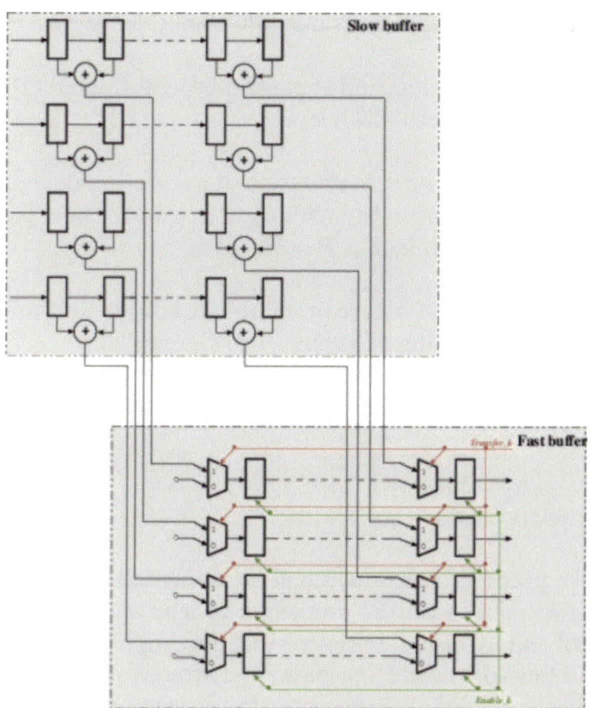

Figure 9.
General view of buffer block structure (in stage $\mathbf{k}$ *and degree of parallelization* $\mathbf{P} = 4$*) of parallel IDWPT architecture.*

slow buffer sub-block. Also, we used a control block unit to manage the control signals. The structure of control block is similar to that present in **Figure** 7.

2.3 Implementation results

Following our strategy, we develop a new pipeline and P-parallel architectures for DWPT and IDWPT. These architectures are full reconfigurable at synthesis. The reconfigurable parameters are the wavelet scale or the depth of DWPT and IDWPT, the filter coefficient and data quantization, the order of modified $\boldsymbol{H}/\boldsymbol{G}$ and $\overline{\boldsymbol{H}}/\overline{\boldsymbol{G}}$ filters, and the degree of parallelism.

Also, these architectures are partially reconfigurable after synthesis function the value of filters coefficients (that mean implicitly the order of filters). This feature, we give the possibility to work with different wavelet family without re-synthesis the FPGA carte where we load dynamically after synthesis the filter coefficients of the corresponding wavelet.

Our aim in this part is to study the performance of these architecture to record the impact of different parameters on:

- The speed of data process that mean the clock frequency (given in MHz) of implemented architecture. Where from the degree of parallelism and clock frequency, we can obtain the data sampling rate of our DWPT and IDWPT architectures.

- The hardware consumption, which represented the logic registers $\boldsymbol{l_r}$ and the logic elements $\boldsymbol{l_e}$.

In the following procedure of the implementation of our new architectures of DWPT and IDWPT transforms on the same FPGA split, we respect these constraints:

- These architectures (pipeline and P-parallel DWPT and IDWPT) are designed and modeled in VHDL at the RTL level.

- Theoretically, we do not have a limitation of parallelism degree but we should take into consideration the exist technology (hardware side) and the value must respect the dyadic rule, i. e. $\boldsymbol{P} = \boldsymbol{2^x}$, $\boldsymbol{x}$ $\mathbb{N}^+$.

- We used Altera Quartus software premium lite edition to synthesis our architectures and Intel/Altera Cyclone-V FPGA as a target technology with a speed grade of -7. For the real implementation, we used an FPGA board from Xilinx product called NexysVideo development board based on Artix-7 FPGA as a target technology.

2.3.1 Real implementation setup

To evaluate the proposed solution, a real implementation setup is depicted in **Figure 10**, where we used the UART connector to send and receive data from PC to NexysVideo board and inversely. Initial verification has been realized by sending the coefficients of Low-pass and High-pass filter after synthesis. Additional verification has been realized when received the reconstructed data.

The different simulations results are shown in **Tables 1–3**.

Based on the results in **Tables 1–3**, we observed that when we increase the quantization order from 5 to 16 this increases linearly the logic and element registers

DOI: http://dx.doi.org/10.5772/intechopen.94858

Figure 10.
Lab implementation setup.

Design parameters (Depth, Filter order, and Quantification)	Clock frequency (MHz)		Resources usage ($l_e l_r$)	
(2, 2, 5)	203.8	*205*	(471,296)	**(109, 186)**
(3, 2, 5)	200.21	*201.82*	(756,510)	**(166,312)**
(4, 2, 5)	197.37	*196.16*	(1204,899)	**(244,505)**
(2, 4, 5)	200.87	*152.88*	(879,456)	**(265,286)**
(3, 4, 5)	185.05	*152.58*	(1299,719)	**(379,442)**
(4, 4, 5)	193.71	*153.37*	(1941,1171)	**(483,665)**
(2, 16, 5)	189.2	*144.03*	(3299,1416)	**(1447,886)**
(3, 16, 5)	192.3	*137.44*	(4794,1924)	**(1983,1222)**
(4, 16, 5)	185.08	*136.24*	(6397,2614)	**(2457,1625)**
(2, 2, 16)	122.62	*132.36*	(2571,905)	**(578,582)**
(3, 2, 16)	119.79	*135.34*	(4216,1599)	**(833,972)**
(4, 2, 16)	123.14	*133.69*	(5850,2853)	**(1102,1572)**
(2, 4, 16)	120.56	*104.57*	(5038,1324)	**(1594,902)**
(3, 4, 16)	118.57	*102.77*	(7521,2260)	**(2174,1388)**
(4, 4, 16)	115.33	*100.61*	(10,374,3636)	**(2772,2084)**
(2, 16, 16)	114.16	*94.14*	(4902,4402)	**(7719,2822)**
(3, 16, 16)	126.16	*92.08*	(6805,5729)	**(10,557,3884)**
(4, 16, 16)	124.23	*90.49*	(9107,7752)	**(13,469,5156)**

Table 1.
Implementation results of pipeline and P = 4 parallel DWPT (italic) and IDWPT (bold) architectures.

Design parameters (Depth, Filter order, and Quantification)	Clock frequency (MHz)		Resources usage ($l_e l_r$)	
(2, 2, 5)	217.31	*207.04*	(1109,504)	**(109, 186)**
(3, 2, 5)	212.45	*195.77*	(1699,935)	**(166,312)**
(4, 2, 5)	213.4	*198.73*	(2754,1531)	**(244,505)**
(2, 4, 5)	217.9	*147.15*	(2120,897)	**(265,286)**
(3, 4, 5)	202.6	*148.39*	(3050,1197)	**(379,442)**
(4, 4, 5)	206.59	*147.65*	(4603,2023)	**(483,665)**
(2, 16, 5)	201.14	*136.44*	(7689,2447)	**(1447,886)**
(3, 16, 5)	202.16	*133.05*	(12,176,3166)	**(1983,1222)**
(4, 16, 5)	196.82	*131.56*	(14,956,4571)	**(2457,1625)**
(2, 2, 16)	95.77	*128.75*	(6079,1696)	**(578,582)**
(3, 2, 16)	97.8	*123.08*	(9279,2735)	**(833,972)**
(4, 2, 16)	98.82	*128.04*	(13,489,5011)	**(1102,1572)**
(2, 4, 16)	97.04	*99.98*	(12,032,2582)	**(1594,902)**
(3, 4, 16)	94.2	*98.87*	(17,549,3965)	**(2174,1388)**
(4, 4, 16)	88.01	*98.95*	(24,311,6363)	**(2772,2084)**
(2, 16, 16)	99.15	*90.16*	(11,263,7856)	**(7719,2822)**
(3, 16, 16)	102.89	*86.02*	(14,750,11,451)	**(10,557,3884)**
(4, 16, 16)	100.24	*86.1*	(21,314,13,091)	**(13,469,5156)**

Table 2.
Implementation results of pipeline and P = 8 parallel DWPT (italic) and IDWPT (bold) architectures.

and decreases logarithmically the clock frequency from around 200 MHz to around 100 MHz. As expected, the impact of depth and order of filters is too weak on the clock frequency and increases linearly the logic and element registers while it was exponential with Mallat binary tree. It is important to notice that the small latency in our architectures give us the possibility to process data in ultra high speed (in the gate of Giga-samples/clock cycle) without requiring any extra memory or DSP blocks.

It is important to notice that the incrementation of the functional frequency is directly proportional to the parallel degree. When we exceed the order of parallelism to 32, the needed resources overcome the capacity of NexysVideo board. To vanquish this problem, we suggest two possible solutions:

i. Under the strategy of minimizing the used hardware of Discrete Wavelet Transform, we look forward to the lifting scheme wavelet transform as a second DWPT transform generation. Section 3 is dictated to describe in details this suggested proposal.

ii. Another possible solution is to upgrade this work with a new FPGA family like Ultra scale. This new FPGA architecture present a high-performance environment which deliver the optimal balance between the required system performance (with 783 k to 5541 k Logic cells) and the smallest power envelope. But it remains a very expensive solution.

Design parameters (Depth, Filter order, and Quantification)	Clock frequency (MHz)		Resources usage $(l_e l_r)$	
(2, 2, 5)	210.36	*197.47*	(3668, 652)	**(389,668)**
(3, 2, 5)	209.23	*195.35*	(6019, 960)	**(573,1096)**
(4, 2, 5)	209.02	*194.97*	(8655, 1243)	**(742,1576)**
(2, 4, 5)	181.83	*147.54*	(5991, 1689)	**(1091,1008)**
(3, 4, 5)	178.58	*142.31*	(8380, 2363)	**(1410,1526)**
(4, 4, 5)	178.37	*141.98*	(11,181, 2601)	**(1552,2036)**
(2, 16, 5)	169.1	*127.6*	(30,012, 4881)	**(5894,3048)**
(3, 16, 5)	167.28	*124.88*	(37,172, 6575)	**(7300,4106)**
(4, 16, 5)	167.43	*125.09*	(38,374, 7680)	**(7536,4796)**
(2, 2, 16)	106.07	*125.25*	(11,116, 3395)	**(2183,2120)**
(3, 2, 16)	105.11	*123.02*	(17,679, 4330)	**(2704,3472)**
(4, 2, 16)	104.98	*122.71*	(25,389, 4830)	**(3016,4986)**
(2, 4, 16)	91.2	*92.6*	(31,336, 5137)	**(6154,3208)**
(3, 4, 16)	90.55	*91.29*	(39,361, 7764)	**(7730,4848)**
(4, 4, 16)	91.85	*93.93*	(42,687, 10,342)	**(8383,6458)**
(2, 16, 16)	86.43	*83.17*	(26,408, 15,572)	**(30,619,9736)**
(3, 16, 16)	86.57	*83.44*	(33,859, 20,959)	**(39,257,13,104)**
(4, 16,16)	85.9	*82.16*	(36,348, 24,456)	**(42,143,15,290)**

Table 3.
Implementation results of pipeline and $P = 16$ parallel DWPT (italic) and IDWPT (bold) architectures.

3. Hardware implementation of second generation

Actually, we find that some new applications especially in modern wireless communication require high throughput but at the same time a low energy consumption. For this reason, we look for the second discrete wavelet generation or lifting scheme wavelet transform. Because the lifting wavelet theory is by nature require less multiplier/adder blocks and consequently low energy.

So, our aim in this section is to conserve the ultra high speed data process and also reduce the hardware conception by introducing the linearization concept in the classic lifting scheme DWPT and IDWPT tree as shown in **Figures 11** and **12**.

A new pipeline and linear lifting scheme DWPT and IDWPT architectures are presented in **Figures 11** and **12**. These new architectures ensure the data speed proceed like the classic lifting scheme transform but with less hardware not affected by the wavelet depth.

The $\boldsymbol{P}/\boldsymbol{U}$ Filter Blocks and $\overline{\boldsymbol{P}}/\overline{\boldsymbol{U}}$ Filter Blocks in linear lifting scheme of DWPT and IDWPT architectures, receptively, are the modified predicted and updated filter. In **Figure 13**, we present the structure of the modified $\boldsymbol{P}/\boldsymbol{U}$ Filter Blocks (the same for $\overline{\boldsymbol{P}}/\overline{\boldsymbol{U}}$ Filter Blocks, we just change the coefficients values) which can process the same amount of data (same functionality) on given stage in the classic lighting scheme tree. This Filter Blocks can process two samplings in one clock cycle.

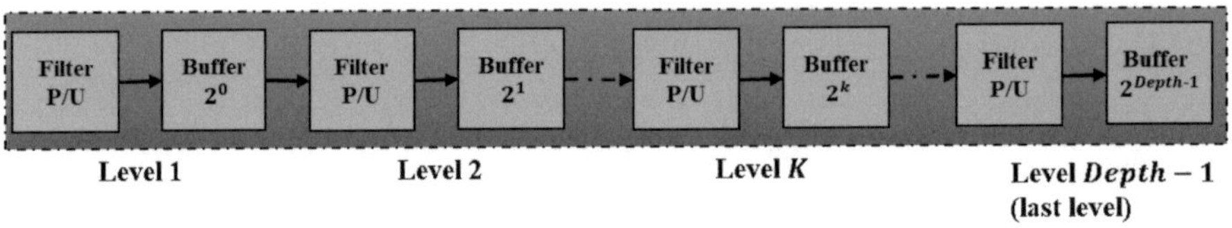

Figure 11.
Data path diagram of linear proposed lifting scheme DWPT architecture.

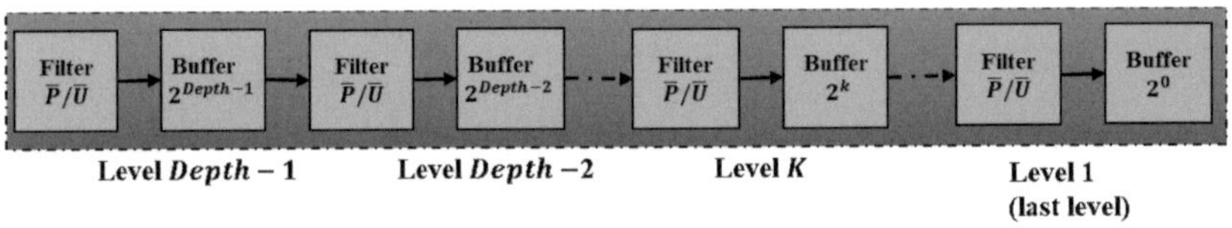

Figure 12.
Data path diagram of linear proposed lifting scheme IDWPT architecture.

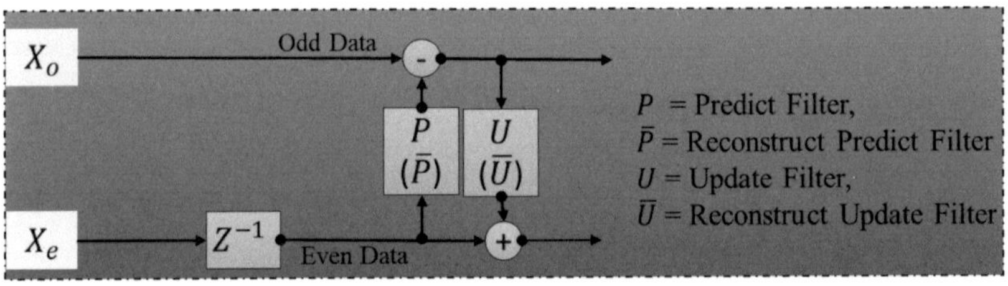

Figure 13.
*Structure of the modified predict/update filters and their conjugate in stage **k**.*

4. Comparison

To evaluate the performance of our architecture, a comparison section is important to prove the potential of our work and to lead us to a new innovated architecture.

In **Table 4**, we present a comparison between our proposed architectures and other achieved architectures of Discrete Wavelet Transform in literature. Without doubt, this table presents the potential of our linear pipeline and parallel architecture where on one hand it ensures a high frequency data processing and on the other hand a full reconfigurable structure using less hardware. Additionally, without missing an important feature, we implemented our architecture without using a memory or DSP blocks which gives us a privilege to more optimization of the used hardware in the next FPGA generation.

5. Conclusion

In this work, we propose ultra-high throughput with low hardware consumption of first generation and second generation of discrete wavelet packet transform. Where title of example, from **Table 3**, with a quantization order = 5, depth order = 2, filter order = 2 and degree of parallelism $\boldsymbol{P}$ = 16, we obtain a clock frequency = 210.36 MHz, theoretically can proceed 3365.52 Mega samplings in one clock cycle with low hardware used $\boldsymbol{l_e}$ = 3668 and $\boldsymbol{l_r}$ = 652.

	Sung et al. [20]	Marino et al. [21]	Mohanty et al. [22]	Madis-hetty et al. [23]	Wang et al. [24]	Wu et al. [5]	Meihua et al. [25]	Proposed parallel architecture
Wavelet	Daub-4	Quadri-filter	Daub-4	Daub-6	Lifting- Db4	Quadri-filter	Quadri-filter	Arbitrary
DWPT/ IDWPT	Both	DWPT	DWPT	DWPT	Both	Both	DWPT	Both
Logic cell	N/A	N/A	426	1040	N/A	30,192 (Logic element)	1835 (Logic element)	3668 (Logic element)
Technology	Xilinx XC2V4000	N/A	CMOS 90 nm	Xilinx Virtex 6	CMOS 180 nm	0.35 μm	Altera EP20K200E	Altera Cyclone IV & V
Max. Freq.(Mhz)/ Bitrate	N/A	N/A	Feq.: 20	Feq.: 306.15	Feq.: 20	Feq.: 100	Feq.: 29	Bitrate:718.13 Feq.: 197.47
Memory	Yes	N/A	Yes	Yes	Yes	N/A	Yes	No
Quantization (bits)	N/A	N/A	N/A	N/A		32	N/A	Test with 5 to 16 (up to the limit of manufacturing technology)
DSP	N/A	N/A	N/A			N/A	N/A	No
Depth	3	N/A	N/A	4	3	3 (up to 6)	3	2, 3 and 4 (up to the limit of manufacturing technology)
Parallel degree	No	No	No	No	No	No	No	4,8 and 16(up to the limit of manufacturing technology)

Table 4.
Comparison of proposed architectures with other works.

Based on the results in **Tables 1–3**, these architectures ensure high operating frequency which is low affected of wavelet depth and filters order because in our structures we maintained a short critical path of effective data path. Furthermore, these architectures are pipelined and P-parallel, modeled in VHDL at the RTL level, generic and fully reconfigurable in pre-synthesis function of the quantization of the filter coefficients and data sampling, the depth of wavelet transform, the order of the filters, and the degree of parallelism.

Last, but not least, our developed architectures are reconfigurable post-synthesis, which is not the case for most of the previous work as shown in the comparison in **Table 4**. Where the values of filters coefficients can be load at run-time which provides a great flexibility in experimental usage in contrary to all previous works.

This work is still in progress where we are making many simulations/verifications in different contexts to verify if the simulation results will agree or not with the implementation results. As perspectives, we work on new version of FIR filter and in parallel another work to create an IP core (Intellectual Property core) FIR to be used with different FPGA boards and in different applications. A natural way of this work is to develop a different parallel version of hardware implementation in FPGA of lifting scheme wavelet transform.

Author details

Mouhamad Chehaitly[1*], Mohamed Tabaa[2], Fabrice Monteiro[3], Safa Saadaoui[2] and Abbas Dandache[3]

1 University of Montpellier, LIRMM Lab, France

2 EMSI Casablanca, LPRI Lab, Morocco

3 University of Lorraine, LGIPM Lab, France

*Address all correspondence to: che.liban.tly@hotmail.com

IntechOpen

© 2020 The Author(s). Licensee IntechOpen. This chapter is distributed under the terms of the Creative Commons Attribution License (http://creativecommons.org/licenses/by/3.0), which permits unrestricted use, distribution, and reproduction in any medium, provided the original work is properly cited.

References

[1] Denk T and Parhi K. Architectures for lattice structure based orthonormal discrete wavelet transforms. IEEE International Conference on Application Specific Array Processors, pp. 259–270, August 1994.

[2] Vishwanath M and Owens R. A common architecture for the DWT and IDWT. IEEE International Conference on Application Specific Systems, Architectures and Processors (ASAP), pp. 193–198.8, August 1996.

[3] Motra A, Bora P and Chakrabarti I. An efficient hardware implementation of DWT and IDWT. IEEE Conference on Convergent Technologies for the Asia-Pacific Region (TENCON), vol. 1, pp. 95–99, October 2003.

[4] Hatem H, El-Matbouly M, Hamdy N and Shehata K-A. VLSI architecture of QMF for DWT integrated system. Circuits and Systems, MWSCAS 2001. Proceedings of the 44th IEEE 2001 Midwest Symposium on (Volume: 2), pp. 560–563, 2001. doi:10.1109/MWSCAS.2001.986253

[5] Wu B-F and Hu Y-Q. An efficient VLSI implementation of the discrete wavelet transform using embedded instruction codes for symmetric filters. IEEE Transactions on Circuits and Systems for Video Technology, vol. 13, no. 9, pp. 936–943, September 2003.

[6] Jing C and Bin H-Y. Efficient wavelet transform on FPGA using advanced distributed arithmetic. 8th IEEE International Conference on Electronic Measurement and Instruments (ICEMI'2007), pp. 2–512-2-515, Aug. 2007.

[7] Wu Z and Wang W. Pipelined architecture for FPGA implementation of lifting-based DWT. 2011 International Conference on Electric Information and Control Engineering, pp. 1535–1538, April 2011.

[8] Palero R, Gironés R and Cortes A. A novel FPGA architecture of a 2-D wavelet transform. Journal of VLSI signal processing systems for signal, image and video technology, vol. 42, no. 3, pp. 273–284, 2006.

[9] Hu Y and Jong C. A memory-efficient high-throughput architecture for lifting-based multi-level 2-D DWT. IEEE Transactions on Signal Processing, vol. 61, no. 20, pp. 4975–4987, Oct. 2013.

[10] Fatemi O and Bolouki S. A Pipeline, Efficient and Programmable Architecture for the 1-D Discrete Wavelet Transform using Lifting Scheme. The Second Conference On Machine Vision, Image Processing & Applications (MVIP 2003), Tehran 2003.

[11] Sowmya K-B and Mathew J. Discrete Wavelet Transform Based on Coextensive Distributive Computation on FPGA. Materials Today: Proceedings, Second International Conference on Large Area Flexible Microelectronics (ILAFM 2016): Wearable Electronics, December 20th-22nd, 2016.

[12] Paya G, Peiro M, Ballester F and Herrero V. A new inverse architecture discrete wavelet packet transform architecture. IEEE, Signal Processing and Its Applications, 7803–7946, 443–446 vol. 2, 2003.

[13] Acharya T. A Systolic Architecture for Discrete Wavelet Transforms. IEEE, Digital Signal Processing Proceedings, 13th International Conference on Volume 2, 571–574 vol. 2, 1997.

[14] Farahani M and Eshghi M. Architecture of a Wavelet Packet Transform Using Parallel Filters.

TENCON 2006 - IEEE Region 10 Conference, 1–4244–0548-3, 1–4, 2006.

[15] Farghaly S-H and Ismail S-M. Floating-Point FIR-Based Convolution Suitable for Discrete Wavelet Transform Implementation on FPGA, 2019 Novel Intelligent and Leading Emerging Sciences Conference (NILES), Giza, Egypt, 2019, pp. 158–161, DOI: 10.1109/NILES.2019.8909290.

[16] Radhakrishnan P and Themozhi G. FPGA implementation of XOR-MUX full adder based DWT for signal processing applications. Microprocessors and Microsystems, Volume 73, 2020, 102961, ISSN 0141–9331, doi:10.1016/j.micpro.2019.102961.

[17] Taha T-B, Ngadiran R and Ehkan P-L. Design and Implementation of Lifting Wavelet Transform Using Field Programmable Gate Arrays. MS&E 767.1 (2020): 012041.

[18] Mohammed Shaaban I et al. High-throughput parallel DWT hardware architecture implemented on an FPGA-based platform. Journal of Real-Time Image Processing 16.6 (2019): 2043–2057.

[19] Mallat S. A wavelet tour of signal processing. Academic Press, 1999.

[20] Sung T-Y et al. Low-power multiplierless 2-D DWT and IDWT architectures using 4-tap Daubechies filters. In Proc. Seventh Int. Conf. PDCAT, pp. 185–190, 2006.

[21] Marino F. Two fast architectures for the direct 2-D discrete wavelet transform. IEEE Trans. Signal Process, vol. 49, no. 6, pp. 1248–1259, 2001.

[22] Mohanty B-K and Meher P-K. Memory-efficient high-speed convolution-based generic structure for multilevel 2-D DWT. IEEE Trans. Circuits Syst. Video Technol., vol. 23, pp. 353–363, 2013.

[23] Madishetty S, Madanayake A, Cintra R and Dimitrov V. Precise VLSI Architecture for AI Based 1-D/2-D Daub-6 Wavelet Filter Banks with Low Adder-Count. IEEE Transactions on circuits and systems-I: regular paper, Vol. 61, No. 7, 1984–1993, July 2014.

[24] Wang C et al. Near-threshold energy-and area-efficient reconfigurable DWPT/DWT processor for healthcare - monitoring Applications. IEEE Transactions on Circuits and Systems II: Express Briefs, 62(1), 70–74, 2015.

[25] Meihua X et al. Architecture research and VLSI implementation for discretewavelet packet transform. High Density Microsystem Design and Packaging and Component Failure Analysis, 2006. HDP'06. Conference on. IEEE, 2006.

Chapter 7

Fault Detection, Diagnosis, and Isolation Strategy in Li-Ion Battery Management Systems of HEVs Using 1-D Wavelet Signal Analysis

Nicolae Tudoroiu, Mohammed Zaheeruddin, Roxana-Elena Tudoroiu and Sorin Mihai Radu

Abstract

Nowadays, the wavelet transformation and the 1-D wavelet technique provide valuable tools for signal processing, design, and analysis, in a wide range of control systems industrial applications, audio image and video compression, signal denoising, interpolation, image zooming, texture analysis, time-scale features extraction, multimedia, electrocardiogram signals analysis, and financial prediction. Based on this awareness of the vast applicability of 1-D wavelet in signal processing applications as a feature extraction tool, this paper aims to take advantage of its ability to extract different patterns from signal data sets collected from healthy and faulty input-output signals. It is beneficial for developing various techniques, such as coding, signal processing (denoising, filtering, reconstruction), prediction, diagnosis, detection and isolation of defects. The proposed case study intends to extend the applicability of these techniques to detect the failures that occur in the battery management control system, such as sensor failures to measure the current, voltage and temperature inside an HEV rechargeable battery, as an alternative to Kalman filtering estimation techniques. The MATLAB simulation results conducted on a MATLAB R2020a software platform demonstrate the effectiveness of the proposed scheme in terms of detection accuracy, computation time, and robustness against measurement uncertainty.

Keywords: battery management system, extended Kalman filter, fault detection and isolation, 1-D wavelet and transform, signals processing analysis, wavelet filters bank

1. Introduction

The most viable way to achieve clean and efficient transport is to boost the automotive industry to be concerned with developing advanced battery technologies, especially lithium-ion (Li-ion), to increase the number of electric and hybrid electric vehicles (EVs/HEVs) to dominate the vehicle market. An essential internal parameter of the Li-ion battery is the state of charge (SOC), defined as the available capacity of the cell that changes according to the current profile of the driving cycle. Due to its crucial role in keeping the battery safe for various operating conditions and significantly extending battery life, SOC is a topic of great interest, as evidenced by an impressive number of research papers published in the literature. In the absence of a measurement sensor, the SOC must be estimated since its calculated value is not accurate enough. The most used model-based Kalman filters can estimate the battery SOC with a high grade of accuracy [1–4]. The Li-ion battery is an important component integrated into battery management system (BMS) that performs tasks regarding the safe operation and reliability of the battery, protecting battery cells and battery systems against damage, as well as battery efficiency and service life [2–4]. The BMS "plays a significant role in fault diagnosis because it houses all diagnostic subsystems and algorithms" [2, 3]; thus it monitors the battery system through sensors and state estimation, such that to detect any abnormalities during the battery system operation" [2, 5]. A signal processing-based method using wavelet transforms proved to be a viable alternative to conventional Kalman filter state estimators, for designing and implementation of real-time FDI strategies. The new FDI approach avoids battery modeling difficulties and is more straightforward with better dynamic performance [7]. The drawback of this method is the difficulty experienced in dealing with the early faults and fault isolation. Its application also requires a large amount of calculations compared to the model-based methods. An intelligent fault detection scheme for microgrid based on wavelet transform and deep neural networks is used in [6] to "provide fast fault type, phase, and location information for microgrid protection and service recovery" [6]. Similar, a wavelet-based transient fault detection and analysis is used successfully in [7] for a microgrid connected power. In this research, our motivation of using 1-D wavelet analysis comes from the preliminary results obtained for similar investigations on the impact of nonlinearities and uncertainties of actuators (electro-pneumatic valves), such as hysteresis, dead zone, dead band, on a healthy pH neutralization plant [8]. An example of multisignal 1-D wavelet analysis is found in [9], and a useful tutorial of using wavelet transforms presented in [10]. In [11] is shown a generic Simscape model of Li-ion Cobalt battery model used to build a SOC AEKF estimator robust to three different driving cycles profile tests, such as UDDS, EPA-UDDS and FTP-75, the last one also used in the case study of this research. For FDI techniques based on 1-D wavelet analysis are used specific MATLAB commands provided by MATLAB Wavelet Toolbox [12]. A strong theoretical background on wavelet transform and their applications is provided by the fundamental work [13]. In [14] is presented an interesting fault isolation technique based on wavelet transform, and a detailed data-based FDI techniques for a nonlinear ship propulsion system are developed in [15]. Several multimedia applications of wavelet transform can be found in [16], and a better understanding of wavelet transform analysis, design and implementation of features extraction methods, for filtering, denoising, decomposition and reconstructing signals is given in [17–23]. From our most recent preliminary results in Li-ion battery field, modeling and SOC estimators disseminated in [11, 24, 25], an interesting state-of-art analysis of similar SOC AEKF estimators performance reported in the literature is done in terms of statistical performance criteria values, such as root mean square error (RMSE), mean square error (MSE), mean absolute error (MAE), standard deviation (std), mean absolute percentage error (MAPE) and R2 (R-squared). Among three SOC Li-ion battery estimators AEKF, adaptive unscented Kalman filter (AUKF) and particle filter SOC estimators the AEKF proved that is the most suitable for HEVs applications.

Let why is used the AEKF SOC estimator of Li-ion battery in the first part of our research for FDI control strategies, excelling by its simplicity, SOC accuracy, real-time implementation capability and robustness. The robustness is tested for four different scenarios, such as to changes in SOC initial values (guess values), from 70–40%, 20%, 90% and 100%, to FTP-75 driving cycle profile test, changes in measurement level noise (from 0.001 to 0.01), to changes in the battery capacity value from 6 Ah to 4.8 Ah due to aging effects, and changes in internal resistance due to temperature effects, and also for simultaneous changes [11, 23]. Based on a rigorous performance analysis of SOC residuals error compared to the similar results reported in the literature with a typically 2% error, in some situations the AEKF estimator SOC residual error reached values smaller than 1%, such as shown in [25]. Since of the lack of data in the literature field for similar situations developed in our research for Li-ion battery, it is not easy to make a state-of-art analysis of the results reported in the literature related to the FDI techniques design and implementation based on 1-D wavelet analysis. The efficiency of 1-D analysis is proved in this paper based on extensive MATLAB simulations to extract the features of input-output signals such as the energy, skewness, kurtosis, and to compute the MSE statistical criteria performance. Finally, the MATLAB simulation results can provide useful information on detection accuracy, computation time, and robustness against measurement uncertainty, thus showing simply the effectiveness of the FDI proposed scheme. The temperature fault is detected without doubt inside the Li-ion battery based on the significant values reached by the details (D1, D2, and D3) and analysis coefficient (A3) of the output terminal battery voltage residual level three decomposition, represented by the following sets of values (4.46, 2.7, 5.349, 87.5) for energy feature, (0.063, −3.92, 13, −1.33) for skewness signal feature, respectively (5.8, 71.4, 389.13, 56) for kurtosis signal feature. Also, the statistic RMSE performance criterion indicates significant values for D1 coefficient in the presence of the of temperature fault for energy feature (4.4654) and skewness and kurtosis features are the same as for current fault. To detect both faults, a multiresolution analysis (MRA) is performed, capable of extracting a smooth trend term, which provides a valuable information to localize transient changes in the fault injection window [500, 1500] seconds [23].

Thus, the presence of the bias current fault and bias temperature fault is detected and localized as a transient significant change in the nonstationary Li-ion output voltage residual signal. For an appropriate choice of the thresholds' values, both faults can be detected with a high accuracy detection times directly from S8 graph; thus, the presence of the false alarms is completely removed compared to Kalman filter FDI estimation techniques. The fault signature and considering the variation trend in SOC residual and internal resistance of the battery also provides a piece of useful information for fault isolation.

2. Li-ion battery model, SOC estimation and fault injection mechanism

This section briefly presents the Rint equivalent circuit model (Rint ECM) as a case study to investigate the effectiveness of the proposed fault detection and isolation (FDI) strategy, using a conventional EKF SOC estimator, as a support for performance analysis comparison, in the first part [1–4], and a 1-D wavelet transformation in the second part [8, 9]. For comparison purpose, an improved adaptive extended conventional Kalman (AEKF) filter algorithm [3, 4, 11] is also briefly presented for estimating the state of charge (SOC) of the adopted Li-ion battery, as well as the faults in Appendix A. Residual methodology is useful to detect and isolate faults. Only three failures of the current, voltage and temperature sensors of the HEV battery management system (BMS) used for the case study are analyzed.

2.1 Li-ion battery model selection

The Rint ECM Li-ion battery model is one of the most common models to describe battery dynamics in many real-time implemented HEV applications with an acceptable range of performance. The reason for using these models is their simplicity, low number of parameters to adjust and easy implementation in a friendly MATLAB simulation environment. Therefore, a compromise we need to make between the accuracy of the battery SOC and the complexity of the model related to the choice of Li-ion battery, so that, for simulation purpose and "proof concept", we adopt a simple Rint ECM Li-ion battery model, as a reasonably simplified version of RC ECM developed in [1], and in [11] for a Li-ion Cobalt battery, as is shown in **Figure 1**.

The Rint ECM Li-ion battery model is an equivalent Thevenin electrical circuit consisting of an open circuit-controlled voltage (OCV) source and an internal resistance designated by Rin. The OCV source strongly depends on the state of charge (SOC), i.e. a dependency described by an extremely nonlinear function OCV = f (SOC), represented by different combinations of models reported in the literature such as Shepherd, Nernst and Unnewehr universal model [1, 3, 4]. The dynamics complexity and the accuracy of ECM increase by adding an RC polarization cell (first-order RC model), two RC cells (second-order ECM) respectively three RC cells (third ECM order model), as those developed in [1–3, 11]. The main input-output and intermediate signals in **Figure 1** are I_{batt} is the input battery instantaneous value of the direct current (DC) flowing through the open circuit controlled-voltage source, and V_{batt} denotes the measured output terminal battery instantaneous value DC voltage that are nonlinear dependent of OCV, as intermediate signal. The internal resistance of the battery is affected by several factors. Still, a significant impact has conductor resistance, electrolyte resistance, ion mobility, separator efficiency, reactive electrode rates, polarization, temperature, and aging effects, and SOC changes, as is mentioned in [11]. Since the SOC of the battery is defined as [1–4, 11]:

$$SOC = \frac{Q - \int idt}{Q}, i = I_{batt} \quad (1)$$

with Q denoting the rating battery capacity, in the schematic shown in **Figure 1**, the controlled voltage source E (open circuit voltage (OCV)) can be modeled by:

$$E = OCV(t) = f\left(E_0, K, \frac{Q - \int idt}{Q}, t\right) = f(E_0, K, SOC, t) = E_0 - Kh(SOC) \quad (2)$$

The battery terminal voltage Vbatt is related to OCV according to following nonlinear equation:

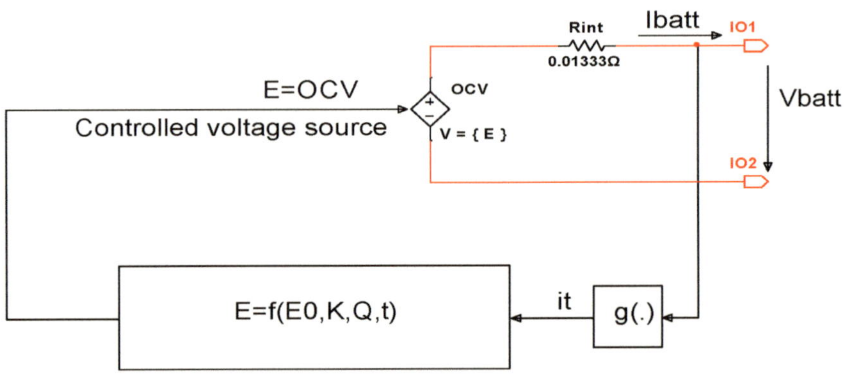

Figure 1.
ECM Rint Li-ion battery model (see [11]).

$$V_{batt} = OCV(t) - R_{in}I_{batt}(t) = E_0 - Kh(SOC) \quad (3)$$

where $K = [K_1 \ K_2 \ K_3 \ K_4]$ is a 1×4 dimensional vector, whose components have the given values suggested in [1], and $h(SOC) = \left[-\frac{1}{SOC} \quad -SOC \ \ln(SOC) \quad \ln(1 - SOC)\right]^T$ is a 4×1 dimensional vector nonlinear function.

However, for the implementation of the proposed FDI techniques, a high-precision model is not required, because the extraction of ECM parameters is beneficial to monitor the battery SOC, rather than to model the battery performance.

2.1.1 Rint ECM Li-ion battery model dynamics represented in continuous and discrete time state space representation

For a discharging current cycle, when $u(t) = i(t) \geq 0$, and for a charging current cycle $u(t) < 0$ the dynamics in continuous time t of an ECM Rint Li-ion battery model is described by the following three Equations [1–4, 11]:

$$\frac{dx_1}{dt} = -\frac{\eta u(t)}{C_{nom}}, x_1 = SOC, x_1(0) = SOC\_ini, \ \ u(t) \geq 0 \quad (4)$$

$$OCV(t) = E_0 - K_2 x_1 - \frac{K_1}{x_1} + K_3 \ln(x_1) + K_4 \ln(1 - x_1) \quad (5)$$

$$y(t) = OCV(t) - R_{in}u(t) \quad (6)$$

where η is the coulombic efficiency that has different values for charging and discharging cycles, E_0, K_1, K_2, K_3, and K_4 are the OCV battery characteristic curve coefficients, whose values are given in [1], "chosen to fit accurately the Li-ion battery model to manufacture's data by using a least squares curve fitting estimation method" [1–3]. Similar, the equivalent battery model in discrete time can be written in the following form:

$$x_1(k+1) = x_1(k) + \frac{\eta T_s u(k)}{C_{nom}}, x_1(k) = SOC(k) = SOC(kT_s) \quad (7)$$

$$OCV(k) = E_0 - K_2 x_1(k) - \frac{K_1}{x_1(k)} + K_3 \ln(x_1(k)) + K_4 \ln(1 - x_1(k)) \quad (8)$$

$$y(k) = OCV(k) - R_{in}u(k) \quad (9)$$

where $\mathbf{T_s = 1(second) - is\ sampling\ time, k \in Z}$ used to denote the discrete time instant $t_k = k\mathbf{T_s}$.

Because the internal resistance Rin is an essential parameter of the battery that is affected much more by the temperature than other parameters of the cell, it is necessary to attach to the Li-ion battery model a thermal model, described in continuous time by a first order differential equation:

$$T_c \frac{dT_{cell}}{dt} = -T_{cell} + R_{th}P_{loss} + T_{amb} \quad (10)$$

where: $\boldsymbol{P_{loss} = R_{amb}u(t)^2, and\ u(t) = i(t)}$.

In discrete time the Eq. (7) becomes:

$$T_{cell}(k+1) = \left(1 - \frac{T_s}{T_c}\right) T_{cell}(k) + \frac{R_{th} T_s P_{loss}(k)}{T_c} + \frac{T_{amb}}{T_c} \tag{11}$$

and $T_c = 2000$ [s] is the thermal resistance, $T_{amb} = 293.15\,[K]$ signifies the ambient temperature in degree Kelvin (T is the most used to denote the absolute temperature in degree Kelvin), and P_{loss} denotes the power losses dissipated on the internal resistance Rin, and T_{cell} is the temperature of the battery cell in degree Kelvin.

2.1.2 The Rint ECM Li-ion battery healthy model: Residual generation and MATLAB simulations

The healthy ECM battery model (free faults model) MATLAB simulations to an input driving cycle Federal Test Procedure (FTP-75) for a city, are shown in **Figure 2**.

In **Figure 2(a)** is shown the FTP-75 driving cycle test profile, **Figure 2(b)** depictures the battery terminal voltage, **Figure 2(c)** reveals the battery SOC, **Figure 2(d)** discloses the temperature profile of the thermal model initiated by an ambient temperature of 20°C, and **Figure 2(e)** exposes the effect of the battery temperature on internal resistance Rin.

2.2 The adaptive extended Kalman filter Li-ion battery SOC estimator for fault detection and isolation

For Li-ion batteries, the aspects such as accuracy performance of the SOC estimation and the prediction of the terminal voltage are essential to be analyzed, thus ensuring the safe operation of the cell, and thus maintaining a long life. Therefore, a brief presentation of an appropriate estimation technique is of real use. Moreover, for any battery, whether it is a Li-ion battery, SOC cannot be measured accurately, so it is necessary to estimate it. The most popular estimation algorithm reported in the literature is the Kalman filter (KF) with its improved version for models with extremely nonlinear dynamics, such as an extended Kalman filter (EKF) /adaptive extended Kalman filter (AEKF) [1–4, 11].

2.2.1 The adaptive extended Kalman filter Li-ion battery SOC estimator- brief presentation

Since the preliminary results obtained in [11] convinced us about the efficiency of applying the AEKF SOC estimator for a Simscape model of Li-ion battery, quite well documented in [4], then the same estimator is used in this paper. For the adopted battery model, the SOC estimator adaptation consists in changing the dimensionality of the state space and the values of the adjustment parameters. For good documentation, the reader can see, in Appendix A, a brief presentation of the steps of AEKF estimation algorithm. Furthermore, the choice of using the AEKF for condition monitoring purposes is explained in this subsection. As is mentioned in the first section, the BMS, through its hardware and software components, plays a vital role in an HEV integrated structure for supervision, control and monitoring all the internal battery parameters. In a BMS, time-based monitoring and FDI techniques based on Kalman filter state and parameters estimators are implementing, and the faults in a system are detected only when measured values exceeded their normal limits [5, 26]. Furthermore, since the Li-ion battery SOC is non-measurable and a critical internal parameter of the battery, the use of AEKF SOC estimator for its estimation is wholly justified.

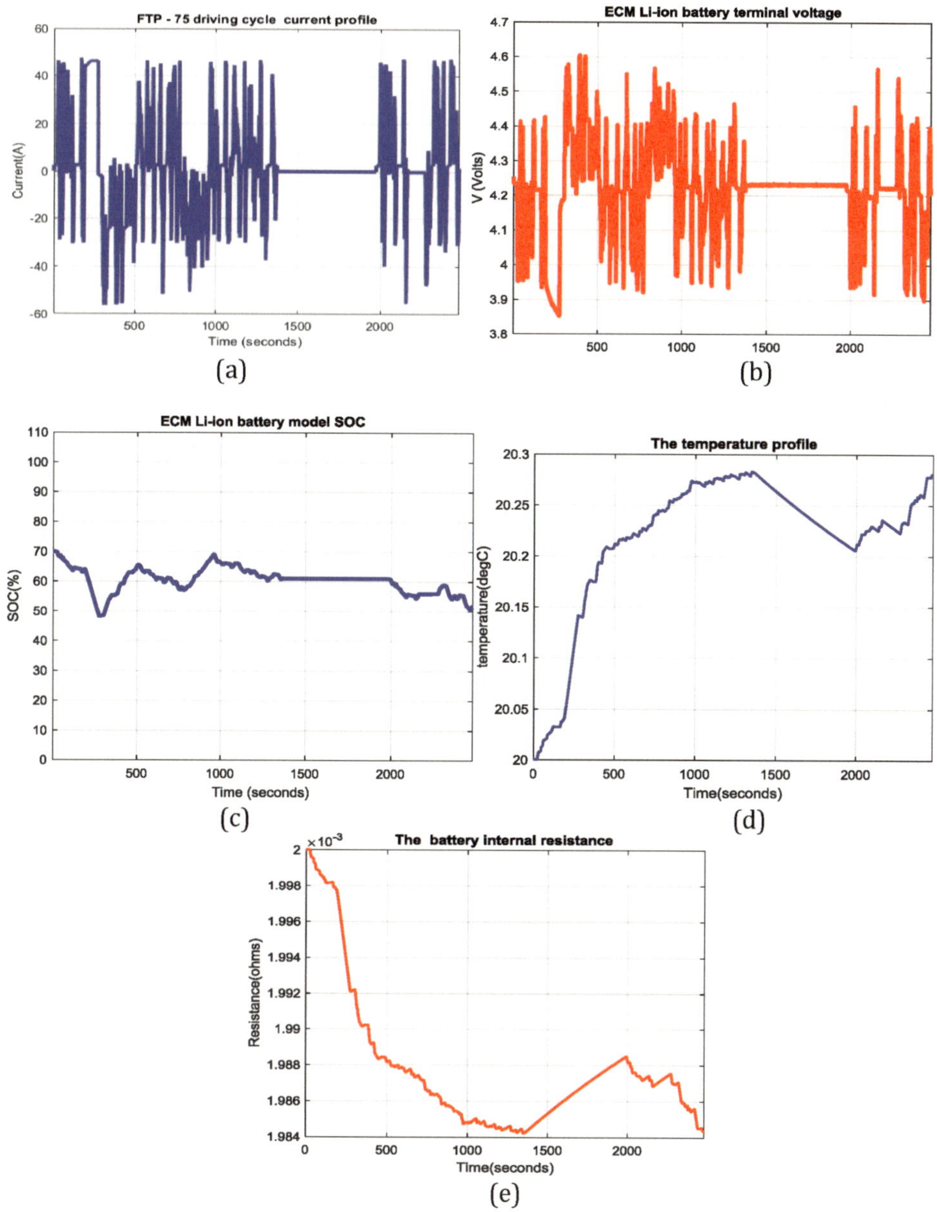

Figure 2.
The ECM Li-ion battery healthy model: (a) FTP-75 driving cycle current profile; (b) output terminal voltage; ECM battery model SOC; (d) temperature profile for changes in ambient temperature; (e) the effect of temperature profile on battery internal resistance.

2.2.2 AEKF SOC estimator of Rint ECM Li-ion battery healthy model: Residue generation and evaluation

For a healthy Li-ion battery (free faults), the MATLAB simulations result of applying AEKF SOC estimator, whose steps are briefly presenting in Appendix A, is shown in **Figure 3**. In **Figure 3(a)** is shown the battery terminal voltage AEKF estimate values versus the Rint ECM Li-ion battery model terminal voltage true values. The MATLAB simulations result reveals an AEKF SOC estimator with an excellent prediction ability for battery terminal voltage. **Figure 3(b)** depictures the residual battery terminal voltage calculated as a difference between the battery

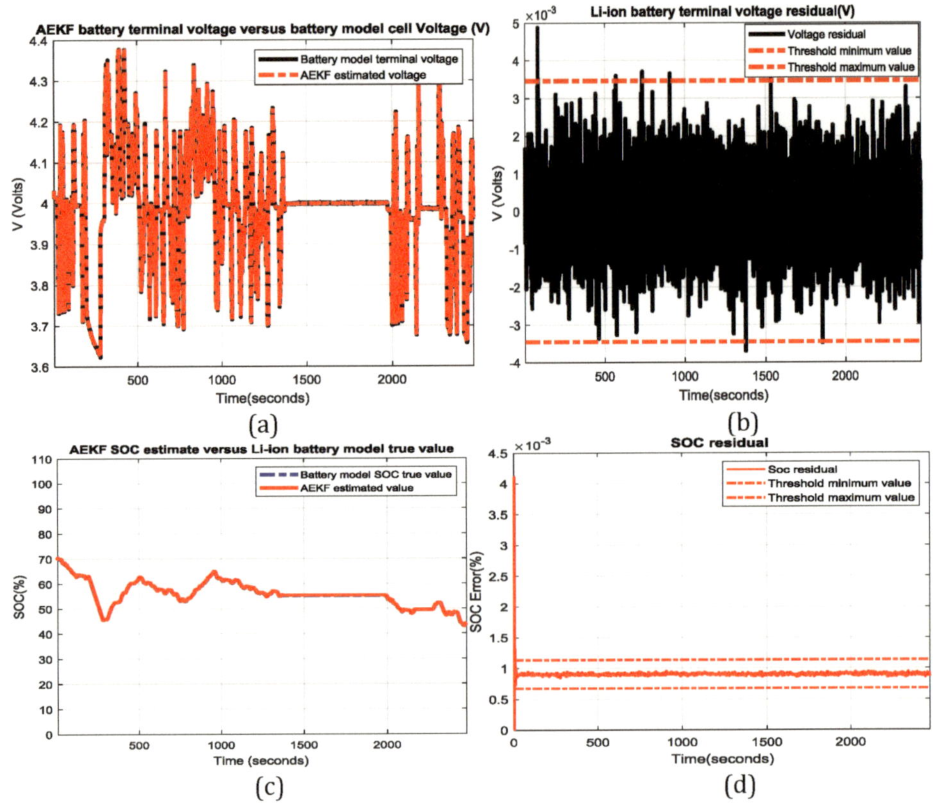

Figure 3.
AEKF estimator and Li-ion ECM battery model – Healthy system: (a) AEKF output terminal voltage estimate versus ECM terminal voltage true value; output terminal voltage residual; (b) terminal voltage residual (c) AEKF SOC estimate versus ECM SOC true value; (d) SOC residual.

terminal voltage true values and the corresponding estimate values of battery terminal voltage, as in Eq. (12).

The residues of battery SOC and for internal resistance are calculated by using the Eqs. (13) and (14):

$$R_y(k) = y(k) - \hat{y}\,(k) \tag{12}$$

$$R_{SOC}(k) = SOC(k) - S\hat{O}C\,(k) \tag{13}$$

$$R_{Rcell}(k) = R_{cell}(k) - R_{cell-fault}(k) \tag{14}$$

For a healthy battery model, the residual is inside the minimum and maximum values of two thresholds, calculated as [5]:

$$\text{Thr}_{ymin} = m_y - 3\sigma_y, \text{Thr}_{ymax} = m_y + 3\sigma_y \tag{15}$$

$$\text{Thr}_{SOCmin} = m_{SOC} - 3\sigma_{SOC}, \text{Thr}_{ymax} = m_{SOC} + 3\sigma_{SOC} \tag{16}$$

$$\text{Thr}_{Rcell,\,min} = m_{Rcell} - 3\sigma_{Rcell}, \text{Thr}_{Rcell,\,max} = m_{Rcell} + 3\sigma_{Rcell} \tag{17}$$

where Thr_{ymin} and Thr_{ymax} denote the minimum value and respectively maximum value of the threshold, m_y is the mean of the clean battery terminal voltage residual values, and σ_y means the standard deviation of the clean battery terminal voltage residual values.

In **Figure 3(c)** is depicted the battery AEKF SOC estimate values versus the battery model SOC true values, and in **Figure 3(d)** is showing the battery SOC residual calculated in the same manner as the battery terminal voltage. The MATLAB simulations result reveals an excellent SOC accuracy, and for a clean battery model the SOC residual is inside the band delimited by the minimum respectively maximum values of the SOC threshold calculated by using a similar formula as in Eq. (9). In **Figure 4(a)** is shown the robustness of AEKF SOC estimator to a change in the initial value of SOC from default value 70% to a SOCini = 40%. A level of the noise in measurements is more realistic in HEVs applications since the initial value of SOC must be guessed, and due to contamination of the measurements with noise. The SOC residual that is showing in **Figure 4(b)** remains inside the band delimited by the same minimum and maximum values of SOC threshold, and in **Figure 4(c)** the battery terminal voltage residual also remains inside the band.

2.2.3 Fault injection mechanism and fault detection based on AEKF SOC estimator - scenarios and residual generation and evaluation

The fault injection mechanism based on AEKF fault estimation and residual generation consists of injecting additive bias sensors faults in the input-output Li-ion battery Rint ECM model, as following:

$$y(k) = OCV(k) - R_{in}\left(u(k) + f_u\right) + f_y \tag{18}$$

$$T_{cell}(k+1) = \left(1 - \frac{T_s}{T_c}\right) T_{cell}(k) + \frac{R_{th} T_s P_{loss}(k)}{T_c} + \frac{T_{amb}(k)}{T_c} + f_T \tag{19}$$

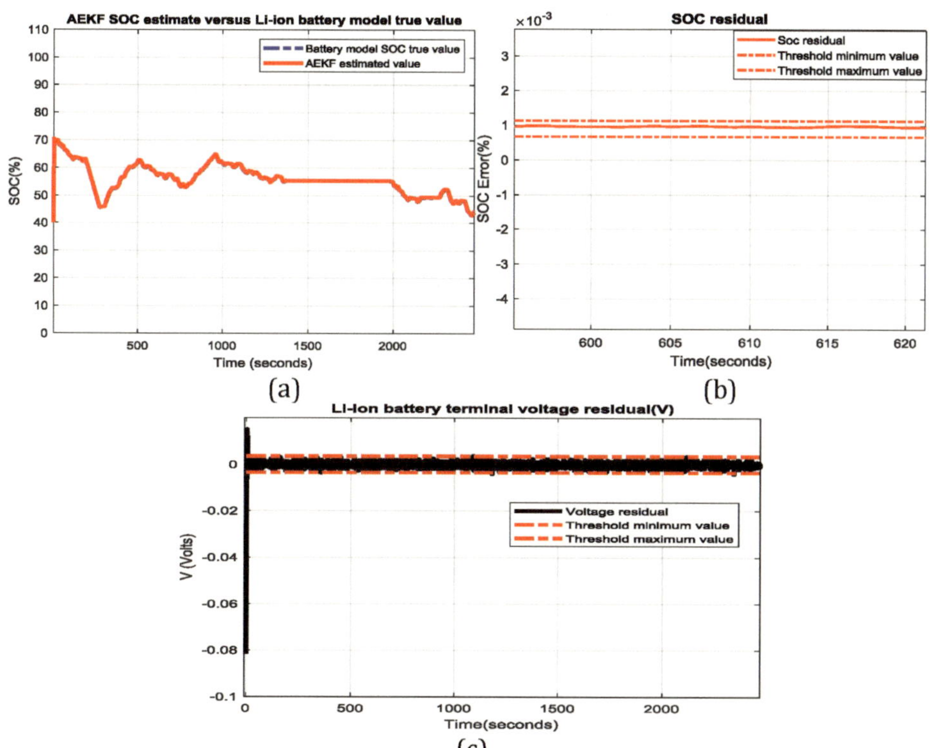

Figure 4.
Robustness of AEKF SOC to changes in SOC initial value, SOCini =70%: (a) robustness to a decrease of 30% from default value SOCini =70% to a SOCini =40%; (b) SOC residual; (c) battery terminal voltage residual.

$$R_{in,cell}(k) = R_{amb}\, exp\left(\alpha\left(\frac{1}{T_{cell}(k)} - \frac{1}{T_{amb}(k)}\right)\right) \tag{20}$$

where f_u denotes the current sensor fault, f_y is the terminal voltage sensor fault, and f_T signifies the temperature fault. The Eq. (12) is useful to measure the impact of f_T on the internal resistance, where $\propto = \frac{E}{RT}$ denotes Arrhenius rate constant, E is the activation energy, $E = 20\ [kJ/mol]$, R signifies the Boltzmann constant, $R = 8.314\ [J/molK]$, as is shown in [26]. As it can be seen in Eqs. (10)–(12) in this research paper are presented only three scenarios for fault injection.

- First scenario - bias sensor fault injection inside the window (500, 1000) seconds.

At the instance 500 seconds is injected a fault in the Voltage measurement sensor of magnitude 1 V, and after 500 seconds the fault is removed, as shown in **Figure 5(a)**.

In **Figure 5(b)** is shown the impact of the injected fault on battery terminal voltage, real and estimated values. The MATLAB simulation result reveals an abnormal behavior of terminal voltage estimate inside the same window of fault injection. The detection of the event is faster at the beginning of the window, persisting only 500 seconds, until the fault is removing. The residual battery terminal voltage is showing in **Figure 5(c)**. It exceeds the band of the clean terminal voltage signal inside the fault window; thus, the same fault is detecting. An abnormal behavior of battery SOC is revealed in **Figure 5(d)** inside the fault window and persists inside the window until the fault is removed at instance 1000. The SOC residual generated by injecting the bias voltage in the Li-ion cell sensor terminal voltage is shown in **Figure 5(e)** that also detects the occurrence of the fault inside the same window. After the fault is removed the SOC residual enters inside of the band and indicates a normal SOC behavior. In **Figure 5(e)** the MATLAB simulations result reveals the fact that the injected fault has not a significant impact on the internal resistance Rin.

- Second scenario: bias current sensor fault injection

Between samples 500 and 1000 is injected a fault in the current measurement sensor of magnitude 2A, such is showing in the **Figure B1(a)** from **Annex B**. Similar as for the first scenario the battery voltage reacts to the fault injection as is shown in **Figure B1(b)**, and its residual depictured in **Figure B1(c)** detects the presence of the fault at the beginning of the window injection. In this scenario, compared to the first scenario, the fault persists until the end of the driving cycle; so its evolution after removing the fault is misclassified and can be considered as a false alarm, that is useful for constructing the FDI logic of fault localization (isolation). A similar situation appears for battery SOC shown in **Figure B1(d)** and for its residual in **Figure B1(e)**. In **Figure B1(f)** the internal resistance Rin has the same evolution as in the first scenario. These last aspects are beneficial also for creating the FDI logic for isolation.

- Third scenario: injection of bias temperature sensor fault

In the temperature sensor, a fault of magnitude 10°C is injected in the same window, similar for first and second scenario, as is shown in **Figure B2(a)** from **Annex B**. The MATLAB simulations result from the impact on the temperature profile of the fault injected is showing in a **Figure B2(b)**. Also, the internal battery resistance is showing in the **Figure B2(c)**, and the battery SOC is disclosed in **Figure B2(d)** together to its residual in **Figure B2(e)**. The battery terminal voltage

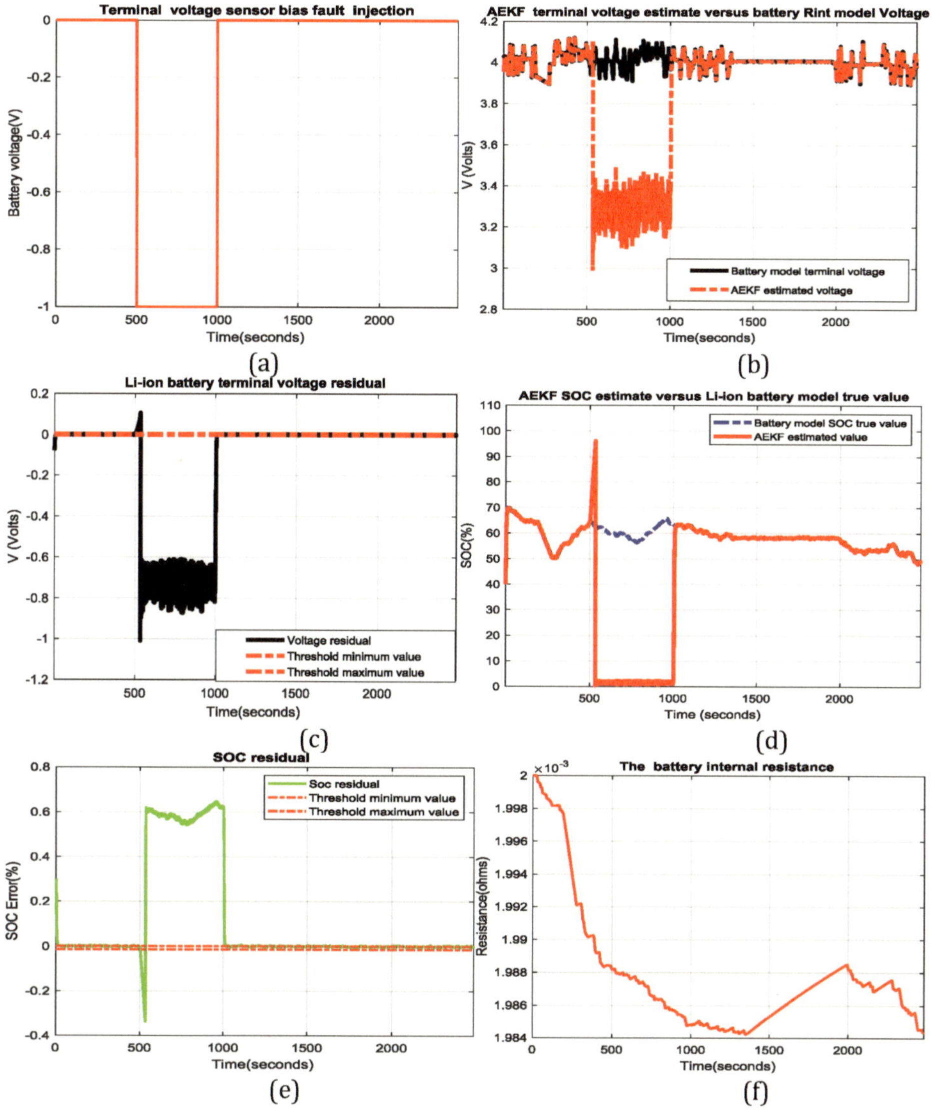

Figure 5.
First scenario of fault injection: (a) bias fault injection of magnitude 1 V in battery cell terminal voltage measurement sensor; (b) AEKF terminal voltage estimate versus EMC battery model terminal voltage true value; (c) battery terminal voltage residual; (d) AEKF SOC estimate versus EMC battery model SOC true value; (e) SOC residual; (f) the battery internal resistance.

is showing in **Figure B2(f)** together to its residual depicted in **Figure B2(g)**. This scenario of point view of fault detection is the same as the first scenario with the fault persistent only inside the window and removed at the end of the same window. Only the internal resistance of the battery withstands a significant impact inside the window, a valuable indication for fault localization.

2.2.4 The residual evaluation and FDI logic isolation

Residual evaluation supposes to define "proper functions for the generated residue evaluation so that fault occurring in the system can be detected correctly", as is stated in [5]. Roughly, in the ideal case, "if no fault occurs, the residue will be

Res_y	Res_SOC	Res_Rcell	Fault signature
1	1 (>0)	0	Voltage sensor fault
1	1 (<0)	0	Current sensor fault, False alarm
1	1	1	Temperature fault sensor

Table 1.
Fault signature for AEKF SOC estimator based diagnostic scheme.

zero and otherwise, it will be non-zero" [5]. More precisely, in a general formulation, the residue evaluation can be defined as:

$$\text{if } R_k = 0, \text{the battery is } \textit{fault free}, \text{otherwise the battery is } \textit{faulty} \quad (21)$$

where R_k might be $R_y(k), R_{SOC}(k)$ *and* $R_{Rcell}(k)$, as defined in Eqs. (9)–(11).

It is possible that in many cases, "the residue might be non-zero even though no fault has occurred; therefore, the evaluation function of Eq. (18) will not be proper. For this purpose, a statistical evaluation function can be defined as [5]:

$$\text{If } m_y - \mu\sigma_y \leq R_k \leq m_y + \mu\sigma_y \text{the Li–ion battery is fault free} \quad (22)$$

otherwise:

$$R_k \leq m_y - \mu\sigma_y \text{ and/or } R_k \geq m_y + \mu\sigma_y, \text{ the Li–ion battery is faulty} \quad (23)$$

for which the values of the parameter μ, i.e. $\mu = 1, 2$ and 3, regarding the evaluation function is a trade-off between maximizing the probability of fault detection and minimizing the probability of wrong fault alarm. If $\mu = 3$ then the probability:

$p\{m_y - 3\sigma_y \leq R_k \leq m_y + 3\sigma_y\} = 98.5\%$ is maximum possible.

The fault signature for AEKF SOC estimator based diagnostic scheme is shown in **Table 1**.

For the second scenario the isolation of the fault can be done based on the tendency of SOC, i.e. for first scenario the SOC increases (Res_SOC >0) after the fault injection, while for second scenario it decreases (Res_SOC < 0) and persists until the end of driving cycle, generating a false alarm.

3. 1-D wavelet transform signal analysis used to extract the faults features in Li-ion batteries

This section investigates the use, in a new approach, of 1-D wave signal analysis, a valuable tool for determining the essential characteristics of faults that occur in a Li-ion battery, a useful basic principle for developing a simple detection of their defects. These techniques are based on detecting changes that occur abruptly in the variation of the residual signal due to a faulty current sensor or a defective temperature measurement sensor, such as those developed in the previous section. Therefore, a similar method of residual generation and evaluation is useful to provide a valuable information to use the wavelet transformation ability to extract the essential features (patterns) of the faults from the output voltage residual of the battery. These faults visibly affect the performance of the Li-ion battery, such as the output voltage and SOC. The dynamics of the battery model under investigation is shown in Section 2. Note that SOC plays a critical role in locating faults (isolation).

3.1 Wavelet transform: Brief presentation and description

Over time, Fourier transform (FT) has proven to be a useful tool for analyzing signal frequency components in a wide variety of applications. However, it has a significant disadvantage, because when it covers the entire time axis, it is impossible to see when a frequency increase. Instead, the short-term Fourier transform (STFT) uses a sliding window to find the spectrogram, which provides complete information on both time and frequency. A small impediment when using STFT in applications is due to the length of the window that limits the frequency resolution [10]. In these situations, the wavelet transforms (WT) seems to be a feasible solution, since it can be applied on a small wavelet of limited duration. Specifically, the wavelet provides local frequency information compared to FT, which captures the global features such as the harmonic components of the entire signal. Besides, the scaled wavelets allow to analyze the signal on different scales. The essential functions designate the "wavelets," which are nothing else than scaled and shifted copies of the same "mother wavelet." With a proper choice of the mother wavelet, the basis wavelets can be orthonormal, or at least linearly independent. Thus, the wavelets form a complete basis, and the wavelet transforms are designed to be reversible.

3.1.1 Continuous and discrete wavelet transforms

A wavelet is a waveform of effectively limited duration that has an average value of zero and nonzero norm, as is stated in [12]. The wavelets compared to sine waves, as the basis of Fourier analysis, "tend to be irregular and asymmetric, while sinusoids are smooth, predictable, and their duration is not limited" [12]. Thus, a wavelet is a wave-like oscillation with an amplitude that starts at zero, increases, and then decreases back to zero. Furthermore, the majority of signals and images of interest "exhibit piecewise smooth behavior punctuated by transients", and the "signals with sharp changes might analyze with an irregular wavelet than with a smooth sinusoid", thus an excellent idea for applying it to develop the detection techniques of the faults [12]. A fundamental work recommended to readers to obtain an excellent theoretical background on the wavelets is the reference [13]. Let us consider the wavelet analyzing function, also called "mother wavelet," and a continuous wavelet transform (CWT). The CWT compares the signal under investigation, denoted by y(t), to shifted and scaling (compressed or stretched) versions of the wavelet function [12]. Since the physical signal y(t), which can be the output of the plant or a residual error, is real-valued, then also the CWT is a real-valued as a function of scale and position. For a scale parameter, $a > 0$, and location, b, a possible representation of a 1-D CWT can be the same as in [12, 13]:

$$CWT(a,b;y(t),\psi(t)) = \frac{1}{\sqrt{a}} \int_{-\infty}^{+\infty} y(t)\psi^*\left(\frac{t-b}{a}\right)dt \tag{24}$$

where $\psi(t)^*$ is the conjugate function of $\psi(t)$, and $CWT(a,b;y(t),\psi(t))$ denotes the coefficients of the wavelet transform CWT. They are affected by the values of scaled and shifting position parameters a, respectively b, as well as by the choice of wavelet function $\psi(t)$. A "mother wavelet" is a waveform for which the most energy is restricted to a finite duration [8]. It is defined as,

$$\psi(t)_{(a,b)} = \frac{1}{\sqrt{a}}\psi\left(\frac{t-b}{a}\right) \tag{25}$$

where $1/a$ denotes the frequency and $1/\sqrt{a}$ is a normalizing constant of each scale parameter. Anyway, there are an infinite number of the functions that can be considered as a "mother wavelet". In Eqs. (24) and (25) the variable a is called also "scale" or "dilation" variable since performs a stretching or compressing action on the "mother wavelet", while the variable b is referred as "time shifting" or "translation" that delays or accelerates the signal start [5]. The $CWT(a,b;y(t),\psi(t)$, as result of wavelet transform on signal y(t) is the wavelet coefficients vector of length L (number of samples) and components $(A_i, D_i, i = \overline{1,N}\ where\ N = length(y)$, as function of scale a, and translation b, i.e., a function:

$$[A\ \ D] = CWT(a,b;y(t),\psi(t)) = cwt(a,b) \tag{26}$$

Each coefficient of the vector $c(a,b)$ $(A_i, D_i, i = \overline{1,N})$ "represents how closely correlated a scaled wavelet is with the portion of the signal y(t) which is determined by translation" [12]. In fact, the c(a, b) coefficients are the time-scale view of the signal y(t), and so the CWT is an important analysis tool capable to "offers insight into both time and frequency domain signal properties" [12]. The results of this interpretation lead to the following useful observations that will be considered for developing the proposed wavelet signals processing and analysis strategy [16]:

- The higher scales correspond to the "most" stretched wavelets, furthermore "the more stretched the wavelet, the longer the portion of the signal with which is compared, and thus the coarser the signal patterns features measured by the wavelet coefficients."

- The coarser features capture the low frequency components $(A_i, i = \overline{1,N})$ called "approximations" that provide basic shapes and properties of the original signal y(t)

- The low scale components $(D_i, i = \overline{1,N})$ are called "details" and capture the high frequency information.

- The CWT is computationally inefficient, since it requires to calculate the c(a, b) coefficients at every single scale, so computationally expensive.

An alternative to the CWT is the discrete wavelet transform DWT, much more efficient and of high accuracy, defined in a similar way that CWT in Eq. (24) [14]:

$$DWT_{j,k}(a,b;y(t),\psi(t)) = \frac{1}{\sqrt{a_0^j}} \int_{-\infty}^{+\infty} y(t)\psi\left(a_0^{-j}t - kb_0\right)dt = \int_{-\infty}^{+\infty} y(t)\psi_{j,k}(t)dt \tag{27}$$

For a parameter (a_0, b_0) setting to the values: $a_0 = 2$, $b_0 = 1$ is obtained a particular dyadic sampling of the time-frequency plane (a set of coefficients per octave), as is mentioned in [14]. Thus, for this particular sampling, it is possible to obtain for the set $\psi_{j,k}$ an orthonormal basis with a "mother wavelet" $\psi(t)$, well localized both in time and frequency, such as the wavelets Morlet, Haar and Daubechies have shown in **Figure 6** [14]. The DWT is based on the wavelet analysis at scales and translations that are power of two, such as 2, 4, 6, 16, and so on [12–14], and wavelet approximations $A_\psi(j_0, k)$ and detailed $D_\psi(j, k)$ coefficients at stage k, are defined as in [10],

$$A_{\psi}(j_0, k) = \frac{1}{\sqrt{N}} \sum_n y(n) \psi_{j_0,k}[n], j_0, k, n\epsilon N, natural\ numbers \tag{28}$$

$$D_{\psi}(j, k) = \frac{1}{\sqrt{N}} \sum_n y(n) \psi_{j,k}[n], j > j_0, j, k, n\epsilon N, natural\ numbers \tag{29}$$

Finally, according to Eqs. (28) and (29) the original signal can be approximated as,

$$y(n) = \sum_k A_{\psi}(j_0, k) + \sum_k D_{\psi}(j, k) \tag{30}$$

or simpler,

$$y = A_N + \sum_{i=1}^{N} D_i \tag{31}$$

starting from last stage N toward the first stage in decomposition, and recursively, at stage level k, it can be writing:

$$A_k = D_{k+1} + A_{k+1}, k = \overline{1, N} \tag{32}$$

In [16] is mentioned the "approximations" of the signals under investigation "provide basic trends and characteristics of the original signals, whereas the details provide the flavor signal". The result of the applying DWT on the original signal y is the so-called wavelet decomposition around both key coefficient vectors, [A] ("approximation" coefficient vector), and [D] ("detail" coefficient vector). The decomposition is repeating on the approximations in each stage. The multiple stage DWT will break down the original signal into many successively lower resolution components, as is described in [15]. According to [15] "at each stage, the approximation coefficients vector [A] represents the basic trends of the original signal characteristics, while the details coefficients vector [D] provides the flavor of the signal". The inverse process opposite to decomposition is the signal reconstruction by using an inverse discrete wavelet transform (IDWT). More details about sample wavelet definitions known as Haar, Morlet and Daubechies wavelets, the reader can find in [8, 13, 17]. As is shown in **Figure 6**, in control systems applications is preferred the Morlet wavelet function for continuous analysis using CWT [13, 14], compared to Haar and the Daubechies wavelet family functions that are very useful for DWT [8–10]. Using the MATLAB/SIMULINK Wavelet and Processing Toolboxes in real-time, the proposed 1-D wavelet analysis strategy is implementing by following the guidelines from [8, 10–12].

3.1.2 Wavelet transform analysis of the faults features extraction in a rechargeable Li-ion battery - setups

Signal processing is a well-known tool to deal with fault diagnosis. It is useful to analyze directly the signals measured online, avoiding system modeling compared to Kalman filter techniques that are model-based. A wave transformation offers a new approach to the analysis of transient regimes that vary over time. It has a specific ability to analyze signals simultaneously in both time and frequency domains. Besides, it can automatically adjust the analysis windows according to frequency, namely, shorter windows for higher frequencies and vice versa. Therefore, the wavelet transform is very suitable for identifying the characteristics of the faults that occur in the Li-ion battery under investigation. However, the identification of such wavelet-based features in HEV Li-ion BMS applications is a novelty. Signal features,

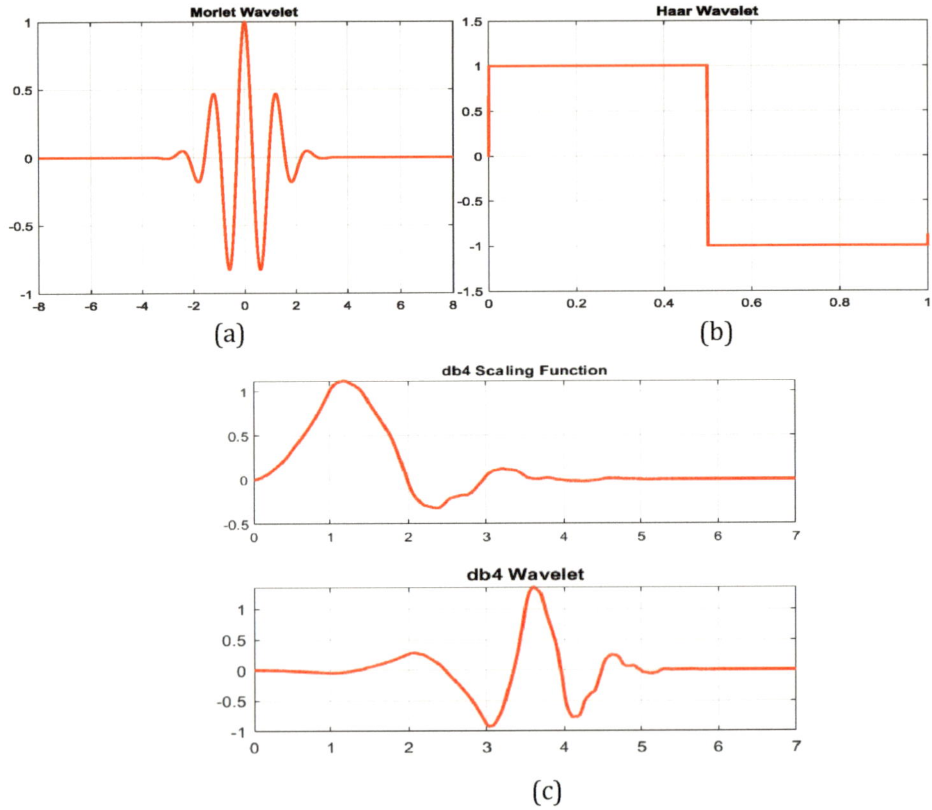

Figure 6.
Wavelet function samples: (a) Morlet wavelet function with 10 vanishes moments; (b) Haar wavelet function; (c) Daubechies wavelet function with 4 vanishes moments and its corresponding scaled function.

such as discontinuity or singularity, are easily detectable through a 1-D wavelet transform. Sudden signal transitions lead to wave coefficients with high absolute values. The changes in the evolution of the signal provide valuable information when something fundamental has occurred in the evolution of the signal. These features suggest an excellent idea in our case study on how to detect measurement sensor errors that often occur in a Li-ion battery used in HEV applications.

Step 1. Simulink model diagram of Li-ion battery and fault injection mechanism setup.

At this stage is investigated the capability of using 1-D wavelet analysis to detect some anomalies in a BMS of the Li-ion battery caused by two faults injected in a current, respectively temperature sensor. **Figure** 7 shows the Simulink diagram of a general model of the Li-ion battery, including the thermal model and fault injection mechanism in both healthy and thermal blocks.

Step 2. Healthy and faulty models of Li-ion battery setup.

The Simulink diagrams of healthy and defective battery cell models are depictured in **Figures** 8 and **9**. In these figures are visible also the fault injection blocks inside the battery (**Figure** 8) and thermal (**Figure** 9) models.

3.1.3 Wavelet transform analysis: MATLAB implementation and simulations

Step 1. Wavelet filter bank decomposition – Biorthogonal wavelet description.

Based on a 1-D DWT signal decomposition, the analysis (decomposition) and synthesis (reconstruction) filters are of more interest than the associated scaling

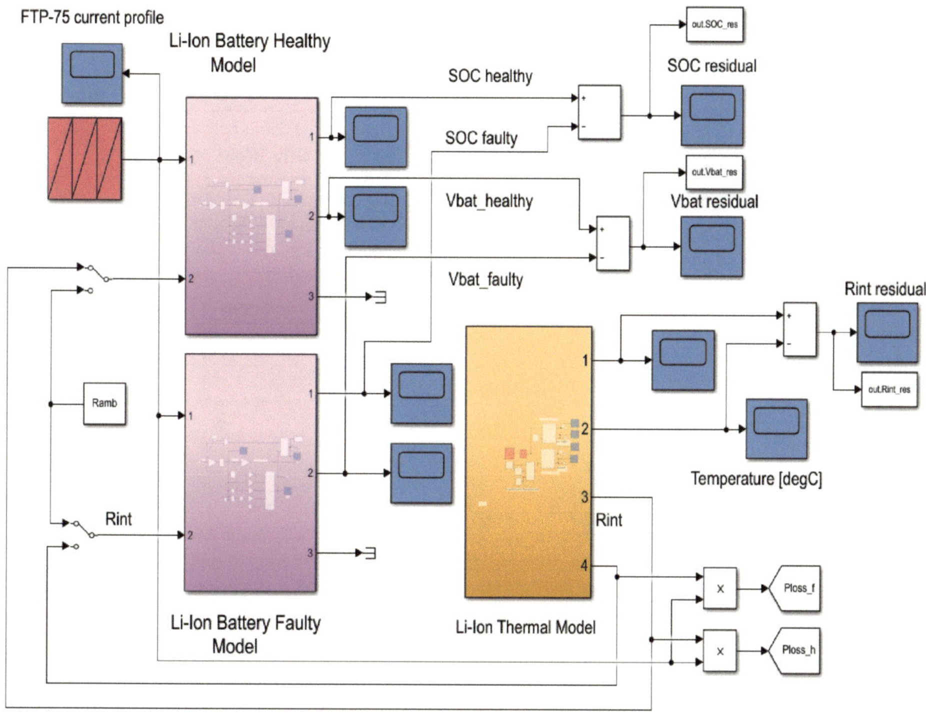

Figure 7.
Simulink diagram of Li-ion battery including the thermal model and fault injection mechanism setup.

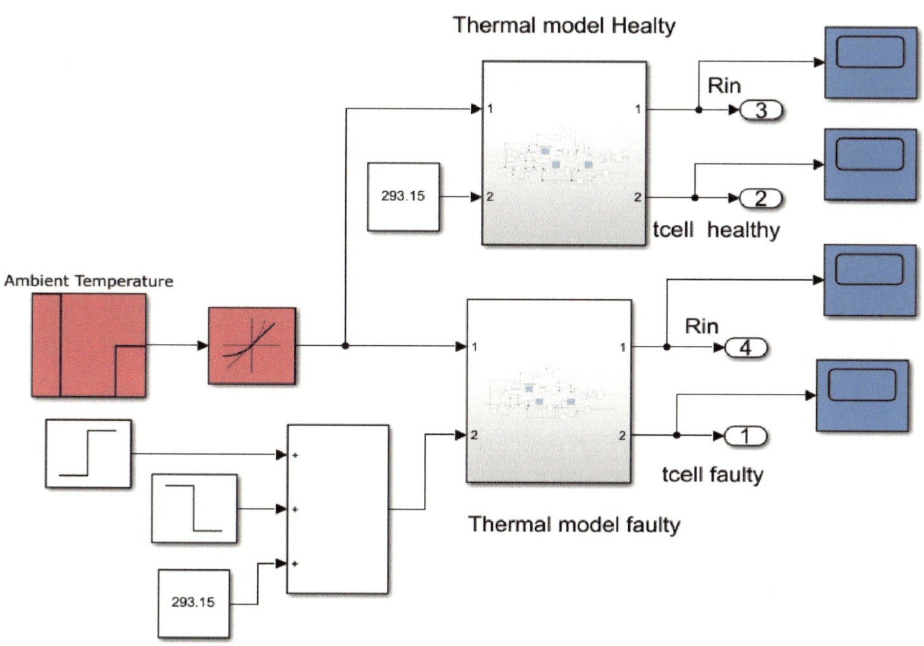

Figure 8.
Simulink diagram of thermal model and fault injection mechanism setup.

function and wavelet for a 1-D CWT. For example, in **Figure 10** are implemented in MATLAB two analysis filters and other two synthesis filters for a B spline biorthogonal wavelet that can reproduce polynomials (vanishing moment property)

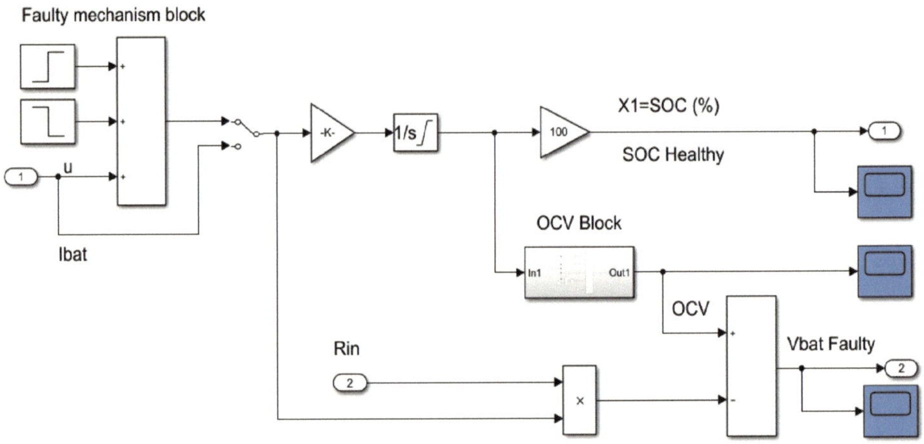

Figure 9.
Simulink diagram of Li-ion battery faulty model setup.

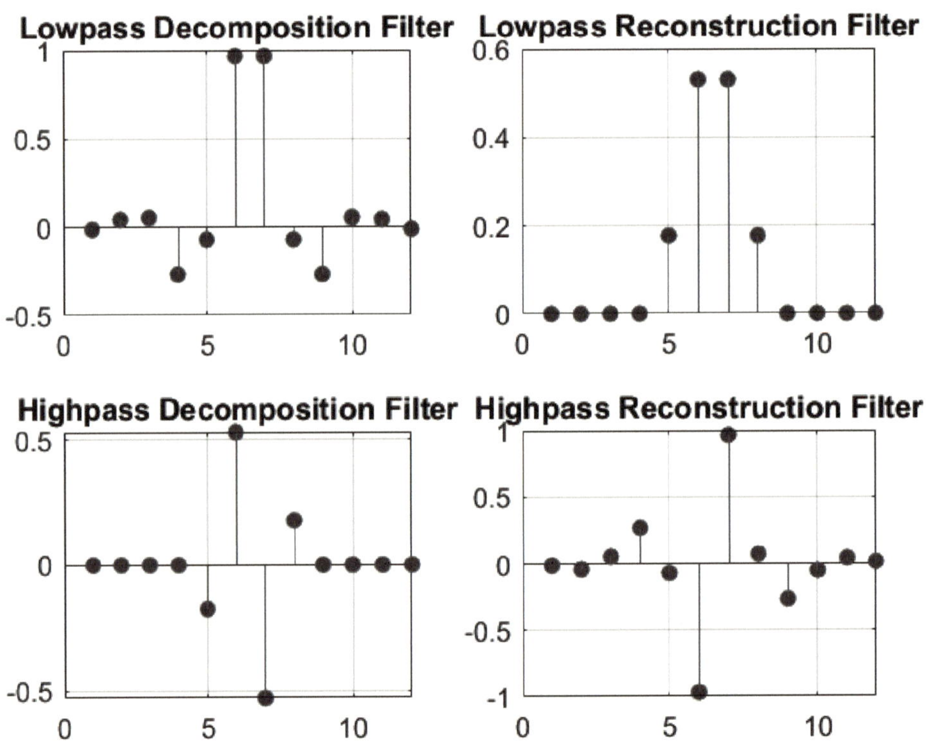

Figure 10.
Analysis and synthesis low pass and high pass decomposition filters, respectively low pass, and high pass reconstruction filters.

with three vanishing moments in the reconstruction filter and five vanishing moments in the decomposition filter, very useful to be used in fault detection. More precisely, both phases analysis and synthesis require two low pass filters (LPF) to filtrate low frequencies signals, respectively two high pass filters (HPF), to filtrate the high frequencies signals [8, 12, 18–21].

Furthermore, the orthogonal and biorthogonal filters banks are an arrangement of low pass, high pass, and bandpass filters that divide the signals data sets into sub-bands [12, 17–21]. If the sub-bands are not modified, these filters enable perfect reconstruction of the original data. In most of applications, the data are processed differently in the different sub-bands and then reconstruct a modified version of the original data. Orthogonal filter banks do not have linear phase, compared to biorthogonal filter banks that have linear phase [12, 18–20]. The wavelet and scaling filters are specifying by the number of the vanishing moments, which allows removing or retaining polynomial behavior in the signals data sets.

In addition, lifting allows designing perfect reconstruction filter banks with specific properties. To obtain and use the most common orthogonal and biorthogonal wavelet filters can be used Wavelet Toolbox™ functions [20]. The design of custom perfect reconstruction filter bank is performing through elementary lifting steps. Besides, can also be added own custom wavelet filters. By using the wavelet filter bank architecture depicted in **Figure 11**, it is possible to obtain residues that change noticeably in order to offer precious information about the timely detection of the faults and its severity [20, 21]. A sub-band model is suggesting in [18, 19] of the form:

$$M(z) = \left(1 - z^{-1}\right)^{-s} \left(a + bz^{-1}\right) \tag{33}$$

where s is an integer number, and a, b are real numbers. In [18] is used the 'db8' wavelet for wavelet filter bank design of level 3 decomposition for a Single-Input Single-Output (SISO) plant extended in [19] for a multiple inputs and multiple outputs (MIMO) plant. Besides, in same reference is developed a wavelet based-frequency sub-band analytical redundancy scheme to calculate the residuals for different faults that uses for wavelet filter bank synthesis and analysis a level three decomposition, as is shown in **Figure 11**. The same wavelet filter bank is adopted in our case study, even if the decomposition resolution can increase by increasing the number of levels. Nevertheless, in our case study, the focus is only on the "concept of proof" and to demonstrate the effectiveness of the proposed error detection technique, based on the use of the multi 1-D signal waveform analysis tool. In **Figure 11**, G(z) and H(z) represent the z-transforms of the low pass filter (LPF) and high pass filter (HPF) respectively. A two-channel critically sampled filters bank play an important role to filtrate the input signal, i.e. the output battery voltage residual, by using a pair of low pass filter (LPF) and high pass filter (HPF) [18, 19, 21]. The subband outputs of the filters are downsampling by two to preserve the overall number of samples. To reconstruct the input, upsampling by two and then interpolate the results using the low pass and high pass synthesis filters. If the filters satisfy specific properties, a perfect reconstruction of the input is achieved [18–21].

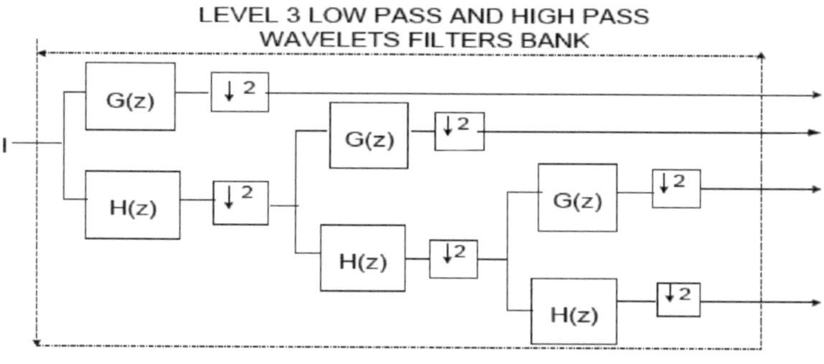

Figure 11.
Wavelet filter bank. Three level decomposition using low and high pass filters for down sampling by two.

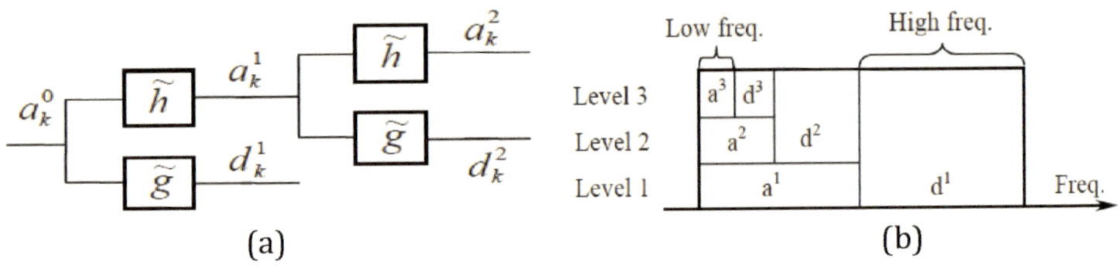

Figure 12.
DWT coefficients interpretation (snapshot from [14]): (a) wavelet filter bank, the approximations (a_k^1, a_k^2) *and details* (d_k^1, d_k^2) *DWT coefficients corresponding to a decomposition level 2; (b) frequency domain.*

In **Figure 12** (a) and (b) are presented the schematic of a Wavelet Filter Bank decomposition on two levels (a), respectively a simple interpretation of the DWT coefficients in frequency domain [14].

The schematic from **Figure 12(a)** give us the idea of a recursive numerical algorithm for the DWT coefficients computation based on digital filters at all levels j = 1:N, which take advantage of using a digital signal processor (DSP):

$$a_k^0 = y(k); a_k^j = \sum_l \tilde{h}_{2k-l} a_l^{j-1}; d_k^j = \sum_l \tilde{g}_{2k-l} a_l^{j-1} \tag{34}$$

Step 2. Fault injection scenarios presentation:

For simulation and "proof-concept" purpose, only two scenarios for error injection are developed in this section, namely a 2A bias fault injected into the current sensor, and a 10°C bias temperature fault injected into the thermal model of Li-ion BMS. The faults are injected separately, in the same window [500,1500] seconds, and their impact on the battery output voltage is analyzed by using the same Li-ion battery residual generation and evaluation method, like in the previous section.

Step 2.1 Scenario 1: Bias current fault MATLAB implementation.

As first scenario is considered a 2A bias fault injected in the current sensor inside the window (500,1500) seconds.

Step 2.1.1 Li-ion output voltage and MATLAB SOC residual generation-original and reconstructed signals.

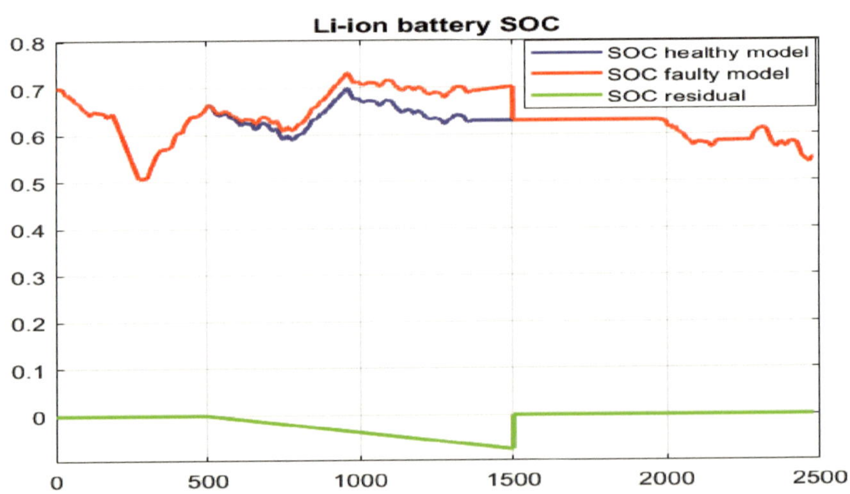

Figure 13.
The impact of the injected bias current fault on the Li-ion battery SOC.

The MATLAB simulations results are shown in the **Figures 13** and **14** for battery SOC (healthy, faulty and residual), respectively for battery voltage residuals, (healthy and faulty) original and reconstructed, using the analysis (approximation) and details wavelets filters (in reconstruction).

Step 2.1.2. Denoising residual signals methods – MATLAB implementation.

In **Figure 15** is used the denoising capability of 1-D wavelet synthesis filters 'sym4' to reduce as much as possible the noise level in the healthy and faulty signals. In [22, 23] is showing how to use wavelets to denoise signals and images. Because wavelets localize features in measurement dataset to different scales, an important signal or image features can be preserved while removing noise [22]. The "basic idea behind wavelet denoising, or wavelet thresholding, is that the wavelet transform leads to a sparse representation for many real-world signals and images" [22]. Thus,

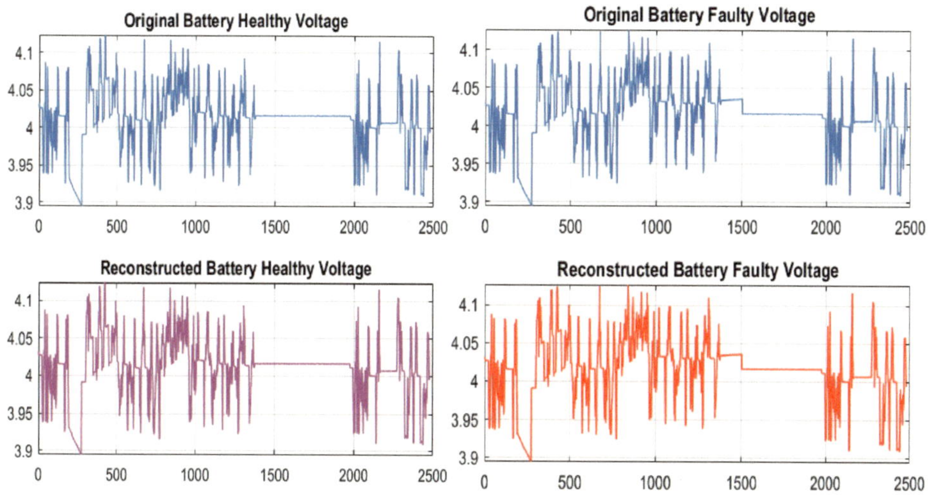

Figure 14.
Li-ion battery terminal output voltage: (a) healthy signal; (b) faulty signal.

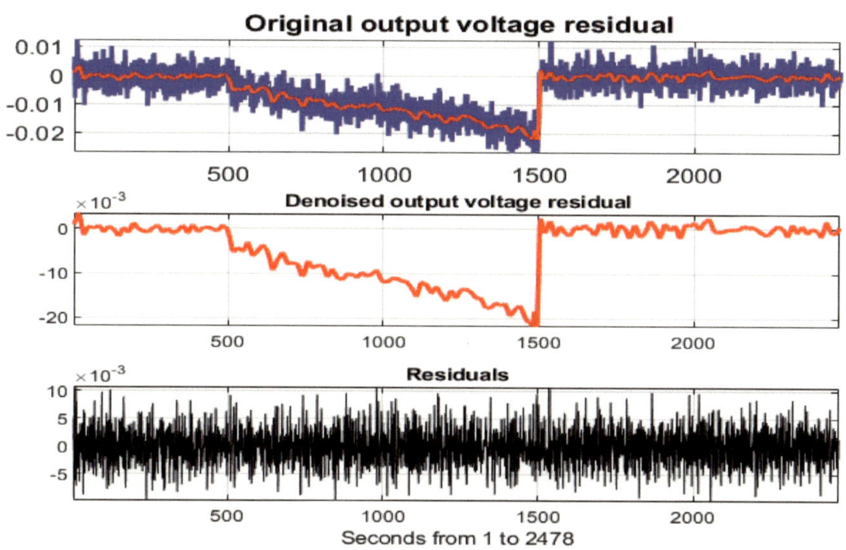

Figure 15.
Li-ion battery output voltage residual – Noisy and denoised signals.

the wavelet transform concentrates the signal and image features in a few large-magnitude wavelet coefficients [22]. Wavelet coefficients which are small in value are typically noise and can be "diminished" those coefficients or much better can be removed without affecting the signal or image quality. Thresholding operation of the coefficients is followed by the reconstruction of the data using the inverse wavelet transform. The denoising operation of the input-output signals can be performed by using an average moving method [23], or decimated ("wdenoise" MATLAB command) and undecimated ("wden" MATLAB command) wavelet transforms [22]. In **Figure 15** is shown the residual between the noisy and denoise signals, where wavelet denoising has removed a considerable amount of the noise while preserving the sharp features in the signal, which is also a challenge for Fourier-based denoising or filtering. The Fourier-based denoising, or filtering, is using a low pass filter (LPF) to remove the noise. However, "when the data has high-frequency features such as spikes in a signal or edges in an image, the low pass filter smooths these out", as is stated in [22]. Moreover, the wavelets can be used to denoise signals in which the noise is nonuniform [22].

Step 2.1.3. Fault detection features:

In **Figure 13**, it easy to see the impact of the injected fault in the windows (500,1500) seconds, where the SOC change by maximum 10%. The information extracted from SOC residual in **Figure 13** and output voltage residual in **Figure 15**, is valuable to detect the incipient moment of the fault, its duration and severity if a threshold value is chosen. The presence of the fault inside the window [500,1500] is visible since sudden changes in the SOC and output voltage of residual levels is easy to visualize. The fault removal at the end of the injected window is noticeable due to a sudden change of the signals' levels in the opposite direction at the initiating time instant of the fault injection. In **Figure 16(a)** and **(b)** is depicted the output voltage residuals noisy and denoised originals and their perfect reconstruction. An impressive result is showing in **Figure 17**, where the presence of the fault inside the injected window is without doubt detecting by analyzing the wavelet variance in signal by scale before injected fault, inside the window and after removing the fault, in bar representation. For the proposed fault detection strategy design, a discrete wave transformation is useful to apply on the output voltage signal of the Li-ion battery. It is equivalent to the analysis branch (with downsampling) of the

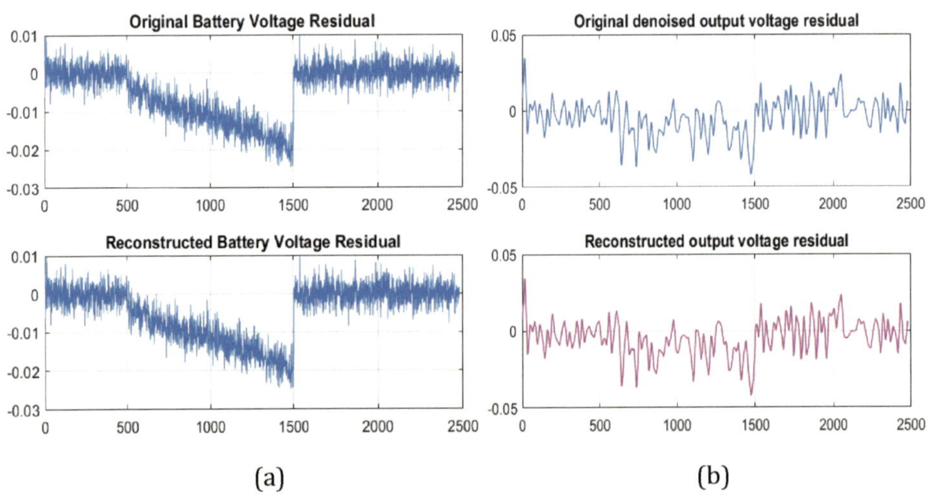

Figure 16.
The Li-ion battery terminal output voltage residual - original and reconstructed waveforms using analysis wavelets filters (reconstruction): (a) contaminated with noise; (b) denoised signals.

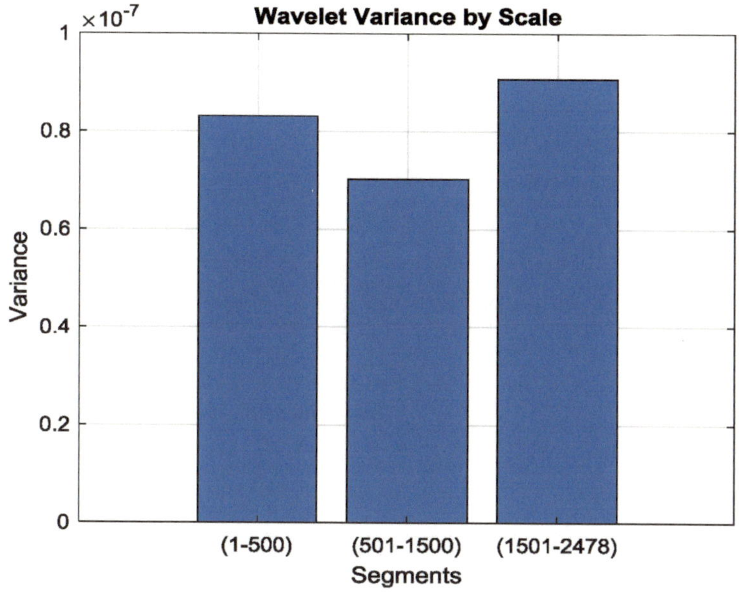

Figure 17.
Li-ion terminal output voltage residual – Wavelet variance in signal by scale before injected fault, inside the window and after removing the fault - bar representation.

two-channel filter bank (decomposition) using LPFs, and HPFs suggested in [17, 18]. They are used for downsampling the input signal up to level 3, as shown in **Figure 18** for all three levels the details of the wave coefficient D1, D2 and D3 and the analysis coefficient A3.

Step 2.1.4. 1-D wavelet transform analysis used for battery voltage residual three levels decomposition – Approximation coefficient A4, and Details coefficients D1, D2 and D3:

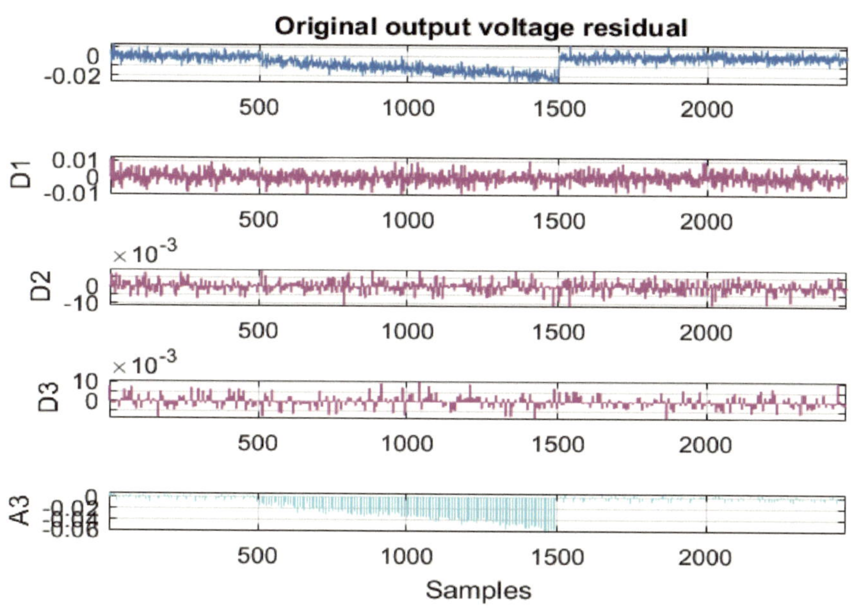

Figure 18.
Li-ion battery output voltage residual decomposition on three levels.

		Details coefficients			Analysis coefficients
Extracted features	Faults	D1	D2	D3	A3
Energy	Current fault	1.7821	0.8357	0.3631	97
	Temperature fault	4.46	2.7	5.349	87.5
Skewness	Current fault	0.063	−0.17	0.15	−4.9
	Temperature Fault	0.063	−3.92	13	−1.33
Kurtosis	Current fault	5.8	11.1	23.11	27.38
	Temperature fault	5.8	71.4	389.13	56
RMSE statistic criterion-performance			D1 coefficient		
Energy	Current fault	1.7821	Remark: Temperature fault features shows significant values.		
	Temperature fault	4.4654			
Skewness	Current fault	0.063			
	Temperature fault	0.063			
Kurtosis	Current fault	5.7581			
	Temperature fault	5.7581			

Table 2.
The main features extracted for faults detection.

In **Figure 18** is presented the MATLAB simulation result of the battery voltage residual decomposition on three level based on the wavelet filter banks shown in **Figures 11** and **12**.

For decomposition is used a Symlet wavelet transform 'sym4' with four vanishing moments. The feature extracted from the wavelet coefficients are summarized in **Table 2** and interpreted at the end of this section, in comparison with the second fault.

Step 2.2. Scenario 2 MATLAB implementation:

As a second fault is investigated a 10°C bias fault injected in the temperature sensor inside the window (500,1500) seconds.

Step 2.2.1 Li-ion output voltage and MATLAB SOC residual generation-original and reconstructed denoised signals:

Like for the first scenario, the same information is extracted from the **Figures 19** and **20**. In **Figure 19(a)** is shown the battery SOC with almost a zero impact of the injected temperature fault since we assumed in this research that SOC does not change significantly if the temperature inside the battery changes. This assumption is not realistic, since in "real life" the SOC and internal resistance of Li-ion battery are dependent on temperature. This assumption was adopted to simplify the Li-ion battery model substantially, since a battery model of high complexity is beyond the topic developed in this research work. Moreover, the assumption is also justified by the fact that the fault detection analysis by using a 1-D wavelet analysis tool is performing online. A model is not required, that is a significant advantage of the new approach compared with the model based Kalman filter technique approach developed in the previous section, for which the SOC accuracy of the battery model is critical. Besides, the main objective of this paper is to provide a "proof concept" and to demonstrate the effectiveness of the use of 1-D wavelet analysis of finding the essential features in the output voltage residual variance for MATLAB design and implementation of the investigated fault detection technique. In **Figure 19** (b), (c) and (d) are visualized the healthy, faulty and the battery temperature residual (b), the healthy, In faulty and the output voltage residual (c), respectively the use

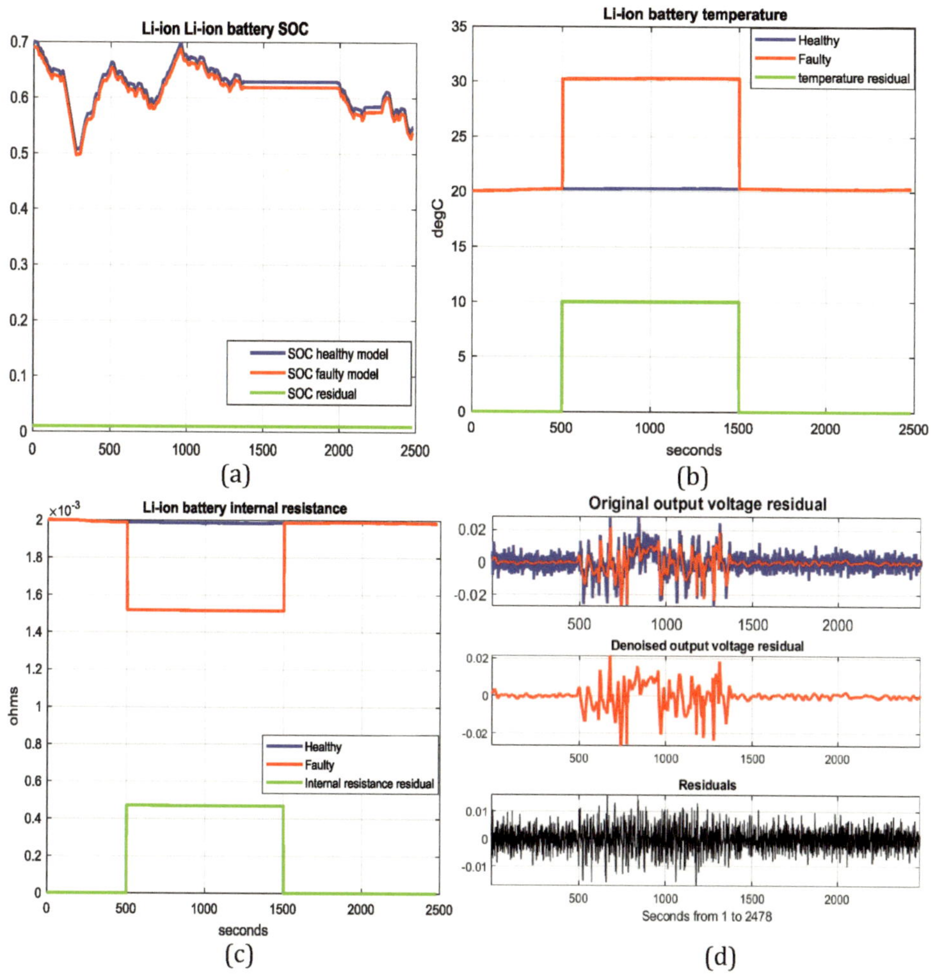

Figure 19.
Li-ion battery temperature fault injected: (a) SOC and its residual; (b) healthy, faulty, and residual temperatures; (c) healthy, faulty, and residual battery internal resistance; (d) original (noisy), denoised and residual output voltage signals.

of 1-D wavelet 'Sym4' for denoising output voltage residual (d). The residual of denoised battery output voltage and its corresponding constructed wave are presenting in **Figure 20(a)**.

Step 2.2.2. Fault detection features.

The MATLAB simulation result shown in **Figure 19** (b), (c) and (d) reveal that the presence of the temperature fault is noticeable by its effect on the output voltage residual at the beginning, inside and at the end of the injected window. and the coefficients D1, D2, D3 and A3 of the 'Sym4' wavelet are shown in **Figure 20 (b)**. The features extracted from **Figure 20(b)** are summarized in **Table 2** and analyzed at the end of this section.

3.1.4 Wavelet transform analysis of fault detection technique. Performance comparison

A rigorous performance analysis of using 1-D wavelet transform tool for fault detection strategy is accomplished based on the information extracted from the details' coefficients of output voltage residual decomposition for both scenarios.

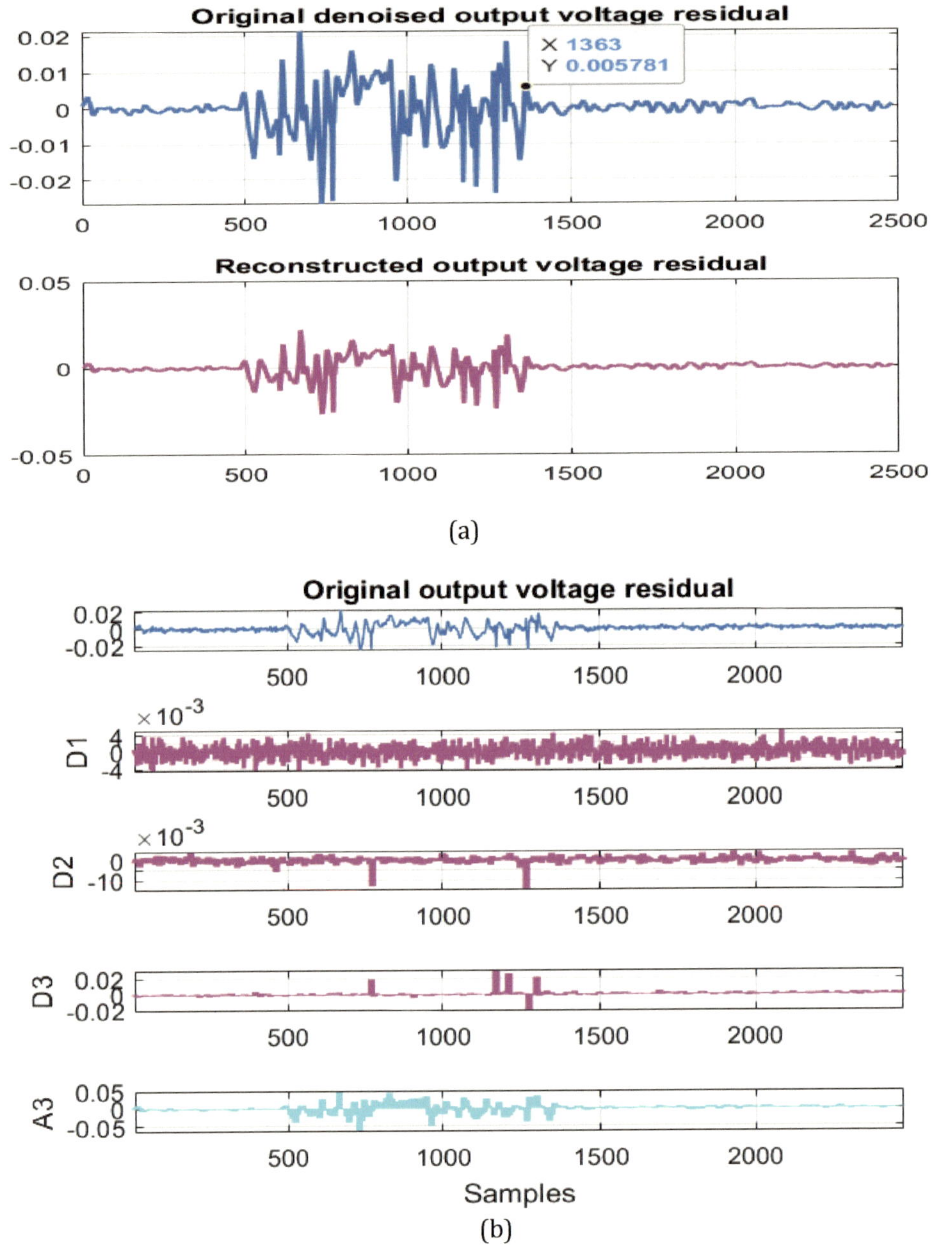

Figure 20.
Li-ion battery output voltage residual second scenario: (a) original denoised output voltage residual; (b) the details (D1, D2, D3) and approximation (A3) Symlet4 (four vanishing moments) wavelets coefficients decomposition at level 3.

From the details coefficients values D1, D2, and D3 can be extracted the wavelet energy, skewness, and kurtosis features. These statistics can identify the types of faults based on their distinct value, as are defined in [14], MATLAB Wavelet Toolbox (for wavelet energy), respectively MATLAB Statistics and Machines Learning Toolbox for skewness and kurtosis).

- The wavelet energy is an important indicator that gives a valuable information about the presence of the fault inside a window that has a concentrated large value of the wavelet energy, defined as,

$$E_{WDT} = \sum_{j=1}^{3} D_j \tag{35}$$

- The skewness is a measure of the asymmetry of the data around the sample mean. If skewness is negative, the data spreads out more to the left of the mean than to the right. If skewness is positive, the data spreads out more to the right. It is defined as,

$$SK_{WTD_i} = \frac{E\{\{D_{ij} - \mu_i)^3\}}{\sigma_i^3} \tag{36}$$

- The kurtosis is a measure of whether the distribution is too peaked, i.e. a very narrow distribution with most of the responses in the center, and is defined as,

$$KT_{WTD_i} = \frac{E\{\{D_{ij} - \mu_i)^4\}}{\sigma_i^4} \tag{37}$$

where μ_i is the mean of D_{ij}, σ_i is the standard deviation of D_{ij}, and E(.) represents the expected value. *The values* of the three indicators defined in Eqs. (31)–(33) are entered in **Table 2** for both error scenarios.

The excess kurtosis and skewness of every coefficient A3, D1, D2 and D3 in the dataset, can be interpreted as follows:

a. For skewness, if the distribution of responses for a variable stretches toward the right or left tail of the distribution, then the distribution is referred to as skewed. A general guideline for skewness is that if the number is greater than +1 or lower than −1, this is an indication of a substantially skewed distribution.

b. For kurtosis, if the number is greater than +1, the distribution is too peaked. Otherwise, a kurtosis of less than −1 indicates a distribution is too flat.

c. When both skewness and kurtosis are zero, the pattern of responses is considered a normal distribution.

Besides, an assessment statistic criterion root mean square error (RMSE) is introduced in **Table 2** for a particular analysis of the high frequency detail component D1 dataset.

The MATLAB simulation results analyzed from the perspective of the fault features extracted from **Table 2** reveal the fact that the temperature fault shows significant values compared to a possible occurrence of current fault in Li-ion battery.

Figures 21 and **22** show how multiresolution decomposition technique, such as 1-D wavelet analysis, allow us to study signal components in relative isolation on the same time scale as the original data [22]. Multiresolution analysis (MRA) refers to "breaking up a signal into components, which produce the original signal exactly when added back together" [22]. The components ideally decompose the variability of the data into physically meaningful and interpretable parts, as is stated also in [22].

The term MRA is often associated with wavelets, and in the "real life" the signals consist of a mixture of different components. Often the interest is focused only in a subset of these components. That is why the MRA allows us to restrict the analysis of the original signal, by separating it into components at different resolutions. Extracting signal components at different resolutions means a decomposition of variations in the data on different time scales, or equivalently in different frequency bands [22]. Consequently, the signal variability at different scales or frequency bands can be seen simultaneously.

In the **Figures 21** and **22**, using a wavelet MRA, the Li-ion battery output voltage residual signal is analyzed in MATLAB at eight resolutions or levels, following the procedure shown in [22] for both faults isolation.

Both graphs from **Figures 21** and **22** starts from the uppermost plot and proceed down until is reached the plot of the original data and is worth noting that the components have become progressively smoother. D2 graph isolates the time-localized high-frequency component, which can be seen and investigated as an essential signal feature practically in isolation. The next two graphs contain the lower frequency oscillation. It is worth to mention that "an important aspect of multiresolution analysis, namely important signal components may not end up isolated in one MRA component, but they are rarely located in more than two" [22]. Finally, from the S8 graph can be extracted a smooth trend term, which provides us a valuable information to localize transient changes, as it can see in the fault injection window [500, 1500] seconds. Thus, the presence of the bias current fault and bias temperature fault is detected and localized as a significant transient change in the nonstationary Li-ion output voltage residual signal. For an appropriate choice of the thresholds' values, both faults can be detected directly from the S8 graph, removing the presence of the false alarms completely.

Besides, the value of the RMSE statistical criterion of the energy feature extracted from the detail coefficient D1, for both faults, shown in **Figures 18** and **20 (b)**, undoubtedly confirms the validation of the results obtained in **Table 3**, adequate to differentiate between the two faults. However, in **Table 3** is shown the Fault signature of 1-D wavelet analysis transform, useful for fault isolation. To distinguish between both faults injected in Li-ion battery, i.e. current sensor bias fault, respectively, temperature bias fault a valuable information is provided by battery SOC and battery internal resistance residuals. It is showing in **Table 3**, like for based model AEKF FDI strategy developed in Section 2. An exciting piece of

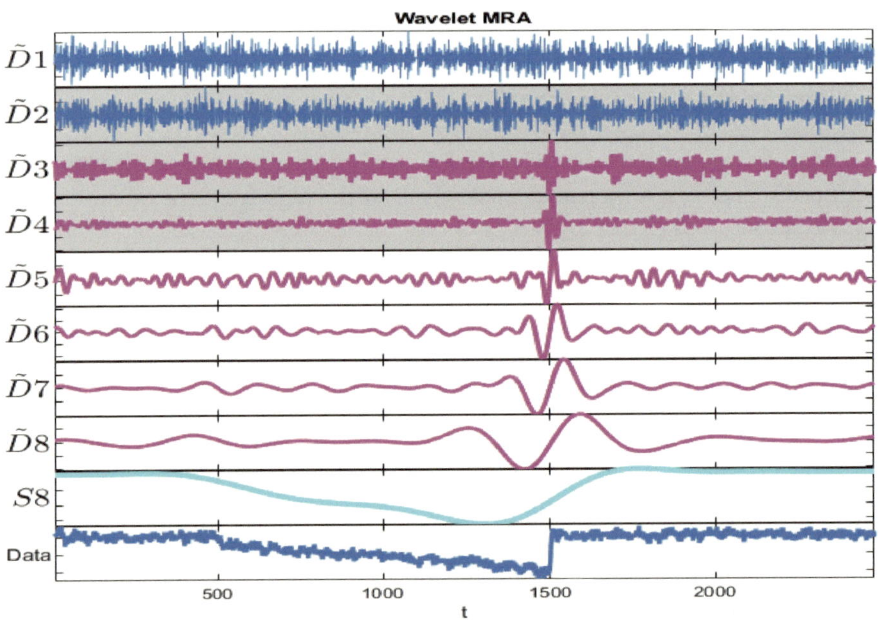

Figure 21.
Li-ion output voltage residual signal using a wavelet MRA for scenario 1 of bias current fault on 8 resolutions (levels) decomposition–extracted smooth trend (S8) and localize transient changes.

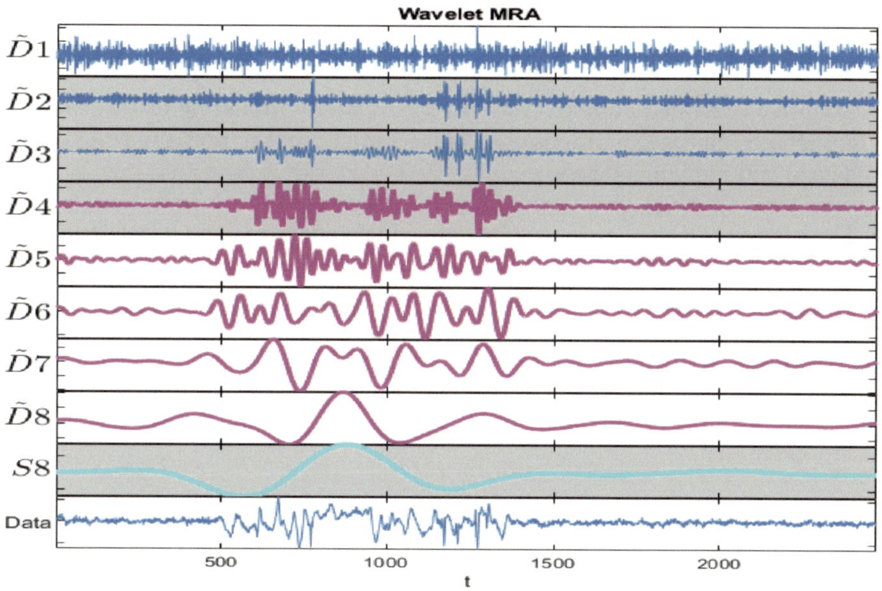

Figure 22.
Synthetic signal using a wavelet MRA on 8 resolutions (levels) decomposition for scenario 2 of bias temperature fault – Extracted smooth trend (S8) and localize transient changes.

Res_y	Res_SOC	Res_Rcell	Fault signature
1	1 (<0)	0	Current fault, no false alarm
1	0	1	Temperature fault, no false alarm

Table 3.
Li-ion battery - fault signature 1-D wavelet analysis transform.

information is related to the "border effects of error injection", clearly visible when the temperature fault is removing, because the healthy signal emerges from the defective one in the window, before the corresponding time tf = 1500 seconds. These "frontier effects" require further investigation in future work.

4. Conclusions

In this research paper is opened a new research direction in HEV BMS applications field by performing many investigations on the use of multisignal 1-D wavelet analysis to improve the accuracy, robustness, the design and the implementation in real-time of Fault detection techniques. Among the most relevant contributions of the authors can be highlighted the following:

- The selection of a suitable and straightforward Li-ion battery model, accurate enough for data generation, and to design and implement a robust adaptive extended Kalman filter SOC estimator to changes in SOC initial values, in the level of measurement noise that contaminate the input-output dataset, to changes in the battery capacity value due to aging effects, and changes in the internal resistance of the battery due to temperature effects

- Representation of the battery model in continuous and discrete time state-space

- Develop the most appropriate thermal model of the battery for data generation and to setup the temperature mechanism fault injection
- Adaptive Extended Kalman Filter SOC estimator with fading feature and covariance matrices of noises correction—brief presentation and MATLAB design and implementation.
- The battery SOC and output voltage residual generation and bias current fault injection mechanism
- The fault detection and isolation estimation technique based on AEKF SOC estimator
- Wavelets transform analysis of the faults features extraction in a rechargeable Li-ion battery
- SOC and output voltage residual generation-original and reconstructed signals
- 1-D wavelet transform analysis used for battery voltage residual three levels decomposition – Approximation coefficient A4, and Details coefficients D1, D2 and D3
- Denoising residual signals methods analysis – MATLAB implementation
- Wavelets transform analysis to extract the fault features for their detection. Performance analysis
- Extracting signal components at different resolutions by using a multiresolution analysis (MRA) method for fault detection
- The use of the fault signature for fault localization (isolation)

These investigations are performed for the case study, principally chosen to evaluate the impact of two bias faults injected in a current and temperature sensor on the output voltage of a BMS Li-ion rechargeable battery used in HEVs applications.

The effectiveness of fault detection strategy is demonstrated through an extensive number of simulations in a MATLAB R2020a software environment. The preliminary simulation results are encouraging, and extensive investigations will be done in future work to extend the applications area. The performance analysis from the last section reveals that 1-D wavelet analysis is a useful tool for signals processing, design and implementation based on wavelet transforms found in a wide range of control systems industrial applications. Compared to AEKF estimation technique described in Section 2, the 1-D wavelet analysis tool has a significant advantage to perform online. Also, it does not require the model of the battery, since it uses directly the input-output signals generated by the battery model. More precisely, it is based only on the measurement input-output dataset collected by a data acquisition (DAQ) system incorporated in BMS of HEVs. Besides, the battery SOC and output voltage signals' accuracy is not affected by noise as long as is using the signals denoising techniques, such in the case of AEKF fault detection and isolation technique during the noise correction step of the algorithm.

Conflict of interest

The authors declare no conflict of interest.

Abbreviations

EV	electric vehicle
HEV	hybrid electric vehicle
BMS	battery management system
FTP-75	Federal test procedure at 75 F
OCV	open-circuit voltage
SOC	state of charge
KF	Kalman filter
EKF	extended Kalman filter
AEKF	adaptive Kalman filter
WCT	wavelet continuous transform
WDT	wavelet discrete transform
LPF	low pass filter
HPF	high pass filter
Sim4	Simlet wavelet with 4 vanishing moments
RMSE	root mean square error
MSE	mean square error
MAE	mean absolute error
MAPE	mean absolute percentage error
std	standard deviation
R^2	R-squared

Appendix A - AEKF SOC estimator steps of Li-ion battery combined model

Step 1. Rint ECM battery nonlinear model represented in discrete time [3, 4, 11]:

$$x_1(k+1) = x_1(k) + \frac{\eta T_s u(k)}{C_{nom}} + w(k) = f(x_1(k), u(k)) + w(k), x_1(k) = SOC(k) \tag{38}$$

$$OCV(k) = K_0 - K_2 x_1(k) - \frac{K_1}{x_1(k)} + K_3 \ln(x_1(k)) + K_4 \ln(1 - x_1(k)) \tag{39}$$

$$y(k) = OCV(k) - R_{in} u(k) + v(k) = g(x_1(k), u(k)) \tag{40}$$

where the process noise $w(k)$ and measurement output noise $v(k)$ are white uncorrelated noises of zero mean and covariance matrices $Q(k)$ and $R(k)$ respectively, i.e.

$$\begin{aligned} & w(k) \sim (0, Q(k)), v(k) \sim (0, R(k)) \\ & E\left(w(k)w(j)^T\right) = Q(k)\delta_{kj}, E\left(v(k)v(j)^T\right) = R(k)\delta_{kj} \\ & \delta_{kj} = \begin{Bmatrix} 0, & k \neq j \\ 1, & k = j \end{Bmatrix} \end{aligned} \tag{41}$$

Step 2. Initialization:

$$\begin{aligned} & \hat{x}_0 = E[x_0] - the\ \ initial\ mean\ \ value \\ & \hat{P}_{x_0} = E\left[(x_0 - \hat{x}_0)(x_0 - \hat{x}_0)^T\right] - \text{the } initial\ state \text{ covariance matrix} \end{aligned} \tag{42}$$

Step 3. Model linearization - The Jacobian matrices of the model linearization are given by:

$$A(k) = \frac{\partial f(k, x(k), u(k))}{\partial x(k)}\bigg|_{\hat{x}(k|k)}$$
$$C(k) = \frac{\partial g(k, x(k), u(k))}{\partial x(k)}\bigg|_{\hat{x}(k|k-1)} \quad (43)$$

For $k \in [1, +\infty)$ do.

Step 4. Prediction phase (forecast or time update from $(k|k)$ to (k + 1)|k):

Step 4.1 Predict the state ahead:

$$\hat{x}(k+1|k) = A(k)\hat{x}(k|k) + B(k)u(k) \quad (44)$$

Step 4.2. Predict the covariance error ahead:

$$\hat{P}(k+1|k) = A(k)\hat{P}(k|k)A(k)^T + \alpha^{-2k}Q(k) \quad (45)$$

Remark. In this phase, the predicted value of the state vector $\hat{x}(k+1|k)$ is calculated based on the previous state estimate $\hat{x}(k|k)$ and the state covariance positive definite matrices $\hat{P}(k|k)$ and $\hat{P}(k+1|k)$ (unidimensional in the case study) are affected by a fading memory coefficient α.

Step 4.3 Compute the updated value of Kalman filter gain:

$$K(k) = \alpha^{2k}\hat{P}(k+1|k)H(k)^T\left(H(k)\alpha^{2k}\hat{P}(k+1|k)H(k)^T + R(k)\right)^{-1} \quad (46)$$

Step 5. Correction phase (analysis or measurement update):

In this phase the Li-ion battery SOC estimated state is updated when an output measurement is available in two steps:

Step 5.1 Update the SOC estimated state covariance matrix with a measurement:

$$\hat{P}(k+1|k+1) = (I - K(k)H(k))\hat{P}(k+1|k)\ (I - K(k)H(k))^T + \alpha^{-2k}K(k)R(k)K(k)^T \quad (47)$$

Step 5.2 Update the SOC estimated state variable with the measurement:

$$\hat{x}(k+1|k+1) = \hat{x}(k+1|k) + K(k)(y(k) - g(\hat{x}(k+1|k), u(k), k) \quad (48)$$

Step 5.3 Update the estimated output (battery terminal voltage):

$$\hat{y}(k|k) = g(\hat{x}(k|k), u(k), k) \quad (49)$$

Step 6. Adaptive noise covariance matrices correction:

For k > = L, the length of the window's samples, compute:

Step 6.1. Output variable error and the correction factor:

$$E_{rr}(k) = y_{mes}(k) - g(\hat{x}(k|k), u_k)$$
$$c(k) = \frac{\sum_{i=k-L+1}^{k} E_{rr}(k)E_{rr}^T(k)}{L} \quad (50)$$

Step 6.2. Measurement noise correction:

$$R(k) = c(k) + H(k)\hat{P}(k|k)H(k)^T \quad (51)$$

Step 6.3. Process noise correction:

$$Q(k) = K(k)c(k)K(k)^T \tag{52}$$

Annex B – Figures

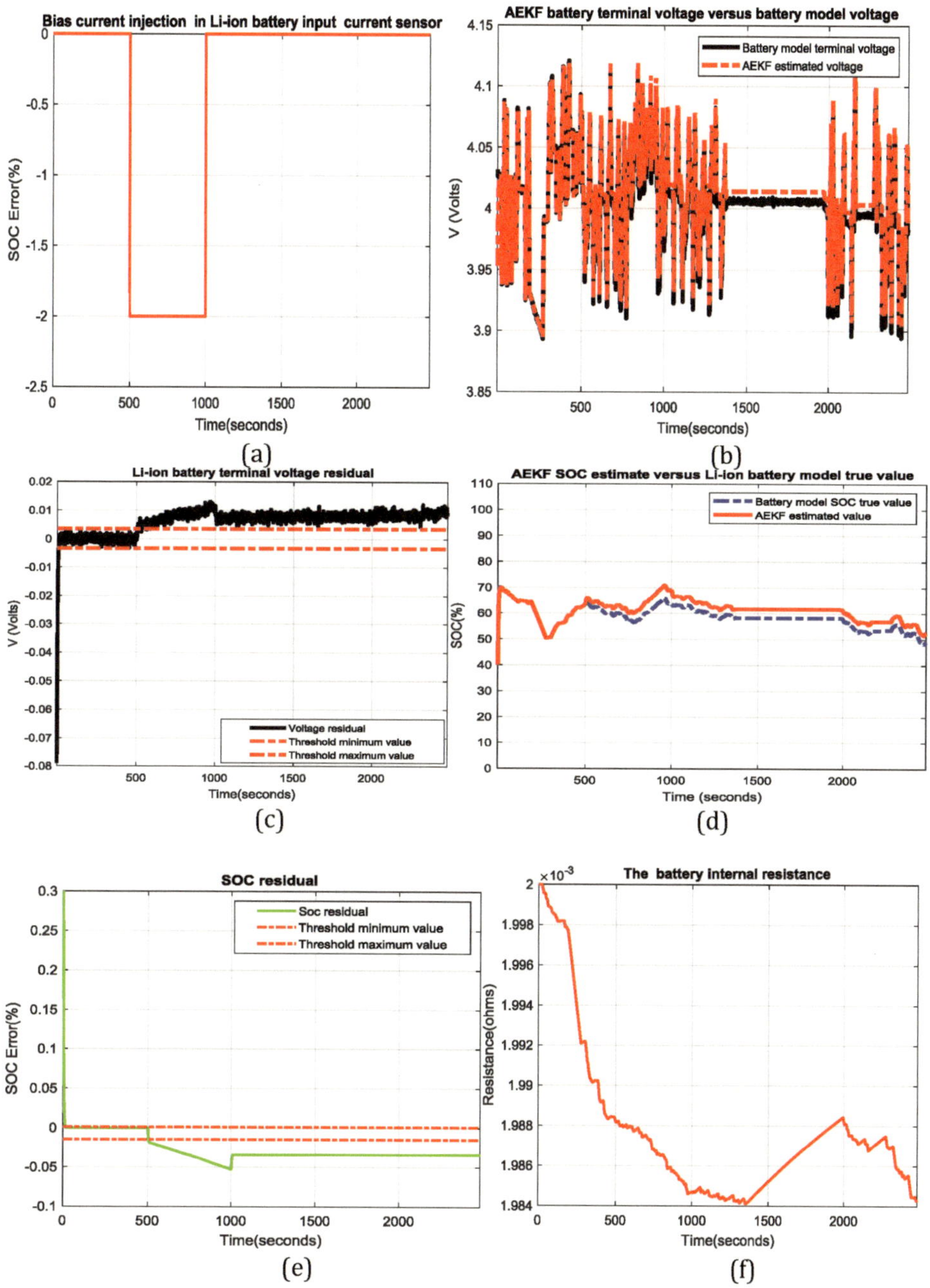

Figure B1.
Second scenario of fault injection: (a) bias fault injection of magnitude 2A in a current measurement sensor; (b) battery terminal voltage residual; (c) AEKF SOC estimate versus EMC battery model SOC true value; (d) SOC residual.

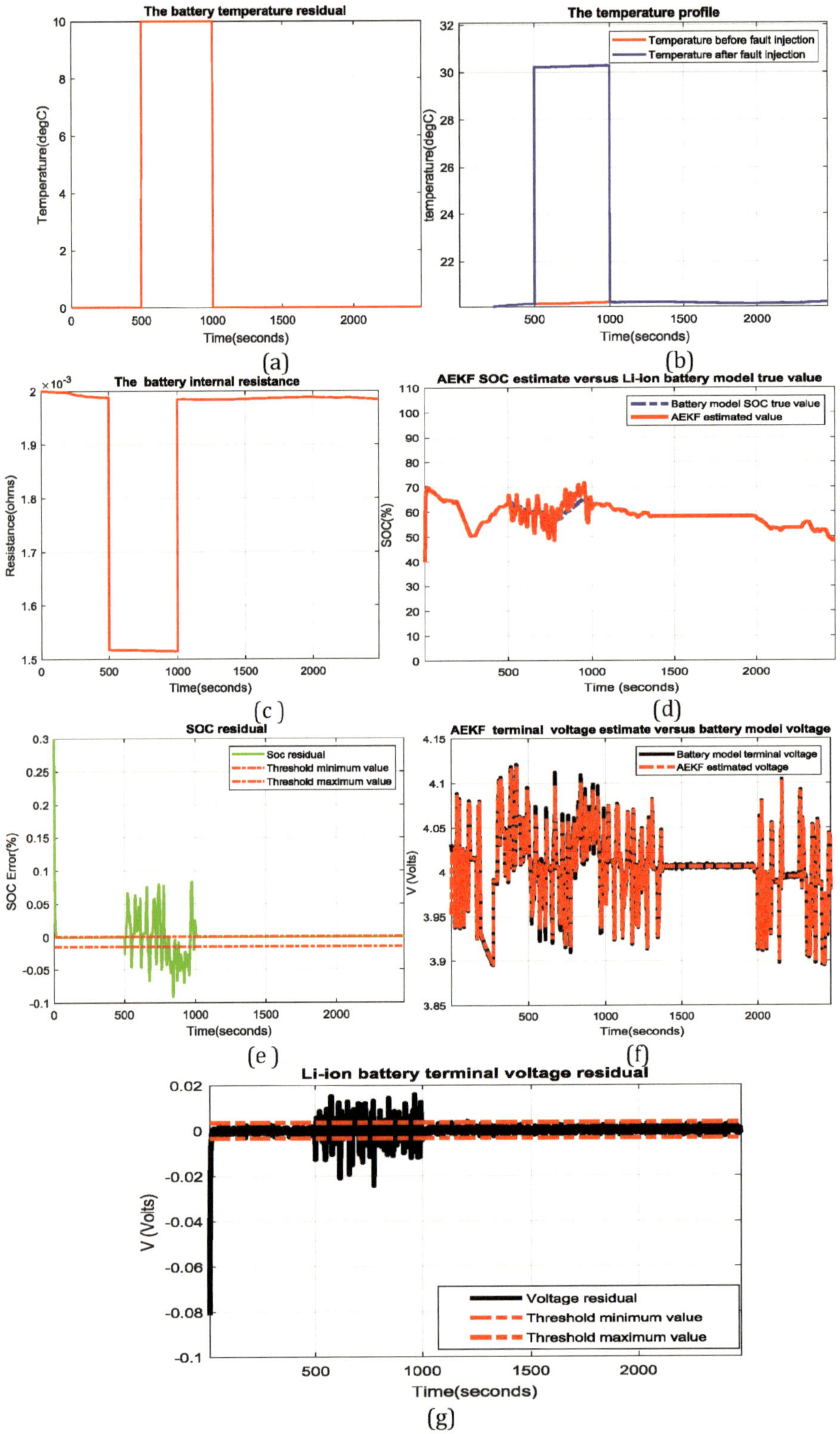

Figure B2.
Third scenario of fault injection: (a) bias fault injection; (b) temperature profile; (c) temperature effect on battery internal resistance Rin; (d) AEKF SOC estimate versus ECM battery model SOC true value; (e) SOC residual; (f) AEKF terminal voltage estimate versus ECM battery model terminal voltage true value; (g) battery terminal output voltage residual.

Author details

Nicolae Tudoroiu[1*], Mohammed Zaheeruddin[2], Roxana-Elena Tudoroiu[3] and Sorin Mihai Radu[3]

1 John Abbott College, Sainte-Anne-de-Bellevue, Quebec, Canada

2 Concordia University from Montreal, Montreal, Canada

3 University of Petrosani, Petrosani, Romania

*Address all correspondence to: ntudoroiu@gmail.com

IntechOpen

© 2020 The Author(s). Licensee IntechOpen. This chapter is distributed under the terms of the Creative Commons Attribution License (http://creativecommons.org/licenses/by/3.0), which permits unrestricted use, distribution, and reproduction in any medium, provided the original work is properly cited.

References

[1] Farag, M. Lithium-Ion Batteries. In: Modeling and State of Charge Estimation, (thesis) Ontario, Canada: McMaster University of Hamilton, 169 p., 2013.

[2] Tudoroiu R-E, Zaheeruddin M, Radu S.M, Tudoroiu N. Louis Romeral Martinez, Miguel Delgado Prieto, editors. In: New Trends in Electrical Vehicle Powertrains. New Trends Electr. Veh. Powertrains 2019, 4, p. 55–81 DOI: 10.5772/intechopen.76230.ch4.

[3] Plett, GL. Extended Kalman Filtering for Battery Management Systems of LiPB-Based HEV Battery Packs - Part 1. In: Modeling and Identification, Power Sources, Elsevier B.V., 2004, 134, p. 252–261, DOI: 10.1016/j.jpowsour.2004.02.031.

[4] Zhang R, Xia B, Li B, Cao L, Lai Y, Zheng W, Wang H, Wang W. State of the Art of Li-ion Battery SOC Estimation for Electrical Vehicles. In: Energies 2018, 11, 1820.

[5] Behrooz Safarinejadian, Elham Kowsari. Fault detection in non-linear systems based on GPEKF and GP-UKF algorithms. In: Systems Science & Control Engineering- An Open Access Journal, 2(1), 610–620, 2014.DOI: 10.1080/21642583.2014.956843.

[6] James J. Q. Yu, Yunhe Hou, Albert Y. S. Lam, Victor O. K. Li. Intelligent Fault Detection Scheme for Microgrids With Wavelet-Based Deep Neural Networks. In: IEEE TRANSACTIONS ON SMART GRID, 2019, 10 (2), p. 1694–1703. DOI: 10.1109/TSG.2017.2776310.

[7] Chandra Shekar S, Ravi Kumar G, Svnl Lalitha. Wavelet Based Transient Fault Detection and Analysis of Microgrid Connected Power System. In: International Journal of Power Systems, 2016, 1, p.46–52.

[8] Tudoroiu, N, Radu S.M, Tudoroiu R-E, Kecs, W, Ilias, N, Elefterie, L. 1-D Wavelet Signal Analysis of the Actuators Nonlinearities Impact on the Healthy Control Systems Performance. In: ASTESJ, 2017, p. 2(3), 1693–1710.

[9] MATLAB R2020a. Multisignal 1-D Wavelet Analysis [Internet].2020. Available from:https://www.mathworks.com/help/wavelet/examples/multisignal-1-d-wavelet-analysis.html [Accessed: 2020 August 4].

[10] Chun-Lin, Liu, A Tutorial of the Wavelet transform. [Internet]. 2013, p.1–71. Available from: http://people.duke.edu/~hpgavin/SystemID/References/Liu-WaveletTransform-2010.pdf [Accessed: 2020 August 3].

[11] Tudoroiu N, Zaheeruddin M, Tudoroiu R-E. Real Time Design, and Implementation of State of Charge Estimators for a Rechargeable Li-ion Cobalt Battery with Applicability in HEVs/EVs-A comparative Study. Energies 2020, 13, 2749; DOI:10.3390/en13112749.

[12] Misity M, Misity Y, Oppenheim G, Poggi J-M. MATLAB Wavelet Toolbox. Getting starting guide. The MathWorks, Inc., 2017.

[13] Mallat St. A Wavelet Tour of Signal Processing. The Sparse Way, 3rd ed. Academic Press: Elsevier, Burlington, MA, USA, 2009, 805 p. ISBN 13: 978–0-12-374370-1.

[14] Lesecq S, Gentil S., Fagarasan I. FAULT ISOLATION BASED ON WAVELET TRANSFORM, Control Engineering and Applied Informatics (CEAI), 9(3), p. 51–58, 2007.

[15] Fang Liu. Data-Based Fault Detection and Isolation Methods for a Nonlinear Ship Propulsion System,

Master of Applied Science Thesis, School of Engineering Science, July 2004, Simon Fraser University, Burnaby, BC, Canada.

[16] Schremmer C. Kerstin. Multimedia Applications on the Wavelet Transform, PhD Thesis, University from Mannheim, 2001.

[17] MATLAB R2020a. Wavelet Toolbox-Examples [Internet]. 2020. Available from: https://www.mathworks.com/examples/wavelet [Accessed 2020 August 5].

[18] Paiva H.M, Duarte M.A.Q, Galvano R.K.H, Hadjiloucas S. Wavelet based detection of changes in the composition of RLC networks, Dielectrics, Journal of Physics: Conferences Series 472, 2013.

[19] Paiva H.M, Galvano R.K.H, Rodrigues L. A wavelet-based multivariable approach for fault detection in dynamic systems, Sba Controle & Automação, Natal, 20(4), p .455–464, 2009.

[20] MATLAB R2020a Wavelet Toolbox Documentation. Orthogonal and Biorthogonal Filter Bank [Internet]. 2020. Available from : https://www.mathworks.com/help/wavelet/ug/orthogonal-and-biorthogonal-filter-banks.html [Accessed 2020 August 6].

[21] Asman S H, Aziz N F Ab, Kadir M Z A Ab, Amirulddin U A Ungku, Izadi M. Determination of Different Fault Features in Power Distribution System Based on Wavelet Transform, International Journal of Recent Technology and Engineering (IJRTE) , 8 (4), 2019. DOI: 10.35940/ijrte.D5104.118419.

[22] MATLAB R2020a Wavelet Toolbox Documentation. Wavelet Denoising. Available from: https://www.mathworks.com/help/wavelet/ug/wavelet-denoising.html [Accessed 2020 August 9].

[23] MATLAB R2020a Wavelet Toolbox Documentation. Practical Introduction on Multiresolution Analysis. 2020. Available from: https://www.mathworks.com/help/wavelet/examples/practical-introduction-to-multiresolution-analysis.html [Accessed 2020 August 10].

[24] Tudoroiu R-E, Zaheeruddin M, Tudoroiu N, Radu S-M, SOC Estimation of a Rechargeable Li-Ion Battery Used in Fuel-Cell Hybrid Electric Vehicles—Comparative Study of Accuracy and Robustness Performance Based on Statistical Criteria. Part I: Equivalent Models, Batteries 2020, 6(3), 42; DOI: https://doi.org/10.3390/batteries6030042.

[25] Tudoroiu R-E, Zaheeruddin M, Tudoroiu N, Radu S-M, SOC Estimation of a Rechargeable Li-Ion Battery Used in Fuel-Cell Hybrid Electric Vehicles—Comparative Study of Accuracy and Robustness Performance Based on Statistical Criteria. Part II: SOC Estimators, Batteries 2020, 6(3), 42; DOI: Batteries 2020, 6(3), 41; https://doi.org/10.3390/batteries6030041.

[26] Chao Wu, Chunbo Zhu, Yunwang Ge, Yongping Zhao. A Review on Fault Mechanism and Diagnosis Approach for Li-Ion Batteries. In: Hindawi Publishing Corporation, Journal of Nanomaterials, 2015, http://dx.doi.org/10.1155/2015/631263.

Section 4

Wavelet Theory and Internet of Things (IoT)

Chapter 8

Industrial IoT Using Wavelet Transform

Mohamed Tabaa, Safa Saadaoui, Mouhamad Chehaitly, Aamre Khalil, Fabrice Monteiro and Abbas Dandache

Abstract

For many years now, communication in the industrial sector has been characterized by a new trend of integrating the wireless concept through cyber-physical systems (CPS). This emergence, known as the Smart Factory, is based on the convergence of industrial trades and digital applications to create an intelligent manufacturing system. This will ensure high adaptability of production and more efficient resource input. It should be noted that data is the key element in the development of the Internet of Things ecosystem. Thanks to the IoT, the user can act in real time and in a digital way on his industrial environment, to optimize several processes such as production improvement, machine control, or optimization of supply chains in real time. The choice of the connectivity strategy is made according to several criteria and is based on the choice of the sensor. This mainly depends on location (indoor, outdoor, ...), mobility, energy consumption, remote control, amount of data, sending frequency and security. In this chapter, we present an Industrial IoT architecture with two operating modes: MtO (Many-to-One) and OtM (One-to-Many). An optimal choice of the wavelet in terms of bit error rate is made to perform simulations in an industrial channel. A model of this channel is developed in order to simulate the performance of the communication architecture in an environment very close to industry. The optimization of the communication systems is ensured by error correcting codes.

Keywords: industrial IoT, wireless communication, DWPT, IDWPT, many-to-one, one-to-many, industrial channel, ECC

1. Introduction

In recent years, technological developments in wireless communication systems have improved user needs in terms of accessibility, data quantity, intelligent decision making and energy consumption. These technologies are still evolving, thanks to the integration of new techniques to improve the connectivity of billions of objects. These connected objects, whether sensors or actuators, are by nature autonomous physical devices with a limited energy source [1, 2]. They are able to communicate with each other, creating a technological revolution. This revolution is bringing more ambitious innovations in a variety of application areas: medicine, industry, energy, security and others.

For industrial applications, research is focused on creating connected, robotic and intelligent factories to improve current production systems. This interconnection of factories is achieved through the connected systems, in which employees, machines and products collaborate with each other to form the new revolution. At the heart of this revolution, the Industrial Internet of Things (IIoT) plays a key role in the development of connectivity for this revolution (**Figure 1**) [3, 4]. In such an industrial environment, propagation differs from other conventional indoor means of communication through its large dimensions and the nature of the objects and obstacles inside. Thus, the industrial environment can be modeled as a fading channel affected by impulsive and Gaussian noise [5, 6].

Given the major advantage of connectivity in the industrial environment, it is necessary to propose wireless, robust and efficient communication architectures inside the factory. The design of these systems differs for each application, taking into account the constraints of the propagation environment. Unlike other traditional indoor environments such as residential buildings or offices, this environment is characterized by its large dimensions and also by the nature of its elements and obstacles. The complexity of the industrial context as well as the noise present in the propagation environment make it necessary to offer a robust wireless communication system to cope with the various disturbances during transmission [5, 7].

In this chapter we focus on applications of industrial communication in a high noise industrial environment. In this work, a multi-user wireless communication system is proposed, characterized by two distinct modes of operation. The first mode provides "Many-To-One" (MtO) communication between several transmitters and a single receiver. The second mode allows one transmitter sensor to send to several receivers in One-To-Many (OtM) mode. These modes of communication illustrate the links between the first three levels of the CIM (Computer-Integrated Manufacturing) pyramid, of which this pyramid illustrates the industrial model on 5 levels. The proposed communication architecture is based on the transformation of wavelet packets The use of the wavelet transform in this context consists, on the one hand, in generating several forms of impulses via their synthesis and being able to simply assign them to each user, and, on the other hand, in the reconstitution of these impulses by the receiver, thus providing an analysis method that is simple to implement and effective. These techniques of analysis and synthesis constitute the major advantage of the wavelet transform for pulse modulations.

An optimal choice of the wavelet in terms of binary error rate is made to perform simulations in an industrial channel. A model of this channel is developed in order to simulate the performance of the communication architecture in an

Figure 1.
IIoT communication in the context of smart factory.

environment very close to industry. Naturally, the optimization of the communication systems is ensured by error correcting codes, this is how we proceeded. We have optimized the performance of our system architecture through conventional channel coding.

This chapter will be presented as follows: in the second part, the state of the art concerning Industry 4.0 and IIoT is developed. The study concerning the discrete wavelet packet transform is discussed in the third part. The part describes the functioning of the proposed architecture in two modes MtO and OtM. the industrial channel model is discussed in the fifth part. The sixth part presents the simulation results and towards the end a conclusion.

2. State of art

2.1 Industry 4.0

Industry 4.0 is characterized by a new way of organizing plants to put an end to complex hierarchical structures. Therefore, ICT techniques must be merged with industrial technologies. In Industry 4.0, embedded systems, IoT and CPS technologies link virtual space to the physical world to give birth to a new connected generation of so-called "intelligent" factories. These factories are capable of more efficient allocation of production resources, with the main objectives of customizing products, minimizing time to market and improving business performance. This opens the way to a new mode of industrial transformation. The concept of Industry 4.0 was first introduced at the Hanover Industrial Technology Fair in 2011, the world's largest technology and industrial trade fair. In 2013, Germany officially adopts the implementation of the concept by the German government's identification of Industry 4.0 in its future projects within its action plan "High-Tech Strategy 2020" (**Figure 2**).

It has rapidly evolved as a German national strategy based on 4 aspects: Building the CPS network, addressing two main themes based on the plant and intelligent production, thus achieving 3 types of integration: Horizontal, vertical and point-to-point. The result is that German industry has welcomed the initiative with open arms. Small, medium and large companies from all sectors participated in the creation of this new era. However, the boost from the government has helped to internationalize the concept of Industry 4.0. In 2014, the State Council of China unveiled its national plan, Made-in-China 2025, inspired by Industry 4.0 and

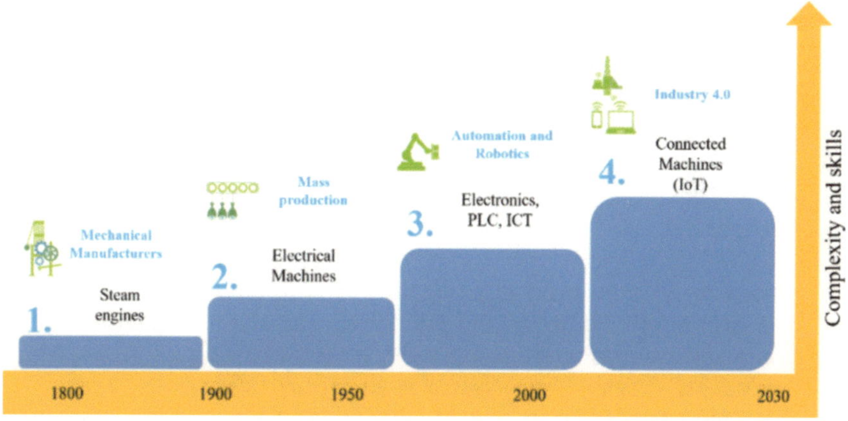

Figure 2.
Evolution of the industry.

designed to improve China's industry globally, integrating digital and industrial technologies. At the same time, several countries have adopted this concept, we cite as an example "the new industrial France" by France, "Industrial internet and advanced manufacturing partnership in USA" by the United States [8, 9].

2.2 IIoT vs. IoT

In the last few decades, technological developments in wireless communication systems have improved user needs in terms of accessibility, data quantity, intelligent decision making and energy consumption. These technologies are still evolving, thanks to the integration of new techniques to improve the connectivity of billions of objects. These connected objects, whether sensors or actuators, are by nature autonomous physical devices with a limited energy source. They are able to communicate with each other, creating a technological revolution. This revolution is bringing more ambitious innovations in a diverse range of applications: medicine, industry, energy, security and others [10, 11].

For industrial applications, research is focused on creating connected, robotic and smart factories to improve current production systems. This interconnection of factories is achieved through the connected systems, in which employees, machines and products collaborate with each other to form the new revolution. At the heart of this revolution, the IIoT plays a key role in the development of connectivity for this revolution. Based on the same concept of IoT, IIoT is based on the use of connected sensors or actuators to improve industrial processes and manufacturing. It integrates intelligence in data processing and analysis to ensure better M2M (Machine-to-Machine) communication. This has been in existence since the integration of electronics in the industrial sector during the third "industry 3.0" revolution. It is now necessary to work on robust communication architectures allowing objects and the technological choice of communication technologies and protocols, in a highly noisy industrial environment, to communicate easily in order to build reliable information for better decision making [12, 13].

General Electric presents the Industrial Internet as a term meaning the integration of complex physical machines with networked sensors and software. The Industrial Internet brings together areas such as IoT, Big Data, machine learning and M2M (Machine to Machine) communication to collect and analyze machine data and use it to adjust operations.

According to the Industrial Internet Consortium IIC, the Industrial Internet connects intelligent devices and machines with people at work, enabling better decisions through advanced analysis that leads to transformational business outcomes. The Industrial Internet covers the non-consumer side of the IoT and applies "internet thinking" to industrial environments.

The Industrial Internet consists of three key elements that together represent the essence of the idea:

- Smart machines: this means connecting machines, fleets, facilities and networks around the world with advanced controls, sensors and software applications.

- Advanced analysis: means combining the power of physics-based analysis, domain expertise, automation and predictive algorithms to understand how machines and systems work.

- People at work: essentially means connecting people at all times to support smarter operations, design and maintenance, as well as high quality of service and safety.

The connection and combination of these three key elements allows companies and economies to benefit from many new opportunities and efficiency gains in several areas. The industrial internet will accelerate productivity growth in the same way that the industrial revolution and the internet revolution have done in the past.

3. Wavelet packet discrete transform

As a matter of principle, the multiresolution analysis in $L^2(R)$ space of the continuous functions of a real variable and an integrable square can be extended to subspaces of it. That is, the same scheme can be applied to the W_j subspaces generated by the previous analysis. Figure shows the hierarchy of the wavelet packet decomposition: it illustrates the principle of wavelet packet decomposition through the analysis of all subspaces [14]. **Figures 3** and **4** illustrate the principle of this decomposition by the discrete wavelet transform.

The analysis used in the wavelet packet transform leads to a decomposition into frequency sub-bands of the input signal. This analysis can be carried out either by the same scale and wavelet functions, which is usually the case, or by different functions. This makes it possible to change the basic functions at each scale. It can be said that perfect reconstruction is ensured by reusing during synthesis, and for a precise resolution of the base functions combined with those used during the analysis at this same resolution. The procedure of analyzing subspaces of signal detail in addition to the approximation subspaces is generally referred to as "wavelet packet analysis."

Because this transform presents a good symmetry of structure which results in identical sampling frequencies on all the inputs of the synthesis filter bank, and on all the outputs of the analysis filter bank, our choice was directed towards an implementation of the discrete wavelet packet transform. This approach will facilitate the generation of the pulses and be able to identify their content (nature of the transmitter, or value of the transported data) [15].

3.1 Packet wavelet transform

The main objective of the wavelet packet decomposition is to extend the construction of a new base from all generated subspaces. By definition, the

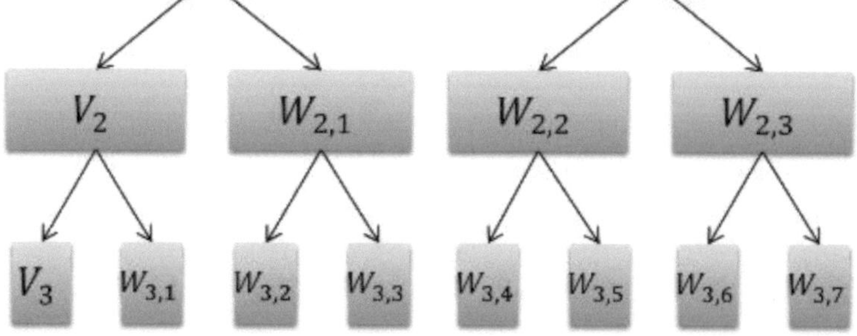

Figure 3.
Discrete wavelet transform decomposition principle.

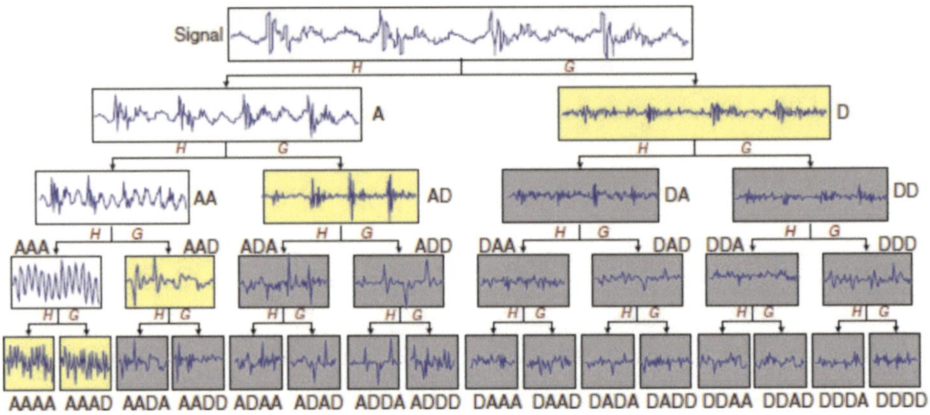

Figure 4.
Scale 4 decomposition procedure by discrete wavelet packet transform. H represents the low-pass filter and G the high-pass filter.

multiresolution analysis of an approximation space V_j is decomposed into two lower resolution spaces V_{j+1} and W_{j+1} [16]. Therefore, this division is obtained by transforming the base $\left\{\phi_j(2^{j+1}t-k)\right\}_{k\in Z}$ de V_j in two orthogonal bases: $\left\{\phi_{j+1}(2^{j+1}t-k)\right\}_{k\in Z}$ de V_{j+1} et $\left\{\psi_{j+1}(2^{j+1}t-k)\right\}_{k\in Z}$ for W_{j+1}.

Note P this tree in which each node corresponds to a subspace P_j^n which admits an orthogonal base $\left\{P_j^n(t-k)\right\}_{k\in Z}$. At a resolution level j we will have:

$$P_j^n = P_{j+1}^{2n} \oplus P_{j+1}^{2n+1.} \tag{1}$$

The functions obtained are wavelet packets that are recursively determined by:

$$P_{j+1}^{2n}(t) = \sqrt{2}\sum_k h(k)p_{j+1}^{2n}(2t-k) \tag{2}$$

$$P_{j+1}^{2n+1}(t) = \sqrt{2}\sum_k g(k)p_{j+1}^{2n}(2t-k) \tag{3}$$

It should be noted that:

p_0^0 represents the scaling function and p_0^1 the associated wavelet via multiresolution analysis and noted respectively ϕ and ψ.

The filters h_n and g_n are respectively the low-pass and high-pass filters represented by quadrature mirror filters, and linked by the following equation:

$$\mathrm{G(n)} = (-1)^{\mathrm{n}}\mathrm{h}(1-\mathrm{n}) \tag{4}$$

The impulse response of the filters satisfies the following conditions:

$$\sum_n h(n-2k)h(n-2l) = \delta_{kl} \,\&\sum_n h(n) = \sqrt{2} \tag{5}$$

$$\sum_n g(n-2k)g(n-2l) = \delta_{kl} \,\&\sum_n g(n) = 0 \tag{6}$$

DOI: http://dx.doi.org/10.5772/intechopen.93879

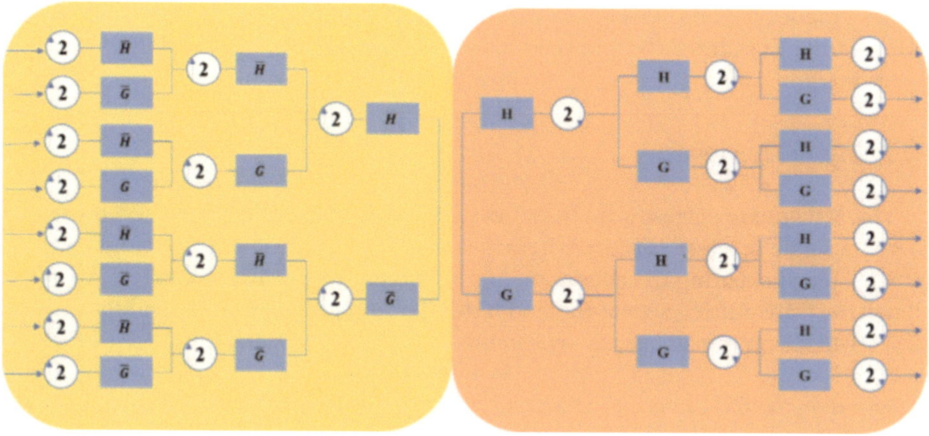

Figure 5.
IDWPT-based transmitter and DWPT-based receiver.

3.2 Decomposition and reconstruction

Recall that each function f(t) in the $L^2(R)$ space can be decomposed on the basis of functions $\{p^n_{j,k}(t)$, where (j,k)ϵZ*Z} as follows:

$$F(t) = \sum_{n,k} a^n_{j,k} p^n_{j,k}(t) \tag{7}$$

with j the depth of decomposition, k the time index, and n the frequency index equivalent to the wavelet number.

The coefficients $a^n_{j,k}$ at a given scale j are expressed as a scalar product of the signal to be analyzed and the analyzing function:

$$a^n_{j,k} = \left\langle f, p^n_{j,k} \right\rangle = \int_{-\propto}^{+\propto} f(t) p^n_{j,k}(t) dt \tag{8}$$

The wavelet packet decomposition is shown in **Figure 5**. In this example, the wavelet packet analysis of the function f is performed with a depth of 4.

The set of coefficients $a^n_{j,k}$ constitutes the discrete wavelet packet transform (DWPT) of f(t) and its inverse transform (IDWPT) is given by:

$$a^n_{j,k} = \sum_{i \in Z} h_{k-2i} a^{2n}_{j,k} + \sum_{i \in Z} g_{k-2i} a^{2n+1}_{j,k} \tag{9}$$

The wavelet packet transform simply consists of filtering the signal using a low-pass filter h_n and a high-pass filter g_n. As for synthesis, it is a regrouping of the signals into a single signal that represents the signal already analyzed. These two approaches give rise to filter banks that check the following conditions:

$$\overline{h}_n = h_{-n} \text{ et } \overline{g}_n = g_{-n} \tag{10}$$

4. Proposed communication architecture

In this chapter, two multi-user operating modes have been studied and tested: the "Many-to-One" mode (MtO) and the "One-to-Many" mode (OtM). The choice of these modes depends essentially on the existence in the current communication architectures (master-slave, bidirectional), in order to facilitate adaptation for a better integration [17–19].

4.1 Many-to-One mode

The MtO mode corresponds to multi-sensor communication from several sensors to a single receiver (**Figure 6**). Each transmitting sensor is in the form of an IDWPT block ensuring the activation of a single input for this transmitter, which allows the transmitting sensor to be identified already. Therefore, each input of the IDWPT block on transmission corresponds to a single output of the DWPT block on reception.

Based on the CIM pyramid (**Figure 7**), this mode of communication corresponds to a communication from level 0 and 1 to level 2. In this mode, data from one or more low flow sensors are transmitted at the same time to the same receiver, and

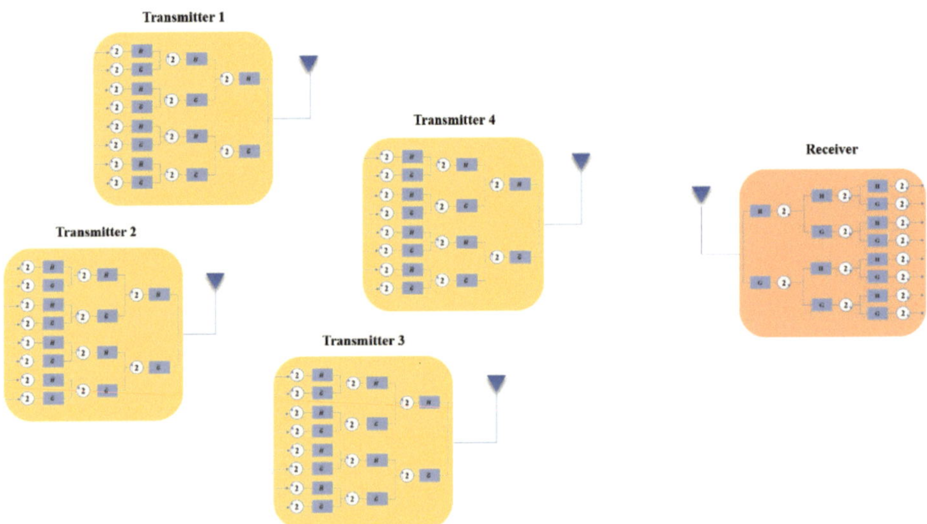

Figure 6.
Many-to-One mode.

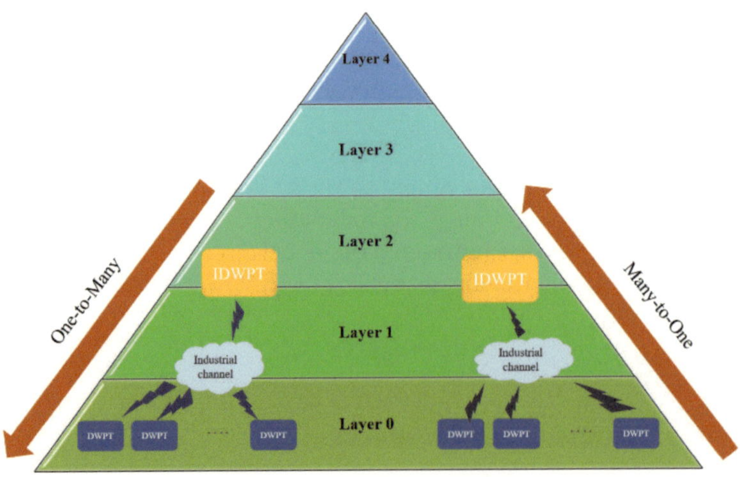

Figure 7.
Operation of MTO and MtO modes in the CIM pyramid.

DOI: http://dx.doi.org/10.5772/intechopen.93879

the activation of one of the inputs generates the activation of a user. **Figure 8** illustrates an 8-input architecture corresponding to 8 potential sensors (scale 3). Therefore, each transmitter uses a single input that is different from the other inputs. The pulse shape of each activated input is different from the waveforms of the other inputs, the other non-activated inputs will be set to zero.

The DWPT-based receiver receives the data stream from all transmitters at the same time. However, each sensor is identified by a unique filter output at the receiver that represents the same input at the receiver. This mode has a higher bandwidth occupancy than single user mode because each user (input enabled) will occupy a separate sub-band. This will result in frequency selectivity of the channel due to interference between users, for which it will be necessary to protect the

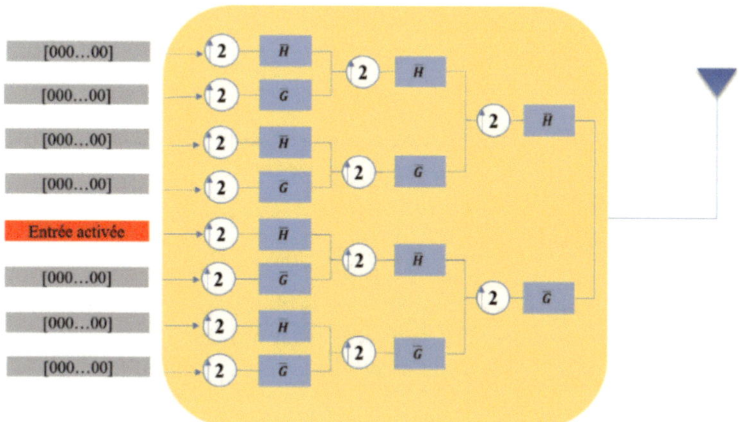

Figure 8.
Transmitter operation in MtO mode.

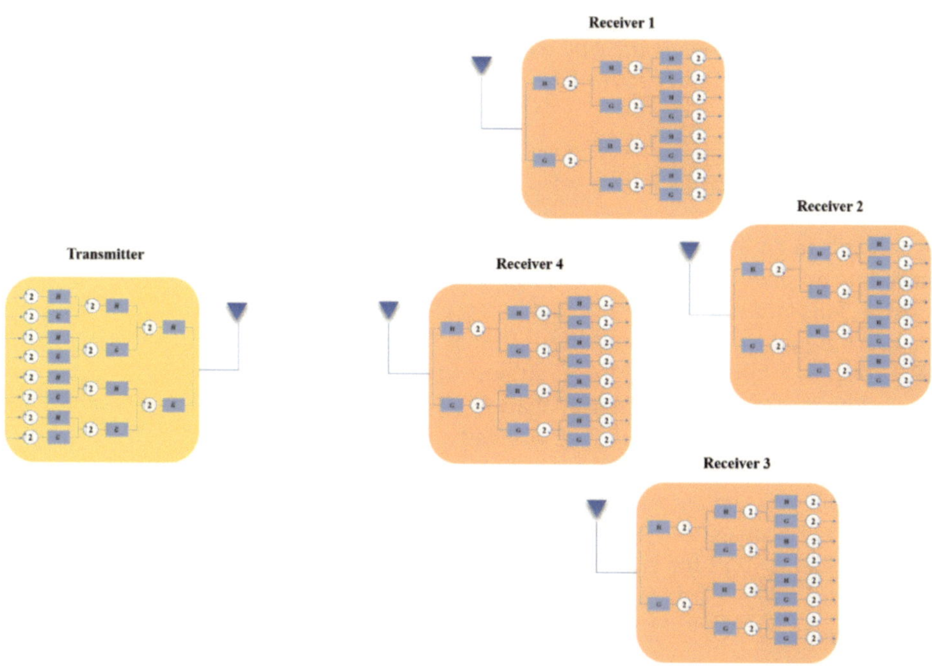

Figure 9.
One-to-Many mode.

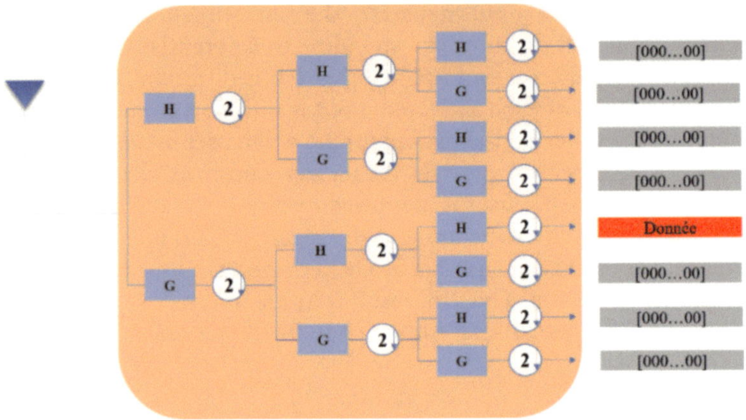

Figure 10.
Receiver in one-to-many mode.

transmitted data as much as possible. Nevertheless, this will allow synchronous communication of several sensors to the same receiver.

4.2 One-to-Many mode

For the OtM mode, an IDWPT transmitter is characterized by n inputs, capable of sending information to m DWPT receivers with n outputs each.

Note that the information sent via input (i) is retrieved at output (i). This is the inverse mode of the MtO mode where the equipment's of levels 1 and 2 of the CIM pyramid send the same information to the level sensors. This mode is equivalent to the master-slave architecture in a conventional industrial communication system. Although the data rate of the transmitted data is generally low, the reception of information from several sensors creates spatial diversity that allows the data to be retrieved by at least one receiver. **Figure 9** illustrates the transmission of data from a single sensor to 4 receivers. The data sent will be detected in the 5th output of the 4 receivers, as shown in **Figure 10**.

5. Industrial channel

Signals in an industrial environment are subject to several disturbances due to propagation phenomena. These disturbances significantly degrade system performance. This environment is affected by very complex noise and interference caused by machine temperatures, vibrations, metal structures and heavy machinery [SHA09]. In addition, the signal is subject to attenuation and shadowing effects caused by abstractions in the propagation channel. The mobility of equipment and people in the wireless medium can also cause time-varying effects. These effects can significantly destroy the information exchanged and thus degrade any communication system performance in the industry [CHE16]. Therefore, it is necessary to estimate the propagation channel in order to design and evaluate the entire wireless transmission system for industrial applications.

5.1 Fading

For wireless propagation in an industrial context, the received information is subject to attenuation and fading effects, of which the expression of the received signal is:

$$y(t) = h(t) * s(t) + n(t) \tag{11}$$

Where, h(t) is the channel impulse response, s(t) is the transmitted signal and n(t) is the additive noise.

In a factory, sensors/actuators are usually arranged according to the production system configuration. Measurements of narrowband and broadband indoor channels have been performed through research in several industrial environments [20], and have shown that the time impulse response h(t) at a fixed location in an industrial context follows a reduced exponential distribution [21, 22]. This distribution depends mainly on the delay and power of each channel, which is shown in Saleh Valenzuela's model [23]. The delay spread of the channels can be determined from the impulse response as a function of the transmission frequency and the LOS (Line-Of-Sight) or NLOS (Non-Line-Of- Sight) configurations. Thus, the objective of the research work is to validate the IDWPT/DWPT-based architecture under a simulated industrial channel, and we then generated a channel impulse response based on the measurements from the work [24, 25] for both LOS and NLOS configurations at 2.4 GHz. The simulated channel impulse response includes 10 significant paths (**Figure 11**).

In order to represent a channel fading phenomenon, all paths follow the same statistical distribution [26]. The time envelope of the received signal follows the Rician statistical distribution in the LOS scenario and the Rayleigh distribution in the NLOS case.

$$P(x) = \frac{x}{\sigma^2} exp\left(-\frac{x^2 + K^2}{2\sigma^2}\right) I_0\left(\frac{Kx}{\sigma^2}\right) \tag{12}$$

With $I_0(x)$ is the Bessel function changed to zero order. K is the shape parameter called Rician factor. For $K = 0$, P(x) converges to the Rayleigh distribution.

5.2 Noise

In the case of wireless communication systems, the noise added to the received signal is White Gaussian Noise (WGN additive). In an industrial environment, the signals will be affected by noise, which is represented as impulsive noise from motors, regulators, electrical equipment and others. However, the industrial noise $n(t)$ in equation will be modeled as a superposition of AWGN w(t) and impulsive

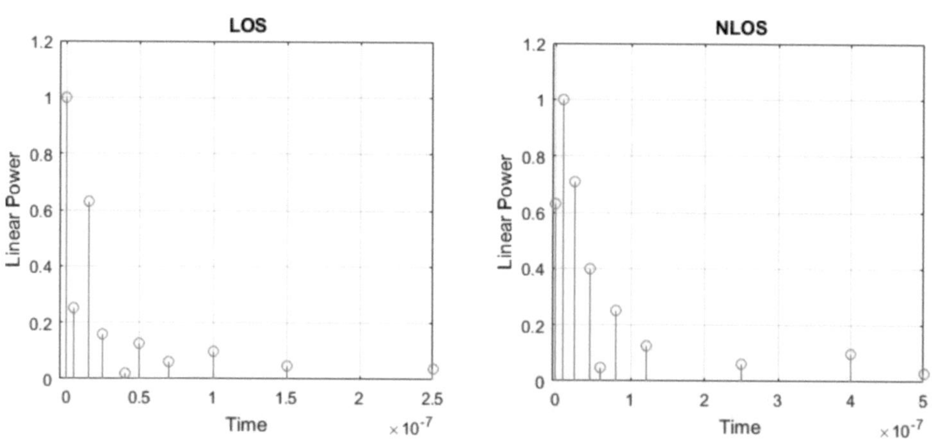

Figure 11.
Simulated channel impulse response.

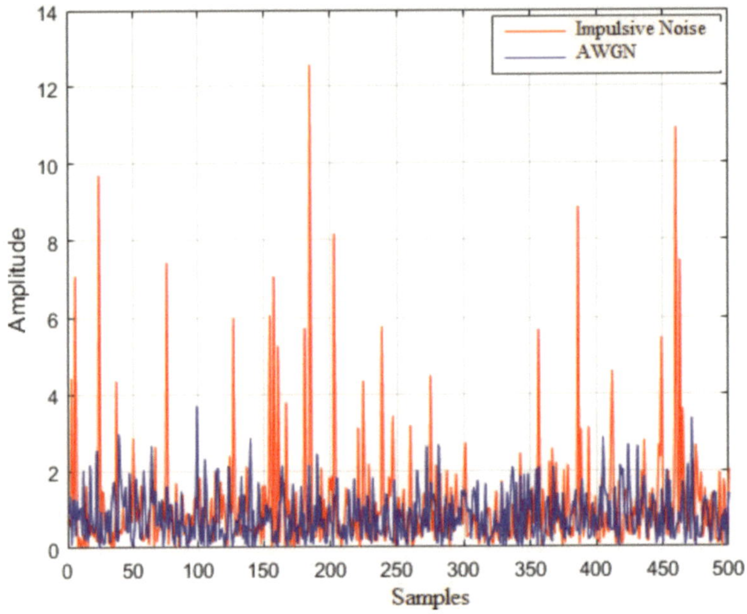

Figure 12.
Industrial noise with a scale factor R = 50.

noise $i(t)$ having a very high variance. Then, $i(t)$ is modeled as a two-state first-order Markov process thus describing the typical impulsive noise [27] (**Figure 12**).

$$n(t) = w(t) + i(t) \tag{13}$$

where $w(t)$ and $i(t)$ are zero-mean Gaussian processes whose probability density functions are respectively:

$$P[w(t)] = \frac{1}{\sqrt{2\pi\sigma^2}} exp\left[-\frac{w(t)^2}{2\sigma^2}\right] \tag{14}$$

$$P[i(t)] = \frac{1}{\sqrt{2\pi R\sigma^2}} exp\left[-\frac{i(t)^2}{2R\sigma^2}\right] \tag{15}$$

With $R \geq 1$ is a scale constant of the amplitude of the impulse noise. The higher the amplitude, the greater the noise. For our simulations, we use R = 50 which corresponds to significant impulse noise.

6. Simulations, results and performances

In this section, we will present the simulation results of the IDWPT/DWPT architecture under a noisy industrial channel. All the simulations presented in this chapter are performed under MATLAB.

6.1 Simulations and results

The proposed system is based on a multi-user IDWPT/DWPT architecture for 2^n sensors/actuators in an industrial environment. The transmitters are based on the IDWPT implementation in the form of synthesis filter banks, and the receivers are based on DWPTs implemented as analysis filter banks. The industrial channel is described as a Rician fading channel for the LOS configuration and a Rayleigh fading channel for the NLOS configuration at the 2.4GHz frequency affected by impulsive noise. In our simulations, we choose the "Symlet" wavelet that has demonstrated the lowest bit error rate for the IDWPT/DWPT architecture under an AWGN channel (**Figure 13**).

In the case of the MTO mode in multi-sensor configuration, the frames for each user are 16 bits long and randomly generated. This data configuration is due to the fact that sensors in industrial environments transmit short data packets. These data frames are pulse modulated and each transmitter is identified by a unique signal. **Figure 14** shows the signals from 4 different sensors (1, 5, 12 and 16) in an architecture with 16 transmitter sensors. The 16 generated signals are all different from each other because the binary data at the input of each filter are different.

Based on the effect of channel fading due to delay propagation in addition to AWGN noise for the LOS and NLOS configurations, it is clear that the multipath effect disturbs the signals of the different users and thus causes interference between them. The proposed architecture allows signal detection at reception for all users as shown in **Figure 15** for a SNR (Signal to Noise Ratio) greater than 20 dB [27–33].

With fading effects, and the addition of industrial noise composed of Gaussian noise and impulse noise, the bit error rate is shown in **Figure 15**. The communication architecture converges more slowly and performance decreases, but it allows the full information of an SNR up to 35 dB. In the case of industrial noise, the data may be completely lost if the effects of the channel are not properly taken into account.

In the case of the OtM mode, a single transmitter based on DWPT with n inputs sends data to m receivers based on DWPT with n outputs each. The principle of this mode is to activate only one input (i) of the transmitter and force the others to zero. When receiving, the data will be detected at output (i) of each receiver. The data are modulated through pulse modulation using a "Symlet" pulse. **Figure 16** shows

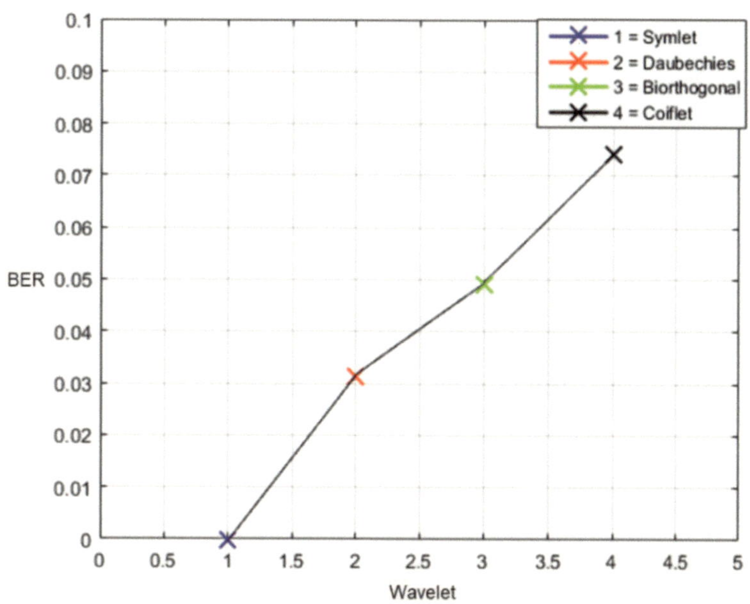

Figure 13.
Performance of four wavelets.

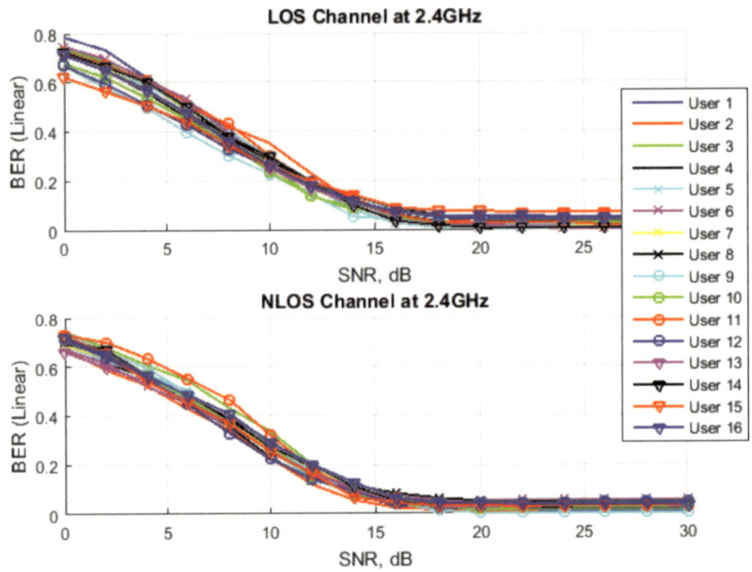

Figure 14.
BER/SNR on a fading channel with AWGN noise for MtO mode.

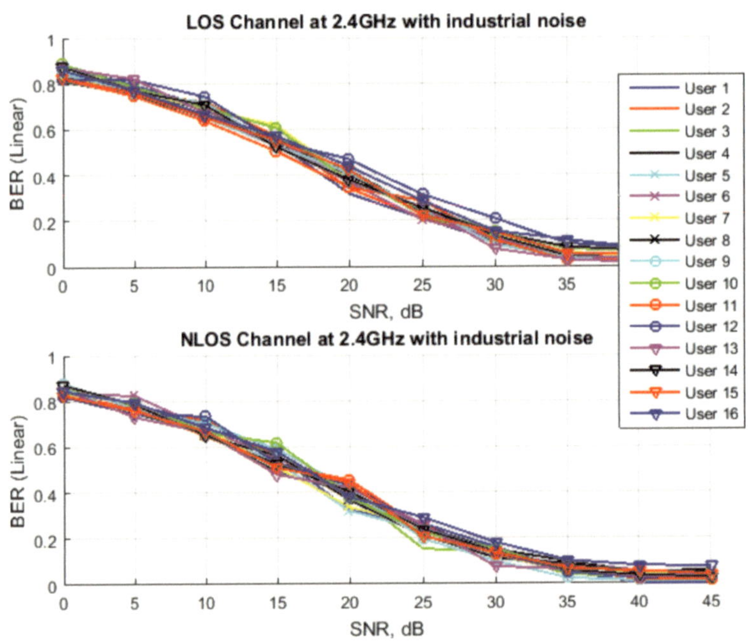

Figure 15.
BER/SNR on a fading channel with industrial noise for MTO mode.

the signals received at the 4 receivers. The data is recovered at the 7th output corresponding to the activated input.

Based on the fading channel and AWGN noise for LOS and NLOS configurations, the architecture detects the signal on reception. According to the simulation results shown in **Figure 16** the transmitted signal is detected at the receiving sensor array for the LOS and NLOS channels at 2.4 GHz. Detection is virtually error-free above 20 dB. Some differences between the LOS and NLOS configurations are

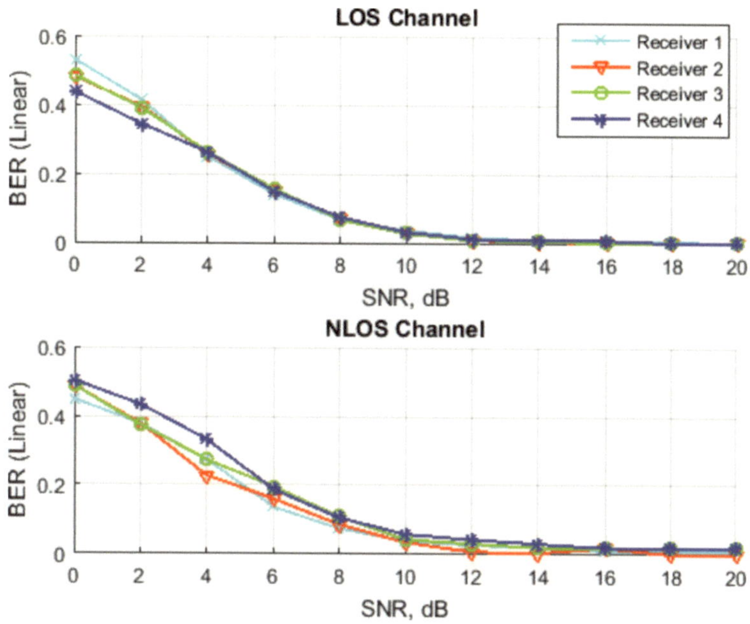

Figure 16.
BER/SNR on a 2.4 GHz fading channel with AWGN noise for OTM mode.

detected from an SNR of 14 dB. This is mainly due to the effects of channel fading and channel dispersion which must be corrected using channel coding during transmission. Taking into account the effect of industrial noise, the communication architecture allows the full detection of 30 dB SNR information as shown in **Figure 17**. The difference in error rate is very large and depends on the propagation channel.

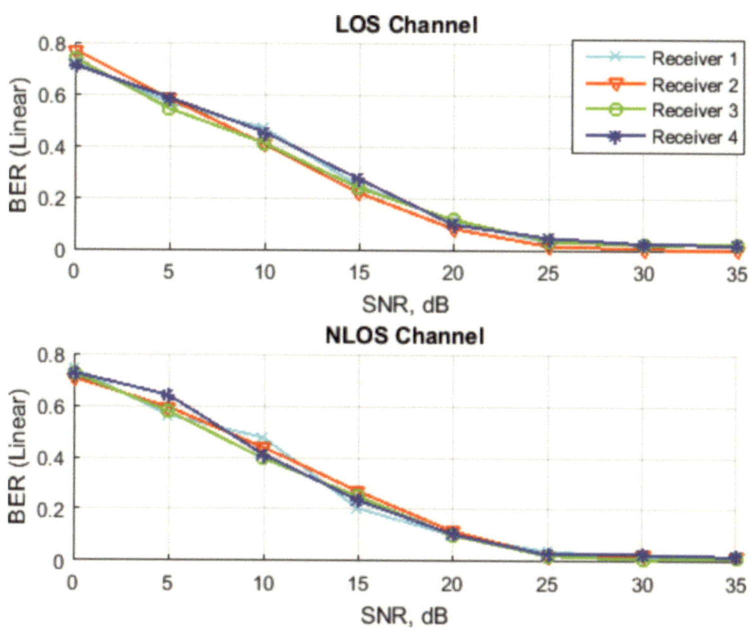

Figure 17.
BER/SNR on a fading channel with industrial noise for OTM mode.

6.2 Performances: ECC

To improve the reliability of the architecture compared to the industrial fading channel, we propose to add an error-correcting channel code on the transmitter side (**Figure 18**). We use two coding techniques: a convolutional code and RS (Reed Solomon) codes. For the convolutional code, we choose an encoder using a trellis diagram with a generating polynomial matrix of having a constraint length of 7 and a code rate = 1/2. On the receiver side, we use a Viterbi decoder [28–30].

As for the Reed Solomon encoder, we use an RS(31,17) with 31 code word symbols and 17 message symbols based on the length of the transmitted data.

As shown in **Figure 19** for an architecture with 8, 16 and 32 users on an industrial channel with AWG noise fading, the error correction code improves the robustness of the architecture against channel fading as a function of the number of sensors used. For a better graphical representation, we have shown the results for only 4 users in each case; for 8 users (user 1, 3, 5 and 7), for 16 users (user 1, 6, 12, 16) and for 32 users (user 1, 12, 22 and 30).

For a fading channel with AWG noise, the SNR is reduced by 2 dB using both convolutional and RS code for an 8-user (or sensor) architecture, and by 4 dB for RS

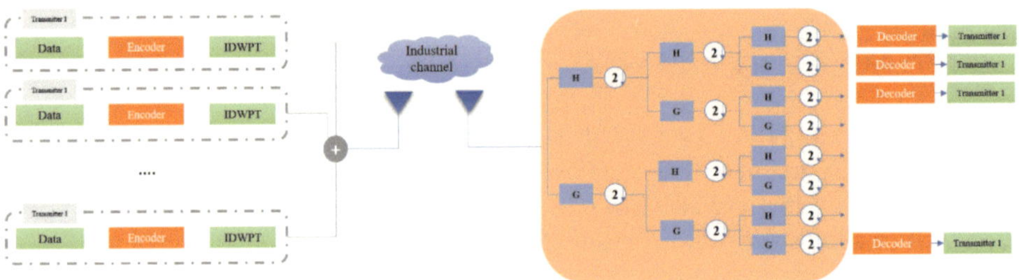

Figure 18.
Architecture for 8 sensors with channel coding.

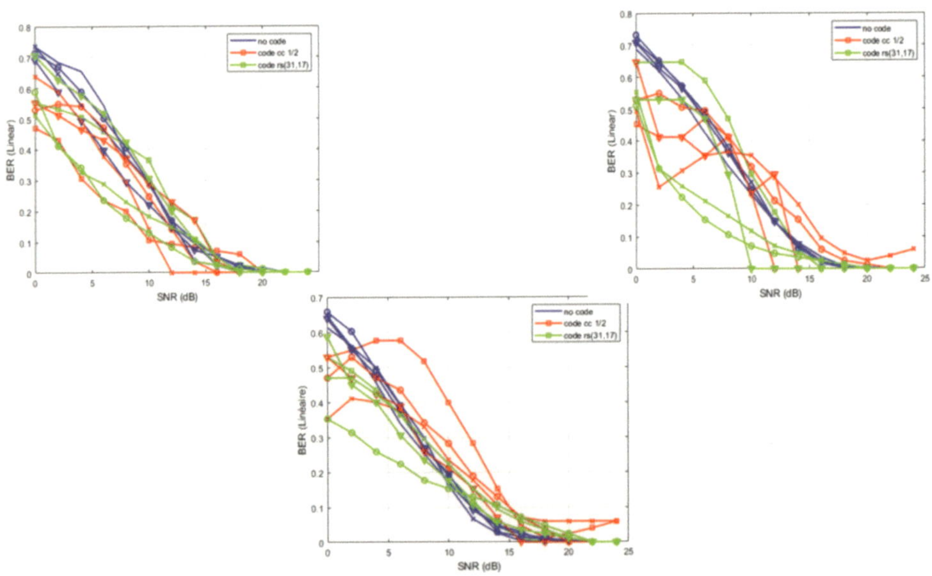

Figure 19.
BER/SNR on a fading channel with AWGN noise for MTO mode.

code in the case of a 32-user use. For a fading channel with industrial noise, the signal-to-noise ratio is reduced by 8 dB for an architecture with 16 users using a convolutional code and by 5 dB for 32 users using an RS code, as shown in **Figure 20**. For a better illustration, **Table 1** shows the different SNR values for a fixed linear bit error rate of 0.1 with or without error-correcting coding [31, 32].

We conclude that for communication over an industrial fading channel, RS coding is optimal for a 32-user architecture. However, convolutional coding is optimal for a 16-user architecture. In the case of an 8-user architecture, the convolutional and RS codes are equal.

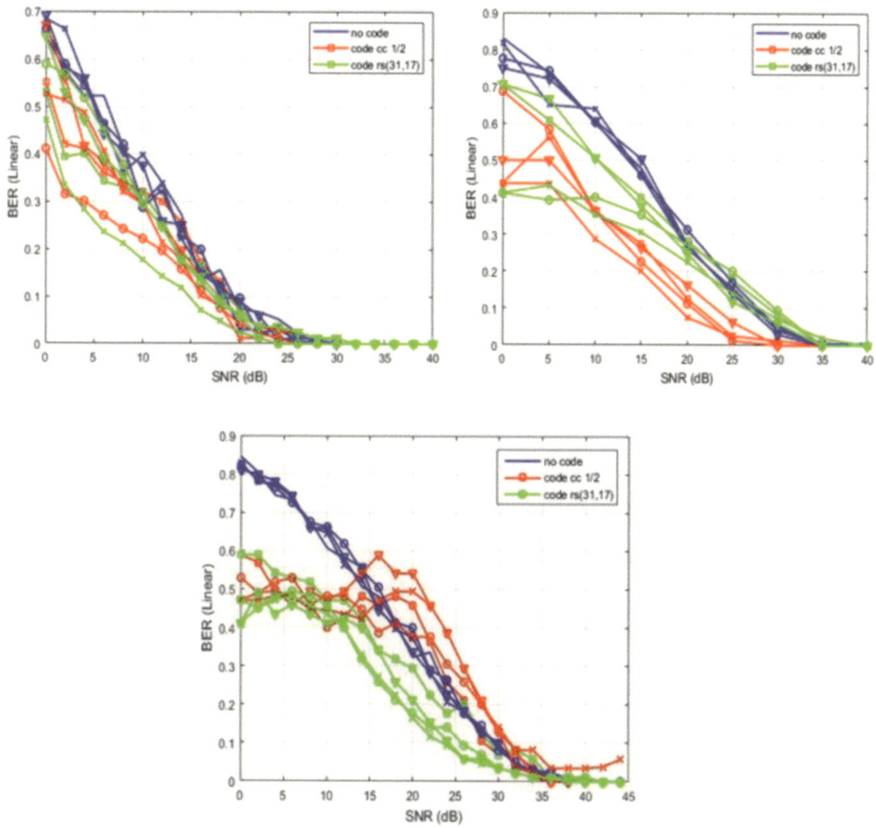

Figure 20.
BER/SNR on a fading channel with industrial noise for MTO mode.

	Number of sensors	No code	Convolutional code 1/2	RS (31,17)
Fading channel with AWGN noise	8	14 dB	12 dB	12 dB
	16	12 dB	14 dB	14 dB
	32	14 dB	12 dB	10 dB
Fading channel with industrial noise	8	20 dB	18 dB	18 dB
	16	28 dB	20 dB	26 dB
	32	30 dB	28 dB	25 dB

Table 1.
System parameters with coding.

7. Conclusion

A robust IIoT multi-user architecture based on IDWPT in transmitter and DWPT in receiver under an industrial channel has been presented in this chapter. The industrial channel was modeled as a fading channel affected by impulse noise combined with AWGN. The wireless sensor network architecture presented, with its two communication modes MtO and OtM, provides better data reception results for a noisy industrial environment. The robustness of the architecture can be improved by using channel coding or industrial noise thresholding at reception. By using a conventional error correction code with a rate of 1/4, the robustness of the MtO mode has been greatly improved and all signals are fully decoded from an 8 dB SNR on a fading channel. In MtO mode, signals are decoded from 6 dB on the same channel. Using an optimal threshold receiver, errors are eliminated by about 25 dB for MtO and OtM modes on a noisy industrial channel. As a perspective, we wish to compare the performance of the proposed architecture with the conventional OFDM communication system.

Author details

Mohamed Tabaa[2*], Safa Saadaoui[2], Mouhamad Chehaitly[1], Aamre Khalil[3], Fabrice Monteiro[3] and Abbas Dandache[3]

1 University of Montpellier, LIRMM Lab, France

2 EMSI Casablanca, LPRI Lab, Morocco

3 University of Lorraine, LGIPM Lab, France

*Address all correspondence to: med.tabaa@gmail.com

IntechOpen

© 2020 The Author(s). Licensee IntechOpen. This chapter is distributed under the terms of the Creative Commons Attribution License (http://creativecommons.org/licenses/by/3.0), which permits unrestricted use, distribution, and reproduction in any medium, provided the original work is properly cited.

References

[1] Wollschlaeger, M.; Sauter, T.; Jasperneite, J. The future of industrial communication: Automation networks in the era of the internet of things and industry 4.0. IEEE Ind. Electron. Mag. 2017, 11, 17–27.

[2] Sasajima, H.; Ishikuma, T.; Hayashi, H. Future IIOT in process automation—Latest trends of standardization in industrial automation, IEC/TC65. In Proceedings of the 54th Annual Conference of the Society of Instrument and Control Engineers of Japan (SICE), Hangzhou, China, 28–30 July 2015; pp. 963–967.

[3] Tabaa, M., Chouri, B., Saadaoui, S., & Alami, K. (2018). Industrial communication based on modbus and node-RED. Procedia computer science, 130, 583-588.

[4] RHOLAM, Oussama, TABAA, Mohamed, et al. Smart Device for Multi-band Industrial IoT Communications. Procedia computer science, 2019, vol. 155, p. 660-665.

[5] LU, Yang. Industry 4.0: A survey on technologies, applications and open research issues. Journal of industrial information integration, 2017, vol. 6, p. 1-10.

[6] LASI, Heiner, FETTKE, Peter, KEMPER, Hans-Georg, et al. Industry 4.0. Business & information systems engineering, 2014, vol. 6, no 4, p. 239-242.

[7] Büchi, G., Cugno, M., & Castagnoli, R. (2020). Smart factory performance and Industry 4.0. Technological Forecasting and Social Change, 150, 119790.

[8] OSTERRIEDER, Philipp, BUDDE, Lukas, et FRIEDLI, Thomas. The smart factory as a key construct of industry 4.0: A systematic literature review. International Journal of Production Economics, 2020, vol. 221, p. 107476.

[9] Skobelev, P. O., and S. Yu Borovik. "On the way from Industry 4.0 to Industry 5.0: From digital manufacturing to digital society." Industry 4.0 2.6 (2017): 307-311.

[10] ZHU, Chunsheng, LEUNG, Victor CM, SHU, Lei, *et al.* Green internet of things for smart world. *IEEE access*, 2015, vol. 3, p. 2151-2162.

[11] ZHANG, Chaofeng, DONG, Mianxiong, et OTA, Kaoru. Enabling Computational Intelligence for Green Internet of Things: Data-Driven Adaptation in LPWA Networking. *IEEE Computational Intelligence Magazine*, 2020, vol. 15, no 1, p. 32-43.

[12] Sisinni, E., Saifullah, A., Han, S., Jennehag, U., & Gidlund, M. (2018). Industrial internet of things: Challenges, opportunities, and directions. *IEEE Transactions on Industrial Informatics*, *14* (11), 4724-4734.

[13] Younan, M., Houssein, E. H., Elhoseny, M., & Ali, A. A. (2020). Challenges and recommended technologies for the industrial internet of things: A comprehensive review. *Measurement*, *151*, 107198.

[14] R.W. Gao, R. Yan, "Wavelets theory and applications for manufacturing", Springer, 2011, ISBN: 978-1-4419-1544-3.

[15] Daubechies, I. Ten Lectures on Wavelets; Society for Industrial and Applied Mathematics: Philadelphia, PA, USA, 1992; 357p.

[16] Mallat, S. A Wavelet Tour of Signal Processing; Academic Press: Cambridge, MA, USA, 1989.

[17] Tabaa, M. A novel transceiver architecture based on wavelet packet

modulation for UWB-IR WSN applications. Wirel. Sens. Netw. 2016, 8, 191–209.

[18] Lakshmanan, M.K.; Nikookar, H. A review of wavelets for digital wireless communication. Wirel. Pers. Commun. 2006, 37, 387–420.

[19] Sauter, T. The three generations of field-level networks—Evolution and compatibility issues. IEEE Trans. Ind. Electron. 2010, 57, 3585–3595.

[20] Shan, Q.; Bhatti, S.; Glover, I.A.; Atkinson, R.; Portugues, I.E.; Moore, P. J.; Rutherford, R. Characteristics of impulsive noise in electricity substations. In Proceedings of the 2009 17th European Signal Processing Conference, Glasgow, UK, 24–28 August 2009; pp. 2136–2140.

[21] Sexton, D.; Mahony, M.; Lapinski, M. Radio channel quality in industrial wireless sensor networks. In Proceedings of the 2005 Sensors for Industry Conference, Houston, TX, USA, 8–10 February 2005; pp. 88–94.

[22] Luo, S.; Polu, N.; Chen, Z.; Slipp, J. RF channel modeling of a WSN testbed for industrial environment. In Proceedings of the 2011 IEEE Radio and Wireless Symposium, Phoenix, AZ, USA, 16–19 January 2011; pp. 375–378.

[23] Saleh, A.A.M.; Valenzuela, R. A statistical model for indoor multipath propagation. IEEE J. Sel. Areas Commun. 1987, 5, 128–137.

[24] Karedal, J. A measurement-based statistical model for industrial ultra-wideband channels. IEEE Trans. Wirel. Commun. 2007, 6, 8.

[25] Cheffena, M. Industrial wireless sensor networks: Channel modeling and performance evaluation. EURASIP J. Wirel. Commun. Netw. 2012, 297, 1–8.

[26] Cheffena, M. Propagation channel characteristics of industrial wireless sensor networks [wireless corner]. IEEE Antennas Propag. Mag. 2016, 58, 66–73.

[27] Saadaoui, S., Tabaa, M., Monteiro, F., Chehaitly, M., & Dandache, A. (2019). Discrete Wavelet Packet Transform-Based Industrial Digital Wireless Communication Systems. Information, 10(3), 104.

[28] Li, L. Energy-Efficient Design and Implementation of Turbo Codes for Wireless Sensor Network. Ph.D. Thesis, University of Southampton, Southampton, UK, 2012.

[29] Schmidt, D.; Berning, M.; Wehn, N. Error correction in single-hop wireless sensor networks: A case study. In Proceedings of the Conference on Design, Automation and Test, Nice, France, 20–24 April 2009; pp. 1296–1301.

[30] Oh, H.; Nam, H.; Park, S. Adaptive threshold blanker in an impulsive noise environment. IEEE Trans. Electromagn. Compat. 2014, 56, 1045–1052.

[31] Hakimi, S.; Hodtani, G.A. Generalized maximum correntropy detector for non-Gaussian environments. Int. J. Adapt. Control Signal Process. 2018, 32, 83–97.

[32] Khalil, A., Saadoui, S., Tabaa, M., Chehaitly, M., Monteiro, F., Oukaira, A., & Dandache, A. (2019). Combined Reed-Solomon and Convolutional codes for IWSN based on IDWPT/DWPT Architecture. Procedia Computer Science, 155, 666-671.

[33] Saadaoui, S., Khalil, A., Tabaa, M., Chehaitly, M., Monteiro, F., & Dandache, A. (2020). Improved many-to-one architecture based on discrete wavelet packet transform for industrial IoT applications using channel coding. Journal of Ambient Intelligence and Humanized Computing, 1-9.

Chapter 9

Wavelet Transform for Signal Processing in Internet-of-Things (IoT)

Indrakshi Dey and Shama Siddiqui

Abstract

The primary contribution of this chapter is to provide an overview of different denoising methods used for signal processing in IoT networks from the perspectives of physical layer in the network. The chapter starts with the introduction to different kinds of noise that can be encountered in any kind of wireless communication networks, different kinds of wavelet transform and wavelet packet transform methods that can be used for denoising sensor signals in IoT networks and the different processing steps that are needed to be followed to accomplish wavelet packet transform for the sensor signals. Finally, a universal framework based on energy correlation analysis has been presented for denoising sensor signals in IoT networks, and such a framework can achieve considerable improvement in denoising performance reducing the effective noise correlation coefficient to 0.00001 or lower. Moreover, this method is found to be equally effective for Gaussian or impact noise or both.

Keywords: denoising, sensor signals, Internet-of-Things (IoT), wavelet transform, wavelet packet transform, energy correlation analysis

1. Introduction

Internet of Things (IoT) refers to a network of diverse range of smart devices used in the domains of healthcare, industry, vehicles, homes, agriculture, retail, poultry and farming, and many more. Typical equipment supporting the IoT functionality include lightning, thermostats, TVs, sensors, mobile phones, speakers, voice assistants, cameras, video cameras, etc. These devices are basically deployed to facilitate the processes of monitoring and automation by transmitting and receiving information via internet. Undoubtedly, IoT has emerged as a rapidly growing ecosystem that promises to deliver unmatched global coverage, quality-of-service (QoS), scalability, security and flexibility to handle different requirements for a comprehensive list of use-cases. This has resulted in increasing number of IoT devices (relays, sensors, transceiver, actuators etc.) being deployed in in all types of urban, suburban and rural environments to cater to the innovative and emerging applications.

Since more devices and appliances have been transforming into their smarter version, we now have the applications such as smart cars with features of smart dashboards, GPS, smart doors and auto-route designed to reduce the accidents. Such applications clearly require high number of connected devices; in fact, it has

IntechOpen

been forecasted by International Energy Agency that the estimated number of connected devices which was 15 billion in 2018 shall reach 46 billion in 2030 [1]. In addition to the IoT devices, the evolution of IoT networking technologies has also been remarkable over the past decade, where more and more IoT devices have been shifting from using Long Term Evolution (LTE) to Narrowband-IoT (NB-IoT) which offers a cost-effective and energy efficient solution for continued operation of these systems. Naturally, the connected devices are expected to transmit large volumes of heterogeneous data at high data rates, and we will be required to deal with ever-increasing radio frequency noise.

The signals carrying IoT data are highly likely to face numerous obstacles and can be corrupted by significant amount of noise present in the environment. White Gaussian model has been commonly been used to quantify the noise faced by [1]. The types of noise which have been found to degrading the quality of IoT signals vary from the impact noise resulting from high frequency interference and instantaneous disturbance on the initialization of large equipment to changing connections around the participating IoT devices [2]. All these kinds of noise negatively influence the multi-device information fusion system [3]. Such noises should be filtered out and the transmitted signal should be reconstructed back to its actual form to ensure the accuracy and reliability of the transmitted information. Here, accuracy of IoT solutions is measured in terms of the number of packets reporting correct information, deviation between the reported and actual results and the delivery to correct destination timely. Similarly, the reliability of IoT is measured using information such as failure rate of the IoT devices, average time between two consecutive failures, average repair time and probability for needing to change a component within a certain time-frame.

Although this chapter mainly deals with algorithms for signal denoising, they can be also be applied for image denoising, as images can be represented as two-dimensional signals. Consequently, signal processing techniques applicable to signals can be modified for images.

2. Noise consideration

The process of removing the noise while retaining and not distorting the quality of the received signal or image is referred to as denoising. The traditional way of denoising is to use a low or band-pass filter with cut-off frequencies. However, the traditional filtering techniques are able to remove out-of-band noise. Therefore many denoising techniques are proposed to overcome this problem.

Denoising is also an indispensable link in speech signal processing owing to the varying origins and non-stationarity, and difficulty in modeling the noise affecting the signal. Assuming that the received signal is affected by white additive Gaussian noise (AWGN) which is also stationary in nature, the received signal $y(i)$ can be represented as,

$$y(i) = x(i) + \sigma\varepsilon(i), \quad i = 0, 1, \ldots, n-1 \tag{1}$$

where $x(i)$ is the noise-free transmitted signal, $\varepsilon(i)$ representing independent normally distributed random variable and σ representing the intensity of the noise affecting $y(i)$. Reconstruction of the original signal $x(i)$ from the instantaneous set of $y(i)$ values without actual assuming a specific model for $x(i)$ or $y(i)$ is the primary aim of the process of 'Denoising'. The most common approach is to recognize noise components as the high frequency components present in the corrupted received signal, apply Fourier transform and then filter out the high frequency components.

Therefore, the most traditional way of denoising signals is based on Fourier analysis and Fourier transform.

Another common denoising method is the modulus maxima method [4]. It is based on the concept that signal and noise exhibit different characteristics when projected to their maxima in space divided in multiple scales. Magnitude scales increasing with decreasing extreme value points are filtered out to remove noise and the extreme value points themselves are reconstructed back [5]. The modulus maxima method in addition to the noise effect is better than any other method when mixed with white noise and singular information is significant, but the computational complexity is quite high. However, Fourier transform based denoising is restricted due to its weakness in obtaining partial characteristic of the transmit signals and possible Gibbs phenomenon [6]. If the signal has the same frequency as the noise, filtering out those frequency components will cause noticeable loss of information of the desired signal when considering the frequency representation of the signal.

3. Wavelet transform

Wavelet Transform (WT) has emerged as a powerful tool for signal and image denoising and processing, that have been successfully used in many scientific fields such as signal processing, image compression, computer graphics and pattern recognition [7, 8]. On contrary to the traditional Fourier transform, WT is particularly suitable for application of non-stationary signals which may instantaneously vary in time. Primarily, the received signal is divided into different frequency components using wavelets. The basis function of WT is scaled based on frequency and a subset of small waves (known as mother wavelet) is used for implementing WT [9]. The mother wavelet is a time-varying window function used for decomposition of $x(i)$ into weighted sets of scaled versions of $y(i)$. Consequently, using wavelet transform in signal processing is the process of the partial transformation of the spatial domain and the frequency domain, in order to get useful information accurately from it though corrupted with noise.

Since different frequency levels are used for WT, it is quite convenient for analyzing the signal characteristics at different frequencies and detecting removing corrupting noise. Broadly, there are two types of WT, Continuous WT (CWT) and Discrete WT (DWT).

3.1 Continuous wavelet transform (CWT)

CWT measures the congruence between an analyzing function and actual signal by calculating the inner product and then integrating the product. The mother wavelet window function can be shifted and moved over the time-axis by changing scale and position parameters, thereby including different frequency components at the different locations. Mathematically CWT can be represented as,

$$\mathrm{CWT}(a, b; x(i), \psi(i)) = \int_{-\infty}^{\infty} x(i) \frac{1}{a} \psi^* \left(\frac{i-b}{a} \right) \mathrm{d}i \tag{2}$$

where $x(i)$ is the transmit signal, $\psi(i)$ is the analyzing function (wavelet), a is the scale parameter, b is a position in time and $*$ represents complex conjugate. Considering $\psi(i)$ as the band-pass impulse response, scaling the wavelet varies the bandwidth of the band-pass filter. CWT allows changing the support of the wavelet to get better resolution in frequency domain. CWT can be realized on computer and the computation time can be significantly reduced if the redundant samples are removed after using the sampling theorem.

3.2 Discrete wavelet transform (DWT)

If suitable transformation is applied to a group of selected wavelet, a collection of orthogonal real-valued wavelets will be generated, a representation of the received signal referred to as wavelet expansion. In this case, the properties of the generated wavelets depend on the features of the mother wavelet. Since the newly generated wavelets are a group of orthogonal wavelets, they provide a time-frequency localization of the actual input signal, thereby concentrating the signal energy over a few frequency coefficients. Scaling and translation of the mother wavelet generated. If the scaling factor is a power of two, the wavelet transform technique is referred to as the dyadic-orthonormal wavelet transform [10]. If the chosen mother wavelet has orthonormal properties, there is no redundancy in the discrete wavelet transforms. In addition, this provides the multiresolution algorithm decomposing a signal into scales with different time and frequency resolution [9].

DWT is an implementation of WT using mutually orthogonal set of wavelets defined by carefully chosen scaling and translation parameters (a and b), such that the normalized area between the analyzing functions is unity, leading to a very simple and efficient iterative scheme for doing the transformation [11]. The translation equation can be expressed as,

$$\mathrm{DWT}\left[n, a^j\right] = \sum_{m=0}^{N-1} x[m]\psi_j^*[m-n]; \quad \psi_j[n] = \frac{1}{\sqrt{a^j}}\psi\left(\frac{n}{a^j}\right) \tag{3}$$

where n is the time delay introduced, N is the signal length and ψ is the discrete mother wavelet windowing function. DWT operates on discrete wavelet sets thereby yielding signal compression and reducing the computational complexity considerably. Moreover, DWT provides better spatial and frequency localization, as compared to other multi-scale signal maxima representation, thereby eliminating redundancy. In DWT, signal is decomposed into 'approximation' and 'detail' coefficients at each level [12].

The process is repeated at multiple levels, a technique equivalent to consecutive iterations of low pass and high pass filtering. As a result, the low frequency and high frequency components of $x(t)$ yield the approximation and detailed coefficients respectively, which can be mathematically expressed as,

$$x(t) = \sum_{m=1}^{L}\left[\sum_{k=-\infty}^{\infty} D_m(k)\psi_{m,k}(t) + \sum_{k=-\infty}^{\infty} A_l(k)\phi_{l,k}(t)\right] \tag{4}$$

Where $D_m(k)$ is the detailed coefficient, $A_l(k)$ is the approximation coefficient, $\psi_{m,k}(t)$ is 2^m-scale discrete analyzing function, and $\phi_{l,k}(t)$ is the 2^l-scale scaling function. After scaling and wavelet filtering, we get [13].

$$\begin{aligned} h(n) &= 2^{-1/2} < \phi(t), \phi(2t-n) > \\ g(n) &= 2^{-1/2} < \psi(t), \phi(2t-n) > = (-1)^n h(1-n) \end{aligned} \tag{5}$$

The approximation and the detailed coefficients are compared by applying FIR filter bank. The filter bank uses a low-pass filtering h for generating the approximation coefficients and high-pass filtering g for generating the detailed coefficients,

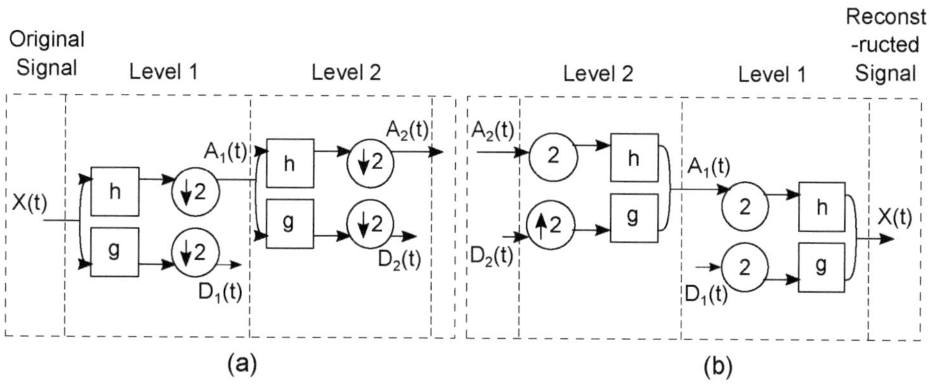

Figure 1.
The DWT decomposition and reconstruction steps of a 1D signal for level of 2; (a) decomposition, (b) reconstruction.

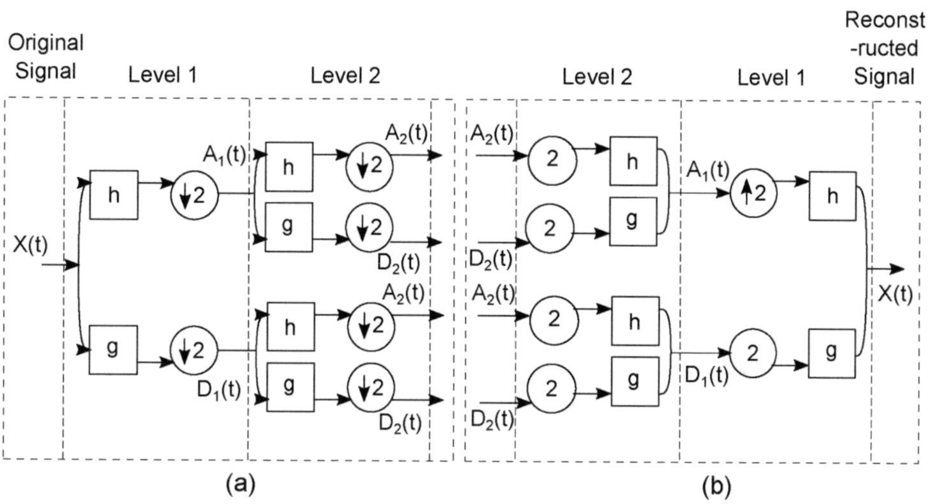

Figure 2.
The wavelet packet decomposition and reconstruction steps of a 1D signal for level of 2; (a) decomposition, (b) reconstruction.

followed by down-sampling by a factor of 2 at each scale level. The entire process is referred to as sub-band coding. The resultant tree structure is presented in **Figure 1**, where, $\downarrow 2$ and $\uparrow 2$ represents the processes of down-sampling and up-sampling respectively. The DWT decomposition process can be applied on both sub part of the signal, approximation coefficients and detail coefficients. This kind of decomposition is referred to as wavelet packet transform or wavelet packet tree decomposition. **Figure 2** represents the wavelet packet decomposition and reconstruction process.

3.3 Wavelet packet transform

Wavelet Packet Transform (WPT) is another powerful denoising tool. WPT is a generalized form of DWT, in which both smooth and details parts are subject to further transforms. A full transformed matrix contains $j(=\log_2 N)$ transform levels for searching for the best basis. The best basis can be chosen using different criteria. Shannon entropy is a very common one, which is defined as,

$$S = -\sum_j p_j \log\left(p_j\right) \quad (6)$$

for which $p_j = |x_j|^2/\|x\|^2$ and $p \log p = 0$ for $p = 0$. The optimized basis function will be a combination of both approximated and detailed coefficients and minimum entropy which can be obtained by comparing all the possible combinations of wavelet coefficients at different levels, minimizing $\sum \log|x_j|$, numbers larger than t and Stein's unbiased estimate of risk (SURE) [14].

Wavelet packet transform (WPT) has several advantages over WT (continuous and discrete) as it sets no requirements of mother wavelet windowing function [15], wavelet packet basis function [16], and selection of the number of decomposition levels [17] and threshold [18]. WPT is introduced in [19] for denoising and harmonic detection by computing the difference between the noise and the desired signal. The effectiveness is also experimentally verified in [20] and tested against dynamics of Electro-encephalogram (EEG) and Electro-cardiogram (ECG) measurements in [21]. Image denoising is implemented by using an adaptive anisotropic dual-tree complex WPT on a bivariate stochastic signal model in [21].

DWT has become a powerful tool for denoising experimental data over the past few years. Original data is decomposed into a series of wavelets at different scales and intensities. Using WT, where the signal is multiplied by a transformation matrix; the detailed and the smooth parts are separated and the process is repeated over $\log_2 N$ iterations. Depending on the length of the filtering steps, we can have different types of wavelets. If the number of steps vary from 4 to 20, the wavelets are referred to as Daublets. The Haar transform is a special case of Daublet 2. There can also be multiple filters, each with different filter lengths. If there are 5 filters, the wavelets are known as Coiflets, where each filter length is a multiple of 6. If there are 7 filters, the wavelets are known as Symmlets, where each filter length is a multiple of 2.

4. DWT for denoising data

The DWT denoising procedure consists of three steps. In the first step, if the length of the data stream is of length of the order of power of two, it is transformed to the wavelet domain. In the second step, coefficients with either zero magnitude or criterion-based minimized values are selected. In the third or final step, the minimized coefficients are reverted back to the original domain from the wavelet domain to extract the denoised data. DWT-based denoising techniques can be broadly classified into two categories - linear and non-linear. In linear DWT, signal and noise are assumed to be belonging to the smooth and the detailed part of the wavelet domain, where high frequency components are attenuated. While in non-linear DWT, the filter removes the coefficients selected in the second stage with amplitudes less than the threshold. In practicality, non-linear DWT is always preferred over linear DWT, as linear DWT introduces error due to the retention of noise components and loss of signal components owing to wavelet filtering.

Whether linear or non-linear DWT denoising technique is used, performance depends on the choice of the wavelet family and the length of the filter. The traditional way for making this choice is based on visual inspection of the data, for example, daublets are implemented when the data appears smooth in the wavelet domain, while Haar or other wavelets are used when the data appears bursty and discontinuous in the wavelet domain. In order to overcome the problems with DWT denoising, correlation denoising method was introduced in [11]. Correlation denoising method implements wavelet transformation and filtering in a way such that the correlation between wavelet coefficients of the signal part and the noise part is different at each level. However, correlation denoising in its original form is computationally complex. In order to reduce computational complexity, wavelet threshold denoising method was proposed by [12]. The method is simple to calculate and the noise can be suppressed to a large extent. At the same time, singular information of the original signal can be preferred well, so it is a simple and effective method. A brief overview of what happens when DWT is applied for denoising is demonstrated in **Figure 3**.

The four major components of the DWT denoising technique are: wavelet-type selection, threshold selection, threshold function selection and threshold application to the wavelet coefficients.

1. **Wavelet Selection** - There is a wide variety of wavelets that can be used for denoising. Selecting the optimum one depends on the selection of the matching wavelet filter. Out of different wavelet transform based denoising methods, only minimum description length (MDL) method has the flexibility of choosing the filter type.

2. **Threshold Selection** - There are four basic types of threshold selection, minimax, Stein's unbiased estimate of risk (SURE), and minimum description length (MDL). The ***Universal*** threshold is computed using,

$$t = \sigma \times \sqrt{2 \times \ln(N)} \tag{7}$$

for which N is the length of the signal data array, and σ is the standard deviation of noise. In practicality, in most cases, σ is unknown, but can be estimated using the first detailed part of the wavelet coefficient x_i through the expression,

$$\sigma_{\text{estimate}} \approx \frac{\text{median}(|x_i|)}{0.6745}. \tag{8}$$

In the case of ***Minimax*** criterion using the estimates of the minimax risk bounds for the transformed wavelets, a table is generated for threshold values corresponding to each set of given data lengths. These threshold values are always smaller than the universal threshold. The noise level estimates are calculated using (8) and signal components are retained along with a few number of noise components.

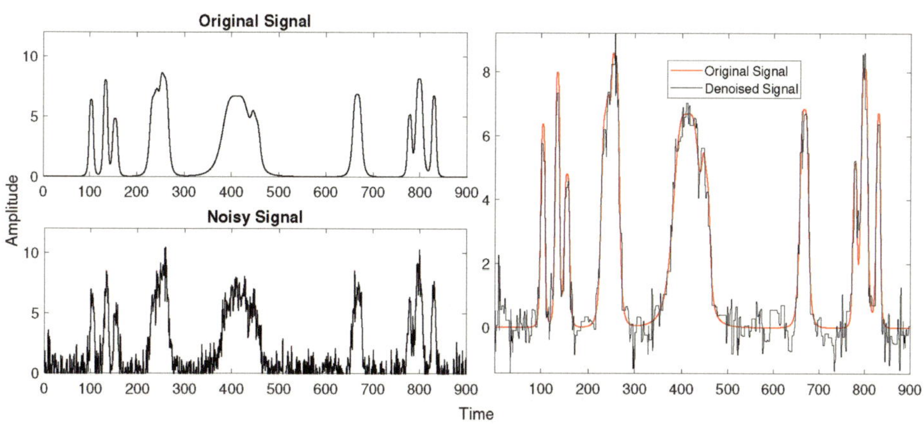

Figure 3.
Denoising with DWT.

Stein's unbiased estimate of risk (SURE) is used to obtain an unbiased estimate of the variance between the filtered and unfiltered data. SURE is defined as

$$\mathrm{SURE}(t,x) = N - 2 \times M_{i:|x_i| \le t} + \sum_{i=1}^{N} \left(|x_i|^t\right)^2 \tag{9}$$

for which t, x_i, N and M refer to the candidate threshold, wavelet coefficient, data length and number of data points less than t. The value of t that minimizes the SURE value is selected as the threshold value while the final term of the SURE function represents the residual energy left after thresholding. The SURE threshold can be modified to yield global thresholds rather than local ones by combining SURE method with cycle-spinning technique; a method referred to as SPINSURE.

The ***Minimum description length (MDL)*** method for threshold computation can be expressed as,

$$\mathrm{MDL}(k*, m*) = \min\left(\frac{3}{2}k\log(N) + \frac{N}{2}\log\left(\sum\left(x_m^2 - x_{mk}^2\right)\right)\right) \tag{10}$$

for which k, m, x_m, and x_{mk} represent the number of largest coefficients retained after filtering, the filter type, the wavelet coefficients from m-type wavelet transform, and the k largest coefficients in amplitude respectively. Here k^* and m^* are the optimized values for the MDL criterion for threshold selection, where k^* is selected as the threshold for the corresponding wavelet coefficient. The $3/2k\log(N)$ term represents the penalty function with value proportional to the number of retained wavelet coefficients. The $\frac{N}{2}\left(\sum\left(x_m^2 - x_{mk}^2\right)\right)$ characterizes the error between the reconstructed and the original signal components.

3. **Selecting threshold function** - whether wavelet threshold denoising method is good or bad depends on two decisive factors; one is the threshold λ and the other important factor is the selection of the threshold function. The most basic threshold functions are the hard and soft threshold functions, comparative performance of which is presented in **Figure 4**.

The ***Hard Threshold Function (HTF)*** nullifies the decomposition coefficients to zero if they are less than the threshold and retains the coefficients if they are more than the threshold [22]. The HTF preserves the local properties of a signal with a few discontinuities introduced by the variations in the reconstructed signals. HTF can be expressed as,

$$\overline{\omega_{j,k}} = \begin{cases} \omega_{j,k}, & |\omega_{j,k}| \ge \lambda \\ 0, & |\omega_{j,k}| < \lambda \end{cases} \tag{11}$$

The ***Soft Threshold Function (STF)*** [23] selects the threshold value such that all decomposition coefficients are nullified to zero. A major drawback with this technique is that a part of the high frequency components is lost owing to their location above threshold. STF can be mathematically expressed as,

$$\overline{\omega_{j,k}} = \begin{cases} \mathrm{sgn}\left(\omega_{j,k}\right)\left(|\omega_{j,k}| - \lambda\right), & |\omega_{j,k}| \ge \lambda \\ 0, & |\omega_{j,k}| < \lambda \end{cases} \tag{12}$$

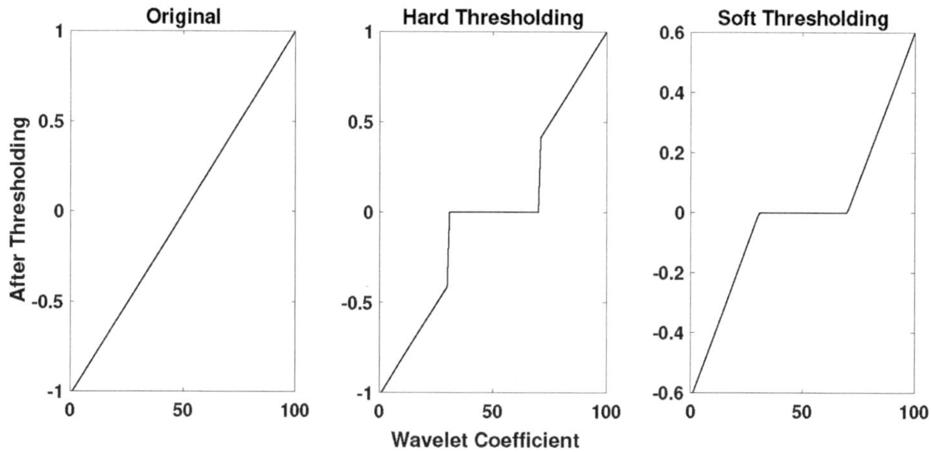

Figure 4.
Comparative hard and soft thresholding when implemented for DWT.

where $\overline{\omega_{j,k}}$, $\omega_{j,k}$, λ, and sgn() denotes the estimated wavelet coefficients, post-decomposition wavelet coefficients, threshold and symbolic piece-wise function respectively [24].

Garrote Threshold Function is proposed in [25] to improve the drawbacks of HTF and STF, whose denoising effect is better than the above two methods with respect to continuity of expressions,

$$\overline{\omega_{j,k}} = \begin{cases} \omega_{j,k} - \dfrac{\lambda^2}{\omega_{j,k}}, & |\omega_{j,k}| \geq \lambda \\ 0, & |\omega_{j,k}| < \lambda \end{cases} \tag{13}$$

The continuity in the soft threshold function is much better, but it has a constant deviation. So, in order to overcome its shortcomings, the soft and hard threshold algorithms are compromised process by the literature; the ***semisoft threshold function*** [26].

$$\overline{\omega_{j,k}} = \begin{cases} \text{sgn}\,(\omega_{j,k})\left(|\omega_{j,k}| - T\lambda\right), & |\omega_{j,k}| \geq \lambda \\ 0, & |\omega_{j,k}| < \lambda \end{cases} \tag{14}$$

It is worth-mentioning here, that the values of the threshold T is fixed with values between 0 and 1 in the case of HTF, STF, Garrote Threshold Function and semi-threshold function.

Another variation is the ***Improved Threshold Function*** which can be given by,

$$\overline{\omega_{j,k}} = \begin{cases} \text{sgn}\,(\omega_{j,k})\left(|\omega_{j,k}| - \dfrac{\lambda}{\exp^{-3}\left[\alpha\left(|\omega_{j,k}| - \lambda\right)/\lambda\right]}\right), & |\omega_{j,k}| \geq \lambda \\ 0, & |\omega_{j,k}| < \lambda \end{cases} \tag{15}$$

The adjustment factor of the new function is different from the semisoft threshold function. It consists of a complex exponential function $\exp^{-3}\left[\alpha\left(|\omega_{j,k}| - \lambda\right)/\lambda\right]$ which has more adaptability; α is the normal number which can be adjusted freely and the values of α are different with the different signal. When $|\omega_{j,k}| = \lambda$, $\overline{\omega_{j,k}} \to \lambda$, $\overline{\omega_{j,k}} \to 0$.Therefore, continuously in place of

λ, the improved threshold function has the characteristics of soft threshold function; when $\overline{\omega_{j,k}} \to \infty$, $\overline{\omega_{j,k}} \to \omega_{j,k}$ improved threshold function based on $\overline{\omega_{j,k}} = \omega_{j,k}$ as the asymptotic line; it can be seen that, with the increase of $\omega_{j,k}$, $\omega_{j,k}$ will gradually be close to $\omega_{j,k}$; when $\omega_{j,k}$ becomes infinite, $\omega_{j,k} \approx \omega_{j,k}$. The choice of α is crucial for the success of the technique and the variation in α affects the denoising effect. When $\alpha = 0$, improved threshold function reduces to STF and when $\alpha = \infty$, improved threshold function reduces to HTF.

4. ***Thresholding or threshold application*** - thresholding is defined as the ways in which threshold is applied for modifying wavelet coefficients. DWT is a multi-level wavelet transform technique with different thresholds being applied at different level of coefficients
Global Thresholding - This technique assumes the corrupting noise as Gaussian distributed with amplitude and frequency distributions same for all orthogonal bases for the entire data space. Global thresholding can be implemented using either hard, soft, Garrote or firm-threshold functions, expressed as,

- Hard:

$$x_i^* = \begin{cases} 0, & if\ |x_i| \le t \\ x_i, & if\ |x_i| > t \end{cases} \tag{16}$$

- Soft:

$$x_i^* = \begin{cases} 0, & if\ |x_i| \le t \\ \text{sign}(x_i)(|x_i| - t), & if\ |x_i| > t \end{cases} \tag{17}$$

- Garrote:

$$x_i^* = \begin{cases} 0, & if\ |x_i| \le t \\ x_i - t^2/x_i, & if\ |x_i| > t \end{cases} \tag{18}$$

- Firm:

$$x_i^* = \begin{cases} 0, & if\ |x_i| \le t_1 \\ \text{sign}(x_i)t_2(|x_i| - t_1)/(t_2 - t_1), & if\ t_1 < |x_i| \le t_2 \\ x_i & if\ |x_i| > t_2 \end{cases} \tag{19}$$

for which x_i and x_i^* represents the wavelet coefficients pre- and post-thresholding respectively. HTF partitions the wavelet coefficients into two parts by the selected threshold eliminating coefficients with low magnitude. STF reduces all coefficients by a factor equal to the threshold eliminating smaller coefficients. Similarly, Garrote thresholding reduces all large coefficients by a factor of a non-linear continuous function. Firm thresholding reduces only the middle coefficients while eliminating small and retaining large coefficients.
Level-Dependent Thresholding - This technique uses different thresholds at each level of wavelet transformations. It uses a combination of SURE and global thresholding techniques to initiate a hybrid method. In this case, if the sample variance at each level is sparse, global thresholding is applied, while SURE thresholding is applied otherwise.

Data-Dependent Thresholding - A Data-dependent threshold (DDT) technique selects a threshold such that empirical wavelet coefficients are shrunk. The thresholding is achieved through statistical tests of hypotheses like linear regression. The level of this statistical test is adjusted to control the smoothness of the resulting estimator such that a good mean-squared error (MSE) performance is achieved for different data analysis settings with smoothness in estimator response. The main aim of this technique is to eliminate a group of wavelet coefficients that exhibit characteristics of pure noise.

Cycle-Spin Thresholding - It combines the process of subspace identification, projecting denoising and averaging of the projections. The subspace mentioned here refers to the region where most of the energy of the signal is concentrated and signal corrupted with noise is projected on to this subspace.

5. Signal denoising for IoT networks

The huge amount of sensor data generated in an IoT network are used to take decisions on a certain observation/ phenomenon based on real-time processing. The decision-making procedure often involves detecting the signal energy level transmitted from the sensors. If the received energy level is higher than a predefined threshold, the target is detected to be present phenomenon and vice-versa. However, the sensor data gets crippled with noise contributed by the wireless environment and the internal electronics of the sensors, on its way to the data center for processing. The WPT method will be the best option in this case for denoising the sensor data, where the original signal coefficients are preserved while removing the noise within the signal. The WPT method can decompose a signal in both scale and wavelet space thereby revealing more details about both the sensor signals and the crippling noise. If energy correlation analysis is used in conjunction with WPT, signal energy from the sensor data can be analyzed and noise can be eliminated by zooming into the signal characteristics at different time scales. Advantages of WPT over WT is evident in **Figure 5**. Hence, in this section, a universal framework is presented for denoising sensor signals in IoT networks. The framework is based on energy correlation analysis and combines the processes of WP decomposition, coefficient modification and WP reconstruction. The functional block diagram for this framework is presented in **Figure 6**.

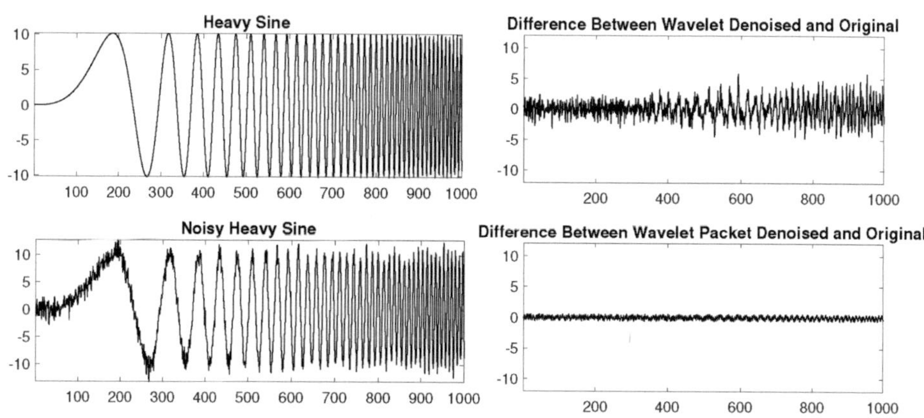

Figure 5.
Comparative performance of WPT and WT.

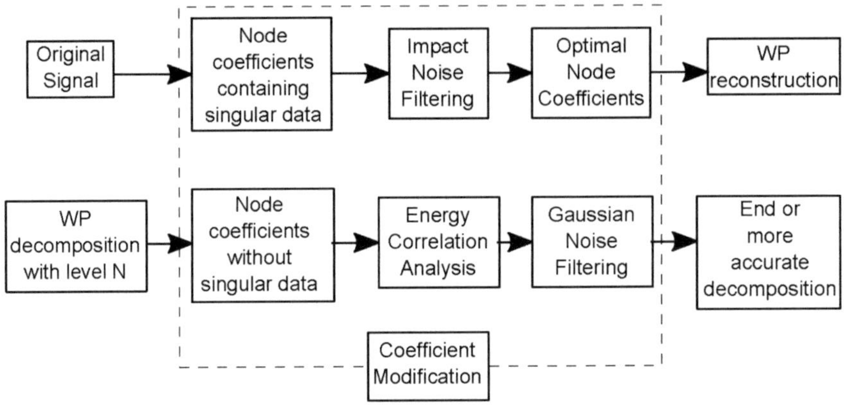

Figure 6.
Architecture of the universal framework.

5.1 Wavelet packet transfer for IoT

In WPT for IoT networks, for a given for a given orthonormal scaling function $\phi(t)$ and wavelet function $\psi(t)$ the double scale Eq. [14] can be described as follows:

$$\phi(t) = \sqrt{2}\sum_{k} h_{0k}\phi(2t-k), \psi(t) = \sqrt{2}\sum_{k} h_{1k}\phi(2t-k) \tag{20}$$

where h_{0k} and h_{1k} are a pair of conjugate orthogonal filter coefficients. WP functions for $n = 0, 1, \ldots$ can be defined as follows,

$$w_{2n}(t) = \sqrt{2}\sum_{k\in Z} h_{0k}w_n(2t-k), w_{2n+1}(t) = \sqrt{2}\sum_{k\in Z} h_{1k}w_n(2t-k) \tag{21}$$

When $n = 0, w_0(t) = \phi(t), w_1(t) = \psi(t)$. $\{w_n(t)\}_{n\in Z}$ represents the wavelet packet assuming standard orthogonal wavelet basis can be constructed from the scaling function. Scaling and wavelet functions generated as a result of this process satisfy the properties of orthogonality over both scale and translation,

$$\begin{aligned} &\langle w_n(t-k) \cdot w_n(t-l)\rangle = \delta_{kl} \quad k, l\in Z \\ &\langle w_{2n}(t-k) \cdot w_{2n+1}(t-l)\rangle = 0 \quad n = 1, 2, \ldots \end{aligned} \tag{22}$$

In the process of WP decomposition, scale space $\{V_j\}_{j\in Z}$ composed of scaling functions and wavelet space $\{W_j\}_{j\in Z}$ composed of wavelet functions can be expressed in a unified way as follow:

$$U_j^0 = V_j, U_j^1 = W_j \quad j\in Z \tag{23}$$

From $V_j = V_{j+1} \oplus W_{j+1}$, then

$$U_j^0 = U_{j+1}^0 \oplus U_{j+1}^1, U_j^n = U_{j+1}^{2n} \oplus U_{j+1}^{2n+1} \quad j\in Z, n\in Z^+ \tag{24}$$

where, U_j^n denotes the closed subspace of square and integrable space $L^2(R)$ generated by the linear combination of wavelet packet w_n after translation and scaling operation.

During the procedure of multi-resolution analysis, objective function is decomposed into the subspace $\{V_j\}_{j\in Z}$, $\{W_j\}_{j\in Z}$ in $L^2(R)$ carried out further decomposition according to binary mode as follows:

$$\begin{aligned} W_j &= U_j^1 = U_{j+1}^2 \oplus U_{j+1}^3, U_{j+1}^2 = U_{j+2}^4 \oplus U_{j+2}^5, U_{j+1}^3 = U_{j+2}^6 \oplus U_{j+2}^7 \\ W_j &= U_{j+2}^4 \oplus U_{j+2}^5 \oplus U_{j+2}^6 \oplus U_{j+2}^7 \end{aligned} \tag{25}$$

Consequently,

$$W_j = U_{j+k}^{2^k+1} \oplus U_{j+k}^{2^k+2} \oplus \dots \oplus U_{j+k}^{2^{k+1}-1} \tag{26}$$

Finally, the wavelet packet coefficients can be computed [27] as follows:

$$d_k^{j+1,2n} = \sum_l h_{0(2l-k)} d_l^{j,n}, d_k^{j+1,2n+1} = \sum_l h_{1(2l-k)} d_l^{j,n} \tag{27}$$

where

$$d_k^{j+1,n} = \sum_k \left[h_{0(2l-k)} d_k^{j,2n} + h_{1(2l-k)} d_k^{j,2n+1} \right] \tag{28}$$

Following this technique of WPT, the efficiency of the denoising process improves quite a bit over the case where just WT is used for denoising the signals, as is evident in **Figure 5**.

5.2 Energy correlation analysis

Digital signal energy computation is achieved by extracting and squaring signal amplitude at different locations in the time domain and then adding them together [28]. The influence of relative large energy is eliminated using normalization technique [29]. This normalization can be avoided by selecting the sum of absolute values of amplitudes at each sampling points as approximations for evaluating energy; the mathematical formulation for which can be represented as:

$$e = \sum_{n=1}^{N} |f(n)|, \quad n = 1, 2, \dots, N. \tag{29}$$

Any kind of non-deterministic relationship existing between two or more variables can be exploited and formalized using correlation analysis. Thus, different kinds of signals can be differentiated by exploring the internal relation with correlation analysis. x_i and y_i denote two random variables, respectively; the calculation formula of correlation coefficient can be given as follows:

$$r = S_{xy} / \sqrt{S_{xx} S_{xy}}, \quad -1 \le r \le 1, \tag{30}$$

where $S_{xx} = \sum_{i=1}^{N} (x_i - \overline{x})^2$, $S_{yy} = \sum_{i=1}^{N} (y_i - \overline{y})^2$ and $S_{xy} = \sum_{i=1}^{N} (x_i - \overline{x})(y_i - \overline{y})$.

The correlation coefficient r is referred to as "Pearson product-moment correlation coefficient," or Pearson's r and is used to estimate the relative relationship between variables using the following principles.

1. The closer the absolute value of Pearson's r to 1, more is the correlation and closer is the Pearson's r to 0, less is the correlation between the variables.

2. The polarity of the coefficient determines the direction of correlation, with plus-sign representing positive and minus-sign representing negative correlation.

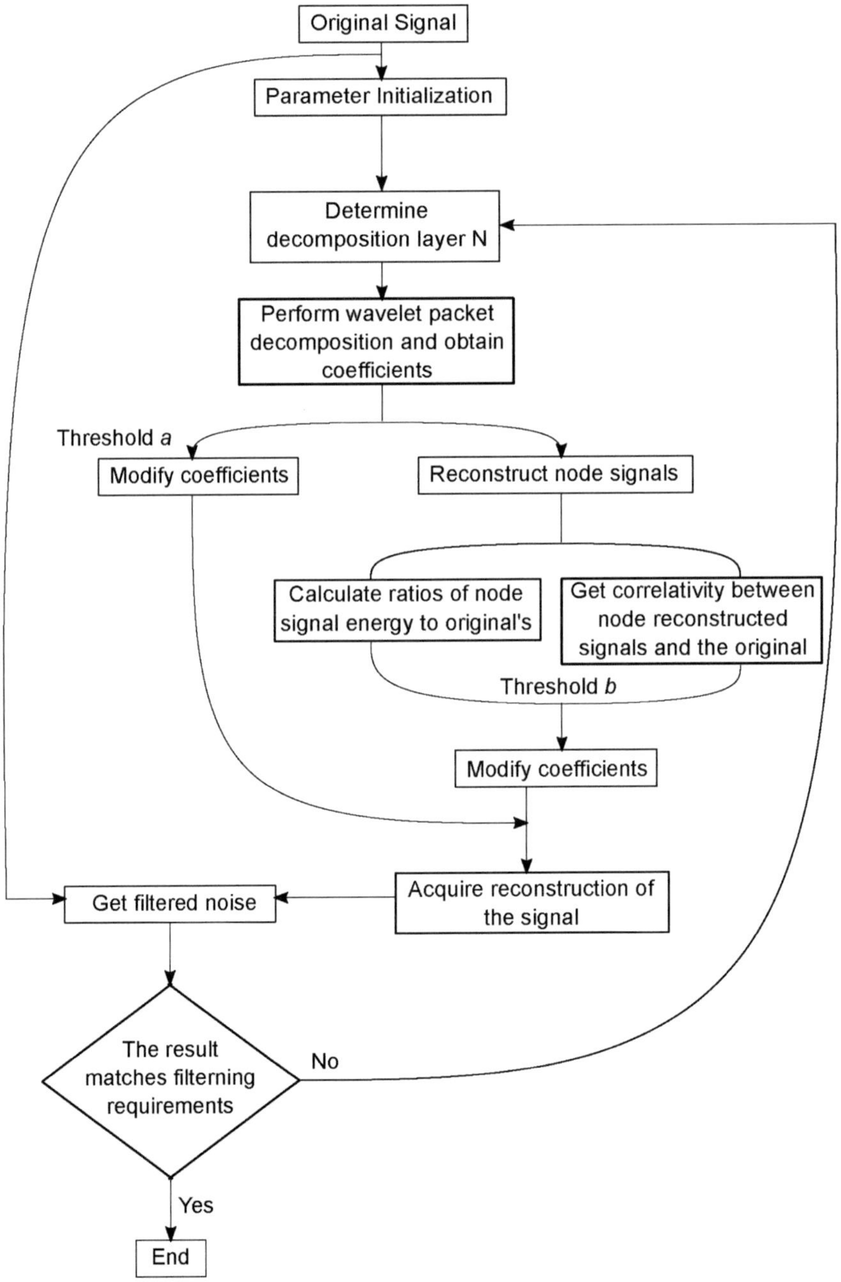

Figure 7.
Flowchart of wavelet packet coefficients based on energy-correlation analysis.

5.3 Processing method for WP coefficients based on energy-correlation analysis

An online filtering process capable of denoising both Gaussian and impact noise is presented below based on the energy correlation between signal components reconstructed from WP coefficients.

Step 1 - Obtain WP decomposition coefficients through the application of appropriate decomposition level and mother wavelet.

Step 2 - Compare WP coefficients in each subspace to eliminate singular data based on a pre-selected threshold through the application of multi-resolution analysis.

Step 3 - After reconstructing WP node signals from real coefficients, compute the ratios of the energy of the reconstructed signal components to the actual signal components to obtain the correlation between them. Subspace unsatisfied coefficients are processed through the use of a different threshold resulting in a series of new coefficients.

Step 4 - Using the new set of modified coefficients on each node, signal components are reconstructed and noise is eliminated. If the filtering requirements are not satisfied, repeat steps to step 4 after increasing the decomposition level. A flow-diagram for energy correlation analysis based WP coefficient processing is depicted in **Figure 7**.

6. Performance analysis of denoising techniques

The best way to denoise a signal is to assume that the noise signal is Gaussian distributed with values that are independent and identical real values. The performance of the denoising process can be evaluated by comparing the quality of the denoised signal with that of the original transmit signal. A variety of methods have been proposed over years to measure the performance of denoising; the most common of which are the metrics of SNR and the peak SNR (PSNR), generally accepted to measure the quality of signal and images respectively. For 1-D signal, measuring the performance of the denoising method by calculating the residual SNR is given by, $SNR = 10\log_{10}\left(\sum_{n=0}^{N-1} x^2(n) / \sum_{n=0}^{N-1} (\overline{x}(n) - x^r(n))^2\right)$ where $x(n)$ is the original signal, $x^r(n)$ is the denoised signal and $\overline{x}(n)$ refers to the mean value of $x(n)$.

In order to measure the quality of image, PSNR is generally used, which is given by $PSNR = 10\log_{10}\left(L / \sum_{n=0}^{N-1}\sum_{m=0}^{M-1} (\overline{x}(n,m) - x^r(n,m))^2\right)$, where L, $x(n)$, $\overline{x}(n,m)$ and $x^r(n,m)$ refer to the quantized gray level of images, original image, mean value of $x(n)$ and the reconstructed image respectively. However, the choice of the noise power is absolutely crucial for visible performance difference. SNR is more important as compared to noise power when evaluating performance and with SNR above 3 dB, it is quite easy to isolate visible corruption.

7. Conclusions

Decomposition in time and frequency domain for Fourier Transform is replaced by decomposition in space domain for WT thereby removing any ambiguity related to time and frequency and offering high flexibility and quality to the overall denoising process. Different threshold estimation methods, wavelet types, threshold types and thresholding functions can be used for implementing WT depending on the application scenario, network architecture, the kind of signal transmitted and the kind of noise commonly observed in the considered application scenario. However, comparing performances of different thresholding methods, wavelet types or threshold types when applied for the WT reveal that the number of decomposition levels are more crucial to the denoising performance than the types of wavelets or thresholds.

If the application scenario is considered to be an industrial IoT network, WPT method is preferred over simple WT for denoising sensor signals. This is because in WPT, signal is decomposed into an approximation and a detail component at each layer of each decomposition level, therefore resulting in 2^n number of components at n decomposition levels in contrast to just 2 components at each of the n decomposition levels of WT. Moreover, WT decomposes only the low frequency components in contrast to WPT which considers both low and high frequency components at each level. If WPT is combined with energy correlation analysis, effectiveness of the denoising process increases manifold owing to its immunity to diversity of signals in an IoT network. Integration of energy and correlation can be used to modify wavelet packet coefficients for eliminating Gaussian and impact noise efficiently.

A. Appendix A

Example MATLAB codeset for signal denoising

Signal Generation

```
N = 2048*2;
name = 'piece-regular';
f0 = load,ignal(name, N);
f0 = rescale(f0,.05,.95);
sigma = 0.05;
f = f0 + randn(size(f0))*sigma;
figure(1)
subplot(2,1,1); plot(f0); axis([1 N 0 1]);
title('Clean signal');
subplot(2,1,2);
plot(f); axis([1 N 0 1]);
title('Noisy signal');
```

Thresholding

```
Theta0 = @(x,T)x.* (abs(x)¿T);
Theta1 = @(x,T)max(0, 1-T./max(abs(x),1e-9)).* x;
t = linspace(-3,3,1024)'; T = 1;
figure(2)
plot( t, [Theta0(t,T), Theta1(t,T)], 'LineWidth', 2 );
axis('equal'); axis('tight');
legend('Θ^0', 'Θ^1');
```

Wavelet-Threholding

```
options.ti = 0; Jmin = 4;
```

DOI: http://dx.doi.org/10.5772/intechopen.95384

```
W = @(f) perform_wavelet_transf(f,Jmin,+1,options);
Wi = @(fw)perform_wavelet_transf(fw,Jmin,-1,options);
x = W(f);
x1 = Theta0(x, 3*sigma);
figure(3)
subplot(2,1,1);
plot_wavelet(x,Jmin); axis([1N -11]);
title('W(f)');
subplot(2,1,2);
plot_wavelet(Theta0(W(f),T),Jmin); axis([1N -11]);
title('Θ^0(W(f))');
f1 = Wi(x1);
figure(4)
subplot(2,1,1);
plot(f); axis([1 N 0 1]);
title('f');
subplot(2,1,2);
plot(f1); axis([1 N 0 1]);
title('f'_1);
x = W(f);
reinject = @(x1)assign(x1, 1:2^Jmin, x(1:2^Jmin));
Theta0W = @(f,T)Wi(Theta0(W(f),T));
Theta1W = @(f,T)Wi(reinject(Theta1(W(f),T)));
```

TI WT

```
options.ti = 1;
W = @(f) perform_wavelet_transf(f,Jmin,+1,options);
Wi = @(fw)perform_wavelet_transf(fw,Jmin,-1,options);
fw = W(f);
nJ = size(fw,3)-4;
figure(5)
subplot(5,1, 1);
plot(f0); axis('tight');
title('Signal');
i = 0;
for j=1:3
i = i+1;
subplot(5,1,i+1);
plot(fw(:,1,nJ-i+1)); axis('tight');
title(strcat(['Scale=' num2str(j)]));
end
subplot(5,1, 5);
plot(fw(:,1,1)); axis('tight');
title('Low scale');
```

Author details

Indrakshi Dey[1,2*] and Shama Siddiqui[3]

1 National University of Ireland, Maynooth, Ireland

2 Trinity College Dublin, University of Dublin, Ireland

3 DHA Suffa University, Karachi, Pakistan

*Address all correspondence to: deyi@tcd.ie; indrakshi.dey@mu.ie

IntechOpen

© 2021 The Author(s). Licensee IntechOpen. This chapter is distributed under the terms of the Creative Commons Attribution License (http://creativecommons.org/licenses/by/3.0), which permits unrestricted use, distribution, and reproduction in any medium, provided the original work is properly cited.

References

[1] A. A. Brincat, F. Pacifici and F. Mazzola. IoT as a Service for Smart Cities and Nations. IEEE Internet of Things Magazine, 2(1), pp. 28–31, 2019.

[2] Z.-C. Liu, X.-G. Chen, and Y.-F. Li. Detection and identification of abrupt changes for on-line sensor output signal. Transaction of Beijing Institute of Technology, 26(12), pp. 1104–1108, 2006.

[3] A.M. Rao and D. L. Jones. A denoising approach to multisensor signal estimation. IEEE Transactions on Signal Processing, 48(5), pp. 1225–1234, 2000.

[4] B. L. Jin, H. Li, N. J. Zhao et al. A new denoising algorithm for wavelet thresholding. Journal of Missile and Missile, 31 (1), pp. 167–169, 2011 (Chinese).

[5] D. L. Donoho. De-noising by soft-thresholding. IEEE Transactions on Information Theory, 41(3), pp. 613–627, 1995.

[6] M. Ding and H. Zhu. Two-Dimensional gibbs phenomenon for fractional fourier series and its resolution. Artificial Intelligence and Computational Intelligence, vol. 7530 of Lecture Notes in Computer Science, pp. 530–538, Springer, Berlin, Germany, 2012.

[7] I. Daubechies. The wavelet transform, time-frequency localization and signal analysis. IEEE Transactions on Information Theory, 36(5), pp. 961–1005, 1990.

[8] A. S. Lewis and G. Knowles. Image compression using the 2-D wavelet transform. IEEE Transactions on Image Processing, 1(2), pp. 244–250, 1992.

[9] S. Mallat and Z. Zhang. Matching pursuits with time-frequency dictionaries. IEEE Transactions on signal processing, 41(12), pp. 3397–3415, 1993.

[10] X. He and M. S. Scordilis. Psychoacoustic music analysis based on the discrete wavelet packet transform. Research Letters in Signal Processing 2008, pp. 1–5.

[11] D. L. Donoho and J. M. Johnstone. Ideal spatial adaptation by wavelet shrinkage. Biometrika, 81(3), pp. 425, 1994.

[12] D. L. Donoho. Denoising by soft-thresholding. IEEE Trans. Inform. Theory, 41(3), pp. 613–627, 1995.

[13] S. G. Mallat. A wavelet tour of signal processing. Academic Pr., 1999.

[14] J. Gubbi, A. Khandoker, and M. Palaniswami. Classification of sleep apnea types using wavelet packet analysis of short-term ECGsignals. Journal of Clinical Monitoring and Computing, 26(1), pp. 1–11, 2012.

[15] F. Adamo, G. Andria, F. Attivissimo, A. M. L. Lanzolla, and M. Spadavecchia. A comparative study on mother wavelet selection in ultrasound image denoising. Measurement, 46(8), pp. 2447–2456, 2013.

[16] R. R. Coifman and M. V. Wickerhauser. Entropy-based algorithms for best basis selection. IEEE Transactions on Information Theory, 38 (2), pp. 713–718, 1992.

[17] A. M. Hasan, K. Samsudm, A. R. Ramli, and R. S. Azmir. Wavelet-based pre-filtering for low cost inertial sensors. Journal of Applied Sciences, 10 (19), pp. 2217–2230, 2010.

[18] D. L. Donoho. De-noising by soft-thresholding. IEEE Transactions on Information Theory, 41(3), pp. 613–627, 1995.

[19] P. Mercorelli. Denoising and harmonic detection using non-orthogonal wavelet packets in industrial applications. Journal of Systems Science Complexity, 20(3), pp. 325–343, 2007.

[20] Y. Li, T. Zhang, L. Deng, B. Wang and M.Nakamura. Denoising and rhythms extraction of EEG under +Gz acceleration based on wavelet packet transform. Proceedings of the 7th ICME International Conference on Complex Medical Engineering (CME '13), pp. 642–647, Beijing, China, May 2013.

[21] J. Yang, W. Xu, Y.Wang, and Q. Dai. 2-D anisotropic dual-tree complex wavelet packets and its application to image denoising. Proceedings of the 15th IEEE International Conference on Image Processing (ICIP '08), pp. 2328–2331, October 2008.

[22] C. He, J. C. Xing, and Q. L. Yang. Optimal wavelet basis selection for wavelet denoising of structural vibration signal. Applied Mechanics and Materials, 578-579, pp. 1059–1063, 2014.

[23] J.-Y. Tang, W.-T. Chen, S.-Y. Chen, and W. Zhou. Wavelet-based vibration signal denoising with a new adaptive thresholding function. Journal of Vibration and Shock, 28(7), pp. 118–121, 2009 (Chinese).

[24] S. Badiezadegan and R. C. Rose. A wavelet-based thresholding approach to reconstructing unreliable spectrogram components. Speech Communication, 67, pp. 129–142, 2015.

[25] X. Chen, S. Li, and W. Wang. New de-noising method for speech signal based on wavelet entropy and adaptive threshold. Journal of Information and Computational Science, 12(3), pp. 1257–1265, 2015.

[26] K. L. Yuan. Wavelet denoising based on threshold optimization method. Engineering Journal of Wuhan University, 48(1), pp. 74–80, 2015.

[27] J. Portilla, V. Strela, M. J. Wainwright, and E. P. Simoncelli. Image denoising using scale mixtures of Gaussians in the wavelet domain. IEEE Transactions on Image Processing, 12 (11), pp. 1338–1351, 2003.

[28] K. Zhang, B.-J. Pang, and M. Lin. Wavelet packet analysis for acoustic emission signals caused by debris cloud impact. Journal of Vibration and Shock, 31(12), pp. 125–128, 2012.

[29] X.-H. Gu, G.-X. Zhang, D.-B. Hou, and Z.-K. Zhou. Detection of water pipe leak location using wavelet packet decomposition and power feature extraction. Journal of Sichuan University, 37(6), pp. 145–149, 2005.

Section 5

Wavelet Transform and Computations

Chapter 10

The Discrete Quincunx Wavelet Packet Transform

Abdesselam Bassou

Abstract

This chapter aims to present an efficient compression algorithm based on quincunx wavelet packet transform that can be applied on any image of size 128 × 128 or bigger. Therefore, a division process into sub-images of size 128 × 128 was applied on three gray-scale image databases, then pass each sub-image through the wavelet transform and a bit-level encoder, to finally compress the sub-image with respect to a fixed bit rate. The quality of the reconstructed image is evaluated using several parameters at a given bit rate. In order to improve the quality in sense of the evaluation quality, an exhaustive search has led to the best packet decomposition base. Two versions of the proposed compression scheme were performed; the optimal version is able to decrease the effect of block boundary artifacts (caused by the image division process) by 27.70% considering a natural image. This optimal version of the compression scheme was compared with JPEG standard using the quality evaluation parameters and visual observation. As a result, the proposed compression scheme presents a competitive performance to JPEG standard; where the proposed scheme performs a peak signal to noise ratio of 0.88 *dB* over JPEG standard at a bit rate of 0.50 *bpp* for a satellite image.

Keywords: quincunx wavelet transform, wavelet packet, quality evaluation parameters, reduction factor, JPEG standard

1. Introduction

Wavelet is defined as a small wave that can be the base of all physical phenomena; which means that a time and/or space variation of a phenomenon is a sum of multiple wavelets. As examples, the wavelet transform was applied on an electrocardiogram (ECG) signal in order to extract the QRS complex [1] (time variation), on a video sequence in order to implement a hidden watermark [2] (time and space variation) and on a 2D image in order to reduce its size (compression) [3, 4] (space variation). In this chapter, one considers the application of the wavelet on 2D image compression.

An image is one of the most important sources of information; it provides a visual comprehension of a phenomenon. The image can take several natures as medical, natural, textural or satellite image, each nature is characterized by a proper amount of details. For a digital image, the size in bytes is as bigger as the amount of details; this applies the use of image compression process.

In other words, if one considers a gray-scale image of size 512 × 512, that means a bit rate of 8 bits per pixel ($R_c = 8\ bpp$) and a file size of 512 × 512 × 8 bits (256 Kbytes). Compressing this image leads to reduce its file size (without changing the image size); for example, to reduce the file size by a factor of 10 (25.6 Kbytes), one have to consider a bit rate of $R_c = \frac{25.6\times1024\times8}{512\times512} = 0.8\ bpp$.

Because of the conservation of all details in the image after decompression, the lossless[1] compression algorithms, as Run Length Coding (RLE), Lempel-Ziv-Welch (LZW) and Huffman [5, 6], are by far the ideal methods. However, such a compression algorithms does not provide a significant reduction of image's file size, and therefore the lossy[2] compression algorithms may be more appropriate.

The most known lossy compression algorithm is the standard JPEG (Joint Photographic Experts Group) [7]; it is based, as lossy algorithm, on a discrete transform (Cosine Discrete Transform, DCT in this case). The Discrete Wavelet Transform (DWT) and Quincunx Wavelet transform (QWT) are two other discrete transforms that can be found in the literature [8, 9]; they apply a progressive transformation on the image followed by an encoding process (like Embedded Zerotree Wavelet, EZW or Set Partitioning In Hierarchical Trees, SPIHT [10]) to give the image a bit-level representation.

This chapter aims to propose a QWT-based compression algorithm that can be applied on any image of size 128 × 128 or bigger. Therefore, the following structure is adopted: In Section 2, the discrete wavelet transform is introduced and the progressive presentation of an image is exposed. Section 3 is dedicated to the quincunx wavelet transform, the QWT extension to wavelet packet (PQWT) and the encoding process employing SPIHT algorithm. The PQWT-based compression algorithm is presented in Section 4, and the results and discussions in Section 5.

2. Discrete wavelet transform

2.1 Definition

As discrete sine and cosine, the DWT is used to represent a digital signal (as an image) with sum of projections over orthogonal functions; these functions are called "wavelet". Several wavelets are described in the literature; among them, one can find dyadic Daubechies family (represented with scaling and wavelet functions in **Figure 1** for four examples [8]).

In order to improve JPEG compression performances (in sense of evaluation parameters presented in Section 4), the researchers have proposed the JPEG 2000 compression algorithm based on a wavelet called CDF 9/7 (Cohen-Daubechies-Feauveau 9-tap/7-tap) [11, 12]. The scaling and wavelet functions, and decomposition and reconstruction low and high filters are shown in **Figure 2**.

2.2 Wavelet decomposition

As it is mentioned above, a wavelet applies a progressive transformation on the image. This process (called filter bank analysis) is realized by passing an image with coefficients $a_0[k]$, at time k, through a decomposition low filter $\overline{h}_0$, a decomposition high filter $\overline{h}_1$ and a decimation function ($\downarrow 2$). As a result of level 1 decomposition, one obtain an approximation image of coefficients $a_1[k]$ and a detail image of coefficients $d_1[k]$. The same process is applied, at level j, on the approximation

---

1 The term "lossless" refers to the conservation of all details in the image after reconstruction, which means that the original and reconstructed images are identical.

2 The term "lossy" refers to the loss of details in the image after reconstruction by quantification or truncation, which means that the original image differs from the reconstructed one.

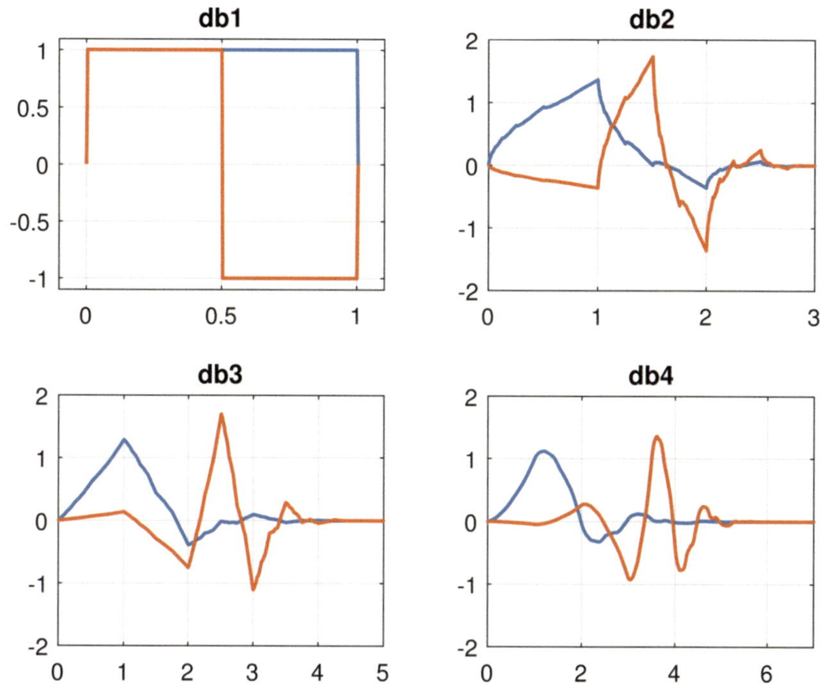

Figure 1.
Four examples of dyadic daubechies wavelets. —— Scaling function, —— Wavelet function.

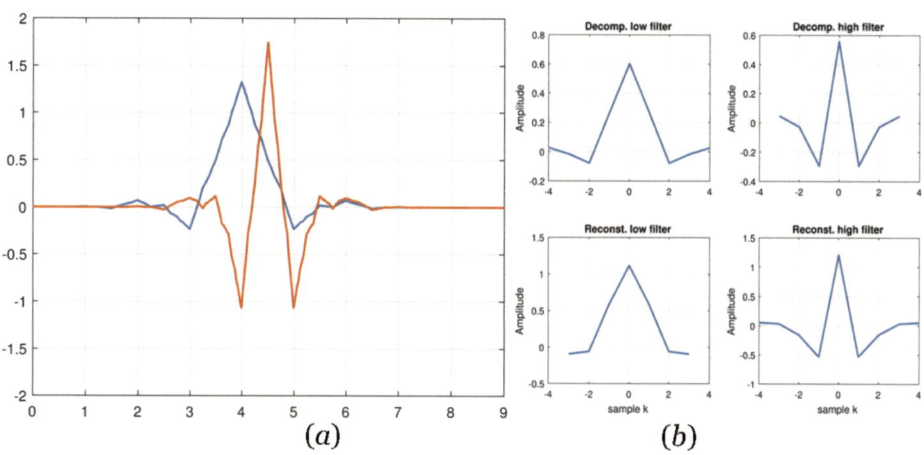

Figure 2.
Wavelet of Cohen-Daubechies-Feauveau 9-tap/7-tap. (a) Scaling and wavelet functions, (b) decomposition and reconstruction filters.

$a_{j-1}[k]$ to get an approximation $a_j[k]$ and a detail $d_j[k]$. **Figure 3** shows a wavelet 3-level decomposition.

2.3 Wavelet reconstruction

The reconstruction process (called filter bank synthesis) follows the inverse order of decomposition process, which means that, at level j and time k, an approximation $a_j[k]$ and a detail $d_j[k]$ are oversampled ($\uparrow 2$) and passed, respectively, through reconstruction low filter h_0 and reconstruction high filter h_1 to

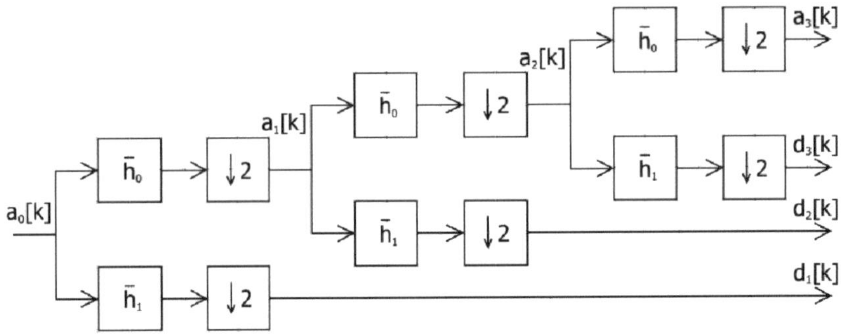

Figure 3.
Wavelet 3-level decomposition.

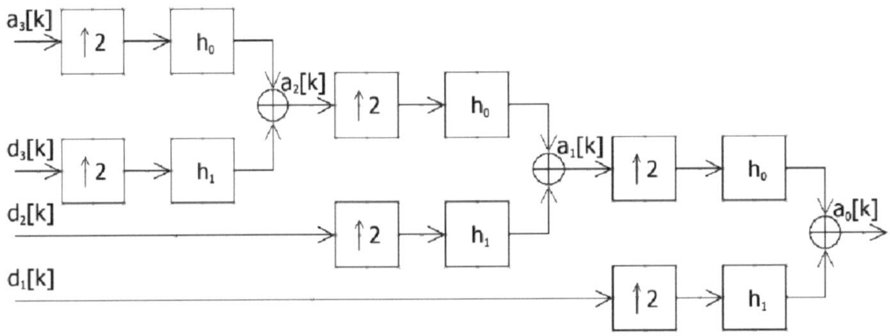

Figure 4.
Wavelet 3-level reconstruction.

generate an approximation image of coefficients $a_{j-1}[k]$. **Figure 4** shows a wavelet 3-level reconstruction.

A perfect reconstruction satisfies the following criteria:

$$\overline{H}_0(f) \bullet H_0(f) + \overline{H}_1(f) \bullet H_1(f) = 2 \tag{1}$$

$$\overline{H}_0(f+1/2) \bullet H_0(f+1/2) + \overline{H}_1(f+1/2) \bullet H_1(f) = 0 \tag{2}$$

where, f is a normolised frequency, $\overline{H}_i(f)$ and $H_i(f)$ $(i = 0, 1)$ are, respectively, the Fourier transform of impulse responses $\overline{h}_i(k)$ and $h_i(k)$.

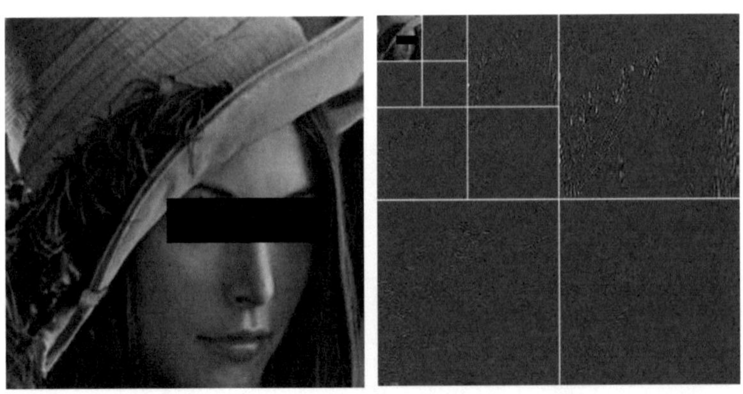

(a) (b)

Figure 5.
3-level decomposition employing CDF 9/7. (a) Original 'Lena' image, (b) decomposed 'Lena' image.

Figure 5 illustrates the result of applying CDF 9/7 3-level decomposition over gray-scale image 'Lena'.

3. Quincunx wavelet transform

3.1 Definition

The decomposition and reconstruction processes using QWT remain the same as DWT; however, there are some differences:

- The diamond McClellan transform [13] is applied to map a 1-D design onto the quincunx structure.

- The decimation factor is $\sqrt{2}$ for each direction.

The 2D quincunx refinement and wavelet filters are given respectively by:

$$H_\lambda\left(e^{j\vec{\omega}}\right) = \frac{\sqrt{2}(2+\cos\omega_1+\cos\omega_2)^{\frac{\lambda}{2}}}{\sqrt{(2+\cos\omega_1+\cos\omega_2)^{\lambda}+(2-\cos\omega_1+\cos\omega_2)^{\lambda}}} \quad (3)$$

$$G_\lambda\left(\vec{z}\right) = z_1 H_\lambda\left(-\vec{z}^{\,-1}\right) \quad (4)$$

where, $\vec{\omega} = (\omega_1, \omega_2)$ is 2D pulse, $\vec{z} = e^{j\vec{\omega}}$ is the discrete Fourier transform parameter and λ is filter order. All simulations in this chapter were performed considering $\lambda = 5$.

The QWT 6-level decomposition of image 'Lena' is given in **Figure 6**.

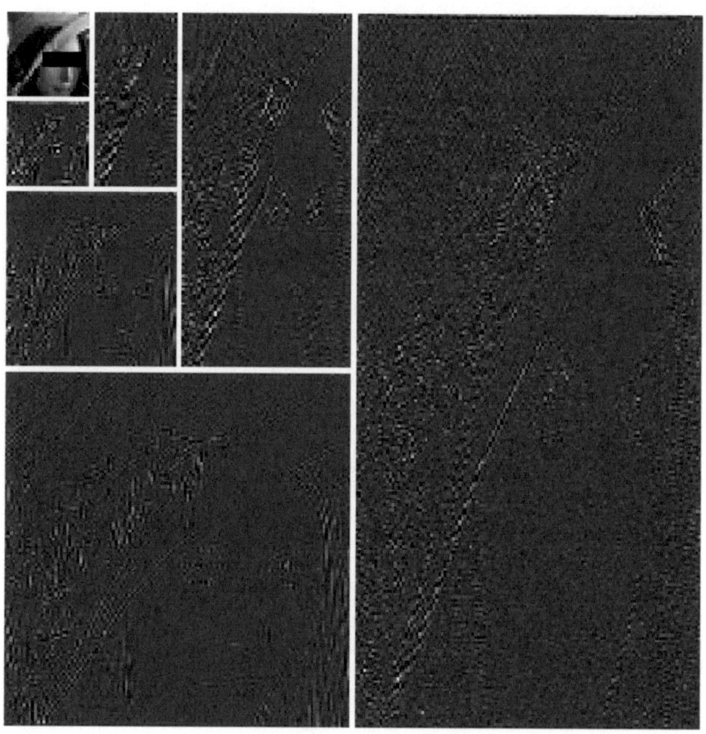

Figure 6.
6-level decomposition of image 'Lena' employing QWT.

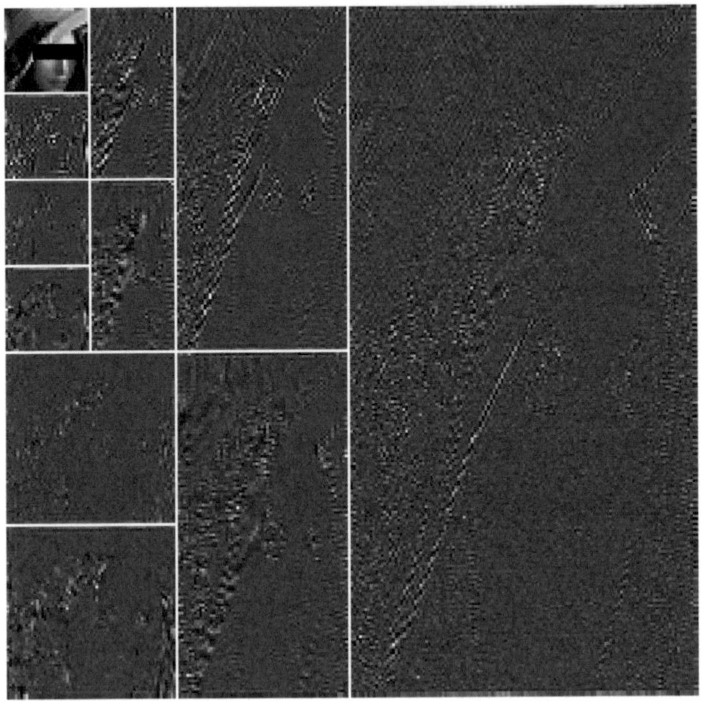

Figure 7.
6-level decomposition of image 'Lena' employing PQWT.

3.2 Quincunx wavelet packet transform

The Wavelet Packet Transform (WP) [14] consists on generalizing the decomposition over all parts of the decomposed images (approximations and details) considering the following condition: a detail image is decomposed if its entropy decreases after decomposition. The literature has shown that this technique is more efficient on textural images.

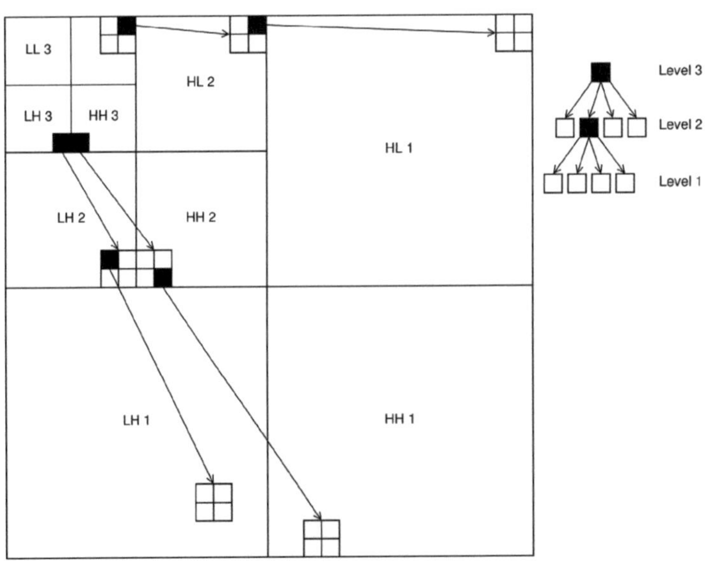

Figure 8.
SPIHT algorithm (L denotes low filtering and H denotes high filtering).

Employing the packet transform on QWT (PQWT) implies that only dyadic parts of QWT decomposition are concerned, which means that the analysis time decreases.

Figure 7 shows the entropy-based PQWT 6-level decomposition of image 'Lena'.

3.3 Set partitioning in hierarchical trees encoder

In order to compress an image employing wavelet-based transform, an encoding step is used to give a bit-level representation to the image. This chapter employs the SPIHT encoder as bit-level representation encoder. **Figure 8** summarizes the relationship between decomposition levels. The authors of [15] had proposed a modified version of SPIHT for the wavelet packet transform; this version is adopted for the PQW transform.

4. Proposed QWT/PQWT-based compression algorithm

4.1 Compression scheme

The JPEG standard is based on dividing an image into sub-images of size 8×8, then applying the DC transform on each image. In the proposed approach, one adopts 8-level PQWT as transform algorithm and a size of $m^2 = 128 \times 128$ for the dividing process. An example of dividing process is given in **Figure 9**.

The proposed compression scheme is summarized in **Figure 10**. It consists on applying, on each image $(I_l, l = 1, 2, \ldots)$ constituting the original image, the QW or

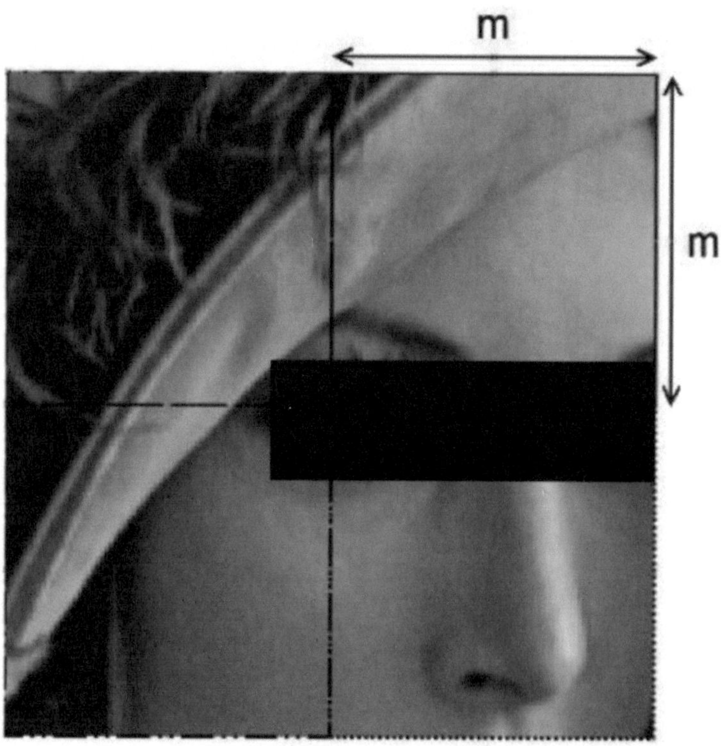

Figure 9.
Example of dividing image 'Lena' into four images of size m^2.

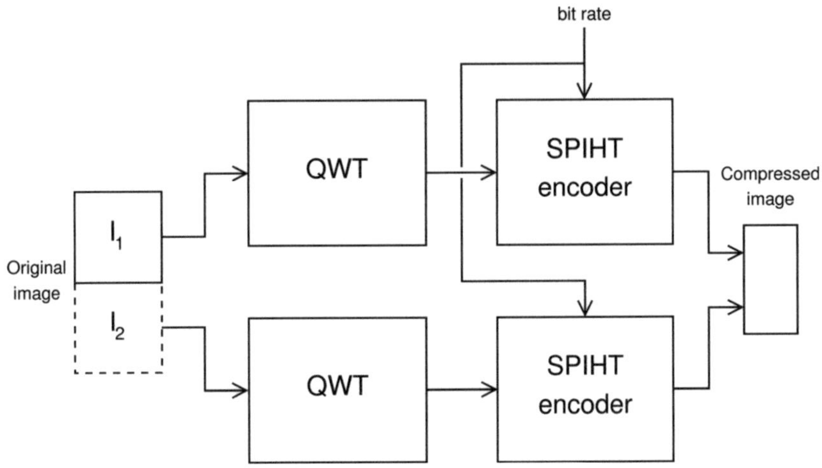

Figure 10.
The proposed compression scheme

PQW transform then the SPIHT algorithm with respect of a compression bit rate. The resulting bit streams are gathered to construct the compressed image.

In order to test the proposed compression algorithm, three gray-scale image databases were employed. The first database consists of 60 images (20 satellite images, 20 natural images and 20 medical images) of size 512×512 [16], the second database consists of 114 textural images of size 512×512 [17] and the third database consists of Shivang Patel 168 fingerprint images of size 256×256. Each image from databases is divided into sub-images of size 128×128 (16 sub-images in case of 512×512 image and 4 sub-images in case of 256×256 image).

Considering packet quincunx wavelet transform, one has tested the 260 possible 8-level decompositions (called decomposition bases) on the sub-images, in order to select the optimal packet decomposition base (in sense of evaluation parameters). The performance of PQWT is compared with QWT and entropy-based PQWT decomposition base (called proper base).

4.2 Reconstruction scheme

The proposed reconstruction scheme is shown in **Figure 11**. The compressed image is divided into bit streams according to the number of sub-images. Each bit

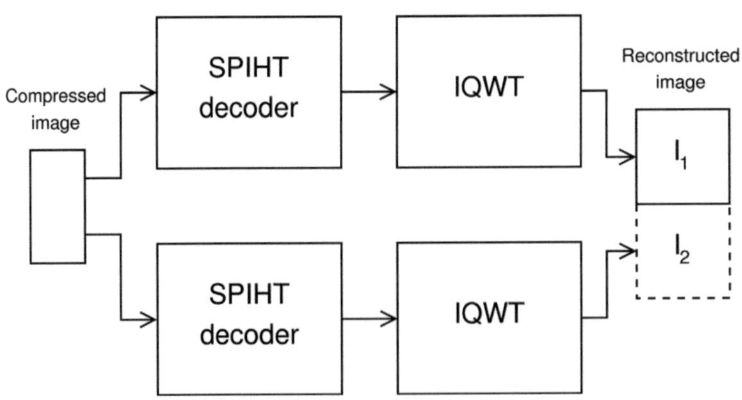

Figure 11.
The proposed reconstruction scheme

stream is decoded and transformed using Inverse QW or Inverse PQW transform, to finally obtain the reconstructed sub-image I_l, ($l = 1, 2, \ldots$). All sub-images are then gathered to construct the reconstructed image.

4.3 Evaluation parameters of compressed image quality

Choosing a compression bit rate $R_c < 8\ bpp$ for gray-level image, leads to a degradation on the original image. This degradation can be measured using the evaluation parameters of compressed image quality. In this chapter, three evaluation parameters are adopted [10]:

- Peak Signal to Noise Ratio: the PSNR parameter is given by $PSNR = 10 \times \log_{10}\left(\frac{(2^R-1)^2}{MSE}\right)$, where, $MSE = \frac{1}{N \times M}\sum_{i=1}^{N}\sum_{j=1}^{N}\left(I(i,j) - \hat{I}(i,j)\right)^2$ is the Mean Square Error, I is the original image, $\hat{I}$ is the reconstructed image and R designates the resolution of a gray-scale image.

- Mean Structural SIMilarity index: the MSSIM index is the average over all local windows of the product of three functions as follow $MSSIM(I, \hat{I}) = \frac{1}{M}\sum_{i=1}^{M} l(I_i, \hat{I}_i) \bullet c(I_i, \hat{I}_i) \bullet s(I_i, \hat{I}_i)$, where, l, c and s are the luminance, contrast and structure comparison functions.

- Visual Information Fidelity: the VIF parameter is a ratio of conditional mutual information measured over all decomposition parts of the image.

5. Results and discussion

The main purpose of this study is to establish a compression strategy using packet quincunx wavelet transform, whatever the type or the size of an image. Therefore, one has begun with applying on the 20 satellite images an exhaustive search among the 260 PQWT decomposition bases. To evaluate the compression quality employing PQWT (for a given bit rate), the relative errors (ε_{PSNR}, ε_{VIF} and ε_{MSSIM}) are used to distinguish between the different performance curves. These parameters are expressed as follow:

$$\varepsilon_X = \frac{m_X^{QWT} - m_X^{PQWT}}{m_X^{QWT}} \times 100\ [\%] \tag{5}$$

where, X designates an evaluation parameter (PSNR, VIF or MSSIM) and m_X is the average of X over all database images. For a negative value of ε_X, the PQWT outperforms the QWT.

Figure 12 illustrates the best 17 decomposition bases that achieve minimum values of ε_X. Base 0 refers to QWT decomposition.

The second step consists on applying these 17 decomposition bases on the other databases, and then evaluates the compressed images quality using the relative errors given in Eq. (5). **Table 1** shows for each database, the top 10 decomposition bases; and in green the five common decomposition bases between all databases.

The evaluation curves in sense of relative error are shown in **Figure 13**. Each Figure compares the evaluation curves of the top 10 decomposition bases, in addition of the PQWT proper decomposition. It can be observed

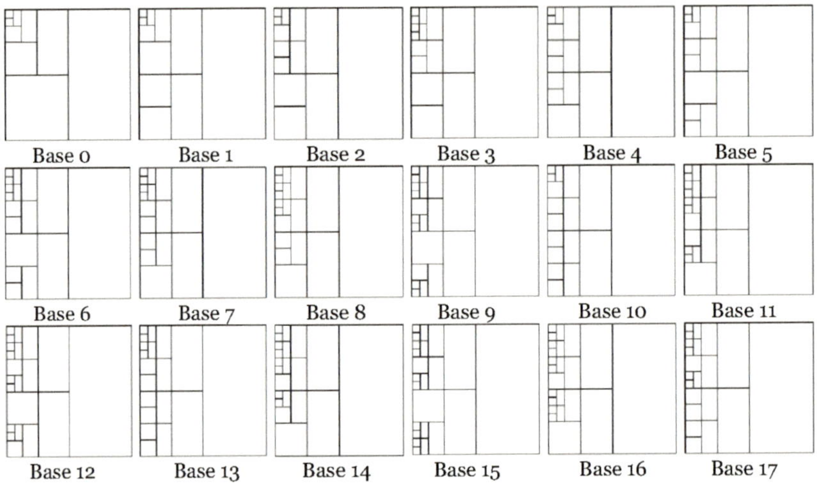

Figure 12.
Best 17 decomposition bases

Base	Satellite images	Natural images	Medical images	Textural Images	Fingerprint images
1	×	×	×		×
2	×	×	×		×
3	×	×	×	×	×
4	×	×	×		
5		×	×		
6	×	×	×	×	×
7	×	×	×	×	×
8	×	×	×	×	×
9	×	×		×	×
10					
11	×	×	×	×	×
12				×	
13			×		×
14	×			×	×
15				×	
16				×	
17					

Table 1.
Top 10 decomposition bases of each database

that the proper decomposition achieves fewer performances in case of satellite, textural and fingerprint images, where a visibility (VIF) average degradation of 8% is measured for fingerprint images. On the other hand, the proper decomposition curves are competitive in comparison with the other decomposition curves in case of natural and medical images, especially for ε_{PSNR} at low bit rate values.

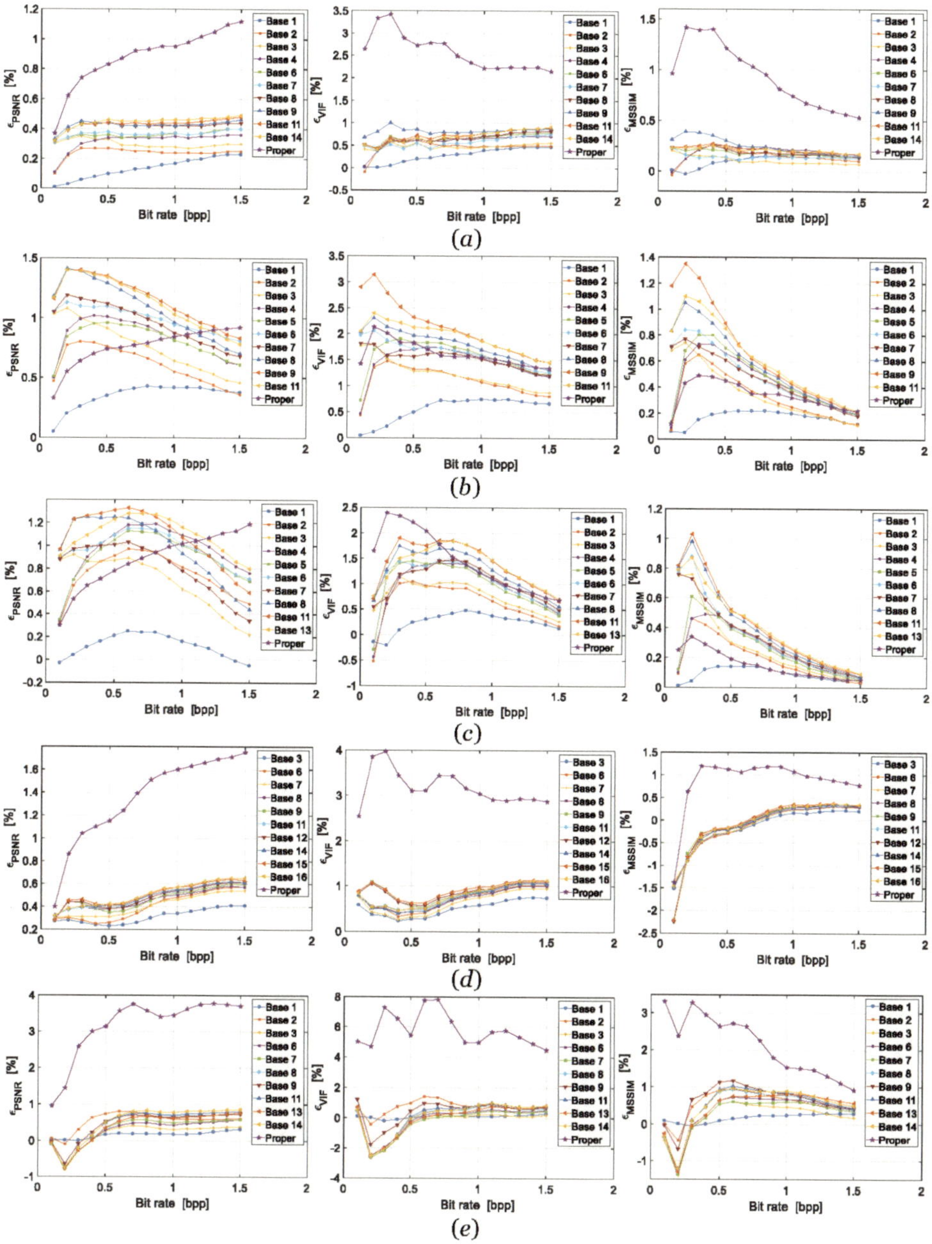

Figure 13.
The evaluation curves vs. bit rate in sense of relative error. (a) Satellite images, (b) natural images, (c) medical images, (d) textural images, (e) fingerprint images

Considering the performance curves of fingerprint images (**Figure 13**.e), negative values of relative errors are observed at low bit rate region; which means that the chosen decomposition bases achieve better performance than base 0 (QWT).

Regarding the five common decomposition bases (marked in green in **Table 1**), the curves of **Figure 13** show that the decomposition base 3 achieves slightly better performance; therefore, in the rest of the chapter, one adopts this decomposition base.

In order to illustrate the compression effect on the database images, one has chosen from each database the image that satisfies the minimum ε_{PSNR}. The chosen images are given in **Figure 14**.

Figure 14.
Original images from databases satisfying the minimum ε_{PSNR}. *(a) satellite image, (b) natural image, (c) medical image, (d) textural image, (e) fingerprint image.*

In **Table 2**, it is given the performance in sense of peak SNR of the five adopted images, for two values of bit rate (0.25 and 2.00 *bpp*). These results show a tiny superiority of decomposition base 3 in comparison with base 0 and the proper decomposition; except for the medical image, where a difference of 9.35 *dB* is observed between base 3 and base 0 at a bit rate of 2.00 *bpp*.

As observed in JPEG compression scheme, the image division into sub-images causes block boundary artifacts; these artifacts are visible at low bit rates values. This phenomenon is clearer for natural, medical and fingerprint image at a bit rate of 0.25 *bpp*.

To remedy to the problem of block boundary artifacts, one propose to add two processes to the compression scheme (as shown in **Figure 15**):

- The two sub-images I_1 and I_2 overlap and have d common pixels (an example of image division with overlapping is given in **Figure 16**),

- Each sub-image is weighted by a 2D Gaussian window defined by the sub-image size m and the minimum amplitude a (as shown in **Figure 17**).

To avoid the pixel redundancy causes by the overlapping, the pixel may have ½ value in case of two overlapped sub-image, and ¼ value in case of four overlapped sub-images. Therefore, as summarized in **Figure 18**, the total size M of a sub-image may be expressed as follow:

- In case of two overlaps, the size of the sub-image equals $M = (m-d)^2 \bullet 1 + 2 \times d \bullet (m-d) \bullet \frac{1}{2} + d^2 \bullet \frac{1}{4} = \left(m - \frac{d}{2}\right)^2$,

- In case of three overlaps, the size of the sub-image equals $M = (m - 2 \bullet d) \bullet (m-d) \bullet 1 + 2 \times d \bullet (m-d) \bullet \frac{1}{2} + d \bullet (m - 2 \bullet d) \bullet \frac{1}{2} + 2 \times d^2 \bullet \frac{1}{4} = (m-d) \bullet \left(m - \frac{d}{2}\right)$,

Base **Bit rate**	**Base 0**	**Base 3**	**Proper decomposition**
0.25 bpp	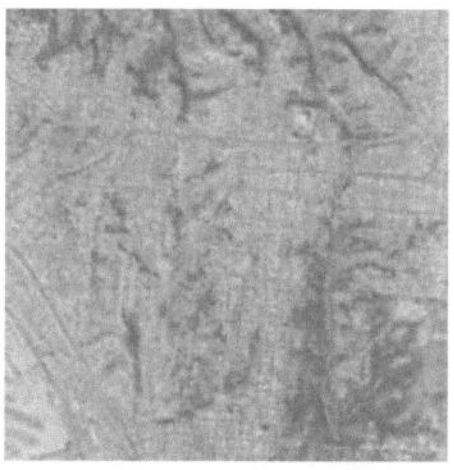 $PSNR = 28.48$ dB	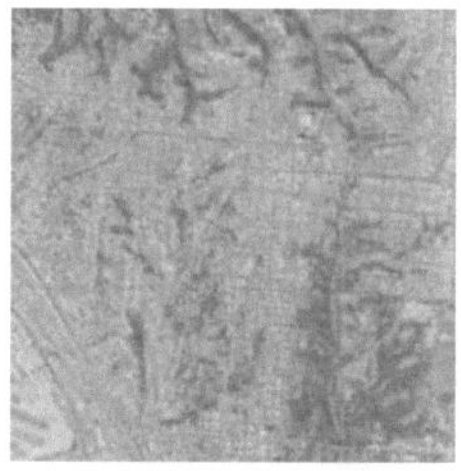 $PSNR = 28.49$ dB	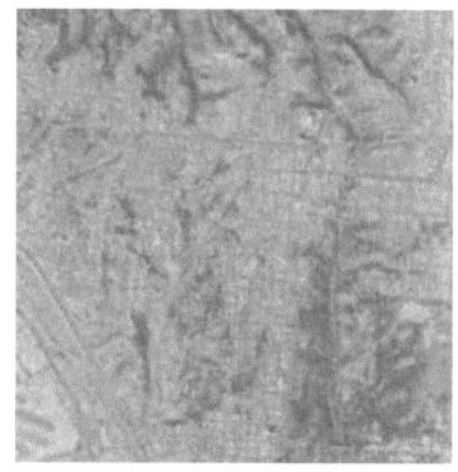 $PSNR = 28.26$ dB
2.00 bpp	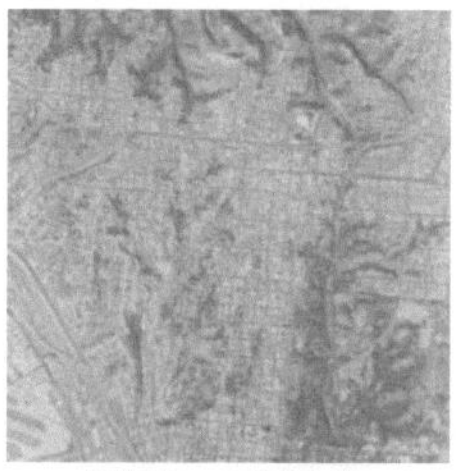 $PSNR = 38.53$ dB	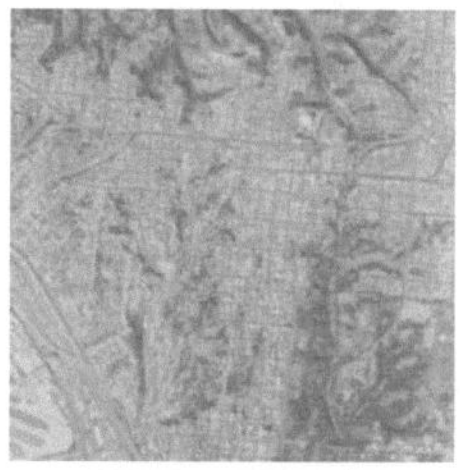 $PSNR = 38.58$ dB	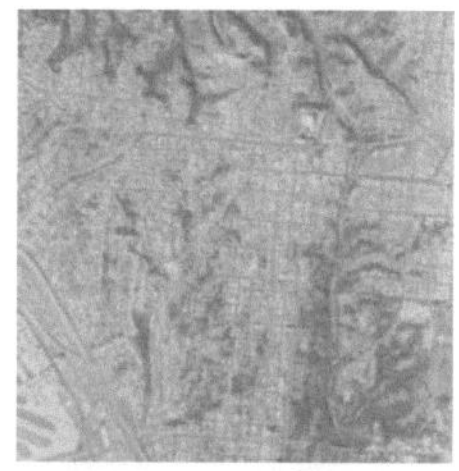 $PSNR = 38.21$ dB
0.25 bpp	 $PSNR = 27.36$ dB	 $PSNR = 27.38$ dB	 $PSNR = 27.16$ dB
2.00 bpp	 $PSNR = 38.91$ dB	 $PSNR = 39.20$ dB	 $PSNR = 38.36$ dB
0.25 bpp	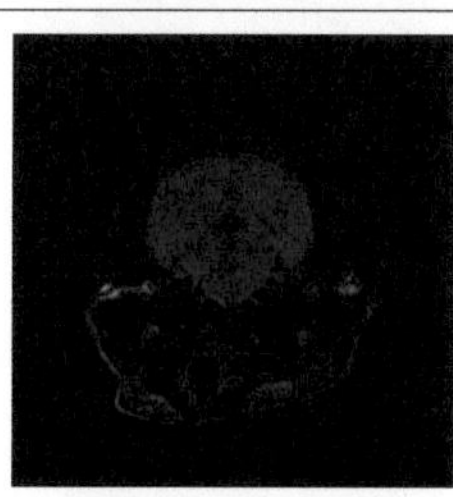 $PSNR = 39.92$ dB	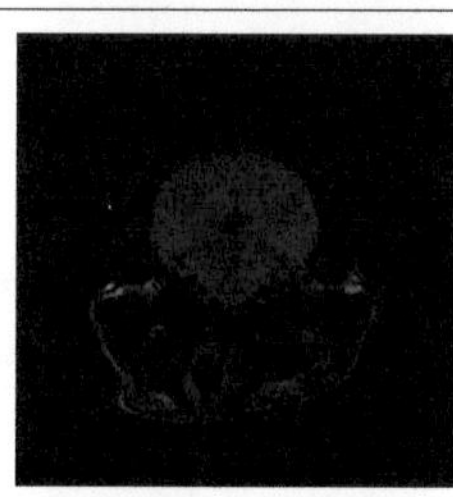 $PSNR = 40.47$ dB	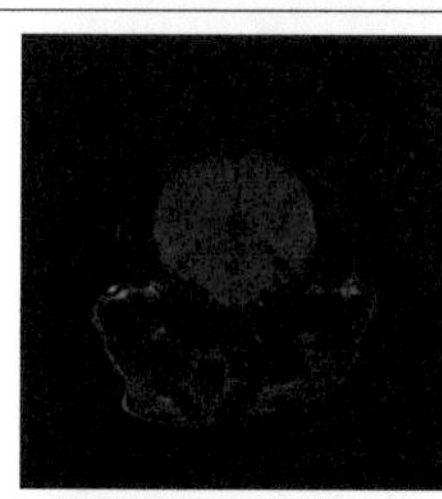 $PSNR = 39.50$ dB
2.00 bpp	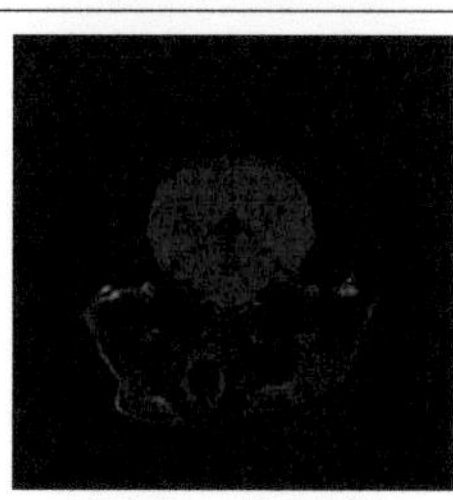 $PSNR = 47.24$ dB	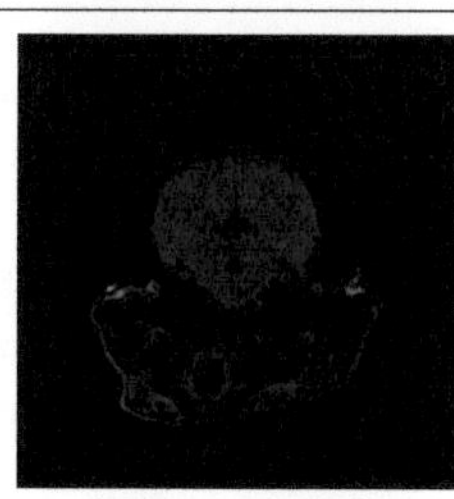 $PSNR =$ **56.59** dB	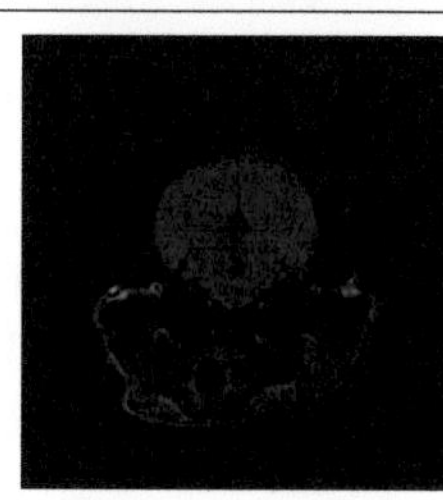 $PSNR = 46.98$ dB

Base Bit rate	Base 0	Base 3	Proper decomposition
0.25 bpp	 $PSNR = 14.62$ dB	 $PSNR = 15.48$ dB	 $PSNR = 15.23$ dB
2.00 bpp	 $PSNR = 23.96$ dB	 $PSNR = 24.55$ dB	 $PSNR = 24.06$ dB
0.25 bpp	 $PSNR = 17.78$ dB	 $PSNR = 17.87$ dB	 $PSNR = 17.04$ dB
2.00 bpp	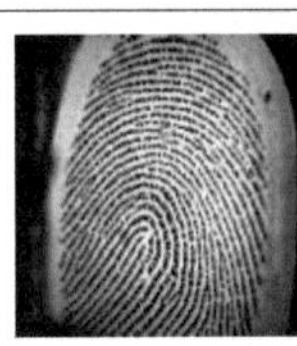 $PSNR = 30.46$ dB	 $PSNR = 31.98$ dB	 $PSNR = 30.06$ dB

Table 2.
Performance in sense of peak SNR of the five adopted images.

- In case of four overlaps, the size of the sub-image equals $M = (m - 2 \bullet d)^2 \bullet 1 + 4 \times d \bullet (m - 2 \bullet d) \bullet ½ + 4 \times d^2 \bullet ¼ = (m - d)^2$.

These new sub-image sizes permit to define a bit rate for each sub-image according to the number of overlaps. In other words, the bit rate of an overlapped sub-image equals:

$$R_c' = R_c \bullet \frac{M}{m^2} \leq R_c \tag{6}$$

where, R_c denotes the bit rate without overlapping and $F_r = \frac{M}{m^2}$ is called the reduction factor of bit rate. The bit rate of the overall image is the average over sub-images bit rates.

In order to demonstrate the effect of the overlapping pixels (d) on bit rates, one has plotted in **Figure 19** the reduction factor curves for $m = 128$. It is clear from these curves that a higher order of reduction factor leads to lower compression performance; therefore, a reduction factor threshold of 0.9 has to be respected.

DOI: http://dx.doi.org/10.5772/intechopen.94970

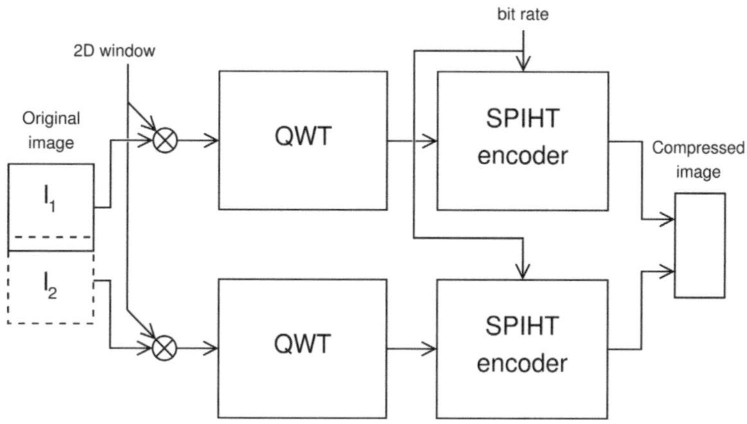

Figure 15.
Proposed compression scheme employing overlapped sub-images.

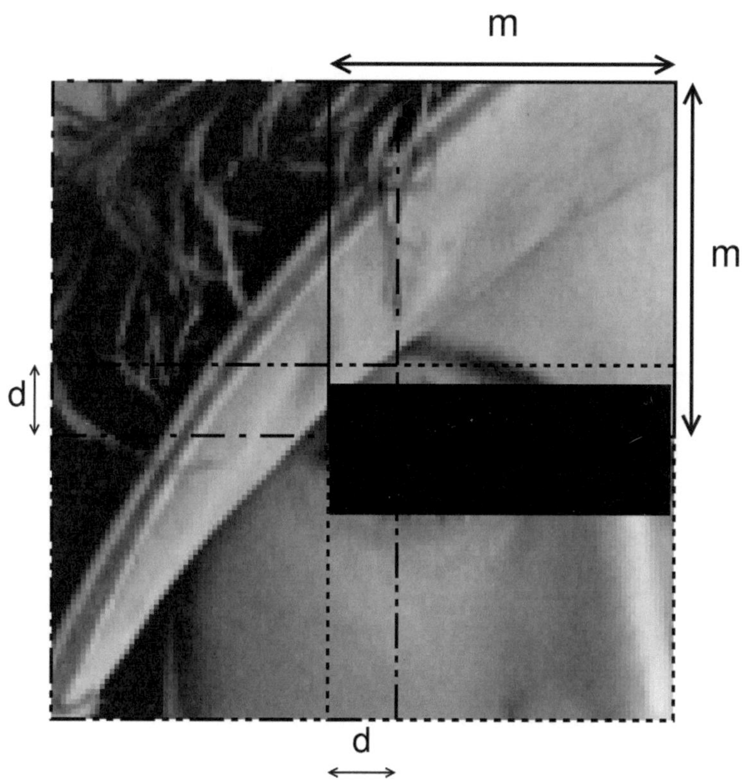

Figure 16.
Example of dividing image 'Lena' with overlapping.

The proposed reconstruction scheme, given in **Figure 20**, divides the output of the inverse QWT or the inverse PQWT by the same 2D Gaussian windows define in the compression process. The reconstructed sub-images are overlapped with the same manner as the compression process, to construct the reconstructed image.

The evaluation curves in sense of PSNR and VIF parameters are shown in **Figure 21**. Each Figure compares the evaluation curves of the PQWT with decomposition base 3 and proper decomposition, in addition of the JPEG compression standard. It can be observed that:

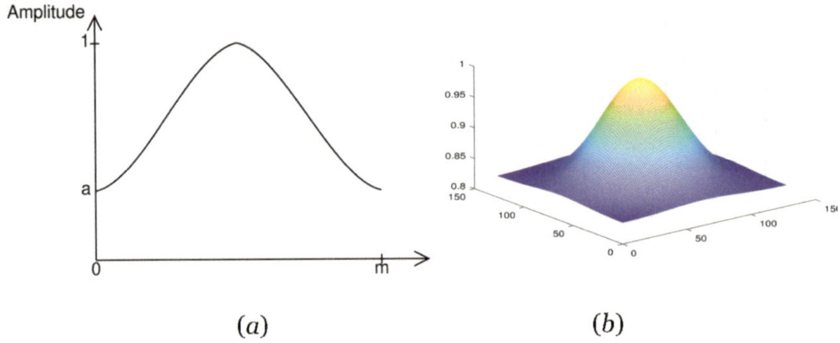

Figure 17.
2D Gaussian window.(a) Gaussian window with parameters m *and* a*.(b) 2D Gaussian window with* m = 128 *and* a = 0.834.

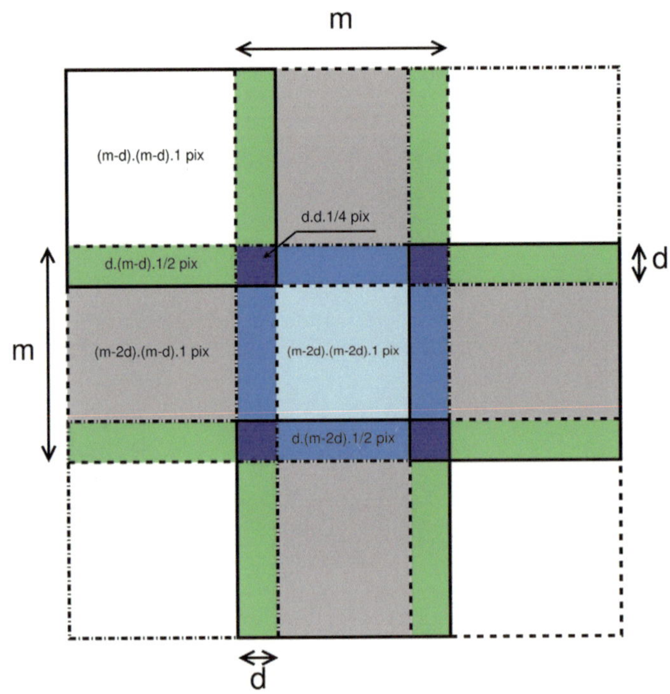

Figure 18.
Computation of pixel sizes for all possible overlapping cases.

- in case of satellite and fingerprint images (**Figure 21**.a and **Figure 21**.e), the proposed PQWT compression scheme presents better performance than JPEG,
- in case of natural and textural images (**Figure 21**.b and **Figure 21**.d), JPEG standard outperforms the proposed compression scheme,
- in case of medical image (**Figure 21**.c), both schemes present slightly the same performance.

In **Table 3**, it is given the performance in sense of PSNR and VIF parameters of the five adopted images, at a bit rate of 0.54 *bpp* for textural image and 0.50 *bpp* for

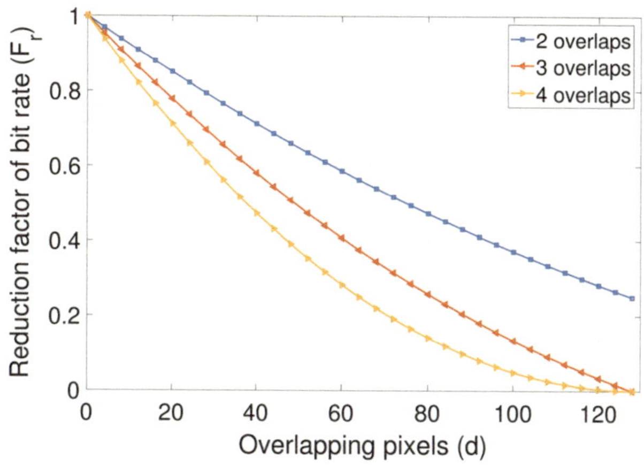

Figure 19.
Reduction factor of bit rate vs. overlapping pixels.

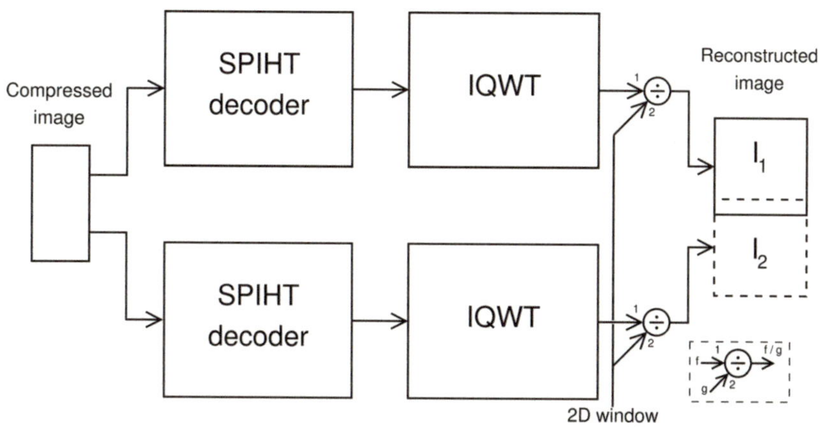

Figure 20.
Proposed reconstructed scheme.

the others. To compare the performance of the proposed scheme (PQWT with a and d parameters), two other schemes are involved: the PQWT with $a = 1$ and $d = 0$ (referring to the first proposed scheme) and JPEG standard.

To fix a and d parameters of PQWT – base 3 and PQWT – proper decomposition, an exhaustive search (with respect of the reduction factor threshold) has been performed to get the maximal value of PSNR parameter. It can be observed in **Table 3** that the a and d parameters differ from an image to another; therefore, these parameters have to be included in the compressed file, as well as the size of the overall image.

The obtained results show a tiny superiority of PQWT – base 3 in comparison with JPEG and PQWT – proper decomposition; except for natural and medical images, where the JPEG standard is slightly better.

Figure 22 compares the visual side of the five adopted compressed images, where details of size 128×128 from original, PQWT – base 3 and JPEG images are magnified. From these figures, it can be observed that PQWT compressed images present lower block boundary artifacts effect in comparison with JPEG images (especially for satellite and textural images), and preserve the continuity of their detail shapes.

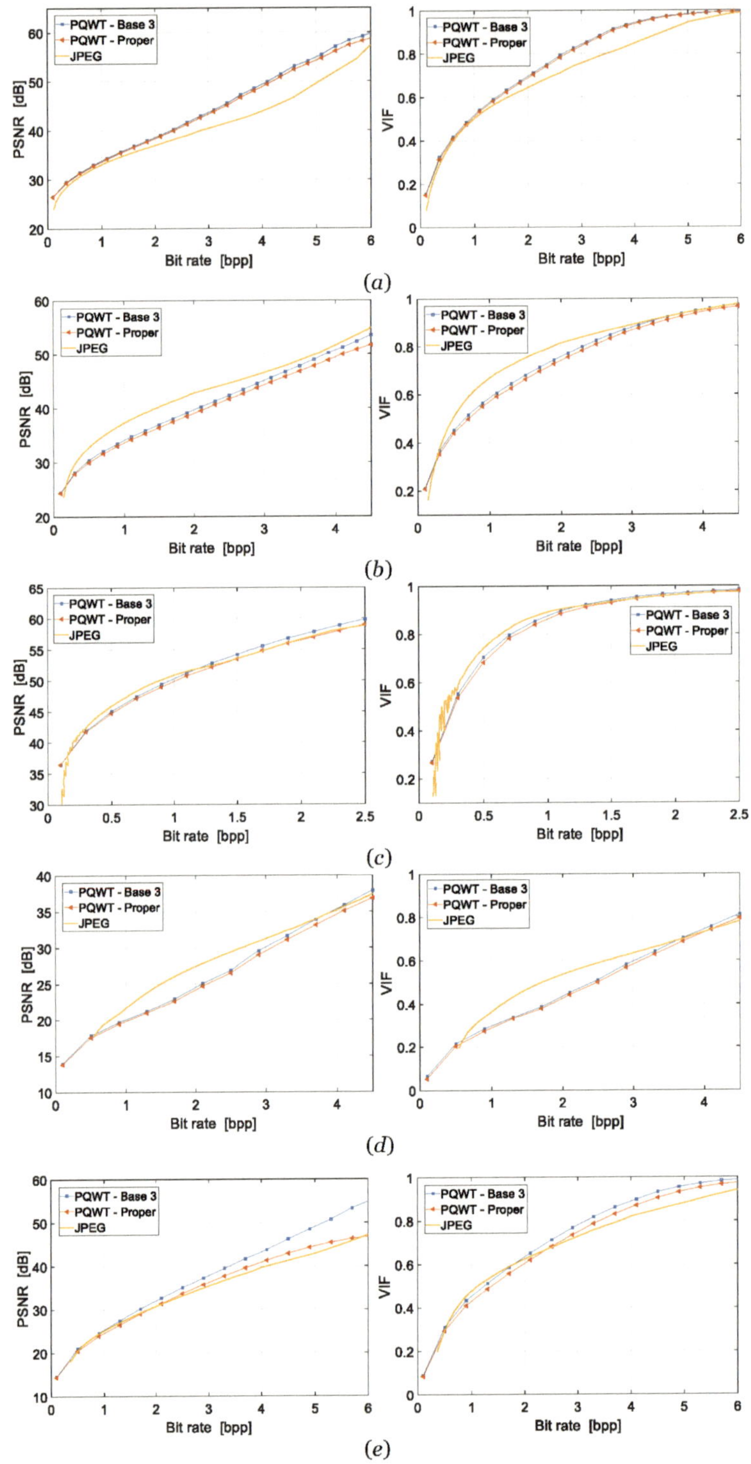

Figure 21.
The evaluation curves vs. bit rate in sense of PSNR and VIF. (a) Satellite image, (b) natural image, (c) medical image, (d) textural image, (e) fingerprint image.

For a deep study of the block boundary artifacts effect between both proposed compression schemes employing PQWT – base 3, one have focused only on the overlapping regions in the five adopted images, where bands of 8 pixels per line and

DOI: http://dx.doi.org/10.5772/intechopen.94970

Bit rate	PQWT – Base 3 $a = 1, d = 0$	PQWT – Base 3	PQWT – Proper decomposition	JPEG
0.50 bpp	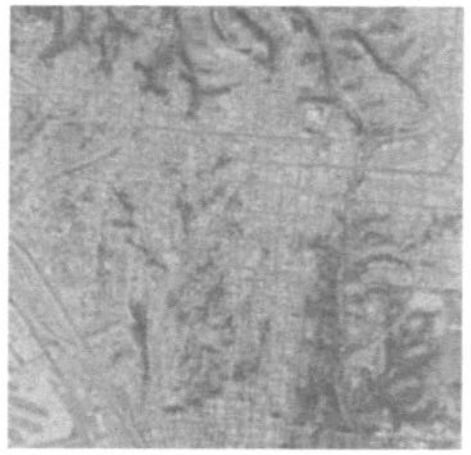 $PSNR = 30.63$ dB, $VIF = 0.38$	 $PSNR = 30.71$ dB, $VIF = 0.39$ $a = 0.969, d = 4, F_r = 0.950$	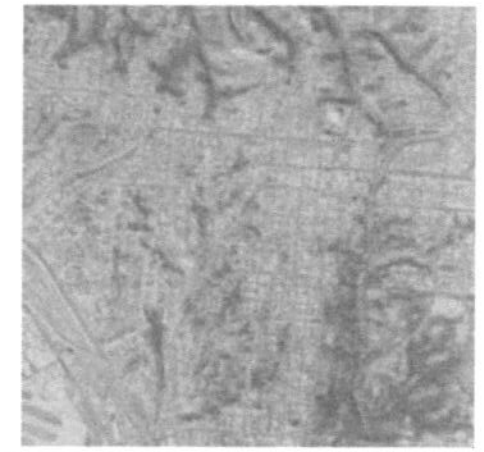 $PSNR = 30.54$ dB, $VIF = 0.38$ $a = 0.962, d = 4, F_r = 0.950$	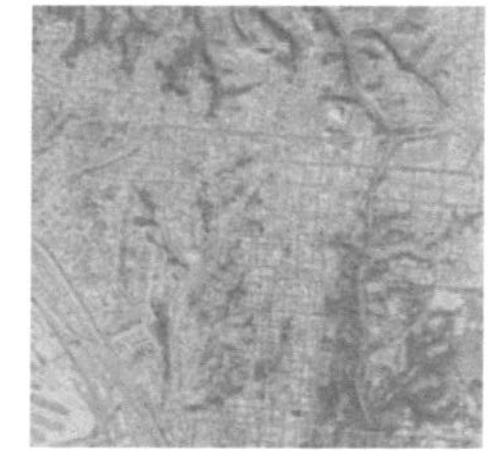 $PSNR = 29.83$ dB, $VIF = 0.36$
0.50 bpp	 $PSNR = 30.24$ dB, $VIF = 0.44$	 $PSNR = 30.43$ dB, $VIF = 0.54$ $a = 0.834, d = 3, F_r = 0.962$	 $PSNR = 30.00$ dB, $VIF = 0.44$ $a = 1, d = 4, F_r = 0.950$	 $PSNR = 32.82$ dB, $VIF = 0.51$
0.50 bpp	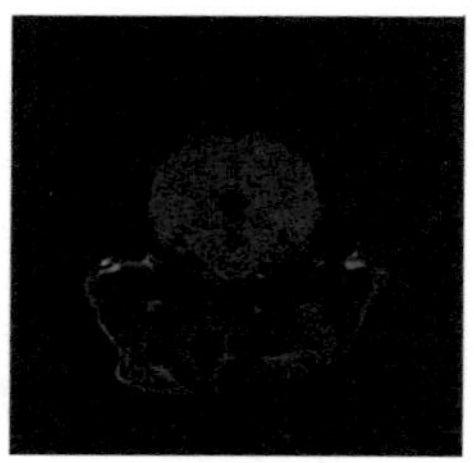 $PSNR = 44.64$ dB, $VIF = 0.69$	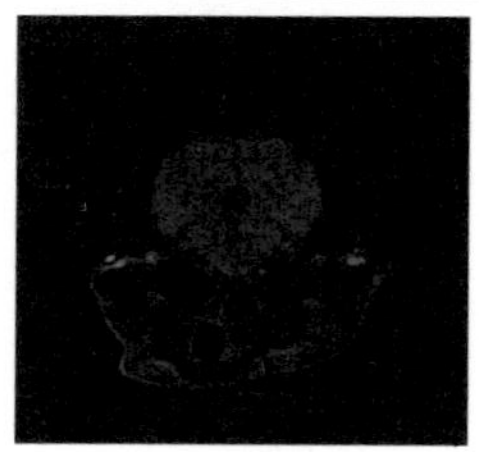 $PSNR = 45.09$ dB, $VIF = 0.70$ $a = 0.938, d = 2, F_r = 0.975$	 $PSNR = 44.76$ dB, $VIF = 0.68$ $a = 0.955, d = 2, F_r = 0.975$	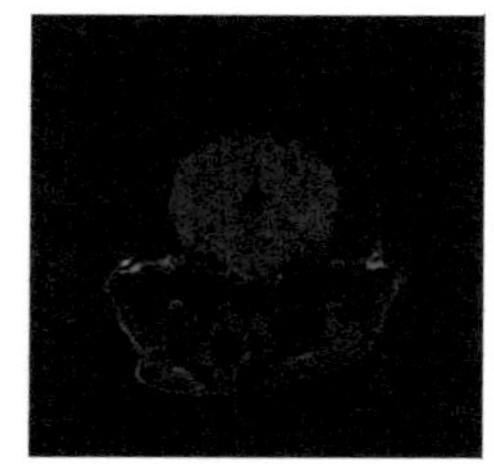 $PSNR = 46.00$ dB, $VIF = 0.74$

Bit rate	PQWT – Base 3 $a = 1, d = 0$	PQWT – Base 3	PQWT – Proper decomposition	JPEG
0.54 bpp	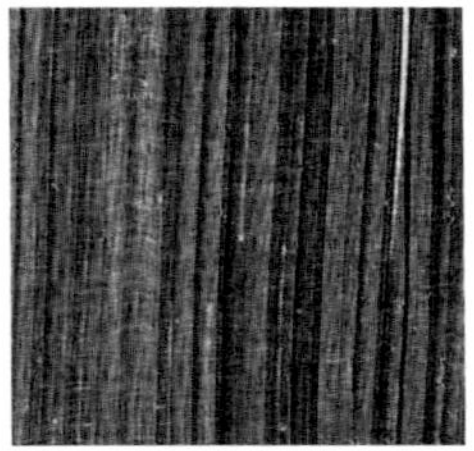 $PSNR = 18.06$ dB, $VIF = 0.22$	 $PSNR = 18.14$ dB, $VIF = 0.22$ $a = 1, d = 4, F_r = 0.950$	 $PSNR = 17.83$ dB, $VIF = 0.21$ $a = 1, d = 2, F_r = 0.975$	 $PSNR = 17.72$ dB, $VIF = 0.19$
0.50 bpp	 $PSNR = 21.02$ dB, $VIF = 0.31$	 $PSNR = 21.12$ dB, $VIF = 0.31$ $a = 1, d = 1, F_r = 0.992$	 $PSNR = 20.48$ dB, $VIF = 0.29$ $a = 0.834, d = 1, F_r = 0.992$	 $PSNR = 20.48$ dB, $VIF = 0.29$

Table 3.
Performance in sense of PSNR and VIF parameters of the five adopted images

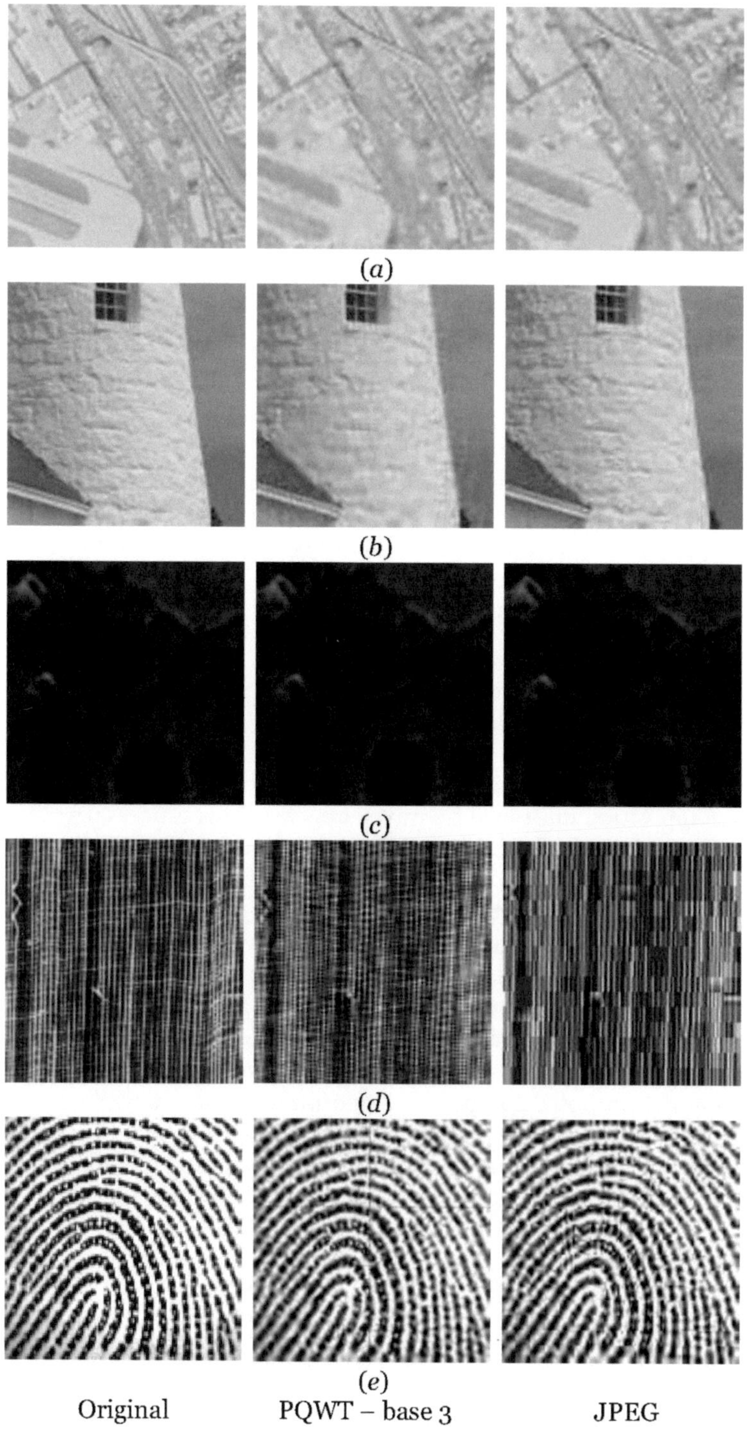

Figure 22.
Magnified detail from the five adopted images. (a) Satellite image – 0.50 bpp, (b) natural image – 0.50 bpp, (c) medical image – 0.50 bpp, (d) textural image – 0.54 bpp, (e) fingerprint image – 0.50 bpp.

column around these regions are extracted in order to measure the effect of block boundary artifacts in sense of PSNR. By denoting $PSNR_1$ and $PSNR_2$ the measured PSNRs of the extracted regions employing, respectively, the first and second proposed schemes, the average block boundary artifacts effect is measured by

Image	Satellite image	Natural image	Medical Image	Textural Image	Fingerprint image
E_{bba} [%]	2.19	27.70	22.87	9.41	0.63

Table 4.
Average block boundary artifacts effect in sense of PSNR for the five adopted images

$$E_{bba} = \sum_{j=1}^{R} \frac{PSNR_2(j) - PSNR_1(j)}{PSNR_1(j)} \times 100\ [\%] \tag{7}$$

where, R values of the bit rate (R_c) from 0.1 bpp up to 8.1 bpp are employed to evaluate the PSNRs.

In **Table 4**, the values of E_{bba} are given for the five adopted images. From these results, it can be concluded that, in comparison with the first compression scheme, the second proposed compression scheme presents a significant reduction of 27.70% and 22.87% of the effect of block boundary artifacts for, respectively, natural and medical images. However, a tiny reduction of E_{bba} for fingerprint image is observed, which means that further processing on sub-images boundaries is necessary for such an image using local 2D filters [18].

6. Conclusion

This chapter introduces an image compression scheme that employs quincunx wavelet decomposition improved by wavelet packet. This process permits to focus on both approximation and detail parts of the image decomposition.

Using the concept of image division into sub-images (employed in JPEG standard compression algorithm), the effect of block boundary artifacts has occurred especially at low range of compression bit rates. To overcome this problem, the sub-images are weighted by a 2D Gaussian window and overlapped with respect to the reduction factor of compression bit rate. This means that, in addition of the overall image size, two parameters have to be included in the compressed file: the minimum amplitude of 2D window and the number of overlapped pixels.

To present the proposed compression algorithm as a standard, its performances were compared, in sense of evaluation parameters, to those of JPEG standard. The main improvement was seen in the capacity of the proposed scheme to provide better image visual quality (detail shapes continuity). This means that, in the first hand, it can be possible to reduce the image file sizes without reducing the image visual quality, and increase the storage capacity in photographic devices in the other hand.

As a result, this compression technique permits to create benchmarks and databases with low capacity whatever its nature (satellite, medical, natural or textural image).

In this work, one have focused on gray-scale images in order to present the proposed compression scheme. It is necessary, in future works, to investigate its efficiency on video and color images compression.

Conflict of interest

The author declare that there is no conflict of interest.

Nomenclature

ECG	Electro-CardioGram
RLE	Run Length Coding
LZW	Lempel-Ziv-Welch
JPEG	Joint Photographic Experts Group
DCT	Cosine Discrete Transform
DWT	Discrete Wavelet Transform
QWT	Quincunx Wavelet transform
EZW	Embedded Zerotree Wavelet
SPIHT	Set Partitioning In Hierarchical Trees
CDF	Cohen-Daubechies-Feauveau
WP	Packet Wavelet
PQWT	Packet-based Quincunx Wavelet transform
IQWT	Inverse Quincunx Wavelet Transform
IPQWT	Inverse Packet-based Quincunx Wavelet Transform
PSNR	Peak Signal to Noise Ratio
MSSIM	Mean Structural SIMilarity index
VIF	Visual Information Fidelity

Author details

Abdesselam Bassou
Information processing and telecommunication laboratory (LTIT), University of Tahri Mohammed Bechar, Algeria

*Address all correspondence to: bassou.abdesselam@univ-bechar.dz

IntechOpen

© 2020 The Author(s). Licensee IntechOpen. This chapter is distributed under the terms of the Creative Commons Attribution License (http://creativecommons.org/licenses/by/3.0), which permits unrestricted use, distribution, and reproduction in any medium, provided the original work is properly cited. CC BY

References

[1] Barmase S, Das S, Mukhopadhyay S. Wavelet transform-based analysis of QRS complex in ECG signals. 2013; arXiv preprint arXiv:1311.6460.

[2] Al-Enizi F, Al-Asmari A. DWT-based data hiding technique for videos ownership protection [Chapter]. In: Baleanu D. Wavelet Transform and Complexity. IntechOpen; 2019. DOI: 10.5772/intechopen.84963

[3] Kanagaraj H, Muneeswaran V. Image compression using Haar discrete wavelet transform. In: Proceedings of 2020 5th International Conference on Devices, Circuits and Systems (ICDCS'20); 5–6 March 2020; Coimbatore, India. p. 271–274. DOI: 10.1109/ICDCS48716.2020.243596.

[4] Patlayenko M, Osharovska O, Pyliavskyi V, Solodka V. Wavelet feature family for image compression. In: Proceedings of 27th National Conference with International Participation (TELECOM'19); 30–31 October. 2019; Sofia, Bulgaria. p. 16–18. DOI: 10.1109/TELECOM48729.2019.8994877.

[5] Ziv J, Lempel A. A universal algorithm for sequential data compression. IEEE trans. inform. theory. 1977;23.3:337–343. DOI: 10.1109/TIT.1977.1055714

[6] Huffman D. A method for the construction of minimum redundancy codes. In: Proceedings of IRE; 1098–1101 September 1952. DOI: 10.1109/JRPROC.1952.273898

[7] JPEG website. Available from http://jpeg.org/ [Accessed: 2020-04-10]

[8] Daubechies I. Ten lectures on wavelets. Society for Industrial and Applied Mathematics; 1992. 357 p. DOI: 10.1137/1.9781611970104

[9] Manuela F, Dimitri VD, Michael U. An orthogonal family of quincunx wavelets with continuously adjustable order. IEEE trans. image proc. 2005;14.4: 499–510. DOI: 10.1109/TIP.2005.843754

[10] Beladgham M. Construction d'une technique d'aide au diagnostic en imagerie médicale: Application à la compression d'images [thesis]. Tlemcen Algeria: Abou Bakr Belkaid University; 2012

[11] Cohen A, Daubechies I, Feauveau JC. Biorthogonal bases of compactly supported wavelets. Communications on pure and applied mathematics. 1992;45:485–560. DOI: 10.1002/cpa.3160450502

[12] Taubman D, Michael M. Image compression fundamentals, standards and practice. Springer; 2002. 777 p. DOI: 10.1007/978-1-4615-0799-4

[13] Shapiro J. Adaptive McClellan transformations for quincunx filter banks. IEEE trans. signal process. 1994; 42.3:642–648. DOI: 10.1109/78.277856

[14] Linderhed A. Adaptive image compression with wavelet packets and empirical mode decomposition [thesis]. Linköping Sweden: Linköping studies in science and technology; 2004

[15] Sprljan N, Grgic S, Mrak M, Grgic M. Modified SPIHT algorithm for wavelet packet image coding. In: Proceedings of the IEEE International Symposium on Video/Image Processing and Multimedia Communications (VIPromCom'02); 16–19 June 2002; Zadar, Croatia. p. 189–194. DOI: 10.1109/VIPROM.2002.1026653

[16] Allaoui CE, Bassou A, Benyahia I, Khelifi M. Selection of compression test images using variance-based statistical method. Indonesian journal of electrical engineering and computer science. 2019;16.1:243–258. DOI: 10.11591/ijeecs.v16.i1.pp243-258

[17] Brodatz P. Textures: a photographic album for artist and designers. Dover Publications, New York; 1966

[18] Lim JS. Two-dimensional signal and image processing. Prentice-Hall International; 1990. 694 p.

Chapter 11

Uncertainty and the Oracle of Market Returns: Evidence from Wavelet Coherence Analysis

Joan Nix and Bruce D. McNevin

Abstract

Wavelet methodology is employed to investigate the statistical relationship between three well-accepted measures of uncertainty and both market and sector returns. Our primary goal is to determine whether uncertainty is sector specific. Although there are periods when the market works effectively as an oracle capturing uncertainty, we also find sector specific uncertainty. The wavelet equivalent of correlation, coherence, is used to determine the presence of sector specific uncertainty. We find that allowing localized information in the time frequency domain is critical for separating out sector specific uncertainty from market uncertainty.

Keywords: finance, sectors, wavelets, uncertainty, coherence

1. Introduction

Uncertainty shocks call the the market's knowledge-gathering role into question. The equity market as an oracle works well when it provides rapid price discovery that reflects the underlying fundamentals of an economy. But when facing uncertainty shocks the equity market's function as a consensus mechanism that reveals economic reality appears at first glance, poorly suited for the environment it faces. An oracle needs a reliable channel for obtaining information. In the face of uncertainty, the equity market turns into a network of pipes where funds flows in ways that leave many skilled observers of market moves caught off guard. The shock filters through to the inter-temporal trade-offs of investors and makes forecasting more of a bet on imagined scenarios than the result of astute modeling that is carefully tested with historic data.

The relevance of wavelet methodology for examining whether the uncertainty measures are correlated at different scales and frequencies with market and sector returns may be more easily imagined with a metaphor.[1] The uncertainty shock operates as a push from behind that a person strolling down the street experiences. The push may be hard and throw the person completely off his path. He may end up face down and in a panic imagining the worse outcome. The push may also be soft from which the person experiences a quick feeling of panic but quickly recovers and

[1] The use of imagery as a guide to economic understanding has a rich history. See for example, the use of a bicycle imagery by Samuelson to explain how a real economic system is capable of resolving indeterminacy even when the path between present and future is far from smooth [1].

continues his walk. It is not that the person is caught completely off guard. The push comes after a signal such as the sound of feet running from behind or the quiet sound of someone walking at a slightly faster rate to catch-up. The relationship between the signal and push is based on short-lived features of the environment. However, very different outcomes are possible. Wavelet methodology is particularly well-suited for capturing these different outcomes because it is designed to capture short-lived features of the environment. Wavelet methodology provides a snapshot of the outcomes in the form of market and sector returns that result from various shocks.

A defining feature of wavelet methodology that makes it particularly well-suited for capturing the economic effects of uncertainty shocks is that at a given point in time the same signal can be analyzed by different wavelets. Most importantly, it is capable of capturing an uncertainty signal that only lasts for a finite period of time. It can also handle non-stationarity which often characterizes uncertainty shocks. Wavelet coherence plots help us discover whether the measures of the shock provide new information that is not reflected in market and sector returns at various scales.

In this chapter, we investigate the statistical relationship between three well-accepted measures of uncertainty and both market and sector returns. The three measures are Macroeconomic and Financial Uncertainty of Jurado, Ludvigson, Ng (JLN) [2] and Economic Policy Uncertainty by Baker, Bloom and Davis (BBD) [3]. We explore the extent to which the impact of uncertainty is sector specific. Employing the wavelet equivalent of correlation, we observe that in the presence of significant coherence, market returns are anti-phase with all three measures of uncertainty. Between market volatility and financial uncertainty, we also observe very high in-phase coherence at low frequencies for prolonged periods of time. However, this is not the case when considering the volatilty of Economic Policy Uncertainty or Macro Uncertainty. For those measures, while there are periods of high coherence, these periods are not as extensive as found with financial uncertainty. One conclusion is that the prolongness of the coherence differs depending on the measure of uncertainty.

Looking at the coherence plots with sector returns and the three measures of uncertainty, we find prolonged high coherence at low frequencies and intermittent coherence at high frequencies. For each coherence plot, we also consider the conditional coherence, after partialing out the effects of the market. By and large, most of the coherence disappears pointing to the question of whether there is any sector-specific uncertainty. Our focus is on six sectors, Telecom, Bus. Equip, Shops. Manufacturing, Energy and Money where each had at least one period of high conditional coherence. For each sector, based on our observation of the conditional coherence plot, we sampled the scales using a Discrete Wavelet Transform (DWT). A DWT is used to run a regression of sector returns against both an uncertainty measure and market returns. A rolling regression is used from which we find the time-pattern of the uncertainty coefficients. It was often the case that the uncertainty had a significant negative impact on sector returns. These snapshots that the wavelet coefficients provide point to the general result that there are significant differences in how uncertainty filters through the sectors that are different from what the market reaction alone tells us.

The remainder of our chapter proceeds as follows: Section 2 highlights research based on wavelet analysis in applied financial economics of particular relevance for our analysis. The important concepts used in wavelet analysis that are applied in our analysis are introduced in Section 3. The data and uncertainty measures are discussed in Section 4. The analysis and results are presented in Section 5. The conclusions follow in Section 6.

2. Literature review

The modern strain of literature relating to uncertainty, and its effects on the economy, grew out concerns in the post credit crisis era that firms were holding off on investments due to uncertainty about the future. Bloom [4] shows that a number of cross-sectional measures of uncertainty are correlated with time series measures of volatility. The cross-sectional measures of uncertainty he considers are the standard deviation of pre-tax profit growth, a stock return measure and the standard deviation of total factor productivity. His time series measure of volatility is stock market volatility. In addition, he evaluates the impact of uncertainty on the real economy using a VAR. He finds that a shock to stock market volatility causes a 1 percent drop in industrial production over a 4 month period. He also reports a similar effect on employment. Bloom identifies 17 major instances of uncertainty based on the stock market volatility measure. Baker, Bloom and Davis (2013) develop a measure of policy uncertainty based on newspaper coverage frequency. They find that their index proxies for movements in policy-related economic uncertainty. Specifically, tight presidential elections, Gulf Wars I and II, the 9/11 attacks, the failure of Lehman Brothers, and the 2011 debt-ceiling dispute are associated with spikes in the index.

Jurado, Ludvigson and Ng [2] develop a measure of uncertainty based on the h-period ahead forecasting error, where h = 1, 3, and 12 months. Using a comprehensive data set of 132 macroeconomic series they aggregate the forecast errors for each series to create a macroeconomic uncertainty index. In contrast to Bloom [4], their analysis finds that there are three major episodes of uncertainty in the 1960–2016 period: 1973–1974, and 1981–1982 recessions, and the Great recession of 2007–2009. Bali, Brown and Tang [5] create an index of macroeconomic uncertainty based on ex-ante measures of cross-sectional dispersion in economic forecasts by the Survey of Professional Forecasters. After controlling for a number of factors, they find a statistically significant negative relationship between their measure of uncertainty and future stock returns. Ludvigson, Ma and Ng [6] examine the question of whether uncertainty is a source of business cycle fluctuations, or an endogenous response. Their analysis distinguishes macroeconomic uncertainty and uncertainty about real economic activity from financial uncertainty. They find that financial uncertainty is primarily an exogenous shock. In addition they find that higher uncertainty about real economic activity is likely to be endogenous, in response to business cycle fluctuations.

3. Wavelet analysis

Prior to the work of Ramsey[2], the use of wavelets in economic and financial analysis was largely non-existent. Today, however, wavelet analysis is a well known, and widely applied tool for any economist who studies time series data.[3] The reason for the rapid increase in wavelet based applications is that the wavelet transform yields a localized decomposition in both time and frequency domain. This stands in sharp contrast to the traditional Fourier transform often used by economists that is global in the sense that there is no time component after the

---

[2] See for instance, [7].

[3] See [8, 9] for an introduction to wavelet methods in economics and finance.

Fourier transform is applied.[4] Since the wavelet transform yields a decomposition that is in localized frequency and time, it has proven to be particularly useful. A clear application where wavelet methodology benefits the analysis is when applied to investment decisions over different time horizons.[5]

The wavelet transform consists of a father wavelet and a set of mother wavelets. Given a function Φ, the father wavelet for the discrete transform is defined as:

$$\Phi_{J,k} 2^{-\frac{J}{2}} \Phi \frac{t - 2^J * k}{2^J} \tag{1}$$

$$\int \Phi(t) dt = 1 \tag{2}$$

The mother wavelets, also in discrete form, are defined as:

$$\Psi_{j,k} 2^{-\frac{j}{2}} \Psi \frac{t - 2^j * k}{2^j}, j = 1, \dots, J \tag{3}$$

$$\int \Psi(t) dt = 0 \tag{4}$$

Where J is the number of scales or levels, 2^J is a scale factor, and k is the time domain index. Note that the father and mother wavelets are each indexed by scale and time. The scale parameter is inversely proportional to frequency.[6] The father wavelet can also be represented as a low pass filter, and the mother wavelets as high pass filters.[7]

Wavelet functions transform a time series, f(t), into a series of wavelet coefficients,

$$S_{J,k} = \int f(t) \Phi_{J,k} \tag{5}$$

and,

$$d_{j,k} = \int f(t) \Psi_{j,k} \quad j = 1, \dots, J \tag{6}$$

Where $S_{J,k}$ are the coefficients for the father wavelet at the maximal scale, 2^J; These coefficients are often referred to as the smooth coefficients. The $d_{j,k}$, or detailed coefficients, are the coefficients of the mother wavelets at the scales from 1 to 2^J.

Applying the transforms results in a time series of length k of smooth coefficients at the maximal scale J, and J time series of detailed coefficients each of length k. If there are 6 scales, the frequency of the first scale is associated with the interval [1/4, 1/2], and the frequency of scale 6 is associated with the interval [1/128, 1/64].

The number of coefficients differs by scale. If the length of the data series is n, and divisible by 2^J, there are $n/2^j$ $d_{j,k}$ coefficients at scale j = 1, ..., J-1. At the

[4] Note that while wavelet transforms are often compared to the Fourier transform are they two differ in a number of fundamental ways. See [10] for a comparison of wavelet versus Fourier transforms.

[5] For the relevance of horizon effects see, for example, [11].

[6] See [7] 99–103 for a complete discussion.

[7] See Ramsey [12].

coarsest scale there are $n/2^J$ $d_{J,k}$ and $s_{J,k}$ coefficients. The wavelet variance at each scale is captured as the wavelet power of each scale.[8]

A time series f(t) can be represented in decomposed form, known as the multi-resolution analysis of f(t), as follows:

$$f(t) = \Sigma_k S_{J,k}\Phi_{J,k}(t) + \Sigma_k d_J, k\Psi_{J,k}(t) + \ldots + \Sigma_k d_{j,k}\Psi_{j,k}(t) + \ldots + \Sigma_k d_{1,k}\Psi_{1,k}(t) \quad (7)$$

Using a more convenient summary notation,

$$f(t) = S_J + D_J + D_{J-1} + \ldots + D_1 \quad (8)$$

The discrete wavelet transform decomposes a time series into orthogonal signal components at different scales. S_j is a smooth signal, and each D_j is a signal of higher detail.

In the case of monthly data, as we use in our analysis, decomposing the series into six scales (D1-D6) corresponds to 2–4, 4–8, 8–16, 16–32, 32–64, and 64–128 months. D1 is the shortest scale (highest frequency) component and D6 is the longest scale (lowest frequency) component. The smooth component (S6) captures the trend of the original series.

The continuous wavelet transform (CWT) is also a useful approach for gaining insight into the localized time-scale decomposition of a time series. One advantage that the CWT has over the DWT is that it produces a powerful visual for detecting time-scale patterns. The CWT, which is based on continuous variations in the scale (λ) and time components (t) is defined as,

$$W(\lambda, t) = \int_{-\infty}^{+\infty} \Psi_{\lambda,t}(u)x(u)du \quad (9)$$

where,

$$\Psi_{\lambda,t}(u) \equiv \frac{1}{\sqrt{\lambda}}\Psi\left(\frac{u-t}{\lambda}\right) \quad (10)$$

The DWT can be viewed as a critical sampling of the CWT with $\lambda = 2^{-j}$ and $t = k2^{-j}$.[9]

The wavelet power spectrum, or squared amplitude, measures the local variance of a time series at different scales. It is defined as $|W(\lambda, t)^2|$, and aids our analysis in terms of understanding how periodic components evolve over time.

In addition to the wavelet power spectrum, we also employ wavelet coherence to measure the co-movement of two time series across time and scale. To define coherence we need to define of two other measures, the cross wavelet transform (XWT), and the cross wavelet power (XWP). The XWT is defined as

$$W_{xy} = W_x(\lambda, t)W_{y*}(\lambda, t) \quad (11)$$

The XWP is the absolute value of the XWT, $|W_{xy}(\lambda, t)|$. It measures the local covariance of 2 series at different time scales. The XWP identifies areas in time-scale space where the two series have high common power.

---

[8] The wavelet power is the amplitude squared.

[9] The DWT can also be derived independently, see [12].

The wavelet coherence, is defined as:

$$R^2(\lambda,t) = \frac{\left|S\left(S^{-1}Wxy(\lambda,t)\right)\right|^2}{S\left(S^{-1}|W_x(\lambda,t)|^2\right) * S\left(S^{-1}|W_{xy}(\lambda,t)|^2\right)} \quad (12)$$

Where S is a smoothing operator in time and scale, and $0 \leq R^2(\lambda,t) \geq 1$. The wavelet coherence is similar to the correlation coefficient, and is typically interpreted as a localized correlation in time-scale space. Note that the coherence between two series may be high even if the XWP is low.

The applicability of wavelet methodology to investigate uncertainty shocks is rooted in the fact that market returns reflect an aggregation of investors' decisions. Investors do not all share the same time horizon. Wavelet methodology is used so that localized information that affects returns is not lost.

4. Data

4.1 Sector returns

The equity return data used for our analysis is from the Kenneth French Data Library (**Table 1**) [13]. The market portfolio (MKT) is a composite portfolio of all stocks traded on the NYSE, AMEX, and NASDAQ. The market is divided into 12 industry groups or sectors defined below.

1 NoDur Consumer NonDurables – Food, Tobacco, Textiles, Apparel, Leather, Toys
2 Durbl Consumer Durables – Cars, TV's, Furniture, Household Appliances
3 Manuf Manufacturing – Machinery, Trucks, Planes, Off Furn, Paper, Com Printing
4 Enrgy Oil, Gas, and Coal Extraction and Products
5 Chems Chemicals and Allied Products
6 BusEq Business Equipment – Computers, Software, and Electronic Equipment
7 Telcm Telephone and Television Transmission
8 Utils Utilities
9 Shops Wholesale, Retail, and Some Services (Laundries, Repair Shops)
10 Hlth Healthcare, Medical Equipment, and Drugs
11 Money Finance
12 Other Other – Mines, Constr, BldMt, Trans, Hotels, Bus Serv, Entertainment

Table 1.
Kenneth French 12 Industry Data Set.

All returns are reported in excess of the risk free rate. The risk-free rate is measured by the yield on the 1-month T-bill.[10] The sample frequency is monthly, and the sample period is July 1960 to Dec. 2019.[11] The sample period includes eight recessions. These are illustrated in **Figure 1**. All but three were less than a year in duration. The 1974–1975 recession was 16 months, this was the time of the first

[10] The 1 month T-bill rate used as a risk free rate is calculated by Ibbotson and Associates, and provided by Kenneth French in his Data Library.

[11] The starting period of the sample is determined by the starting period of the uncertainty indexes.

OPEC price shock, when oil prices quadrupled. The recession starting in July 1981 lasted 16 months. This coincided with Fed interest rate tightening which was implemented to reduce inflation. Finally the Great Recession of 2008–2009 had a duration of 18 months. An examination of the cumulative returns of each sector indicates a high degree of variability across sectors and over time for a given sector. **Figure 2** shows the sectors with cumulative growth that exceeds the market for the sample period. **Figure 3** shows sectors with cumulative growth near or below cumulative market growth. The sector with the highest cumulative growth over the sample period is Consumer Non-durables (NoDur) with growth of almost 6400%, compared with the market as a whole which increased 2378%.The sector with the lowest cumulative returns is Durable Goods (900%). The effect of the technology bubble burst (2000–2001), on Telecom, and BusEq returns is salient. The drop is so precipitous that by the onset of the Great Recession (Dec. 2007), cumulative returns for these sectors was still below peak (March 2000). They did not reach the March 2000 peak until 2016. As a whole, **Figures 2** and **3** indicate a change occurring with the 2001 recession in that when it comes to cumulative returns, sector returns appear to part ways. One result is that some sectors recovered quickly from the 2001 and 2007 recessions and some recovered very slowly.

Figure 4 contains the wavelet power spectrum for market returns. Wavelet power is a measure of variance local to time and scale. The most striking feature of

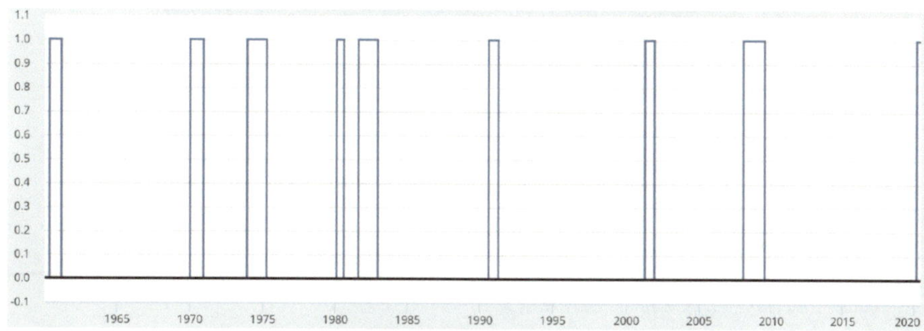

Figure 1.
US Economic Recessions, 1960–2020 - NBER Dating.

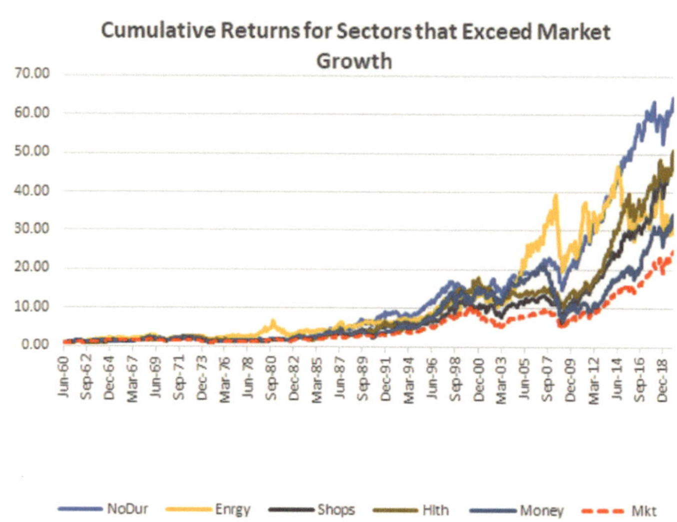

Figure 2.
Cumulative Returns - High Growth Sectors.

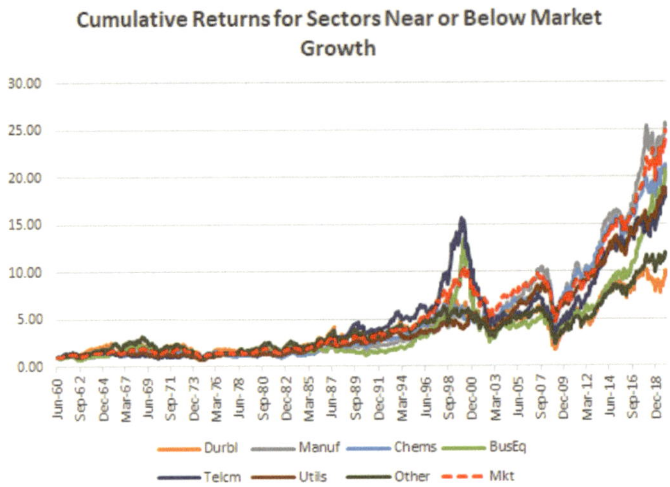

Figure 3.
Cumulative Returns - Low Growth Sectors.

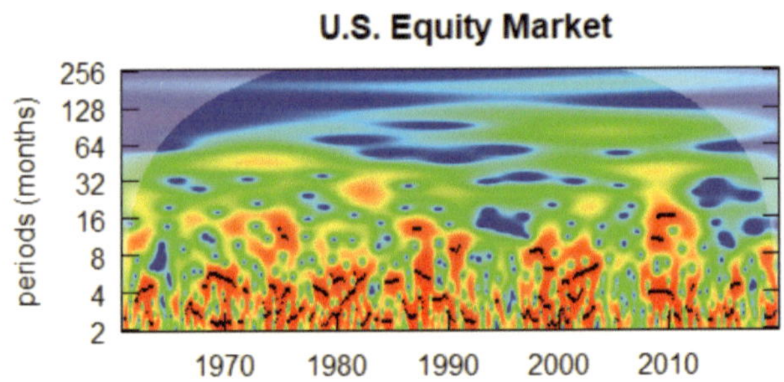

Figure 4.
Wavelet Power Spectrum - U.S. Equity Market.

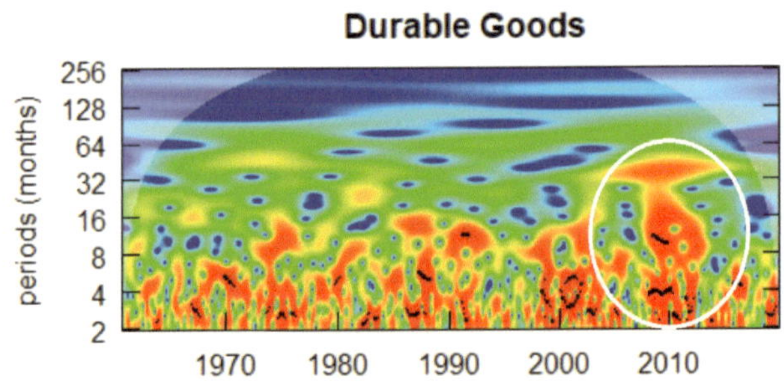

Figure 5.
Wavelet Power Spectrum -Durable Goods.

this chart is that most of the power occurs intermittently at high frequencies. The wavelet power spectrum for the Durables sector (the lowest growth sector) is shown in **Figure 5**, and the power spectrum for the Consumer Non-durables sector (the highest growth sector) is shown in **Figure 6**. Both sectors look similar to the market at high frequencies. At intermediate frequencies (16–32 months) the Durable goods sector shows high power during the Great Recession, but the the Non-durables goods sector does not. This is outlined in white for expository purposes. Consumer non-durables have relatively high power at the 32–64 month frequency

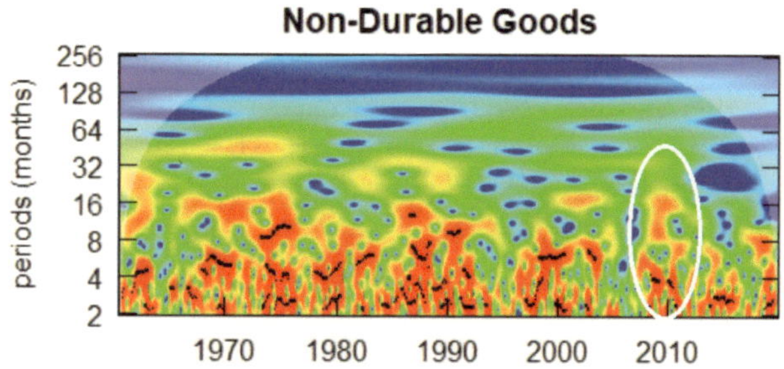

Figure 6.
Wavelet Power Spectrum - Consumer Non-durables.

during the 1970s, while the Durable Goods sector has less variability associated with this frequency band.[12]

A set of descriptive statistics for the monthly excess returns (%) is reported in **Table 2**. Monthly returns range from a high of 42.6% for Durable goods (Apr. 2009) to a low of minus 32.7% also for Durable goods (Oct. 2008). Skewness is negative for most sectors, the exceptions being except Durable goods (0.13%); Excess kurtosis is positive (leptokurtic) for all of the sectors, suggesting that the distribution of returns has fatter tails than a Normal distribution. It ranges from 1.0 for Utilities to 4.8 for Durables.

4.2 Uncertainty measures

We use three measures of uncertainty in our analysis, macroeconomic and financial uncertainty from Jurado, Ludvigson, and Ng [2], and economic policy

Sector	Mean	Std Dev	Median	Minimum	Maximum	Skewness	Kurtosis
NoDur	0.68	4.24	0.76	−21.63	18.3	−0.33	2.03
Durbl	0.51	6.15	0.46	−32.71	42.62	0.13	4.77
Manuf	0.59	5.22	0.98	−29.18	21.07	−0.49	2.52
Enrgy	0.63	5.4	0.66	−19.01	23.6	−0.03	1.23
Chems	0.53	4.55	0.74	−25.19	19.71	−0.26	2.11
BusEq	0.63	6.36	0.66	−26.41	20.32	−0.24	1.3
Telcm	0.51	4.59	0.62	−16.43	21.22	−0.18	1.14
Utils	0.49	3.95	0.64	−12.94	18.26	−0.15	1.01
Shops	0.67	5.08	0.75	−28.83	25.28	−0.31	2.4
Hlth	0.67	4.87	0.77	−21.06	29.01	−0.03	2.29
Money	0.64	5.38	0.91	−22.53	20.59	−0.38	1.59
Other	0.49	5.34	0.82	−29.81	18.85	−0.5	2.16
Note: n = 714							

Table 2.
Summary Statistics - Sector Returns.

[12] The highest power level is shown in red, and the lowest is blue. For the U.S. equity market the highest power level is 4. For the Durable goods sector the highest power level is 6.8, and for the non-durable goods sector it is 5.0.

uncertainty from and Baker, Bloom and Davis [3]. These are arguably the most common measures of uncertainty, and they are both updated on a regular basis and available online.[13] JLN and BBD are different measures of uncertainty constructed using very different methodologies. We summarize both approaches in this section.

JLN define uncertainty for variable y_{jt} as the volatility of the unforecastable part of the future value of y_{jt}.

$$U_{jt}^{y}(h) \equiv \sqrt{E\left[\left(y_{jt+h} - E\left[y_{jt+h}|I_t\right]\right)^2 |I\right]} \tag{13}$$

where, $E(.|I_t)$ is the expectation conditional on information at time t, and h is the number of time periods for the projection. An increase in the squared forecasting error of y_{jt} indicates an increase in uncertainty at time of y_j at t. The JLN methodology computes financial and macroeconomic indexes by aggregating uncertainty measures of the individual economic series.

$$U_t^{y}(h) = \sum_{j=1}^{N_y} w_j U_{jt}^{y}(h) \tag{14}$$

where w_j are the aggregation weights.

JLN used a total of 132 economic series to estimate macroeconomic uncertainty. The series span the following categories: real output and income, employment and hours, real retail, manufacturing and trade sales, consumer spending, housing starts, inventories and inventory sales ratios, orders and unfilled orders, compensation and labor costs, capacity utilization measures, price indexes, bond and stock market indexes, and foreign exchange measures.

The financial uncertainty series is comprised of uncertainty measures for 147 financial series. These series include valuation ratios such as the dividend-price ratio and earnings-price ratio, growth rates of aggregate dividends and prices, default and term spreads, yields on corporate bonds of different ratings grades, yields on Treasuries and yield spreads, and a broad cross-section of industry equity returns. In addition, returns on 100 portfolios of equities sorted into 10 size and 10 book-market categories are included. The data set also includes excess return on the market, small-minus-big and high-minus-low portfolio returns, a momentum factor, a measure of the bond risk premium, and also a small stock value spread.

JLN provide measures of financial and macroeconomic uncertainty based on 1, 3, and 12 month forecast horizons. Our analysis focuses on the one month horizon series, denoted as h = 1. Their macroeconomic uncertainty series is shown in **Figure** 7. In general the peaks of the series align with the NBER recession dates (shaded areas). The three highest peaks are in the mid-1970's during the first OPEC oil shock, the early 1980's when there were back to back recessions, and the recession of 2008–2009. **Figure 8** shows the wavelet power spectrum for this series. The power is highest for periods of 32 to 128 months. Unlike the time plot of uncertainty which peaks at each recession the power spectrum has two basic clusters of uncertainty. It does not distinguish among the first four recessionary periods, and instead shows one extended period of uncertainty from the early 1970's to the late 1980's. The second period of uncertainty is the Great Recession which is notable for the

[13] [2] is available at https://www.sydneyludvigson.com/macro-and-financial-uncertainty-indexes, and [3] is available at https://www.policyuncertainty.com/index.html.

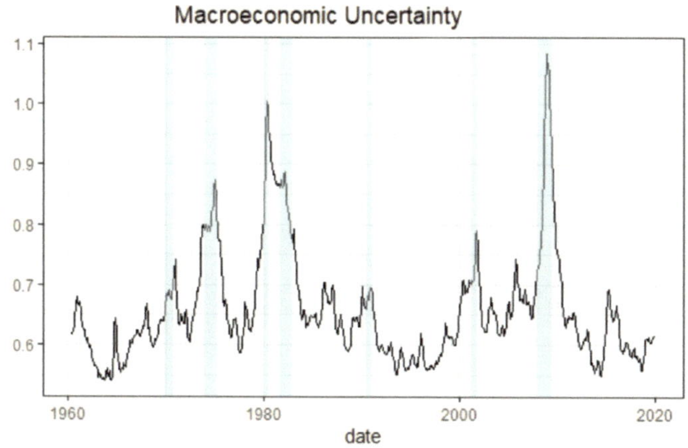

Figure 7.
Macroeconomic Uncertainty, Jurado, Ludvigson and Ng, (Ph1), horizon = 1 month.

range of scale which is from 8 to 256 months. The Great Moderation in the 1990's (outlined in white) is also apparent as a break in the low frequency power. In contrast power spectrum for the market returns 4 macroeconomic uncertainty tends to have low power at high frequencies.

Figure 9 displays the financial uncertainty series for h = 1. The peaks of this series do not align as closely with recessions as does the macroeconomic uncertainty. For instance there is a noteworthy spike in Oct. 1987 (outlined in red) when the equity market dropped, and there was no recession. Events such as the 1997 Asian financial crisis, the 1998 Russian financial crisis and the 2000 Tech. bubble bust are all apparent prior to the 2001 recession. Also, the magnitude of the peaks in the financial uncertainty index are greater than those of the macroeconomic uncertainty index. Alignment of the peaks of the macroeconomic index with recessions, and to a lesser extent the financial uncertainty index, is intuitive as forecasting a turning point is difficult if not impossible. One way useful to think about this is with regard to asset returns which can be written as,

$$r_t = sign * |r_t| \tag{15}$$

It is generally possible to forecast the absolute value of returns but not returns themselves. The reason is that one cannot forecast the sign.

The wavelet power spectrum or variance (**Figure 10**) for financial uncertainty is generally highest at low frequencies. The 1987 stock market crash (outlined in

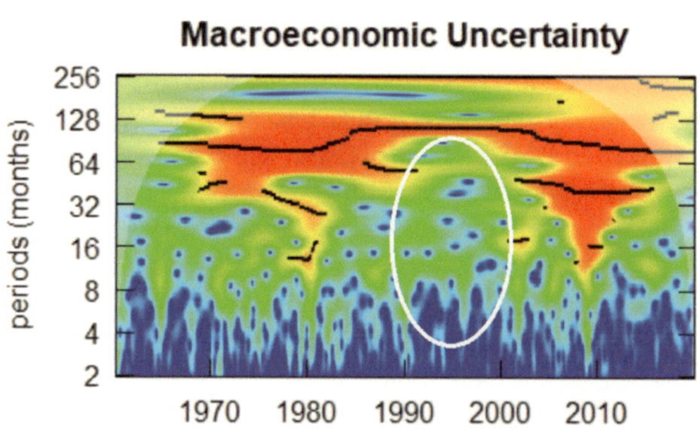

Figure 8.
Wavelet Power Spectrum - Macroeconomic Uncertainty (EP1).

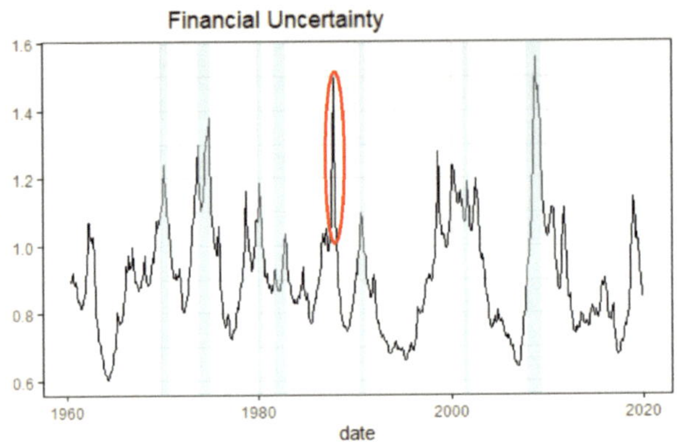

Figure 9.
Financial Uncertainty, Jurado, Ludvigson and Ng, (Fh1), (bottom), horizon = 1 month.

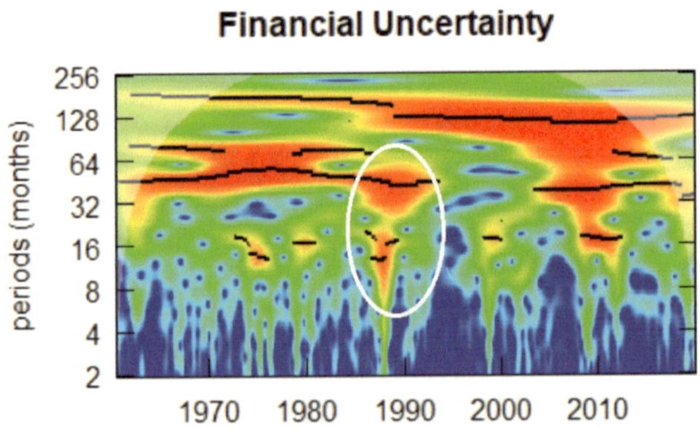

Figure 10.
Wavelet Power Spectrum - Financial Uncertainty (Fh1).

white) is one of several instances where high variance can be also be observed at medium (8 to 32 month) frequencies.[14] The scale of uncertainty is lower from 1960 to 1990 (32–64 months) than it is from 1990 to 2020 (up to 128 months). In effect low frequency financial uncertainty exists through the sample period.

Baker, Bloom and Davis (BBD) construct a measure of economic policy uncertainty using three major components. For the first component, they search ten major newspapers and create an index based on the volume of news relating to economic policy uncertainty. The second component, which is designed to capture uncertainty in the federal tax code, is derived from Congressional Budget Office reports on temporary tax code due to expire over the next ten years. The third component uses the dispersion of opinions among professional forecasters regarding the future of the Consumer Price Index, Federal expenditures and, State and Local Government Expenditures. The forecasts are from the Philadelphia Federal Reserves Survey of Professional Forecasters.[15] These three components are combined to create the Index of Economic Policy Uncertainty. The index is shown in **Figure 11**. Unfortunately the series starts in 1985, so we are unable to study this series over the same sample period as the JLN uncertainty indices. The BBD index shown in **Figure 11**, appears to have a local maxima during each of the 3 recessions,

[14] The Wavelet power and coherence analysis was done using MatLab 2020B.

[15] https://www.philadelphiafed.org/research-and-data/real-time-center/survey-of-professional-forecasters/historical-data/individual-forecasts.

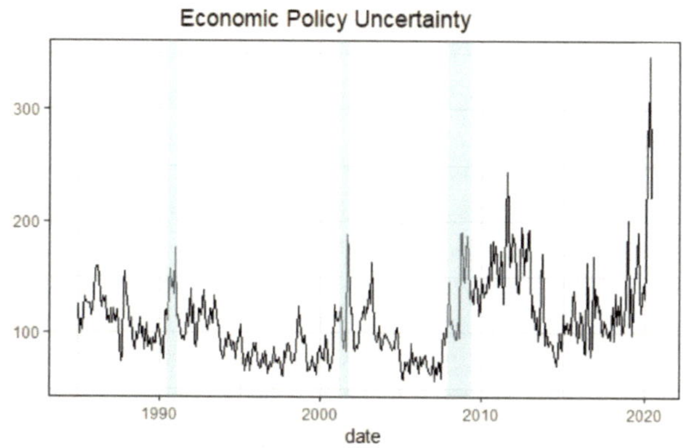

Figure 11.
Economic Policy Uncertainty Index - Baker, Bloom and Davis (EC1).

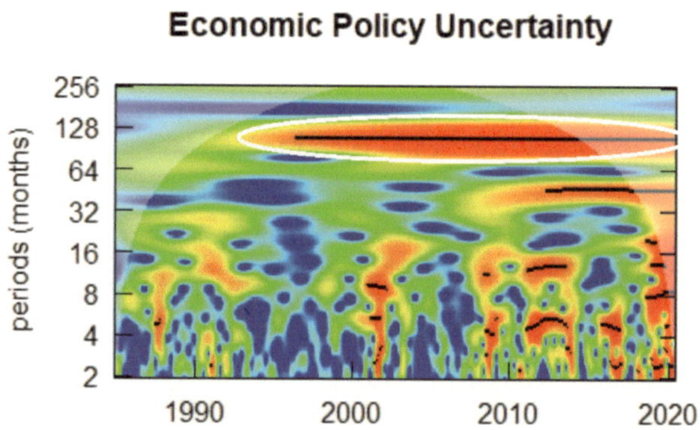

Figure 12.
Wavelet Power Spectrum - Economic Policy Uncertainty.

but it tends to be volatile with peaks at other important dates such as the two Gulf wars (1991, and 2003), the debt ceiling debates (2011–2012), the fiscal cliff (2013), the government shutdowns (1995,2013,2018), and the election of Trump (2016).[16] The wavelet power spectrum for the BBD series is shown in **Figure 12**. High power is found at the 128 month scale from the mid-1990's to the present (outlined in white). There are also a series of high frequency (2–16 months) spikes in the wavelet power which are not present in the other two uncertainty indices.

5. Uncertainty & the market portfolio

We begin our analysis by examining the relationship between the three measures of uncertainty, and the market portfolio. **Figure 13** shows the coherence between macroeconomic uncertainty and excess market returns. Red indicates high coherence and blue indicate no coherence. Coherence is a measure of co-movement between the two series, similar to a correlation. Note that high coherence does imply high power. The heavy black lines around the outside of the red areas indicates statistical significance at the 95% level of confidence. The frequency is inverted in the coherence

[16] [3] provide an annotated version of the index at https://www.policyuncertainty.com/media/US_Annotated_Series.pdf

charts compared with the power spectrum charts. The coherency charts also contain phase arrows which are explained in **Table 3**. There are two basic categories of coherence in **Figure 13**. Sporadic high coherence at the 8 to 64 month scale, and prolonged coherence at the 64–128 month scale. The high coherence at the 64–128 month scales lasts from 1960 until the mid-1980s, breaks for about 15 years (outlined in white), and reoccurs from 2000 to 2019. The period of the break in coherence is shorter than the typical time frame known as the Great Moderation (mid-1980's to 2007). The phase arrows are pointing left indicating that the two series are out of phase. The sporadic high coherence at the 16 to 32 month scales occurs in the middle 1970s, and to a much greater extent during the Great Recession. The phase arrows indicate that the two series are in anti-phase.

In addition to the monthly returns, we also examine coherence of uncertainty with the absolute value of market returns that we use as a measure of market volatility. **Figure 14** shows the coherence plot for the absolute value of market returns with macroeconomic uncertainty. As was the case in **Figure 13** there as two periods of high coherence at a low frequency, but in this instance the scale is lower (32–64 month), the break (outlined in white) begins in the early 1990s and lasts for about 5 years. During the Great Recession, uncertainty and market volatility are in phase.[17]

Figure 15 shows the coherence of the market returns with the financial uncertainty. Coherence occurs at lower scales than the coherence of the market with the

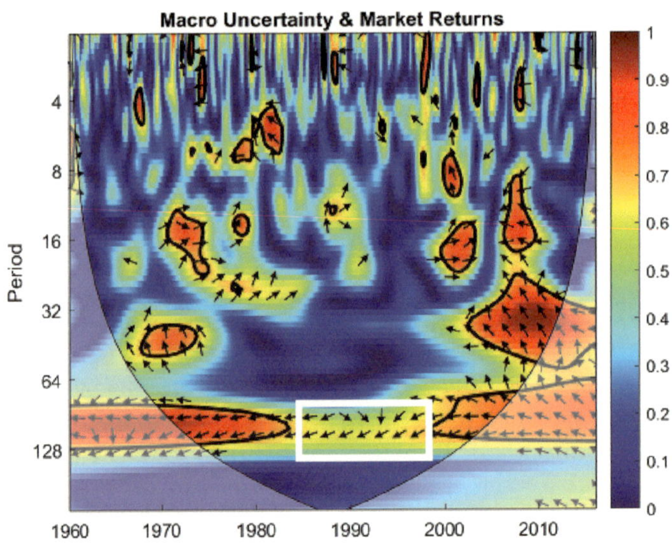

Figure 13.
Wavelet Coherence - Macroeconomic Uncertainty and U.S. Equity Market Returns.

Left arrow: anti-phase
Right arrow: in-phase
Down arrow: X leading Y by 90deg
Up arrow: Y leading X by 90deg

Table 3.
Phase arrow definitions.

[17] Note: interpreting the phase as a lead(/lag) should always be done with care. A lead of 90 degrees can also be interpreted as a lag of 270 degrees or a lag of 90 degrees relative to the anti-phase (opposite sign).

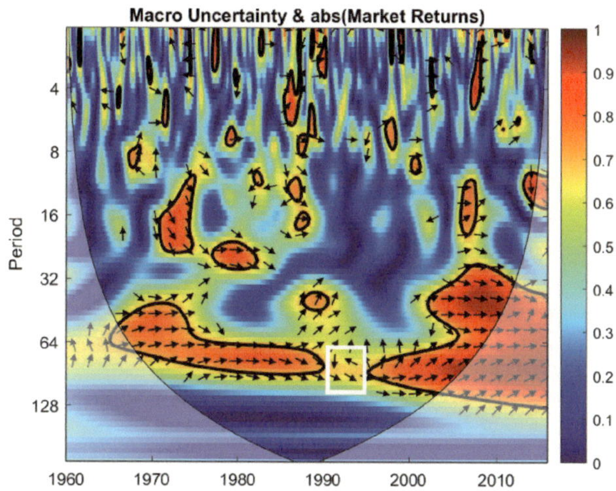

Figure 14.
Wavelet Coherence - Macroeconomic Uncertainty and Absolute Value of U.S. Equity Market Returns.

macroeconomic uncertainty. There is a high degree of coherence in the 1960s and 1970s at the 16–32 month scale, then there is a break (outlined in white) of 20 years when coherence at these scales is non-existent. Beginning in 2000, coherence is high once again at the 16–32 month scales. The two series are out of phase during these periods of high coherence.

The coherence of financial uncertainty and market volatility is shown in **Figure 16**. The coherence is high and in-phase throughout the entire sample period for scales above 32 months. There also are numerous low scale periods when the two series are in phase and have high coherence.

As shown in **Figure 17** the coherence of economic policy uncertainty and market returns generally occurs at a lower scale then coherence of the two BLN indices. Statistically signficiant coherence never exceeds the 64 month scale, but it is high and nearly continuous at the 8 to 16 month scale from 1993 to 2003 (outlined in white). These periods of high coherence at lower scales appear to coincide with financial crises. During the Great Recession there is a very clear distinction in coherence between the 8–16 month scales (2008–2012), and the coherence at the

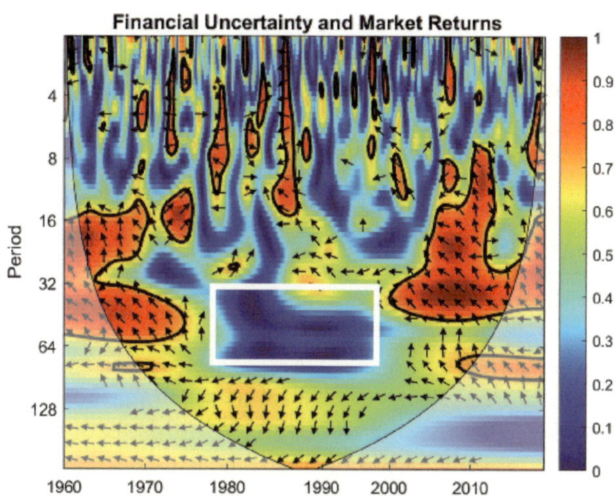

Figure 15.
Wavelet Coherence - Financial Uncertainty and U.S. Equity Market Returns.

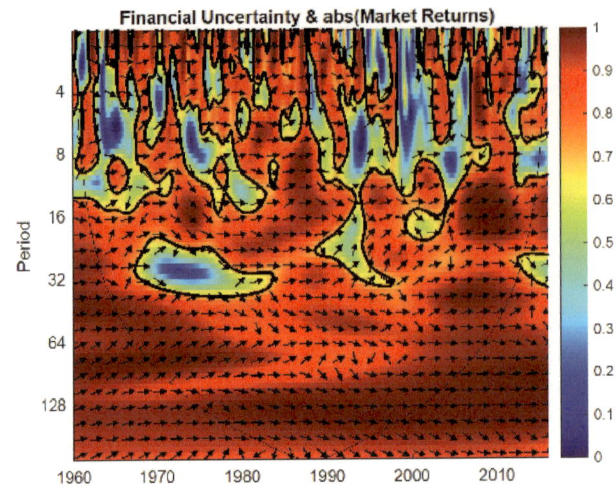

Figure 16.
Wavelet Coherence - Financial Uncertainty and Absolute Value of U.S. Equity Market Returns).

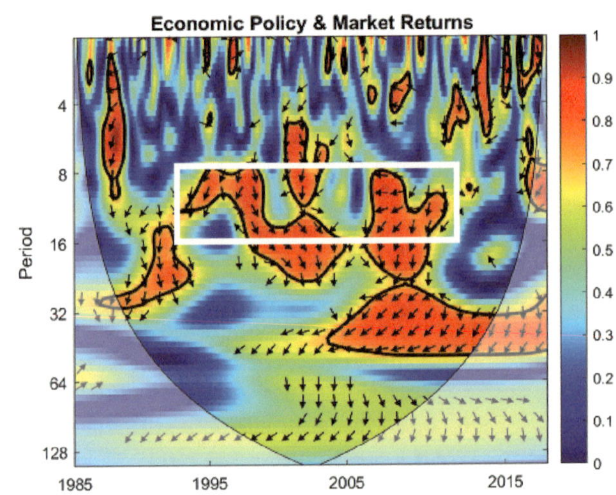

Figure 17.
Wavelet Coherence -Economic Policy Uncertainty and U.S. Equity Market Returns.

32 months scale from 2003 to 2018. The phase arrows generally point upwards indicating a lead–lag relationship.

Figure 18 shows the coherence between economic policy uncertainty and the absolute value of market returns. Coherence is generally quite low, except for the 128 month scale beginning in the mid -1990s and going to 2019.

5.1 Uncertainty and sector returns

In the previous section, we showed that regardless of the uncertainty measure employed, there is evidence of high coherence between market returns and uncertainty, and also market volatility and uncertainty. In this section, we examine the coherence between sector returns and the three measures of uncertainty. Our goal is to characterize the extent to which uncertainty has impacted individual sectors. To do this we examine the coherence of each measure of uncertainty with sector returns. However, uncertainty affects the market and we want to find the

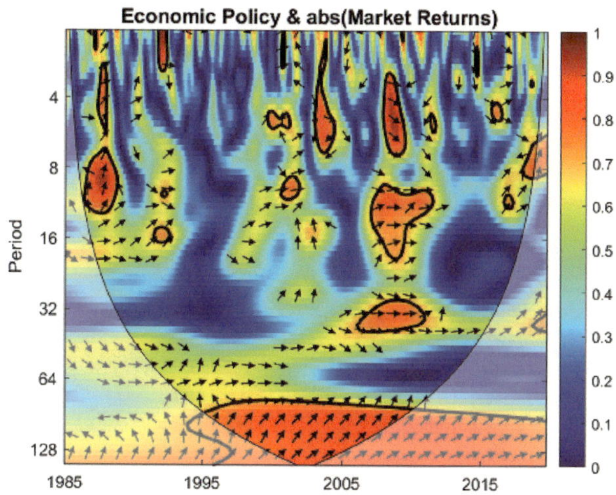

Figure 18.
Wavelet Coherence -Economic Policy Uncertainty and absolute Value of U.S. Equity Market Returns).

relationship between uncertainty and sector returns after removing the relationship of uncertainty with the market. In order to accomplish this, there is a "before and after." We refer to the "after" as the conditional coherence. Conditional coherence is the coherence of sector returns with uncertainty conditional on the market returns.[18] In addition, to further illustrate the extent to which uncertainty impacts a sector independently of the impact it has vis-a-vis the market portfolio, we estimate a rolling regression. The rolling regression is of sector returns against the market return and the uncertainty index for one or more scales using the discrete wavelet transform.[19] Selection of the scales was based on high conditional coherence of sector returns and uncertainty measures. The functional form of the regression is,

$$r^s_{sector,t} = \beta^s_{0,t} + \beta^s_{1,t} * r^s_{mkt,t} + \beta^s_{2,t} * U^s_{i,t} \tag{16}$$

where $U_{i,t}$ for $i = 1, 2, or\ 3$ is a of the measure of uncertainty, and $s = 1, .., 6$ is the scale. The estimation window is 60 months.

As there are three measures of uncertainty and 12 sectors, it would be cumbersome to show all of the coherence plots for the 12 sectors. Instead, we have chosen to discuss a subset consisting of six sectors where each sector has at least one period of high coherence with one measure of uncertainty. While the other sectors may also have periods of high coherence, including them would add little to our analysis.[20]

5.1.1 Financial sector returns and uncertainty

The first sector examined is the financial sector. **Figures 19–21** show the coherence of the financial sector with all three types of uncertainty both before and after the market has been partialed out. Most of the coherence with all three types of uncertainty is subsumed by the market. However, financial uncertainty is one measure of uncertainty that does matter. As shown, in **Figure 20** the right hand side chart shows several areas of significant conditional coherence. Conditional

[18] The partial CWT was estimated using code from [14] updated to run on MatLab 2020B.

[19] The rolling regressions were estimated using R. DWT was estimated using the Waveslim package [15] in R.

[20] The other sectors are available from the authors upon request.

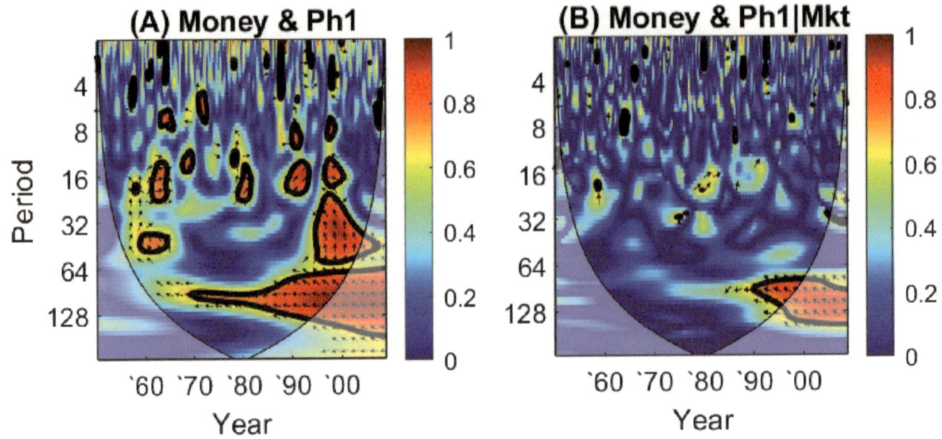

Figure 19.
Wavelet Coherence -Macroeconomic Uncertainty and the Money Sector.

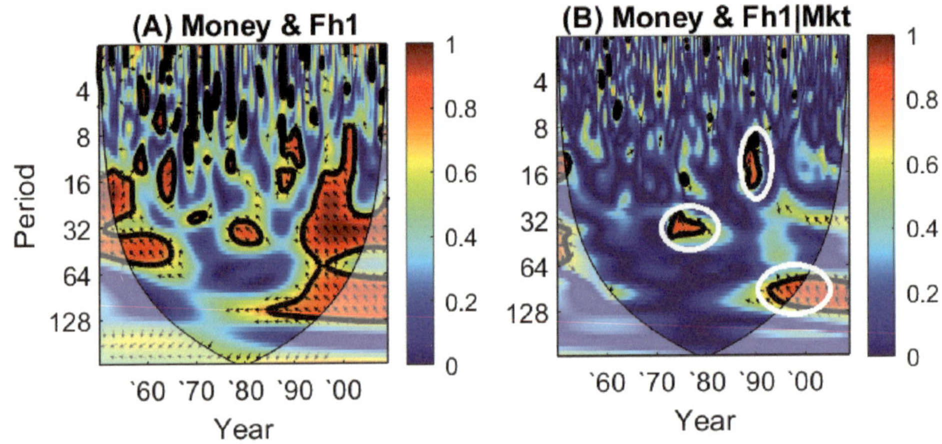

Figure 20.
Wavelet Coherence - Financial Uncertainty and the Money Sector.

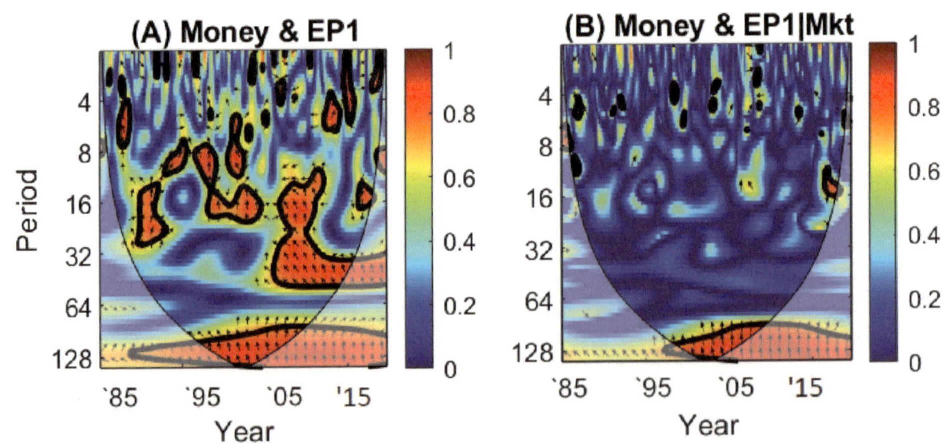

Figure 21.
Wavelet Coherence -Economic Policy Uncertainty and the Money Sector.

coherence exists for the 32 month scale in the 1970s (circled in white). There is also a small area of conditional coherence around 1990 at the 16 month scale and it is also present at the 64–128 month scale beginning around in the mid 1990s. It is interesting to note that there is no apparent conditional coherence with Economic Policy Uncertainty in 2010 when Dodd-Frank was passed.

Figure 22 shows the financial uncertainty coefficient estimates (top) and corresponding t-statistics (bottom) for scales 4 and 5 using the rolling window regression given in Eq. (16). The scales were chosen based on the conditional coherence charts. Scales 4 and 5 for financial uncertainty suggest by the presence of red hot spots that there would be significance for the uncertainty measure. The dashed lines in the t-stat chart indicate +/− 2. Although both scales are presented, scale 5 stands out. Scale 5 (32 months) is negative and statistically significant for three segments of time beginning in the late 1970s. The longest of these three periods begins in the late 1990s and ends in 2010. Given the large number of shocks all of which effected equity values, it is interesting to find that the financial uncertainty index had significance in a rolling window regression that includes both the market and the uncertainty index. Some of the important financial events include the Asian financial crisis (1997), the Russian financial crisis (1998), the collapse of Long Term Capital Management (1998), the repeal of the Glass-Steagall Act (1999), the 2001 recession, and the 2008 recession. The significant negative coefficients for this time period are an indication that there was sector specific uncertainty over and above that which impacted the market as a whole.

5.1.2 Energy sector returns and uncertainty

Figures 23–25 contain the coherence of the Energy sector returns with all three type of uncertainty before (left) and after (right) the market returns are partialed out. Each of the three right hand side charts display several short periods of high conditional coherence indicating uncertainty specific to the energy sector. The longest period of conditional coherence is at scale 6 (64 months) for the Macroeconomic Uncertainty Index (circled in white). There are two small period of high

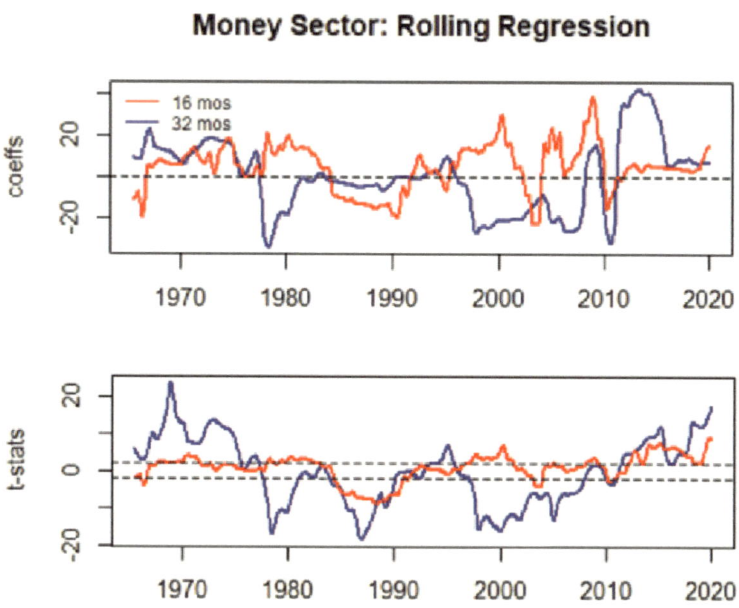

Figure 22.
Rolling Regression Coefficients -Financial Uncertainty and the Money Sector.

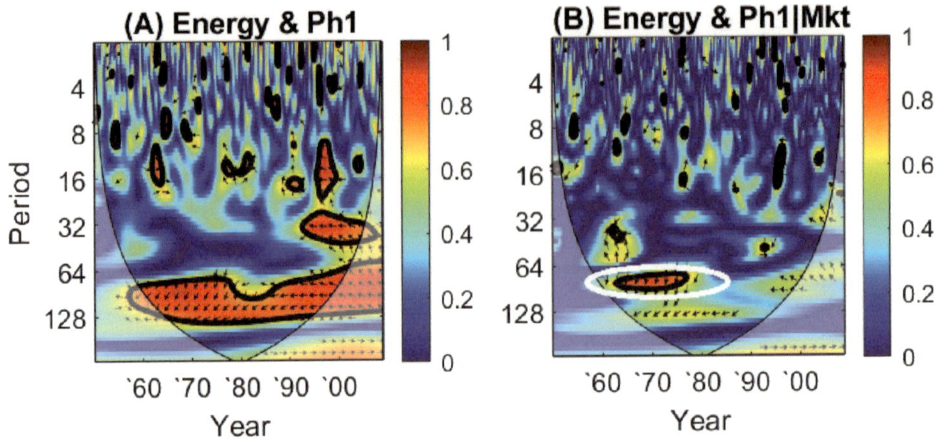

Figure 23.
Wavelet Coherence - Macroeconomic Uncertainty and the Energy Sector.

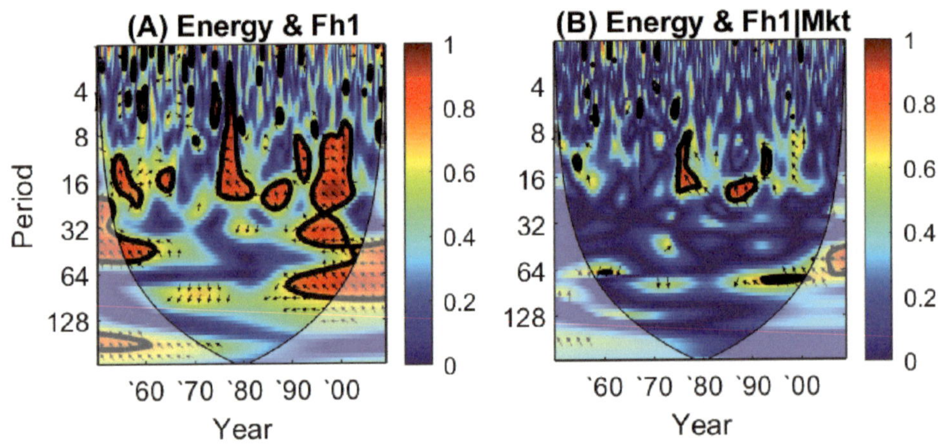

Figure 24.
Wavelet Coherence - Financial Uncertainty and the Energy Sector.

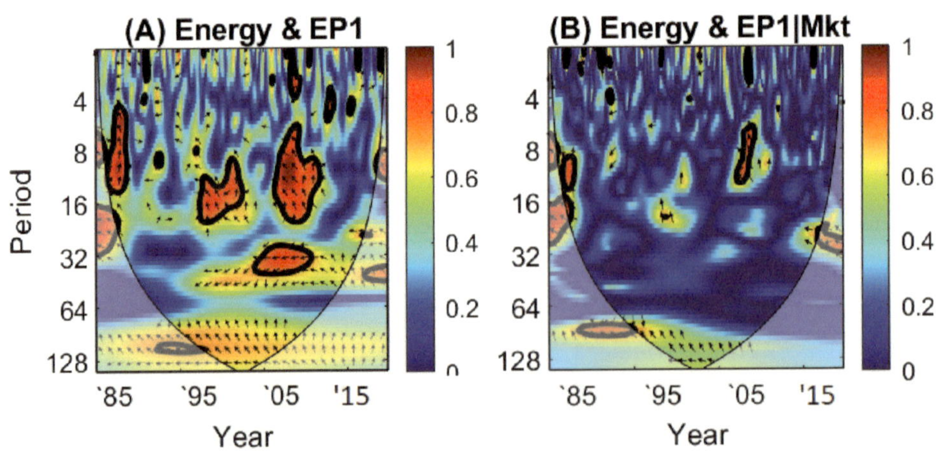

Figure 25.
Wavelet Coherence -Economic Policy Uncertainty and the Energy Sector.

conditional coherence at the 16 month scale with financial uncertainty. There is moderate conditional coherence with policy uncertainty from the mid-1980s until the mid-1990s, but the scale (128 months) is almost completely outside the cone of influence (COI) and is not statistically significant. The 2014 oil glut that led to a steep decline in oil prices has a very small area of significance in the conditional coherence charts. Note also that there is little or no sign of the boom in hydraulic fracking which began in the mid 1990s, nor the environmental backlash that began in 2013–2014 in the conditional coherence charts.

Figure 26 shows rolling window regression coefficients at scales 5 (32 months) and 6 (64 months) for macroeconomic uncertainty. Again, the scales were chosen based on observations from the conditional coherence charts. Scale 6 is negative and statistically significant from the mid-1960s until the mid-1970s and clearly shows the impact of the 1973 OPEC embargo. The uncertainty coefficients for scale 6 are also negative and statistically significant from 2010 to 2015 and may reflect the environmental backlash to fracking.

In summary, although the conditional coherence charts show surprisingly little sector specific uncertainty relating to the OPEC II (1979) oil shocks or the 2014 oil glut, the DWT regression shows a strong negative sector specific impact.

5.1.3 Telecommunications sector returns and uncertainty

Over the course of the sample period, the Telecommunications Industry evolved from a heavily regulated monopoly to a more competitive industry with at least six large firms. In addition, technological changes in communications broadened the scope of services offered by the industry. This resulted in redefining communications providers as content providers and making cell phone usage an imperative for the vast majority of adults. An examination of the coherence charts (**Figures 27–29**) does show some periods of sector specific uncertainty. The conditional coherence chart for financial uncertainty shows high coherence at the 32 months scale throughout the 1980s (highlighted with white) which is coincident with the restructuring of AT&T. The conditional economic policy chart shows four high

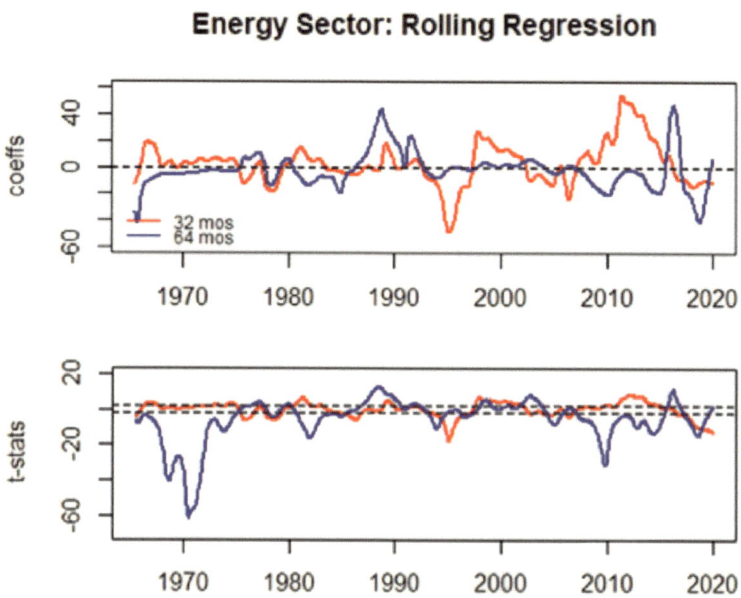

Figure 26.
Rolling Regression Coefficients -Macroeconomic Uncertainty and the Energy Sector.

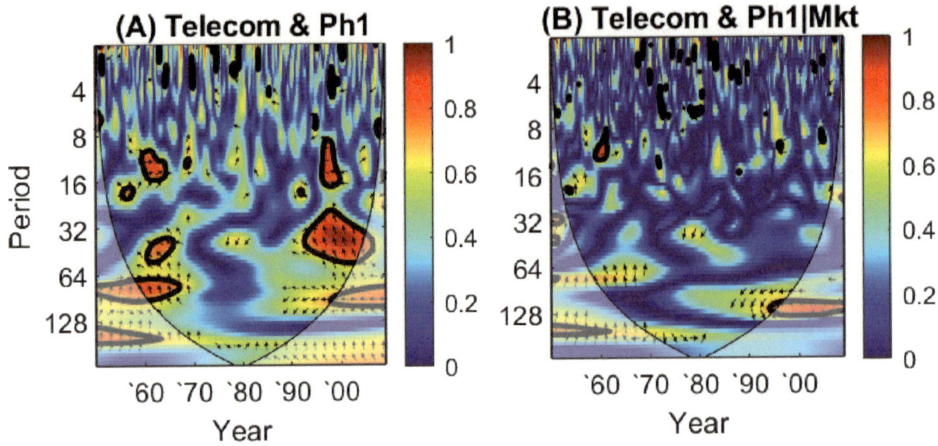

Figure 27.
Wavelet Coherence -Macroeconomic Uncertainty and the Telecom. Sector.

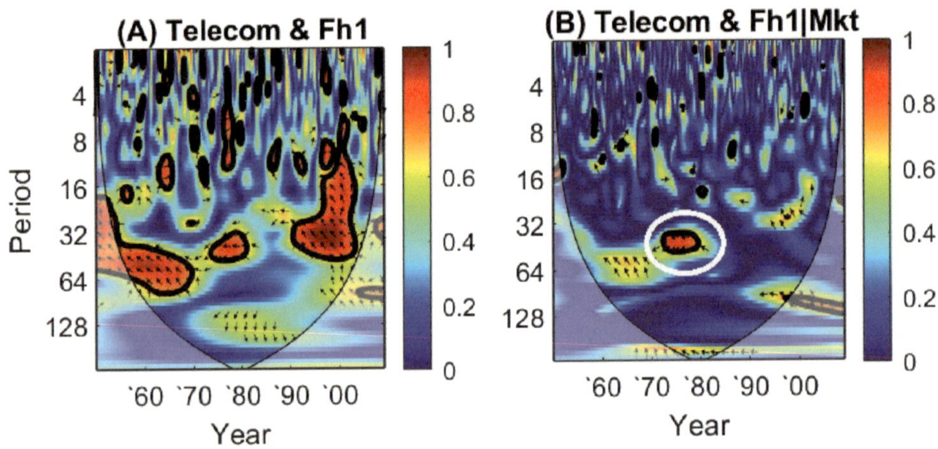

Figure 28.
Wavelet Coherence - Financial Uncertainty and the Telecom. Sector.

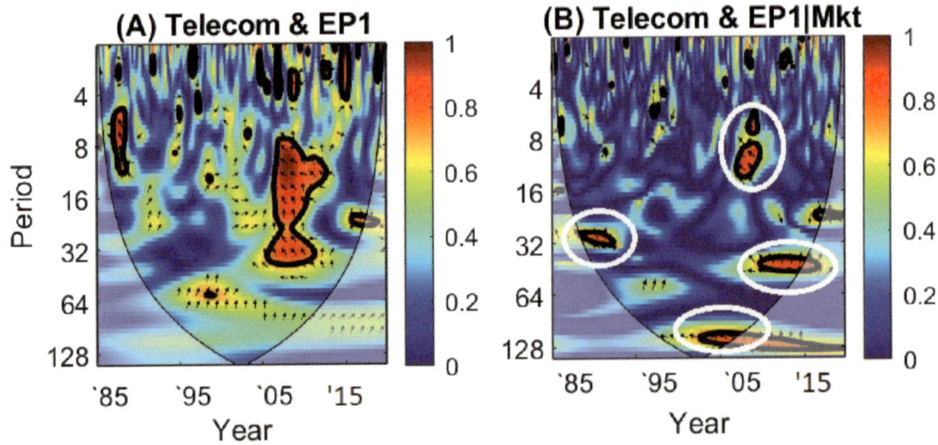

Figure 29.
Wavelet Coherence -Economic Policy Uncertainty and the Telecom. Sector.

periods of coherence. Two are at the 32 month scale, one in the early 1990s and the second starting in 2010. There is also high coherence during the Great Recession at the 8 to 16 month scale. The fourth period of high coherence is at the 64 to 128 month scale beginning in 2005 and ending in 2010.

Based on the conditional coherence charts, rolling regressions were run for both Economic Policy Uncertainty and Financial Uncertainty indices. Significant coefficients were found for both indices. **Figure 30** shows the uncertainty coefficient estimates and t-statistics for the regression of sector returns against market returns and Economic Policy Uncertainty for scales 16 and 32 months. Policy Uncertainty at scale 32 is negative and statistically significant from 1995 to 2000, and from from 2007 to 2015. The Telecommunications Act of 1996 may be partially responsible for the large drop in the 32 month coefficient in 1996–1997.

Coefficient estimates for Financial Uncertainty at 16 and 32 month scales are shown in **Figure 31**. This is a longer time series than the policy uncertainty index and it shows a modest negative impact in the 1980s at the 32 month scale. There is a much larger negative impact at the 32 month scale starting in the late 1990s and extending until 2018. The significance coefficients are consistent with the explanation that the Telecommunications Sector exhibits sector specific uncertainty.

5.1.4 Business equipment sector returns and uncertainty

Coherence plots for the business equipment sector, which is comprised of computer, electronic and software firms are shown in **Figures 32–34**. There appears to be very little conditional coherence for Macroeconomic Uncertainty. The 32 month scale for Financial Uncertainty shows some coherence in the early 1990's and from 2000 to 2010. Conditional coherence with Economic Policy Uncertainty is low except in the 2018–2919 period at the 8 month scale (highlighted in white). This may be the result of the policy change in favor of repealing net neutrality on the part of the Trump Administration.

Figure 35 shows the coefficient estimates for Economic Policy Uncertainty at the 8 and 16 month scales. The coefficients for the 16 month scale are negative and

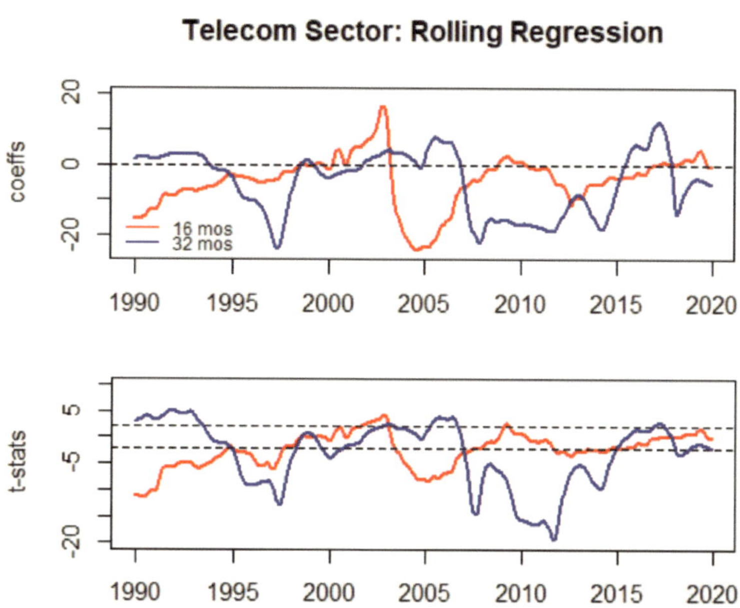

Figure 30.
Rolling Regression Coefficients-Economic Policy Uncertainty and the Telecom. Sector.

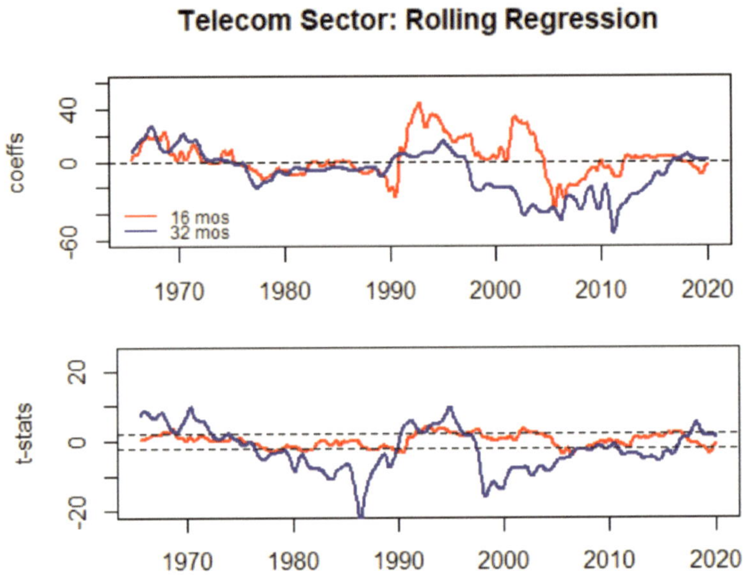

Figure 31.
Rolling Regression Coefficients-Financial Uncertainty and the Telecom. Sector.

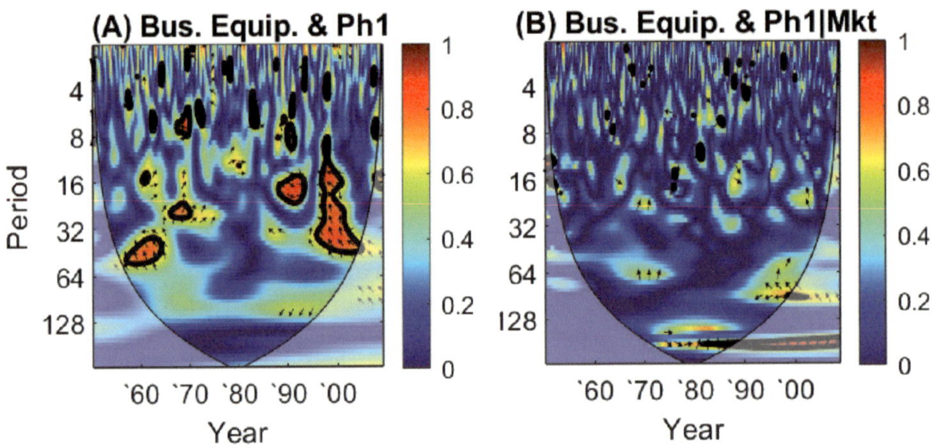

Figure 32.
Wavelet Coherence -Macroeconomic Uncertainty and the Business Equipment Sector.

significant from the late 1990s until 2008. This is a time of rapid growth for the internet. The tech bubble burst after partialing out the market effect clearly has a stand alone component. The significant coefficients are consistent with difficulties encountered when introducing a new technology.

5.1.5 Shops sector returns and uncertainty

There are several instances of significant conditional coherence for the shops sector (**Figures 36–38**). Notable is the coherence with Macroeconomic Uncertainty, and to a lesser extent Financial Uncertainty, at the 64 month scale from the mid 1970s to the late 1980s. The Shops sector, which consists of Wholesale, Retail, and Some Services, shows very little conditional coherence with the JLN indices after the late 1980s. This is consistent with the rise and expansion of big box retailers such as Walmart and Target and the demise of small Mom & Pop stores. Big Guns

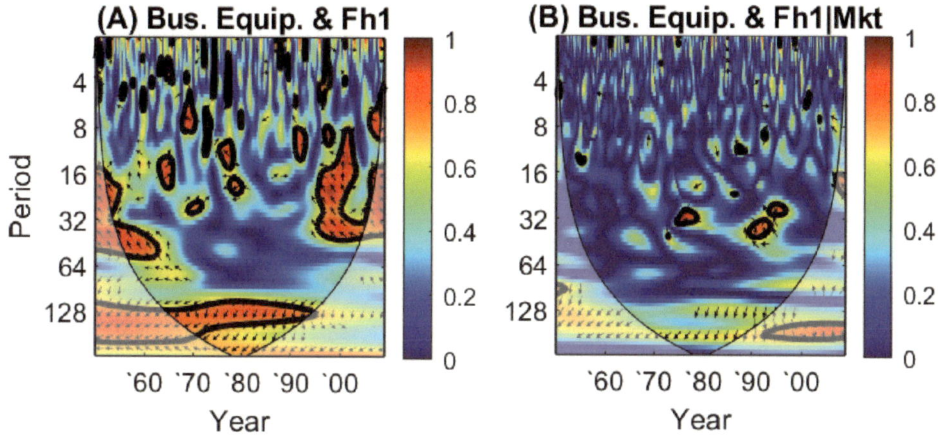

Figure 33.
Wavelet Coherence - Financial Uncertainty and the Business Equipment Sector.

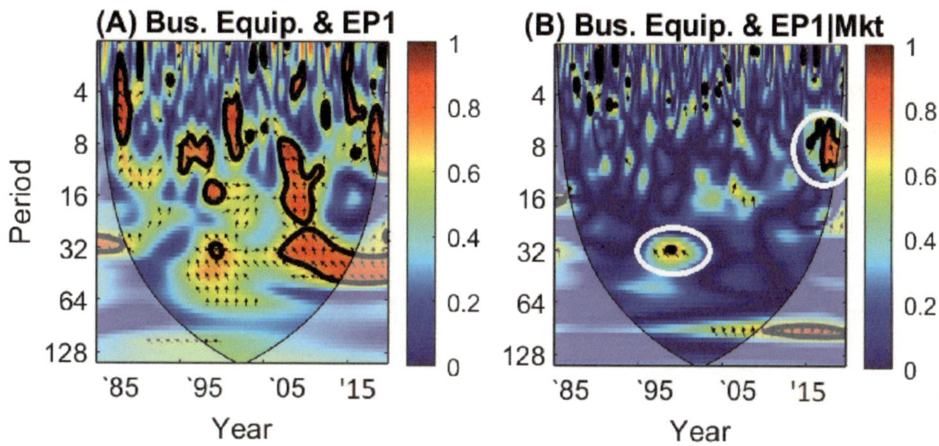

Figure 34.
Wavelet Coherence -Economic Policy Uncertainty and the Business Equipment Sector.

are better able to weather uncertainty storms. The rise of internet retail captured by Amazon's IPO in 1997 does show up in the conditional coherence for Economic Policy (highlighted in white). This is suggestive that policy treatment regarding internet retail mattered, especially considering taxes.

The rolling regression coefficients for the shops sector with economic policy uncertainty are shown in **Figure 39** for 8 and 16 month scales. The uncertainty coefficients for the 16 month scale are negative starting in 1996 and remain negative until 2015. This seems consistent with the war between internet shopping and brick and mortar retail businesses.

5.1.6 Manufacturing sector returns and uncertainty

Coherence plots for the manufacturing sector are shown in **Figures 40-42**. There is high conditional coherence with Macroeconomic Uncertainty from 1990 to 2019 at the 128 month scale (highlighted in white). However, at least half of this coherence is outside the cone of influence (COI). The onset of this uncertainty coincides with the signing of NAFTA in 1994. There appears to be less conditional coherence with Financial Uncertainty, although there is a significant patch at the

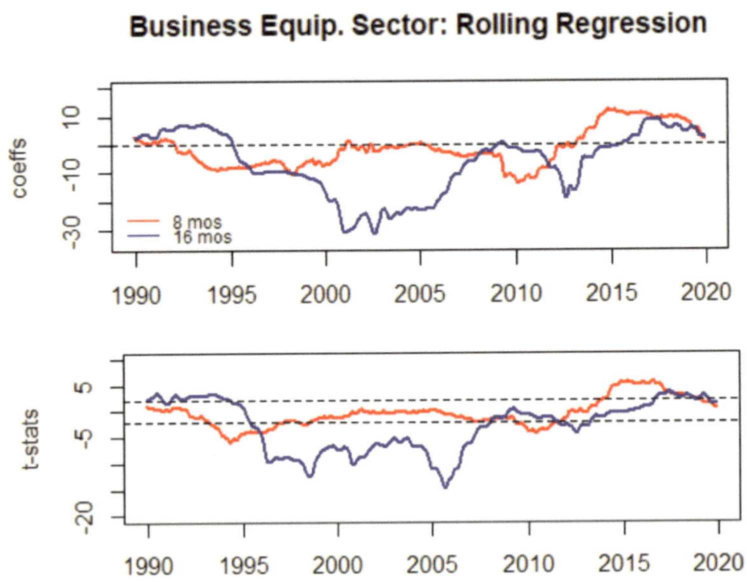

Figure 35.
Rolling Regression Coefficients - Economic Policy Uncertainty and the Business Equipment Sector.

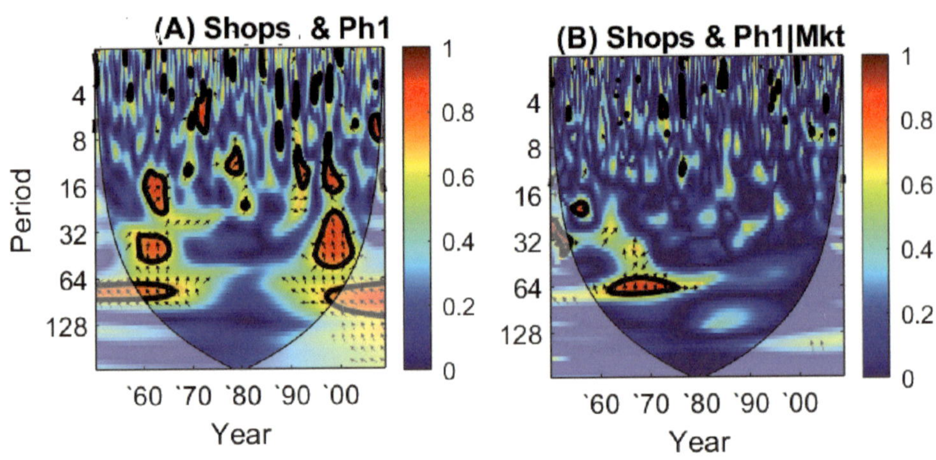

Figure 36.
Wavelet Coherence -Macroeconomic Uncertainty and the Shops Sector.

64 month scale during the Great Recession. Conditional coherence with Economic Policy Uncertainty occurs in the mid to late 1990s at the 8 to 16 month scale (highlighted in white). These findings suggest that the market bears the risk to manufacturing captured by the indices of Macroeconomic and Financial Uncertainty. However, since Economic Policy Uncertainty is showing some hot spots of conditional coherence this suggests that the market responded to news and information regarding this sector, especially the effects of trade on manufacturing. The stand alone uncertainty that is captured may be associated with policy uncertainty regarding trade.

Figure 43 shows the rolling regression coefficients for Macroeconomic Uncertainty at the 64 and 128 month scale. The 128 month scale coefficients are negative and significant throughout the 1970s and 1980s, and from 2000 to 2010. **Figure 44** shows the coefficients for Economic Policy Uncertainty at the 8 and 16 month

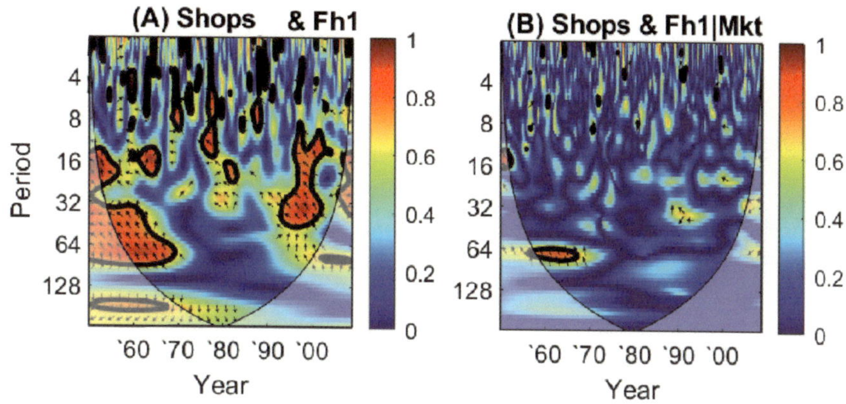

Figure 37.
Wavelet Coherence - Financial Uncertainty and the Shops Sector.

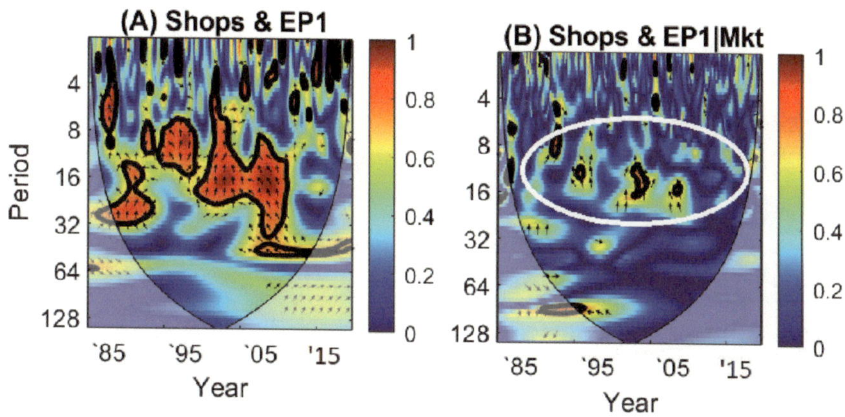

Figure 38.
Wavelet Coherence -Economic Policy Uncertainty and the Shops Sector.

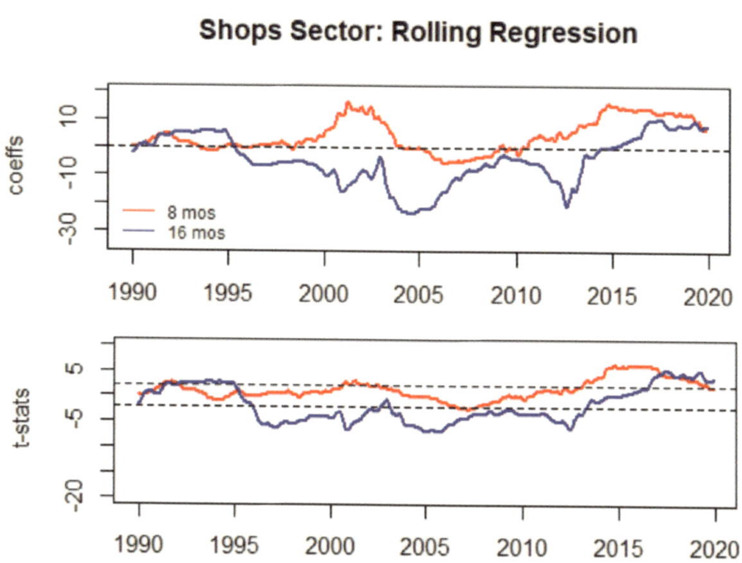

Figure 39.
Rolling Regression Coefficients - Economic Policy Uncertainty and the shops Sector.

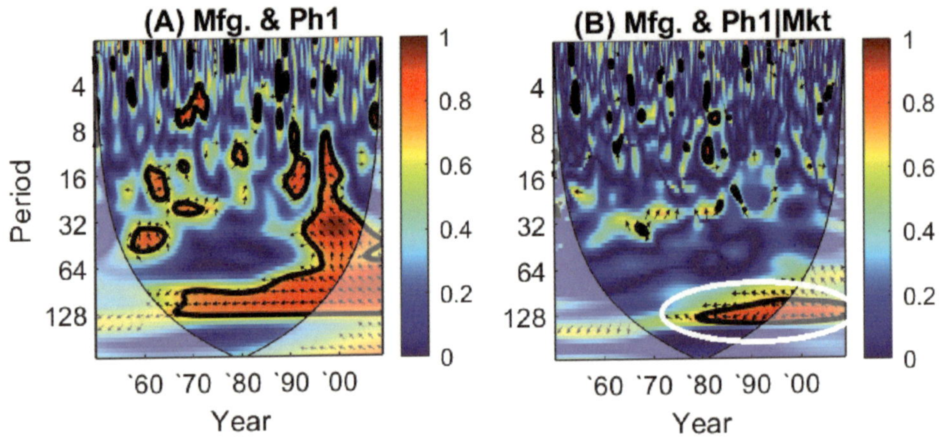

Figure 40.
Wavelet Coherence -Macroeconomic Uncertainty and the Manufacturing Sector.

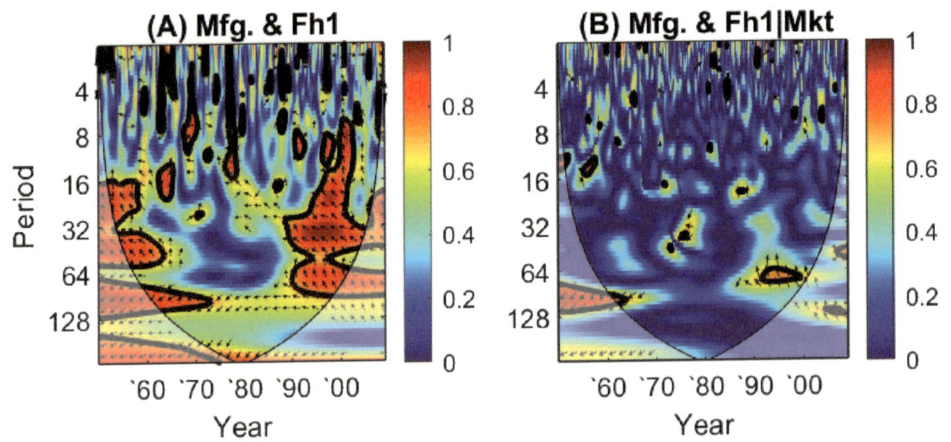

Figure 41.
Wavelet Coherence - Financial Uncertainty and the Manufacturing Sector.

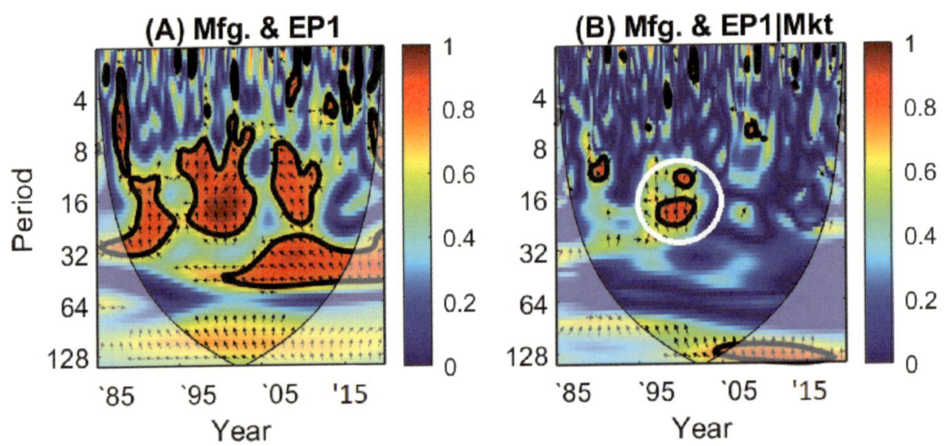

Figure 42.
Wavelet Coherence -Economic Policy Uncertainty and the Manufacturing Sector.

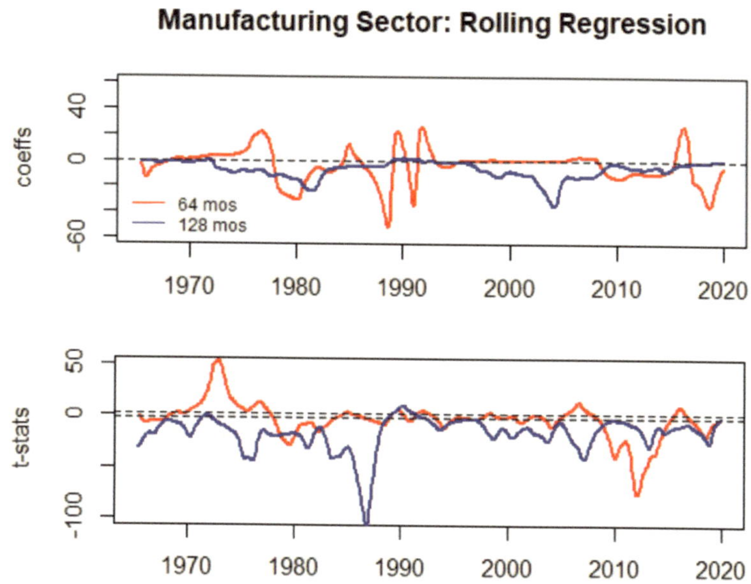

Figure 43.
Rolling Regression Coefficients - Macroeconomic Uncertainty and the Mfg Sector.

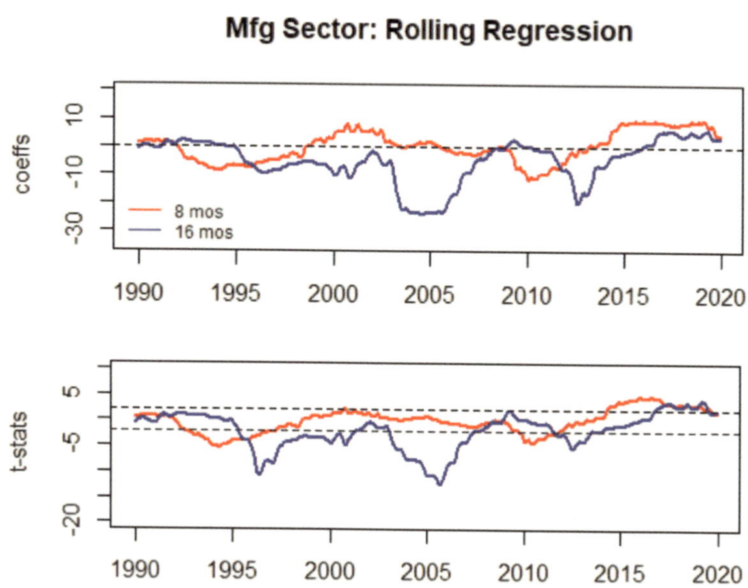

Figure 44.
Rolling Regression Coefficients - Economic Policy Uncertainty and the Mfg Sector.

scales. The impact of concerns about NAFTA is apparent at the 16 month scale with negative coefficients in 1996 and ending in 1998.

6. Conclusions

Wavelet methodology, by allowing local features of the environment to be captured took the lead in our exploration of uncertainty shocks. We examine changes in coherence between each uncertainty measure and the returns of each sector both before and after partialing out the coherence of uncertainty with the market portfolio. Rolling regressions were used to identify sector-specific uncertainty that is not captured by the overall market. Such uncertainty was found for the Money Sector, Energy sector, Telecommunications sector, and Manufacturing sector. These findings suggest that there are periods when the market reaction to shocks is not reflecting all the information captured by the uncertainty indices. One interpretation of our results is that an industry like Telecommunications, Money, Energy, and Manufacturing undergoing significant technological or regulatory changes will have a great reaction to shocks than the overall market response captures. These sectors have a greater sensitivity to uncertainty shocks when the design of the uncertainty metric is largely macro in orientation.

Our finding that there are episodes of uncertainty when there is increased comovements across frequency and over time for specific sectors helps paint a more complete picture of how uncertainty affects the economy through its transmission across sectors. When local features of the return environment are considered, we conclude that in the face of uncertainty shocks the market's knowledge-gathering role could be improved by introducing uncertainty measures that in terms of information-gathering are less global and more local. Localized or micro measures of uncertainty shocks should be of direct relevance to traders and portfolio managers who must respond to such shocks in a ways that are wealth-preserving for their clients.

Conflict of interest

"The authors declare no conflict of interest."

Disclaimer

The material presented in thsi Chapter gives the views of the author, and not necessarily Queens College or the Bank of America.

Author details

Joan Nix[1*†] and Bruce D. McNevin[2†]

1 Queens College, New York, USA

2 Bank of America, New York, USA

*Address all correspondence to: joan.nix@qc.cuny.edu

† These authors contributed equally.

IntechOpen

© 2020 The Author(s). Licensee IntechOpen. This chapter is distributed under the terms of the Creative Commons Attribution License (http://creativecommons.org/licenses/by/3.0), which permits unrestricted use, distribution, and reproduction in any medium, provided the original work is properly cited.

References

[1] Gram, H, G.C. Harcourt 2019. Keynesian Uncertainty: The Great Divide Between Joan Robinson and Paul Samuelson in Their Correspondence and Public Exchanges. In Remaking Economics: Eminent Post-War Economists, R. A. Cord et al.(eds.), SpringerLink.

[2] Jurado, Kyle. Sydney Lugdvison, and Serena Ng, 2015. Measuring Uncertainty. *American Economic Review* 105(3), 1177–1216.

[3] Baker, Scott R., Nicholas Bloom, Steven J. Davis, Measuring Economic Policy Uncertainty, The Quarterly Journal of Economics, Volume 131, Issue 4, November 2016, Pages 1593–1636,

[4] Bloom, Nicholas, 2009. The Impact of Uncertainty Shocks, *Econometrica* 77.3, 623–685.

[5] Bali, Turan G., Stephen J. Brown, and Yi Tang, 2014. Macroeconomic uncertainty and expected stock returns. *Working Paper*.

[6] Ludvigson, Sydney, Sai Ma, and Serena Ng, 2017,"Uncertainty and Business Cycles: Exogenous Impulse or Endogenous Response?" American Economic Journal: Macroeconomics, forthcoming.

[7] Ramsey, James B., Daniel Usikov and George M. Zasslavskyan, 1995, "An Analysis of U.S. Stock Price Behavior Using Wavelets", Fractals, Vol. 3, No. 2, pp 377–389.

[8] Gencay, Ramazan, Faruk Selcuk, and Brandon Whitcher, 2010. *An Introduction to Wavelets and Other Filtering Methods in Finance and Economics.* (Academic Press, New York).

[9] Crowley, Patrick M., A GUIDE TO WAVELETS FOR ECONOMISTS*, Journal of Economic Surveys, Vol.21, No. 2,pages = 207–267.

[10] Strang, Gilbert, 1993. Wavelet transforms versus Fourier transforms, *Bulletin of the American Mathematical Society* 28, 288–305.

[11] Kamara, Avraham and Korajczyk, Robert A. and Lou, Xiaoxia and Sadka, Ronnie, Short-Horizon Beta or Long-Horizon Alpha? (June 11, 2018). Journal of Portfolio Management, Vol. 45, No. 1, (Fall 2018): 96–105.

[12] Ramsey, James, 2002. Wavelets in Economics and Finance: Past and Future, *Studies in Nonlinear Dynamics & Econometrics*, 6, 3 (Online)

[13] French, Kenneth. http://mba.tuck.dartmouth.edu/pages/faculty/ken.french/data_library.html

[14] Ng, E. K. W. and J. C. L. Chan, 2012: Geophysical application of partial wavelet coherence and multiple wavelet coherence. J. Atmos. Oceanic Technol., doi: 10.1175/JTECH-D-12-00056.1.

[15] Waveslim, Version: 1.7.5.1, 2014-12-21,"Basic Wavelet Routines for One-, Two- And Three-Dimensional Signal Processing",Author: Brandon Whitcher, Maintainer: Brandon Whitcher ¡bwhitcher@gmail.com¿

Chapter 12

Case Study: Coefficient Training in Paley-Wiener Space, FFT, and Wavelet Theory

Kayupe Kikodio Patrick

Abstract

Bessel functions form an important class of special functions and are applied almost everywhere in mathematical physics. They are also called cylindrical functions, or cylindrical harmonics. This chapter is devoted to the construction of the generalized coherent state (GCS) and the theory of Bessel wavelets. The GCS is built by replacing the coefficient $z^n/n!$, $z \in \mathbb{C}$ of the canonical CS by the cylindrical Bessel functions. Then, the Paley-Wiener space PW_1 is discussed in the framework of a set of GCS related to the cylindrical Bessel functions and to the Legendre oscillator. We prove that the kernel of the finite Fourier transform (FFT) of L^2-functions supported on $[-1, 1]$ form a set of GCS. Otherwise, the wavelet transform is the special case of CS associated respectively with the Weyl-Heisenberg group (which gives the canonical CS) and with the affine group on the line. We recall the wavelet theory on $\mathbf{R}$. As an application, we discuss the continuous Bessel wavelet. Thus, coherent state transformation (CST) and continuous Bessel wavelet transformation (CBWT) are defined. This chapter is mainly devoted to the application of the Bessel function.

Keywords: coherent state, Hankel transformation, Bessel wavelet transformation

1. Introduction

Coherent state (CS) was originally introduced by Schrödinger in 1926 as a Gaussian wavepacket to describe the evolution of a harmonic oscillator [1].

The notion of coherence associated with these states of physics was first noticed by Glauber [2, 3] and then introduced by Klauder [4, 5]. Because of their important properties these states were then generalized to other systems either from a physical or mathematical point of view. As the electromagnetic field in free space can be regarded as a superposition of many classical modes, each one governed by the equation of simple harmonic oscillator, the CS became significant as the tool for connecting quantum and classical optics. For a review of all of these generalizations see [6–9].

Four main methods are well used in the literature to build CS, the so-called Schrödinger, Klauder-Perelomov, Barut-Girardello and Gazeau-Klauder approaches. The second and third approaches are based directly on the Lie algebra symmetries with their corresponding generators, the first is only established by means of an appropriate infinite superposition of wave functions associated with the harmonic oscillator whatever the Lie algebra symmetries. In [10–12] the authors introduced a new family of CS as a suitable superposition of the associated Bessel functions and in [13–15] the authors also use the generating function approach to construct a new type CS associated with Hermite polynomials and the associated Legendre functions, respectively. The important fact is that we do not use algebraic and group approaches (Barut-Girardello and Klauder-Perelomov) to construct generalized coherent states (GCS).

We first discuss GCS associated with a one-dimensional Schrödinger operator [16, 17] by following the work in [18, 19]. We build a family of GCS through superpositions of the corresponding eigenstates, say $\psi_n, n \in \mathbb{N}$, which are expressed in terms of the Legendre polynomial $P_n(x)$ [16]. The role of coefficients $z^n/\sqrt{n!}$ of the canonical CS is played by

$$\mathfrak{D}_n(\xi) := i^n \left(\frac{\pi(2n+1)}{2\xi}\right)^{\frac{1}{2}} J_{n+\frac{1}{2}}(\xi), \quad n = 0, 1, 2, \ldots, \tag{1}$$

where $\xi \in \mathbb{R}$ and $J_{n+\frac{1}{2}}(.)$ denotes the cylindrical Bessel function [20]. When $n = 0$, Eq. (1) becomes

$$\mathfrak{D}_0 = \mathcal{J}_0(\xi) = \frac{\sin(\xi)}{\xi} \tag{2}$$

where $\mathcal{J}_0(.)$ denotes the spherical Bessel function of order 0. The choosen coefficients (1) and eigenfunctions (27) (see below) have been used in ([21], p. 1625). We proceed by determining the wavefunctions of these GCS in a closed form. The latter gives the kernel of the associated CS transform which makes correspondence between the quantum states Hilbert space $L^2([-1,1], 2^{-1}dx)$ of the Legendre oscillator and a subspace of a Hilbert space of square integrable functions with respect to a suitable measure on the real line. We show that the kernel $e^{ix\xi}, \xi \in \mathbb{R}$, of the L^2-functions that are supported in $[-1,1]$ form a set of GCS.

There are in literature several approach to introducce Bessel Wavelets. We refer for instence to [22, 23]. Note that, for $[-1,1] \ni x \mapsto \cos(y/n),\ n \in \mathbb{N}$, the Legendre polynomial $P_n(x)$ and the Bessel function of order 0 are related by the Hansen's limit

$$\lim_{n\to\infty} P_n\left(\cos\frac{y}{n}\right) = \int_0^{\pi} e^{iy\cos\phi} d\phi = J_0(y),$$

and the integral

$$\int_0^{\infty} J_0(y) J_0(y) dy = \frac{\pi}{2}. \tag{3}$$

Note that in [22, 23] the authors have introduced the Bessel wavelet based on the Hankel transform. The notion of wavelets was first introduced by J. Morlet a French petroleum engineer at ELF-Aquitaine, in connection with his study of seismic traces. The mathematical foundations were given by A. Grossmann and J. Morlet [24]. Harmonic analyst Y. Meyer and other mathematicians understood the importance of this theory and they recognized many classical results within (see [25–27]). Classical wavelets have several applications ranging from geophysical and acoustic signal analysis to quantum theory and pure mathematics. A wavelet base is a family of functions obtained from a function known as mother wavelet, by translation and dilation. This tool permits the representation of L^2-functions in a basis well localized in time and in freqency. Wavelets are special functions with special properties which may not be satisfied by other functions. In the current context, our objective is to make a link between the construction of GCS and the theory of wavelets. Therefore, we will talk about coherent state transformation (CST) and the continuous Bessel wavelet transformation (CBWT).

The rest of this chapter is organized as follows: Section 2 is devoted to the generalized CS formalism that we are going to use. In Section 3, we briefly introduce the Paley-Wiener space PW_Ω and some notions on Legendre's Hamiltonian. We give in Section 4 a summary concept on the continuous wavelet transform on $\mathbb{R}$. In Section 5, we have constructed a class of GCS related to the Bessel cylindrical function for the legendre Hamiltonian. In Section 6, we discuss the theory of CBWT where we show as an example that the function $f \in L^2_\sigma(\mathbb{R}_+)$

$$f(t) := \frac{2w_0 - t^2}{2\left(w_0^2 + t^2\right)^{5/2}}, \quad w_0 > 0, \tag{4}$$

such that $\int_{\mathbb{R}_+} f(t)d\sigma(t) = 0$ is the mother wavelet where $d\sigma(t)$ is an appropriate Legesgue's measure on $\mathbb{R}$. Finally in Section 7. we gives some concluding remarks on the chapter.

2. Generalized coherent states formalism

We follow the generalization of canonical coherent states (CCS) introduced in [18, 19]. The definition of CS as a set of vectors associated with a reproducing kernel is general, it encompasses all the situations encountered in the physical literature. For applications we will work with normalized vectors. Let $(\mathscr{X}, \mu)$ be a measure space and let $\mathfrak{N}^2 \subset L^2(\mathscr{X}, \mu)$ be a sub-closed space of infinite dimension. Let $\{\mathscr{C}_n\}_{n=0}^{\infty}$ be a satisfactory orthogonal basis of $\mathfrak{N}^2$, for arbitrary $x \in \mathscr{X}$

$$\sum_{n=0}^{\infty} \rho_n^{-1} |\mathscr{C}_n(x)|^2 < +\infty \tag{5}$$

where $\rho_n := \|\mathscr{C}_n\|^2_{L^2(\mathscr{X},\mu)}$. Define the kernel

$$K(x,y) := \sum_{n=0}^{\infty} \rho_n^{-1} \mathscr{C}_n(x)\overline{\mathscr{C}_n(y)}, \quad x,y \in \mathscr{X}. \tag{6}$$

Then, the expression $K(x,y)$ is a reproducing kernel, $\mathfrak{N}^2$ is the corresponding kernel Hilbert space and $\mathscr{N}(x) := K(x,x), \ x \in \mathscr{X}$. Define

$$\vartheta_x := (\mathscr{N}(x))^{-1/2} \sum_{n=0}^{\infty} \rho_n^{-1/2} \overline{\mathscr{C}_n(x)} \varphi_n.$$

Therefore,

$$\langle \vartheta_x, \vartheta_x \rangle = \mathscr{N}(x)^{-1} \sum_{n=0}^{\infty} \rho_n^{-1} \mathscr{C}_n(x)\overline{\mathscr{C}_n(x)} = 1,$$

and

$$\mathscr{W} : \mathscr{H} \to \mathfrak{N}^2 \quad with \quad \mathscr{W}\phi = \mathscr{N}^{1/2}\langle \vartheta_x, \phi \rangle$$

is an isometry. For $\phi, \psi \in \mathscr{H}$, whe have

$$\langle \phi, \psi \rangle_{\mathscr{H}} = \langle \mathscr{W}\phi, \mathscr{W}\psi \rangle_{\mathfrak{N}^2} = \int_{\mathscr{X}} \overline{\mathscr{W}\phi(x)}\mathscr{W}\psi(x)d\mu(x) \tag{7}$$

$$= \int_{\mathscr{X}} \langle \phi, \vartheta_x \rangle \langle \vartheta_x, \psi \rangle \mathscr{N}(x)d\mu(x), \tag{8}$$

and

$$\int_{\mathscr{X}} |\vartheta_x\rangle\langle\vartheta_x|\mathscr{N}(x)d\mu(x) = I_{\mathscr{H}}, \tag{9}$$

where $\mathscr{N}(x)$ is a positive weight function.

Definition 1. *Let $\mathscr{H}$ be a Hilbert space with* $\dim \mathscr{H} = \infty$ *and* $\{\varphi_n\}_{n=0}^{\infty}$ *be an orthonormal basis of $\mathscr{H}$.The generalized coherent state (GCS) labeled by point $x \in \mathscr{X}$ are defined as the ket-vector $\vartheta_x \in \mathscr{H}$, such that*

$$\vartheta_x := (\mathscr{N}(x))^{-1/2} \sum_{n=0}^{\infty} \rho_n^{-1/2} \overline{\mathscr{C}_n(x)} \varphi_n. \tag{10}$$

By definition, it is straightforward to show that $\langle \vartheta_x, \vartheta_x \rangle_{\mathscr{H}} = 1$.

Definition 2. *For each function $f \in \mathscr{H}$, the coherent state transform (CST) associated to the set $(\vartheta_x)_{x \in \mathscr{X}}$ is the isometric map*

$$\mathscr{W}[f](x) := (\mathscr{N}(x))^{1/2} \langle f | \vartheta_x \rangle_{\mathscr{H}}. \tag{11}$$

Thereby, we have a resolution of the identity of $\mathscr{H}$ which can be expressed in Dirac's bra-ket notation as

$$\mathbf{1}_{\mathscr{H}} = \int_{\mathscr{X}} T_x \mathscr{N}(x)d\mu(x) \tag{12}$$

where the rank one operator $T_x := |\vartheta_x\rangle\langle\vartheta_x| : \mathscr{H} \to \mathscr{H}$ is define by

$$f \mapsto T_x[f] = \langle \vartheta_x | f \rangle \vartheta_x.$$

$\mathscr{N}(x)$ appears as a weight function.

Next, the reproducing kernel has the additional property of being square integrable, i.e.,

$$\int_{\mathscr{X}} K(x,z)K(z,y)\mathscr{N}(z)d\mu(z) = K(x,y). \tag{13}$$

Note that the formula (10) can be considered as generalization of the series expansion of the CCS [28].

$$\vartheta_z = \sqrt{\pi} e^{-\frac{z\bar{z}}{2}} \sum_{k=0}^{\infty} \frac{z^n}{\sqrt{n!}} \phi_n, \quad z \in \mathbb{C} \tag{14}$$

with $\{\phi_n\}_{n=0}^{\infty}$ being an orthonormal basis of eigenstates of the quantum harmonic oscillator. Then, the space $\mathfrak{N}^2$ is the *Fock space* $\mathcal{F}(\mathbb{C})$ and $\mathcal{N}(z) = \pi^{-1}e^{z\bar{z}},\ z\in\mathbb{C}$.

3. The Paley-wiener space PW_Ω and the Legendre Hamiltonian: a brief overview

3.1 The Paley-wiener space PW_Ω

The Paley-Wiener space is made up of all integer functions of exponential type whose restrictions on the real line is square integrable. We give in this Section a general overview on this notion ([29], pp. 45–47).

Definition 3. *Consider F as an entire function. Then, F is an entire function of exponential type if there exists constants* $A, B > 0$ *such that, for all* $z \in \mathbb{C}$

$$|F(z)| \le Ae^{B|z|}. \tag{15}$$

Note that, if F satisfy Definition 3, we call Ω the type of F where

$$\Omega = \lim_{r\to+\infty} \sup \frac{\log M(r)}{r} \tag{16}$$

and where $M(r) = \sup_{|z|=r}|F(z)|$. The following conditions on an entire function F are verified:

1. For all $\varepsilon > 0$ there exists C_ε such that

$$|F(z)| \le C_\varepsilon e^{(\Omega+\varepsilon)|z|};$$

2. There exists $C > 0$ such that

$$|F(z)| \le Ce^{\Omega|z|};$$

3. as $|z| \to +\infty$

$$|F(z)| = o\left(e^{\Omega|z|}\right).$$

Then cleary, (3) $\Rightarrow$ (2) $\Rightarrow$ (1) $\Rightarrow$ F is of exponential type at most Ω.

Definition 4. *Let* $\Omega > 0$ *and* $1 \le p \le \infty$. *The Paley-Wiener space* PW_Ω^p *is defined as*

$$PW_\Omega^p = \left\{ f \in L^2(\mathbb{R}) : f(x) = \int_{-\Omega}^{\Omega} g(y)e^{-ixy}dy, \text{ where } g \in L^p(-\Omega,\Omega)\right\} \tag{17}$$

and we set

$$\|f\|_{PW_\Omega^p} = 2\pi\|g\|_{L^p}. \tag{18}$$

The Paley-Wiener PW_Ω^p is the image via the *Fourier transform* of the L^p-function that are supported in $[-\Omega, \Omega]$. We will be interested in the case $p = 2$, in which PW_Ω to denote the Paley-Wiener space PW_ω^2. From the *Plancherel formula* we have

$$\|f\|_{PW_\Omega^2} = \|\hat{g}\|_{PW_\Omega^2} = 2\pi\|g\|_{L^2} = \|\hat{f}\|_{L^2} = \|f\|_{L^2}. \tag{19}$$

Hence, by polarization, for $f, \varphi \in PW_\Omega$,

$$\langle f, \varphi \rangle_{PW_\Omega} = \langle f, \varphi \rangle_{L^2}. \tag{20}$$

Theorem 1.1 Let F be an entire function and $\Omega > 0$. Then the following are equivalent

- $F_{|\mathbb{R}} \in L^2(\mathbb{R})$ and

$$|F(z)| = o\left(e^{\Omega|z|}\right) \text{ as } |z| \to +\infty, \tag{21}$$

- there exists $f \in L^2(\mathbb{R})$ with $supp\hat{f} \subseteq [-\Omega, \Omega]$ such that

$$F(z) = \frac{1}{2\pi}\int_{\mathbb{R}} \hat{f}(\xi)e^{iz\xi}d\xi. \tag{22}$$

The function $f \in PW_\Omega$ if and only if $f \in L^2(\mathbb{R})$ and $f = F_{|\mathbb{R}}$ (that is, f is the restriction to the real line of a function F), where F is an entire function of exponential type such that $|F(z)| = o\left(e^{\Omega|z|}\right)$ for $|z| \to +\infty$.

Theorem 1.2 The Paley-Wiener space PW_Ω is a Hilbert space with reproducing kernel w.r.t the inner product (20). Its reproducing kernel is the function

$$K(x,y) = \frac{\Omega}{\pi} sinc(\Omega(x-y)), \tag{23}$$

where $sinct = sint/t$. Hence, for every $f \in PW_\Omega$

$$f(x) = \frac{\Omega}{\pi}\int_{\mathbb{R}} f(y) sinc(\Omega(x-y))dy, \tag{24}$$

where $x \in \mathbb{R}$.

3.2 The Legendre Hamiltonian

The Legendre polynomials $P_n(x)$ and the Legendre function $\psi_n(x)$ are similar to the Hermite polynomials and the Hermite function in standard quantum mechanics. Based on the work of Borzov and Demaskinsky [16, 17] the Legendre Hamiltonian has the form

$$H = X^2 + P^2 = a^+a^- + a^-a^+, \tag{25}$$

where X and P denotes respectively the position and momentum operators, a^+ and a^- are the creation and annihilation operators. The eigenvalues of operators H are equal to

$$\lambda_0 = \frac{2}{3}, \quad \lambda_n = \frac{n(n+1) - \frac{1}{2}}{\left(n+\frac{3}{2}\right)\left(n-\frac{1}{2}\right)}, n = 1, 2, 3, \ldots\ , \tag{26}$$

and the corresponding eigenfunctions reads

$$\psi_n(x) = \sqrt{2n+1}P_n(x), \quad n = 0, 1, 2, 3, \ldots\ , \tag{27}$$

in terms of the Legendre polynomial $P_n(.)$, which form an orthonormal basis $\{\psi_n \equiv |n\rangle\}_{n=0}^{\infty}$ in the Hilbert space $\mathscr{H} := L^2([-1,1], 2^{-1}dx)$. These functions satisfy the recurrence relations

$$x\psi_n(x) = b_{n-1}\psi_{n-1}(x) + b_n\psi_{n+1}(x), \quad \psi_{-1}(x) = 0, \ \psi_0(x) = 1, \tag{28}$$

with coefficients

$$b_n = \sqrt{\frac{(n+1)^2}{(2n+1)(2n+3)}}, \quad n \geq 0. \tag{29}$$

The generalized position operator on the Hilbert space $\mathscr{H}$ connected with the Legendre polynomials $P_n(x)$ is an operator of multiplication by argument $X\psi_n = x\psi_n$. Taking into account of the relation (28), then

$$X\psi_n(x) = b_n\psi_{n+1}(x) + b_{n-1}\psi_{n-1}(x), \tag{30}$$

whee b_n are coefficients defined by Eq. (29). Because $\sum_{n=0}^{\infty} 1/b_n = +\infty$, X is a self-adjoint operator on the Hilbert space $\mathscr{H}$ (see [30–32]). The momentum operator P by the way described in ([17], p. 126) acts on the basis elements in $\mathscr{H}$, by the formula $P\psi_n = i(b_n\psi_{n+1} - b_{n-1}\psi_{n-1})$. The usual commutator of operator X and P on the basis elements reads as

$$[X, P]\psi_n = 2i(b_n^2 - b_{n-1}^2)\psi_n = \frac{2i}{(2n-1)(2n+1)(2n+3)}\psi_n. \tag{31}$$

The creation and annihilation operators (25) are define by relations

$$a^+ = \frac{1}{\sqrt{2}}(X - iP); \quad a^- = \frac{1}{\sqrt{2}}(X + iP), \tag{32}$$

these operators act as $a^+\psi_n = \sqrt{2}b_n\psi_{n+1}$ and $a^-\psi_n = \sqrt{2}b_{n-1}\psi_{n-1}$. They satisfy $[a^-, a^+] = -i[X, P]$, the commutation relations.

4. Wavelet theory on $\mathbb{R}$ and the reproduction of kernels

We briefly describe below some basis definitions and properties of the one-dimensional wavelet transform on $\mathbb{R}_+$, we refer to [22, 23, 33]. In the Hilbert space $\mathfrak{N} = L^2(\mathbb{R}, dx)$, the function ψ satisfying the so-called admissibility condition

$$\mathscr{C}_\psi := \int_{-\infty}^{\infty} \frac{|\hat{\psi}(\xi)|^2}{\xi} d\xi < \infty, \tag{33}$$

where $\hat{\psi}$ being the Hankel transform of ψ. Not every vector in $\mathfrak{N}$ satisfies the above condition. A vector ψ satisfying (33) is called a mother wavelet. Combining dilatation and translation, one gets affine transformation

$$y = (b, a)x \equiv ax + b, \ a > 0, \ b \in \mathbb{R}, \ x \in \mathbb{R}_+. \tag{34}$$

Thus $\{(b,a)\}=:G_{aff}=\mathbb{R}\times(0,\infty)$, the affine group of the line. Specifically, for each pair (a,b) of the real numbers, with $a>0$, from translations and dilatations of the function ψ, we obtain a family of wavelets $\{\psi_{a,b}\}\in\mathfrak{N}$ as

$$\psi_{a,b}(x)=\frac{1}{\sqrt{a}}\psi\left(\frac{x-b}{a}\right), \qquad \psi_{1,0}=\psi. \tag{35}$$

Here a is the parameter of dilation (or scale) and b is the parameter of translation (or position). It is then easily cheked that

$$\left\|\psi_{a,b}(x)\right\|_{\mathfrak{N}}^{2}=\|\psi(x)\|_{\mathfrak{N}}^{2}, \quad for \quad all \quad a>0 \quad and \quad b\in\mathbb{R}. \tag{36}$$

Moreover, in terms of the Dirac's bracket notation it is an easy to show that the resolution of the identity

$$\frac{1}{\mathscr{C}_{\psi}}\int\int_{\mathbb{R}\times\mathbb{R}_{+}^{*}}\left|\psi_{a,b}\right\rangle\left\langle\psi_{a,b}\right|\frac{dbda}{a}=I_{\mathfrak{N}} \tag{37}$$

holds for these vectors (in the weak sense). Here $I_{\mathfrak{N}}$ is the identity operator on $\mathfrak{N}$. The *continuous wavelet transform* of an arbitrary vector (signal) $f\in\mathfrak{N}$ at the scale a and the position b is given by

$$\mathscr{S}_{f}(a,b)=\int_{0}^{\infty}f(t)\psi_{a,b}(t)dt. \tag{38}$$

The wavelet transform $\mathscr{S}_{f}(a,b)$ has several properties [34]:

- It is linear in the sense that:

$$\mathscr{S}_{\alpha f_{1}+\beta f_{2}}(a,b)=\alpha\mathscr{S}_{f_{1}}(a,b)+\beta\mathscr{S}_{f_{2}}(a,b), \quad \forall\alpha,\beta\in\mathbb{R} \ and \quad f_{1},f_{2}\in L^{2}(\mathbb{R}_{+}).$$

- It is translation invariant:

$$\mathscr{S}_{\tau_{b'}f}(a,b)=\mathscr{S}_{f}(a,b-b')$$

where $\tau_{b'}$ refers to the translation of the function f by b' given

$$(\tau_{b'}f)(x)=f(x-b').$$

- It is dilatation-invariant, in the sense that, if f satisfies the invariance dilatation property $f(x)=\lambda f(rx)$ for some $\lambda,r>0$ fixed then

$$\mathscr{S}_{f}(a,b)=\lambda\mathscr{S}_{f}(ra,rb). \tag{39}$$

As in Fourier or Hilbert analysis, wavelet analysis provides a Plancherel type relation which permits itself the reconstruction of the analyzed function from its wavelet transform. More precisely we have

$$\langle f,g\rangle=\frac{1}{\mathscr{C}_{\psi}}\int_{a>0}\int_{b\in\mathbb{R}}\mathscr{S}_{f}(a,b)\overline{\mathscr{S}_{g}(a,b)}\frac{dadb}{a^{2}}, \quad \forall f,g\in L^{2}(\mathbb{R}) \tag{40}$$

which in turns to reconstruct the analyzed function f in the L^2- sense from its wavelet transform as

$$f(x) = \frac{1}{\mathscr{C}_\psi} \int_{a>0} \int_{b\in\mathbb{R}} \mathscr{S}_f(a,b)\psi_{a,b} \frac{dadb}{a^2}, \quad where \quad \mathscr{S}_f(a,b) = \langle \psi_{a,b} | f \rangle. \tag{41}$$

The function $\mathscr{S}_f$ is the continuous wavelet transform of the signal f. The parameter $1/a$ represents the signal frequency of f and b its time. The conservation of the energy of the signal is due to the resolution of the identity (37), so

$$\mathscr{C}_\psi \|f\|^2 = \int\int_{\mathbb{R}\times\mathbb{R}_+} |\mathscr{S}_f(b,a)|^2 \frac{dbda}{a^2}. \tag{42}$$

Then, the transform $\mathscr{S}_f$ is a fonction in the Hilbert space $L^2\left(\mathbb{R}\times\mathbb{R}_+^*, \frac{dbda}{a^2}\right)$. The reproducing kernel associated to the signal is

$$K_\psi(b,a,b',a') = \frac{1}{\mathscr{C}_\psi} \langle \psi_{a,b} | \psi_{a',b'} \rangle. \tag{43}$$

which satisfies the square integrability condition (13) with respect to the measure $dbda/a^2$. The corresponding reproducing kernel Hilbert space $\mathfrak{N}_\psi$, one see that this is the space of all signal transforms, corresponding to the mother wavelet ψ. If ψ and ψ' are two mother wavelets such that $\langle \psi' | \psi \rangle \neq 0$, then

$$\frac{1}{\langle \psi' | \psi \rangle} \int\int_{\mathbb{R}\times\mathbb{R}_+^*} |\psi_{a,b}\rangle\langle\psi'_{a,b}| \frac{dbda}{a^2} = I_{\mathfrak{N}}, \tag{44}$$

The formula (41) generalizes to

$$f = \frac{1}{\langle \psi' | \psi \rangle} \int\int_{\mathbb{R}\times\mathbb{R}_+^*} \mathscr{S}'_f(b,a)\psi_{a,b} \frac{dbda}{a^2}, \quad where \quad \mathscr{S}'_f(a,b) = \langle \psi'_{a,b} | f \rangle. \tag{45}$$

The vector ψ' is called the analyzing wavelet and ψ the reconstructing wavelet. The repoducing kernel Hilbert space $\mathfrak{N} \subset L^2(\mathbb{R}\times\mathbb{R}_+^*)$, consisting of all signal transforms with respect to the mother wavelet ψ'. Then, we have

$$K_{\psi,\psi'}(b,a;b',a') = \frac{1}{[\mathscr{C}_\psi \mathscr{C}_{\psi'}]^{\frac{1}{2}}} \left\langle \psi_{a,b} | \psi'_{a',b'} \right\rangle \tag{46}$$

is the integral kernel of a unitary map between $\mathfrak{N}_{\psi'}$ and $\mathfrak{N}_\psi$. The properties of the wavelet transform can be understood in terms of the unitary irreductible representation of the one-dilensional affine group.It is important to note that the Wavelets built on the basis of the group representation theory have all the properties of CS. There is a wole body of work devoted to the study of CS arising from group representation theory [7, 33, 35].

5. Application 1: GCS for the Legendre Hamiltonian and CS transform

5.1 GCS for the Legendre Hamiltonian

By replacing the coefficients $z^n/\sqrt{n!}$ of the canonical CS by the function $\mathfrak{O}_n(\xi)$ in (1) as mentioned in the introduction. We construct in this section a class of GCS indexed by point $\xi \in \mathbb{R}$.

Definition 5. *The GCS labeled by points $\xi \in \mathbb{R}$ is defined by the following superposition*

$$\vartheta_\xi = \mathcal{N}(\xi)^{-1/2} \sum_{n=0}^{\infty} \mathfrak{D}_n(\xi)\psi_n, \quad \xi \in \mathbb{R} \tag{47}$$

here $\mathcal{N}(\xi)$ is a normalization factor, the function $\mathfrak{D}_n(\xi) := \Phi_n(\xi)\rho_n^{-1/2}$, with

$$\Phi_n(\xi) = i^n \sqrt{\frac{\pi}{2\xi}} J_{n+\frac{1}{2}}(\xi), \tag{48}$$

where $J_{n+1/2}(.)$ is the cylindrical Bessel function ([20], p. 626):

$$J_{n+\frac{1}{2}}(z) = \sum_{s=0}^{\infty} \frac{(-1)^s}{s!(s+n+1/2)!} \left(\frac{z}{2}\right)^{2s+n+\frac{1}{2}}, \quad z \in \mathbb{C} \tag{49}$$

and ρ_n are positive numbers given by

$$\rho_n = \frac{1}{2n+1}, \quad n = 0, 1, 2, \ldots , \tag{50}$$

and $\{\psi_n\}$ is an orthonormal basis of the Hilbert space $\mathcal{H} = L^2([-1,1], 2^{-1}dx)$ defined in (27).

Proposition 1. *The normalization factor defined by the GCS (47) reads as*

$$\mathcal{N}(\xi) = 1, \tag{51}$$

for every $\xi \in \mathbb{R}$.

Proof. From (47) and by using the orthonormality relation of basis elements $\{\psi_n\}_{n=0}^{+\infty}$ in (27), then

$$\langle \vartheta_\xi | \vartheta_\xi \rangle = \pi(\xi\mathcal{N}(\xi))^{-1} \sum_{n=0}^{\infty} \left(n + \frac{1}{2}\right) J_{n+\frac{1}{2}}(\xi) J_{n+\frac{1}{2}}(\xi). \tag{52}$$

In order to identify the above series, we make appeal to the formula ([36], p. 591):

$$\sum_{n=0}^{\infty} \left(n + \frac{1}{2}\right) J_{n+\frac{1}{2}}(\xi) J_{n+\frac{1}{2}}(\xi) = \pi^{-1}\xi, \tag{53}$$

we then obtain the result (51) by using the GCS condition $\langle \vartheta_\xi | \vartheta_\xi \rangle = 1$. □

Proposition 2. *The GCS defined in* (47) *satisfy the following resolution of the identity*

$$\int_{\mathbb{R}} T_\xi d\mu(\xi) = \mathbf{1}_{\mathcal{H}}, \tag{54}$$

(in the weak sense) in terms of an acceptable measure

$$d\mu(\xi) = \frac{1}{\pi} d\xi, \tag{55}$$

where $d\xi$ the Lebesgue's measure on $\mathbb{R}$. The rank one operator $T_\xi = |\vartheta_\xi\rangle\langle\vartheta_\xi| : \mathcal{H} \to \mathcal{H}$ is define as

$$\varphi \mapsto T_\xi[\varphi] = \langle \vartheta_\xi | \varphi \rangle \vartheta_\xi. \tag{56}$$

Proof. We need to determine the function $\sigma(\xi)$. Let

$$d\mu(\xi) = \sigma(\xi) d\xi, \tag{57}$$

where $\sigma(\xi)$ is an auxiliary function. Let us writte $T_{m,n} := |\psi_m\rangle\langle\psi_n|$, defined as in (56). According to (56) and by writing

$$\int_{\mathbb{R}} T_\xi d\mu(\xi)$$

$$= \sum_{n,m=0}^{\infty} \frac{\pi}{2} (-1)^n i^{n+m} \left(\int_{-\infty}^{\infty} \frac{J_{m+\frac{1}{2}}(\xi) J_{n+\frac{1}{2}}(\xi)}{\sqrt{\rho(m)\rho(n)}} \sigma(\xi) \frac{d\xi}{\xi} \right) T_{m,n} \tag{58}$$

$$= \sum_{n,m=0}^{\infty} \frac{\pi}{2} (-1)^n i^{n+m} \sqrt{(2m+1)(2n+1)} \left(\int_{-\infty}^{\infty} J_{m+\frac{1}{2}}(\xi) J_{n+\frac{1}{2}}(\xi) \sigma(\xi) \frac{d\xi}{\xi} \right) T_{m,n}. \tag{59}$$

Hence, we need $\sigma(\xi)$ such that

$$\int_{-\infty}^{\infty} J_{n+\frac{1}{2}}(\xi) J_{m+\frac{1}{2}}(\xi) \sigma(\xi) \frac{d\xi}{\xi} = \frac{2}{\pi(2n+1)} \delta_{m,n}. \tag{60}$$

We make appeal to the integral ([36], p. 211):

$$\int_{-\infty}^{\infty} \frac{1}{y} J_{m+\frac{1}{2}}(cy) J_{n+\frac{1}{2}}(cy) dy = \frac{2}{2n+1} \delta_{m,n}, \tag{61}$$

with condition $c > 0$. Then, for parameters $c = 1$, we have

$$\int_{-\infty}^{\infty} \frac{1}{\xi} J_{m+\frac{1}{2}}(\xi) J_{n+\frac{1}{2}}(\xi) d\xi = \frac{2}{2n+1} \delta_{m,n}. \tag{62}$$

By comparing (62) with (66) we obtain finally the desired weight function $\sigma(\xi) = 1/\pi$. Therefore, the measure (57) has the form (55) [37]. Indeed (59) reduces further to $\sum_{n=0}^{\infty} T_{n,n} = 1_{\mathscr{H}}$, in other words

$$\int_{\mathbb{R}} T_\xi d\mu(\xi) = \mathbf{1}_{\mathscr{H}}. \tag{63}$$

According to this construction, the state ϑ_ξ form an overcomplete basis in the Hilbert space $\mathscr{H}$ (**Figure 1**). □

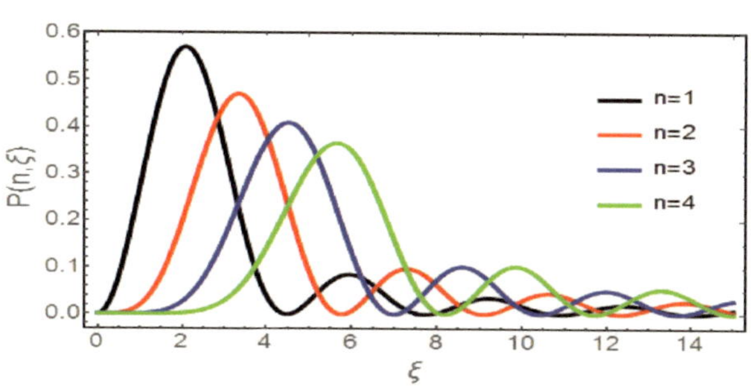

Figure 1.
Plots of the probability distribution $P(n,\xi)$ versus ξ for various values of n.

When the GCS (47) describes a quantum system, the probability of finding the state ψ_n in some normalized state ϑ_ξ of the state Hilbert space $\mathscr{H}$ is given by $P(n,\xi) := \left|\langle \psi_n | \vartheta_\xi \rangle\right|^2$. For the GCS (47) the probability distribution function is given by

$$P(n,\xi) = \frac{\pi(2n+1)}{2|\xi|}\left|J_{n+\frac{1}{2}}(\xi)\right|^2, \quad \xi \in \mathbb{R}_+^*. \tag{64}$$

5.2 Coherent state transform

To discuss coherent state transforms (CST), we will start by establishing the kernel of this transformation by giving the closed form of the GCS (47).

Proposition 3. *For all $x \in [-1,1]$, the wave functions of GCS in (47) can be written as*

$$\vartheta_\xi(x) = e^{-ix\xi}, \tag{65}$$

for all $\xi \in \mathbb{R}$.

Proof. We start by the following expression

$$\vartheta_\xi(x) = \mathcal{N}(\xi)^{-1/2}\mathfrak{S}(x,\xi), \tag{66}$$

where the series

$$\mathfrak{S}(x,\xi) := \sum_{n=0}^{\infty} \mathfrak{O}_n(\xi)\psi_n(x), \tag{67}$$

with the function $\mathfrak{O}_n(\xi) = \Phi_n(\xi)\rho_n^{-1/2}$, mentioned in Definition 5. To do this, we start by replacing the function $\Phi_n(\xi)$ and the positive sequences ρ_n by their expressions in (48) and (50) thus Eq. (67) reads

$$\mathfrak{S}(x,\xi) = \sqrt{\frac{\pi}{2\xi}}\sum_{n=0}^{\infty}(-1)^n i^n \sqrt{2n+1}J_{n+\frac{1}{2}}(\xi)\psi_n(x). \tag{68}$$

Making use the explicit expression (27) of the eigenstates $\psi_n(x)$, then the sum (68) becomes

$$\mathfrak{S}(x,\xi) = \sqrt{\frac{2\pi}{\xi}}\sum_{n=0}^{\infty}(-1)^n i^n \left(n+\frac{1}{2}\right)J_{n+\frac{1}{2}}(\xi)P_n(x). \tag{69}$$

We now appeal to the Gegenbauer's expansion of the plane wave in Gegenbauer polynomials and Bessel functions ([38], p. 116):

$$\mathrm{e}^{\mathrm{i}\xi \mathrm{x}} = \Gamma(\gamma)(\xi 2)^{-\gamma}\sum_{\mathrm{n}=0}^{\infty}\mathrm{i}^{\mathrm{n}}(\mathrm{n}+\gamma)\mathrm{J}_{\mathrm{n}+\gamma}(\xi)\mathrm{C}_{\mathrm{n}}^{\gamma}(\mathrm{x})$$

Then, for $\gamma = 1/2, y = x$ and by using the identity $\Gamma(1/2) = \sqrt{\pi}$, we arrive at (65). □

Corollary 1. *When the variable $\xi \ll 1$, the GCS in (47) becomes*

$$\vartheta_\xi \approx \mathcal{N}(\xi)^{-1/2}\sum_{n=0}^{\infty}\frac{\sqrt{2\pi}(-i\xi)^n}{\sqrt{2^{2n+1}(2n+1)\Gamma\left(n+\frac{1}{2}\right)}}\psi_n. \tag{70}$$

Proof. The result follows immediately by using the formula ([20], p. 647):

$$\mathscr{J}_n(\xi) \approx \frac{\xi^n}{(2n+1)!!}, \quad \xi \ll 1 \tag{71}$$

where

$$\mathscr{J}_n(\xi) = \sqrt{\frac{\pi}{2\xi}} J_{n+\frac{1}{2}}(\xi), \quad n = 0, 1, 2, \ldots \ , \tag{72}$$

is the spherical Bessel function [20]. This ends the proof. □

The careful reader has certainly recognized in (70) the expression of nonlinear coherent states [38].

Let us note that, in view of the formula ([36], p. 667):

$$\sum_{n=0}^{\infty}\left(n+\frac{1}{2}\right) J_{n+\frac{1}{2}}(\eta) J_{n+\frac{1}{2}}(\xi) = \frac{\sqrt{\eta\xi}}{\pi(\eta-\xi)} \sin(\eta-\xi), \tag{73}$$

the reproducing kernel arising from GCS (47) can be written as

$$K(\eta,\xi) := \langle \vartheta_\eta | \vartheta_\xi \rangle \tag{74}$$

$$= \pi \sum_{n=0}^{\infty}\left(n+\frac{1}{2}\right) \frac{J_{n+\frac{1}{2}}(\eta)}{\sqrt{\eta}} \frac{J_{n+\frac{1}{2}}(\xi)}{\sqrt{\xi}} = \frac{\sin(\eta-\xi)}{\eta-\xi}, \tag{75}$$

denotes the Dyson's sine kernel, which is the reproducing kernel of the Paley-Wiener Hilbert space PW_1. Then, the family $\left\{[\pi(n+1/2)/\xi]^{1/2} J_{n+\frac{1}{2}}(\xi)\right\};\ n \in \mathbb{N}_0$, forms an orthonormal basis of PW_1 [39].

Once we have a closed form of GCS, we can look for the associated CST, this transform should map the space $\mathscr{H} = L^2([-1,1], 2^{-1}dx)$ spanned by eigenstates $\{\psi_n\}$ in (27) onto $PW_1 \subset L^2(\mathbb{R}, d\mu)$ as.

Proposition 4. *For* $\varphi \in L^2([-1,1], 2^{-1}dx)$, *the CST is the unitary map*

$$\mathscr{W}\left(L^2([-1,1], 2^{-1}dx)\right) = PW_1, \tag{76}$$

defined by means of (65) *as*

$$\mathscr{W}[\varphi](\xi) = (\mathscr{N}(\xi))^{1/2} \langle \varphi | \xi \rangle_{\mathscr{H}} = \int_{-1}^{1} e^{-ix\xi} \overline{\varphi(x)} \frac{dx}{2}, \tag{77}$$

for all $\xi \in \mathbb{R}$.

Corollary 2. *The following integral*

$$\frac{(-i)^n}{\sqrt{\xi}} J_{n+\frac{1}{2}}(\xi) = \frac{1}{\sqrt{2\pi}} \int_{-1}^{1} P_n(x) e^{-i\xi x} dx, \quad \xi \in \mathbb{R}. \tag{78}$$

holds.

Proof. From (75), the image of the basis vector $\{\psi_n\}$ under the transform $\mathscr{W}$ should exactly be

$$\mathscr{W}[\psi_n](\xi) = (-i)^n \sqrt{\frac{\pi(2n+1)}{2\xi}} J_{n+\frac{1}{2}}(\xi). \tag{79}$$

Now, by writing (75) as

$$\mathscr{W}[\psi_n](\xi) = \int_{-1}^{1} e^{-ix\xi}\psi_n(x)\frac{dx}{2},$$

and replacing ψ_n by their values given in (27), we obtain

$$\mathscr{W}[\psi_n](\xi) = \frac{\sqrt{2n+1}}{2}\int_{-1}^{1} e^{-ix\xi}P_n(x)dx,$$

the integral (78) can be evaluated by the help of the formula ([40], p. 456):

$$\int_{-1}^{1} P_n(x)e^{i\xi x}dx = i^n\sqrt{\frac{2\pi}{\xi}}J_{n+\frac{1}{2}}(\xi), \tag{80}$$

this ends the proof. □

Note that, in view of ([28], p. 29), by considering $h_n(\xi) := \rho_n^{-1/2}\overline{\Phi_n(\xi)}$ and GCS $\mathscr{K}(\xi,x) := \langle x|\vartheta_\xi\rangle$, the basis element $\psi_n \in L^2([-1,1],2^{-1}dx)$ has the integral representation

$$\psi_n(x) = \int_{-\infty}^{\infty} h_n(\xi)\overline{\mathscr{K}(\xi,x)}d\mu(\xi) \tag{81}$$

where the function $\Phi_n(\xi)$ and the positive sequences ρ_n are given in (48) and (50) respectively, the measure $d\mu(\xi)$ is given in (55), then the Legendre polynomial has the following integral representation

$$P_n(x) = \frac{(-i)^n}{\pi}\int_{-\infty}^{\infty} \mathscr{J}_n(t)e^{ix\xi}d\xi, \tag{82}$$

where the function $\mathscr{J}_n(.)$ is given in (72), which is recognized as the Fourier transform of the spherical Bessel function (72) (see [40], p. 267):

$$\int_{-\infty}^{\infty} e^{ixt}\mathscr{J}_n(t)dt = \begin{cases} \pi i^n P_n(x), & -1<x<1 \\ \frac{1}{2}\pi(\pm i)^n, & x=\pm 1, \\ 0, & \pm x>1 \end{cases} \tag{83}$$

where $P_n(.)$ the Legendre's polynomial [40].

Remark 1. *Also note that:*

- *The usefulness expansion of GCS was made very clear in a paper authored by Ismail and Zhang, where it was used to solve the eigenvalue problem for the left inverse of the differential operator, on L^2-spaces with ultraspherical weights [41, 42].*

- *For $x, \xi \in \mathbb{R}$, the function $\varphi_\xi(x) = e^{ix\xi}$, is known as the Gabor's coherent states introduced in signal theory where the property $\psi_\xi = \hat{T}(\xi)\psi$, with $\psi \in L^2(\mathbb{R})$, and $\hat{T}(\xi)$ the unitary transformation, is obtained by using the standard representation of the Heisenberg group in three dimensions, in $L^2(\mathbb{R})$, for more information (see [43]).*

Exercise 1. *Show that the vectors*

$$\vartheta_\xi = \mathcal{N}(\xi)^{-1/2}\sum_{n=0}^{\infty}\frac{\sqrt{2\pi}(-i\xi)^n}{\sqrt{2^{2n+1}(2n+1)\Gamma\left(n+\frac{1}{2}\right)}}\psi_n. \tag{84}$$

forms a set of GCS and gives the associated GCS transform.

6. Application 2: continuous Bessel wavelet transform

The continuous wavelet transform (CWT) is used to decompose a signal into wavelets. In mathematics, the CWT is a formal tool that provides an overcomplete representation of a signal by letting the translation and scale parameter of the wavelets vary continuously. There are several ways to introduce the Bessel wavelet [22, 23]. For $1 \le p \le \infty$ and $\mu > 0$, denote

$$L^p_\sigma(\mathbb{R}_+) := \left\{\psi \ \ such\ \ as\ \ \|\psi\|^p_{p,\sigma} = \int_0^\infty |\psi(x)|^p d\sigma(x) < \infty\right\}$$

and $\|\psi\|_{\infty,\sigma} = \ ess_{0<x<\infty}\ \sup|\psi(x)| < \infty$ and $d\sigma(x)$ is the measure defined as

$$d\sigma(x) = \frac{x^{2\mu}}{2^{\mu+\frac{1}{2}}\Gamma\left(\mu+\frac{3}{2}\right)}dx. \tag{85}$$

Now, let us consider the function

$$j(x) = 2^{\mu-\frac{1}{2}}\Gamma\left(\mu+\frac{1}{2}\right)x^{\frac{1}{2}-\mu}J_{\mu-\frac{1}{2}}(x), \tag{86}$$

where $J_{\mu-\frac{1}{2}}(x)$ is the Bessel function of order $l := \mu - 1/2$ given by

$$J_l(x) = \left(\frac{x}{2}\right)^l \sum_{k=0}^{\infty}\frac{(-1)^k}{k!\Gamma(k+l+1)}\left(\frac{x}{2}\right)^{2k}. \tag{87}$$

For $\mu = 1$, the function $j(x) = \mathfrak{O}_0(x)$ coincides with equation (2) discussed in the introduction. For each function $\phi \in L_{1,\sigma}(0,\infty)$, the Hankel transform of order μ is defined by

$$\hat{\phi}(x) := \int_0^\infty j(xt)\phi(t)d\sigma(t), \quad 0 \le x < \infty. \tag{88}$$

We know that from ([44], p. 316) that $\hat{\phi}(x)$ is bounded and continuous on $[0,\infty)$ and $\|\hat{\phi}\|_{\infty,\sigma} \le \|\phi\|_{1,\sigma}$. If $\phi, \hat{\phi} \in L_{1,\sigma}(0,\infty)$, then by inversion, we have

$$\phi(x) = \int_0^\infty j(xt)\hat{\phi}(t)d\sigma(t). \tag{89}$$

From ([45], p. 127) if $\phi(x)$ and $\Phi(x)$ are in $L_{1,\sigma}(0,\infty)$, then the following Parseval formula also holds

$$\int_0^\infty \hat{\phi}(t)\hat{\Phi}(t)d\sigma(t) = \int_0^\infty \phi(x)\Phi(x)d\sigma(x). \tag{90}$$

Denoting therefore by

$$\mathcal{D}(x,y,z) = \int_0^\infty j(xt)j(yt)j(zt)d\sigma(t). \tag{91}$$

For a 1-variable function $\psi \in L^2_\sigma(\mathbb{R}_+)$, we define the Hankel translation operator

$$\tau_y\psi(x) := \psi(x,y) = \int_0^\infty \mathcal{D}(x,y,z)\psi(z)d\sigma(z), \quad \forall x>0, \ y<\infty. \tag{92}$$

Trime'che ([46], p. 177) has shown that the integral is convergent for almost all y and for each fixed x, and

$$\|\psi(x,.)\|_{2,\sigma} \leq \|\psi\|_{2,\sigma}. \tag{93}$$

The map $y \mapsto \tau_y\psi$ is continuous from $[0,\infty)$ into $(0,\infty)$. For a 2-variables the function ψ, we define a dilatation operator

$$D_a\psi(x,y) = a^{-2\mu-1}\psi\left(\frac{x}{a},\frac{y}{a}\right). \tag{94}$$

From the inversion formula in (89), we have

$$\int_0^\infty j(zt)\mathcal{D}(x,y,z)d\sigma(z) = j(xt)j(yt), \quad \forall 0<x,y<\infty, \quad 0\leq t<\infty,$$

for $t=0$ and $\mu - 1/2 = 0$, we arrive at

$$\int_0^\infty \mathcal{D}(x,y,z)d\sigma(z) = 1. \tag{95}$$

The Bessel Wavelet copy $\psi_{a,b}$ are defined from the Bessel wavelet mother $\psi \in L^2_\sigma(\mathbb{R}_+)$ by

$$\psi_{a,b}(x) := D_a\tau_b\psi(x) = D_a\psi(b,x) \tag{96}$$

$$= a^{-2\mu-1}\int_0^\infty \mathcal{D}\left(\frac{b}{a},\frac{x}{a};z\right)\psi(z)d\sigma(z), \quad \forall\ a>0, \ b\in\mathbb{R}, \tag{97}$$

the integral being convergent by virtue of (92). As in the classical wavelet theory on $\mathbb{R}$, let us define the continuous Bessel Wavelet transform (CBWT) of a function $f \in L^2_\sigma(\mathbb{R}_+)$, at the scale a and the position b by

$$\mathcal{B}(b,a) := (\mathcal{B}_\psi f)(b,a) = \langle f(t), \psi_{b,a}(t)\rangle \tag{98}$$

$$= \int_0^\infty f(t)\overline{\psi_{a,b}(t)}d\sigma(t) \tag{99}$$

$$= a^{-2\mu-1}\int_0^\infty\int_0^\infty f(t)\overline{\psi(z)}\mathcal{D}\left(\frac{b}{a},\frac{t}{a},z\right)d\sigma(z)d\sigma(t). \tag{100}$$

The continuity of the Bessel wavelet follows from the boundedness property of the Hankel translation ([46], (104), p. 177). The following result is due to [22]:

Theorem 1.3 Let $\psi \in L^2_\sigma(\mathbb{R}_+)$ and $f,g \in L^2_\sigma(\mathbb{R}_+)$. Then

$$\int_0^\infty \int_0^\infty (\mathscr{B}_\psi f)(b,a)\overline{(\mathscr{B}_\psi g)(b,a)}d\sigma(a)d\sigma(b) = \mathscr{C}_\psi \langle f, g\rangle \tag{101}$$

whenever

$$\mathscr{C}_\psi = \int_0^\infty t^{-2\mu-1}|\hat{\psi}(t)|^2 d\sigma(t) < \infty. \tag{102}$$

For all $\mu > 0$.

Proof. For the function $f \in L^2_\sigma(\mathbb{R}_+)$, let us write the Bessel wavelet by using Eq. (38) as

$$(\mathscr{B}_\psi f)(b,a) = \int_0^\infty f(t)\psi_{a,b}(t)d\sigma(t) \tag{103}$$

$$= \frac{1}{a^{2\mu+1}}\int_0^\infty \int_0^\infty f(t)\overline{\psi}(z)\mathscr{D}\left(\frac{b}{a}, \frac{t}{a}, z\right)d\sigma(z)d\sigma(t). \tag{104}$$

Now observe that

$$\mathscr{D}\left(\frac{b}{a}, \frac{t}{a}, z\right) = \int_0^\infty j\left(\frac{bu}{a}\right)j\left(\frac{tu}{a}\right)j(zu)d\sigma(u). \tag{105}$$

Hence whe have that

$$(\mathscr{B}_\psi f)(b,a) = \frac{1}{a^{2\mu+1}}\int_{\mathbb{R}^3_+} f(t)\psi(z)j\left(\frac{bu}{a}\right)j\left(\frac{tu}{a}\right)j(zu)d\sigma(u)d\sigma(z)d\sigma(t) \tag{106}$$

$$= \frac{1}{a^{2\mu+1}}\int_{\mathbb{R}^2_+} \hat{f}\left(\frac{u}{a}\right)\psi(z)j\left(\frac{bu}{a}\right)j(zu)d\sigma(u)d\sigma(z) \tag{107}$$

$$= \frac{1}{a^{2\mu+1}}\int_{\mathbb{R}_+} \hat{f}\left(\frac{u}{a}\right)\hat{\psi}(u)j\left(\frac{bu}{a}\right)d\sigma(u) \tag{108}$$

$$= \int_{\mathbb{R}_+} \hat{f}(v)\hat{\psi}(av)j(bv)d\sigma(v) \tag{109}$$

$$= \left(\hat{f}(v)\hat{\psi}(av)\right)^\frown(b). \tag{110}$$

In terms of the Parseval formula (90), we obtain

$$\int_{\mathbb{R}_+} (\mathscr{B}_\psi f)(b,a)\overline{(\mathscr{B}_\psi f)}(b,a)d\sigma(b)$$

$$= \int_0^\infty \left(\hat{f}(v)\hat{\psi}(av)\right)^\frown(b)\overline{\left(\hat{g}(v)\overline{\hat{\psi}(av)}\right)^\frown}(b)d\sigma(u) \tag{111}$$

$$= \int_0^\infty \hat{f}(u)\overline{\hat{\psi}(au)}\overline{\hat{g}(u)\overline{\hat{\psi}(au)}}d\sigma(u) \tag{112}$$

Now multiplying by $a^{-2\mu-1}d\sigma(a)$ and integrating, we get

$$\int_{\mathbb{R}_+}\int_{\mathbb{R}_+} (\mathscr{B}_\psi f)(b,a)\overline{(\mathscr{B}_\psi f)}(b,a)a^{-2\mu-1}d\sigma(a)d\sigma(b) \tag{113}$$

$$= \int\int_0^\infty \hat{f}(u)\overline{\hat{\psi}(au)}\overline{\hat{g}(u)\hat{\psi}(au)}\frac{d\sigma(a)}{a^{2\mu+1}}d\sigma(u) \tag{114}$$

$$= \int_{\mathbb{R}} \hat{f}(u)\overline{\hat{g}(u)}\left(\int_{\mathbb{R}} |\hat{\psi}(au)|^2 \frac{d\sigma(a)}{a^{2\mu+1}}\right)d\sigma(u) = C_\psi \int_{\mathbb{R}} \hat{f}(u)\overline{\hat{g}(u)}d\sigma(u) \tag{115}$$

$$= \mathscr{C}_\psi \langle f, g \rangle. \tag{116}$$

The *admissible condition* (102) requires that $\hat{\psi}(0) = 0$. If $\hat{\psi}$ is continuous then from (88) it follows that

$$\int_0^\infty \psi(x)d\sigma(x) = 0. \tag{117}$$

6.1 Example

Let us consider the function

$$f(t) = \frac{2w_0^2 - t^2}{2(w_0^2 + t^2)^{5/2}}, \quad w_0 > 0, \quad t \in \mathbb{R}_+. \tag{118}$$

In the case $\mu = 1/2$, the measure (85) takes the form

$$d\sigma(t) = \frac{t}{2}dt \tag{119}$$

and the function (86) reduces to

$$j(t) = J_0(t), \tag{120}$$

where $J_0(x)$ the Bessel's function of the first kind. Also note that

$$\int_0^\infty \frac{(2w_0^2 - t^2)^2}{2(w_0^2 + t^2)^5} d\sigma(t) < \infty. \tag{121}$$

The Bessel wavelet transform of $f(t)$ is given by

$$\left\{\mathscr{B}_\psi\left(\frac{2w_0^2 - t^2}{2(w_0^2 + t^2)^{5/2}}\right)\right\}(b, a) = a^{-2}\int_0^\infty \frac{2w_0^2 - t^2}{2(w_0^2 + t^2)^{5/2}}\psi\left(\frac{b}{a}, \frac{t}{a}\right)d\sigma(t) \tag{122}$$

$$= a^{-2}\int_0^\infty \psi(z)\left(\int_0^\infty \frac{2w_0^2 - t^2}{2(w_0^2 + t^2)^{5/2}}\mathscr{D}\left(\frac{b}{a}, \frac{t}{a}, z\right)d\sigma(t)\right)d\sigma(z) \tag{123}$$

Using the representation

$$\mathscr{D}\left(\frac{b}{a}, \frac{t}{a}, z\right) = \int_0^\infty J_0\left(\frac{b}{a}u\right)J_0\left(\frac{t}{a}u\right)J_0(zu)d\sigma(u) \tag{124}$$

then (122) becomes

$$a^{-2}\int_0^\infty \psi(z)\left(\int_0^\infty J_0\left(\frac{b}{a}u\right)J_0(zu)\mathfrak{D}_{a,w_0}(u)d\sigma(u)\right)d\sigma(z)$$

Where the integral

$$\mathfrak{D}_{a,w_0}(u) = \int_0^\infty \frac{2w_0^2 - t^2}{2(w_0^2 + t^2)^{5/2}} J_0\left(\frac{t}{a}u\right) d\sigma(t). \tag{125}$$

In terms of the Legendre polynomial $P_2(t)$, the function

$$\frac{2w_0^2 - t^2}{2(w_0^2 + t^2)^{5/2}} = (w_0^2 + t^2)^{-3/2} P_2\left[w_0 (w_0^2 + t^2)^{-1/2}\right]. \tag{126}$$

Then (125) reads

$$\mathfrak{D}_{a,w_0}(u) = \int_0^\infty (w_0^2 + t^2)^{-3/2} P_2\left[w_0 (w_0^2 + t^2)^{-1/2}\right] J_0\left(\frac{t}{a}u\right) d\sigma(t). \tag{127}$$

The above equation can be evaluated by means of the formula ([47], p. 13):

$$\frac{1}{n!} y^{n-1/2} e^{-py} = \int_0^\infty x^{1/2} (p^2 + x^2)^{-\frac{1}{2}n - \frac{1}{2}} P_n\left[p(p^2 + x^2)^{-1/2}\right] (xy)^{1/2} J_0(xy) dx. \tag{128}$$

For parameters $n = 2$ and $p = w_0$, we find that

$$\mathfrak{D}_{a,w_0}(u) = \frac{1}{4} u \exp\left(-w_0 \frac{u}{a}\right). \tag{129}$$

In terms of the above result, the CBWT read as

$$\left\{\mathscr{B}_\psi\left(\frac{2w_0^2 - t^2}{2(w_0^2 + t^2)^{5/2}}\right)\right\}(b, a) = a^{-2} \int_0^\infty \psi(z) \mathfrak{M}_{a,w_0}(z) d\sigma(z) \tag{130}$$

where

$$\mathfrak{M}_{a,w_0}(z) = \int_0^\infty 8^{-1} u^2 e^{-\frac{w_0}{a}u} J_0\left(\frac{b}{a}u\right) J_0(zu) du. \tag{131}$$

To evaluated (131) we make appeal to the Lipschitz-Hankel integrals ([48], p. 389):

$$\int_0^\infty e^{-pt} J_\nu(qt) J_\nu(rt) t^{\mu-1} dt \tag{132}$$

$$= \frac{(qr)^\nu}{\pi p^{\mu+2\nu}} \frac{\Gamma(\mu + 2\nu)}{2\nu + 1} \int_{{}_2F_1}^{\pi} \left(\frac{\mu + 2\nu}{2}, \frac{\mu + 2\nu + 1}{2}; \nu + 1; -\frac{\zeta^2}{p^2}\right) \sin^{2\nu}\phi d\phi$$

with conditions $\Re(p \pm iq \pm ir) > 0$ and $\Re(\mu + 2\nu) > 0$, while ζ is written in place of $(q^2 + r^2 - 2qr\cos\phi)^{1/2}$, where ${}_2F_1$ denotes the hypergeometric function. For parameters $p = w_0/a, q = \mathrm{b/a}, r = z, \mu = 3$ and $n = 0$, we arrive at

$$\mathfrak{M}_{a,w_0}(z) = \frac{a^3}{4\pi w_0^3} \int_{{}_2F_1}^{\pi} \left(\frac{3}{2}, 2; 1; -(a w_0^{-1} \zeta)^2\right) d\phi \tag{133}$$

where $\zeta = \left[(a^{-1}b)^2 + z^2 - 2a^{-1}bz\cos\phi\right]^{1/2}$.

Next, by using the representation of the hypergeometric ${}_2F_1$-sum ([49], p. 404, Eq. 209) (**Figure 2**):

$$ {}_2F_1\left(\frac{3}{2},2;1;z\right)=\frac{1}{2}(2+z)(1-z)^{-5/2}. \tag{134} $$

Then (131) takes the form

$$ \mathfrak{M}_{a,w_0}(z)=\frac{a^3}{8\pi w_0^3}\int_0^{\pi}\left(2-\left(w_0^{-1}a\zeta\right)^2\right)\left(1+\left(w_0^{-1}a\zeta\right)^2\right)^{-5/2}d\zeta, \tag{135} $$

This leads to the following CBWT

$$ \left\{\mathscr{B}_{\psi}\left(\frac{2w_0^2-t^2}{2\left(w_0^2+t^2\right)^{5/2}}\right)\right\}(b,a)=\frac{a}{4\pi}\int_0^{\infty}\psi(z)\int_0^{\pi}\frac{2w_0^2-(a\zeta)^2}{2\left(w_0^2+(a\zeta)^2\right)^{5/2}}d\phi d\sigma(z). \tag{136} $$

We have given an example of a signal $f(t)\in L_{\sigma}^2(0,\infty)$ such that the CBWT is written as

$$ \left\{\mathscr{B}_{\psi}(f(t))\right\}(b,a)=\frac{a}{4\pi}\int_0^{\pi}\int_0^{\infty}\psi(z)f(a\zeta)d\sigma(z)d\phi. \tag{137} $$

According to Theorem 1.3, let $\psi\in L_{\sigma}^2(\mathbb{R}_+)$ and $f,g\in L_{\sigma}^2(\mathbb{R}_+)$, then

$$ \int_0^{\infty}\int_0^{\infty}\left(\mathscr{B}_{\psi}f\right)(b,a)\overline{\left(\mathscr{B}_{\psi}g\right)(b,a)}d\sigma(a)d\sigma(b)=\frac{1}{128w_0^2}\langle f,g\rangle. \tag{138} $$

Note that, for all $w_0>0$, the given function

$$ f(t)=\frac{2w_0^2-t^2}{2\left(w_0^2+t^2\right)^{5/2}},\quad t\in\mathbb{R}_+, \tag{139} $$

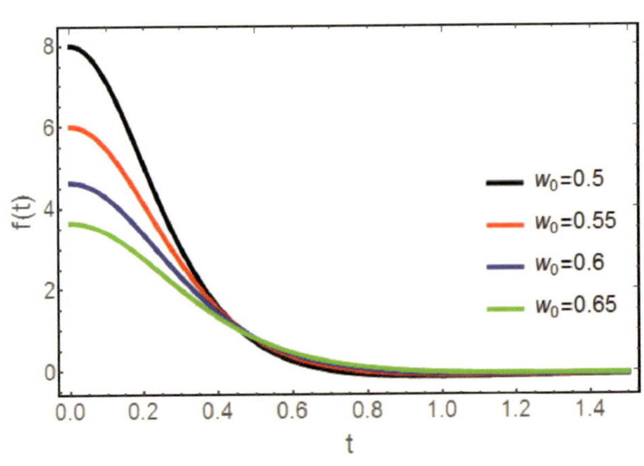

Figure 2.
Plots of the mother wavelet $f(t)$ defined in (6.34) versus t, for various values of the parameters w_0.

is the mother wavelet. The Hankel transform of $f(t)$ is given by

$$\hat{f}(y) = \int_0^\infty \frac{2w_0^2 - t^2}{2(w_0^2 + t^2)^{5/2}} J_0(xy) d\sigma(t) = \frac{1}{4} y e^{-w_0 y}, \quad \forall \quad 0 \leq y < \infty. \tag{140}$$

and satisfy the admissible condition

$$\mathscr{C}_f = \frac{1}{2} \int_0^\infty \frac{\left|\hat{f}(\xi)\right|^2}{\xi} d\xi \tag{141}$$

$$= \frac{1}{128 w_0^2}, \quad w_0 > 0. \tag{142}$$

The Hankel transformation $\hat{f}(0) = 0$, so by the help of (140) we obtain

$$\int_0^\infty \frac{t(2w_0^2 - t^2)}{(w_0^2 + t^2)^{5/2}} dt = 0. \tag{143}$$

Exercise 2
For which numbers $n \in \mathbb{N}$, the following function

$$f_n(t) = (w_0^2 + x^2)^{-\frac{1}{2}n - \frac{1}{2}} P_n\left[w_0 (w_0^2 + x^2)^{-1/2}\right] \tag{144}$$

Is the mother wavelet where $P_n(.)$ the Legendre's polynomial.

7. Conclusions

In this chapter we are interested in the construction of the generalized coherent state (GCS) and the theory of wavelets. As it is well know wavelets constructed on the basis of group representation theory have the same properties as coherent states. In other words, the wavelets can actually be thought of as the coherent state associated with these groups. Coherent state is very important because of three properties they have: coherence, overcompleteness, intrinsic geometrization. We have seen that it is possible to construct coherent states without taking into account the theory of group representation. Throughout this chapter we have used the Bessel function to construct the coherent state transform and Bessel continuous wavelets transform. We have prove that the kernel of the finite Fourier transform (FFT) of L^2-functions supported on $[-1, 1]$ form a set of GCS. We therefore discussed another way of building a set of coherent states based on Wavelet's theory makes it easier.

Building coherent states in this chapter is always not easy because it is necessary to find coefficients which will make it possible to find vectors which will certainly satisfy certain conditions but the procedure based on Wavelet's theory makes it easier.

It should be noted that the theory of classical wavelets finds several applications ranging from the analysis of geophysical and acoustic signals to quantum theory. This theory solves difficult problems in mathematics, physics and engineering, with several modern applications such as data compression, wave propagation, signal processing, computer graphics, pattern recognition, pattern processing. Wavelet analysis is a robust technique used for investigative methods in quantifying the timing of measurements in Hamiltonian systems.

Conflict of interest

The authors declare no conflict of interest.

Author details

Kayupe Kikodio Patrick
Faculty of Sciences, Lubumbashi University, DRC

*Address all correspondence to: kayupepatrick@gmail.com

IntechOpen

© 2021 The Author(s). Licensee IntechOpen. This chapter is distributed under the terms of the Creative Commons Attribution License (http://creativecommons.org/licenses/by/3.0), which permits unrestricted use, distribution, and reproduction in any medium, provided the original work is properly cited.

References

[1] Schrödinger, E.: Der stetige Ã¼bergang von der Mikro- zur Makromechanik. Naturwissenschaften 14, 664–666 (1926)

[2] R. Glauber, The quantum theory of optical coherence, *Phys. Rev.* **130** (1963) 2529

[3] R. Glauber, Coherent and incoherent states of radiation field, *J. Phys. Rev.* **131** (1963) 2766

[4] J. R. Klauder, Continuous representation theory. I. Postulayes of continuous representation, *J. Math. Phys.* **4** (1963) 1055.

[5] J. R. Klauder, Generalized relation between quantum and classical dynamics, *J. Math. Phys.* **4** (1963) 1058.

[6] J. R. Klauder and B-S. Shagertan, Coherent states, (world scientific, Singapore, 1985)

[7] A. M. Perelomov, Generalized coherent states and their Application, (Spring-Verlag Berlin 1986)

[8] Peter. W. Milonni and Michael Martin Nieto, Coherent states, In: Compendium of Quantum Physics, pp.106–108. Springer (2009)

[9] S. T. Ali, J-P. Antoine and J-P. Gazeau, Coherent states, Wavelets and their Gneralization, (Springer-Verlag, New York, 2000)

[10] B. Mojaveri and S. Amiri Faseghandis, Generalized coherent states related to the associated Bessel functions and Morse potential, *Phys. Scr.* **89** (2014) 085204 (8pp)

[11] B. Mojaveri, Klauder-Perelomov and Gazeau-Klauder coherent states for an electron in the Morse-like magnetic field, *Eur. Phys. J. D,* **67** (2013) 105

[12] H. Fakhri, B. Mojaveri and M. A. Gomshi Nobary, Landau Levels as a Limiting case of a model with the Morse-Like Magnetic field, *Rep. Math. Phys*, **66** (2010) 299

[13] B. Mojaveri and A. Dehghani, New Generalized coherent states arising from generating functions: A novel approach, *Rep. Math. Phys*, **75** (2015) 47

[14] A.Dehghani, B.Mojaveri and M. Mahdian, New Even and ODD Coherent States Attached to the Hermite Polynomials, *Rep. Math. Phys*, **75** (2015) 267

[15] H. Fakhri and B. Mojaveri, Generalized Coherent States for the Spherical Harmonics *Int. J. Mod. Phys. A*, **25** (2010) 2165.

[16] V. V. Borzov, E. V. Damaskinsky, Coherent states for the Legendre oscillator, ZNS POMI, 285–52 (2002)

[17] V. V. Borzov, Orthogonal polynomials and generalized oscillator algebras, *Integral Transf. and Special Funct.* **12**,(2001).

[18] K. Thirulogasantar and N. Saad, Coherent states associated with the wavefunctions and the spectrum of the isotonic oscillator, *J. Phys. A, Math. Gen.* **37**, 4567–4577 (2004).

[19] A. Horzela and F. H. Szafraniec, A measure-free approach to coherent states, *J. Phys. A: Math. Theor.* **45** (2012).

[20] G. Arfken, Mathematical methods for physicists, Academic Press Inc, 1985.

[21] Neal C. Gallagher, J. R., and Gary L. Wise, A representation for Band-Limited Functions, *Proceedings of the IEEE*, **63** 11, 1975.

[22] R. S. Pathak, M. M. Dixit, Continuous and discrete Bessel wavelet

transforms, Journal of Computational and Applied Mathematics, 160 (2003), pp. 240–250.

[23] R. S. Pathak, S. K. Upadhyay and R. S. Pandey, The Bessel wavelet convolution product, Rend.Sem.Mat. Univ.Politec.Torino, 96(3) (2011), pp. 267–279

[24] A. Grossmann and J. Morlet, Decomposition of Hardy functions into square integrable wavelets of constant shape, *SIAM J. Math. Anal.*, Vol. 15, No. 4 (1984), 723–736.

[25] C.K. Chui, An Introduction to Wavelets, Academic Press, 1992

[26] T.H. Koornwinder, The continuous wavelet transform. Vol. 1. Wavelets: An Elementary Treatment of Theory and Applications. Edited by T.H. Koornwinder, World Scientific, 1993, 27–48

[27] Y. Meyer, Wavelets and Operators, Cambridge University Press, Cambridge, 1992

[28] S. T. Ali, J. P. Antoine and J. P. Gazeau, Coherent states, Wavelets, and their generalization, Springer Science +Business Media New York 2014

[29] M. M. Pelosso, Classical spaces of holomorphic functions, Technical report, Università di Milano, 2011.

[30] M. S. Birman and M. Z. Solomyak, Spectral theory of self-adjoint operators in Hilbert space, Leningrad Univ. Press, 1980 (in Russian).

[31] N. I. Akhieser, The classical moment problem and some related questions in analysis, Hafner Publ. Co, New York, 1965

[32] J-M. Sixdenierrs, K. A. Penson and A. I. Solomon, Mittag-Leffler coherent states, *J. Phys. A.* **32**, (1999)

[33] Ali, S.T., Reproducing kernels in coherent states, wavelets, and quantization, in Operator Theory, Springer, Basel (2015), p. 14; doi 10.1007/978-3-0348-0692-3 63–1

[34] Imen Rezgui and Anouar Ben Marbrouk, Some Generalized q-Bessel type Wavelets and Associated Transforms, *Anal. Theory Appl.*, **34** (2018), pp. 57–76

[35] Ali, S.T., Antoine, J.-P., Gazeau, J.-P.: Coherent States, Wavelets and their Generalizations. Springer, New York (2014)

[36] A. P. Prudnikov, Y. A. Brychkov and O. I. Marichev, Integrals and series. volume 2, Special functions, OPA (Overseas Publishers Association), Amsterdam 1986

[37] J. R. Higgins, Sampling theory in Fourier and signal analysis: Foundations, Oxford Science Publication, 1996.

[38] Mourad E. H. Ismail, Classical and Quantum Orthogonal Polynomials in one Variable, Encyclopedia of Mathematics and its Applications, Cambridge University Press 2005

[39] J. R. Higgins, Sampling Theory in Fourier and Signal Analysis: Foundations. Oxford, U.K.: Clarendon, 1996.

[40] F. W. J. Olver, D. W. Lozier, R. F. Boisvert and C. W. Clark, editors. NIST Handbook of Mathematical Functions. Cambridge University Press, 2010

[41] M. E. H. Ismail and R. Zhang, Diagonalization of certain integral operators. *Adv. Math.* **109** (1994), no. 1, 1–33

[42] L. D. Abreu, The reproducing kernel structure arising from a combination of continuous and discrete orthogonal

polynomials into Fourier systems, *Constr. Approx.* **28**, 219–235.

[43] D. Robert, La cohérence dans tous ses états. *SMF Gazette* **132** (2012)

[44] I.I.Hirschman Jr., Variation diminishing Hankel transforms, *J.Anal. Math.*8 (1960–1961) 307–336.

[45] A.H. Zemanian, Generalized Integral Transformations, Interscience, New York, 1968

[46] K.Trime′che, Generalized Wavelets and Hypergroups, Gordon and Breach, Amsterdam, 1997

[47] Fritz Oberhettinger, Tables of Bessel Transforms, Springer-Verlag New York Heildelberg Berlin 1972

[48] G.N. Watson, A treatise of Bessel functions, Cambridge at the University Press, Second Edition 1944

[49] A. P. Prudnikov, Yu. A. Brychkov, Integrals and series, More special Functions, volume 3, computing center of the USSR Academy of sciences Moscow, from Russian version

Chapter 13

Wavelet Filter Banks Using Allpass Filters

Xi Zhang

Abstract

Allpass filter is a computationally efficient versatile signal processing building block. The interconnection of allpass filters has found numerous applications in digital filtering and wavelets. In this chapter, we discuss several classes of wavelet filter banks by using allpass filters. Firstly, we describe two classes of orthogonal wavelet filter banks composed of two real allpass filters or a complex allpass filter, and then consider design of orthogonal filter banks without or with symmetry, respectively. Next, we present two classes of filter banks by using allpass filters in lifting scheme. One class is causal stable biorthogonal wavelet filter bank and another class is orthogonal wavelet filter bank, all with approximately linear phase response. We also give several design examples to demonstrate the effectiveness of the proposed method.

Keywords: wavelet, filter bank, allpass filter, perfect reconstruction, symmetry, orthogonality

1. Introduction

The discrete wavelet transform (DWT), which is implemented by a two band perfect reconstruction (PR) filter bank, has been applied extensively to digital signal processing, image processing, medical and health care, economy and so on [1–4]. In many applications such as image processing, wavelets are required to be real since the signal is real-valued in general. We restrict ourselves to real-valued wavelet filter banks in this chapter.

In addition to orthogonality, one desirable property for wavelets is symmetry, which requires all filters in the filter bank to possess exactly linear phase, because the symmetric extension method is generally used to treat the boundaries of images [5, 6]. It is known in [1–4] that finite impulse response (FIR) filters (corresponding to the compactly supported wavelets) can easily realize exactly linear phase. However, it is widely appreciated that the only FIR solution that produces a real orthogonal symmetric wavelet basis is the Haar wavelet, which is not continuous and the corresponding filter is of order 1 only that is not enough for many practical applications. To obtain wavelet filter banks with higher degrees of freedom, infinite impulse response (IIR) filters have been used to construct wavelet filter banks with some of the desired properties [7–12]. Among the existing IIR wavelet filter banks, wavelet filter banks composed of allpass filters are attractive [7, 9, 10, 12], which can realize both of orthogonality and symmetry.

Allpass filter is a computationally efficient versatile signal processing building block and quite useful in many applications [13]. Allpass filter possesses unit

IntechOpen

magnitude at all frequencies (see Appendix) and is a basic scalar lossless building block. The interconnection of allpass filters has found numerous applications in practical filtering problems, such as low sensitivity filter structures, multirate filtering, filter banks and so on [7, 10, 12, 13]. The phase approximation of allpass filters has been also discussed in [13–15].

The lifting scheme proposed by W. Sweldens in [16, 17] is an efficient tool for constructing second generation wavelets, and has advantages such as faster implementation, fully in-place calculation, reversible integer-to-integer transforms, and so on. It has been proved in [18, 19] that every FIR wavelet filter bank can be decomposed into a finite number of lifting steps, thus this allows the construction of an integer version of the wavelet transform. Such integer wavelet transforms are invertible, and then are attractive in lossless coding applications. Due to these properties, the lifting implemantation has been adopted in the international standard JPEG2000 [5]. Conventionally, the lifting scheme is often used to construct a class of biorthogonal wavelet filter banks. It has been shown in [18] that orthogonal wavelet filter banks can also be realized by the lifting scheme. However, it is not always possible for IIR wavelet filter banks to be decomposed into a finite number of lifting steps.

In this chapter, we discuss several classes of wavelet filter banks by using allpass filters. Firstly, we describe two classes of orthogonal wavelet filter banks composed of two real allpass filters or a single complex allpass filter. We consider design of the proposed orthogonal wavelet filter banks without or with symmetry, respectively, and give the maximally flat solutions, where the orthogonal symmetric wavelet filter banks using real or complex allpass filter are corresponding to half sample symmetric (HSS) and whole sample symmetric (WSS) wavelets, respectively. Next, we present two classes of wavelet filter banks based on the lifting scheme with two lifting steps only. By using real allpass filters in the lifting steps, we can obtain one class of causal stable biorthogonal wavelet filter bank and another class of orthogonal wavelet filter bank, all with approximately linear phase response. In addition, we show some design examples to demonstrate the effectiveness of the proposed method.

2. Two band wavelet filter bank

It is well-known [1–4] that wavelet basis can be generated by two band filter bank shown in **Figure 1**. In **Figure 1**, $H_0(z)$ and $H_1(z)$ are analysis filters, and $G_0(z)$ and $G_1(z)$ are synthesis filters. The relationship of input $X(z)$ and output $Y(z)$ of the filter bank is given by

$$\begin{aligned} Y(z) &= \frac{1}{2}\{H_0(z)G_0(z) + H_1(z)G_1(z)\}X(z) \\ &+ \frac{1}{2}\{H_0(-z)G_0(z) + H_1(-z)G_1(z)\}X(-z). \end{aligned} \tag{1}$$

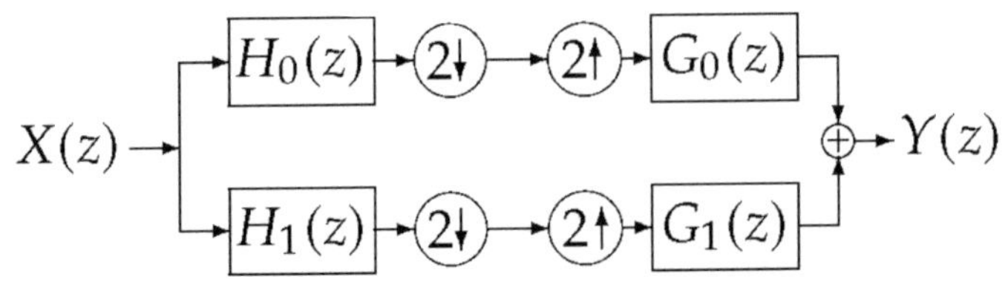

Figure 1.
Two band wavelet filter bank. $X(z)$ *is input and* $Y(z)$ *is output.* $H_0(z), H_1(z)$ *are analysis filters and* $G_0(z), G_1(z)$ *are synthesis filters.*

Therefore the PR condition is

$$\begin{cases} H_0(z)G_0(z) + H_1(z)G_1(z) = cz^{-I} \\ H_0(-z)G_0(z) + H_1(-z)G_1(z) = 0 \end{cases}, \tag{2}$$

where c is constant and I is integer.

One desirable property for wavelets is orthogonality, which requires the filter bank is orthogonal, i.e., $|H_0(e^{j\omega})| = |G_0(e^{j\omega})| = |H_1(e^{j(\pi-\omega)})| = |G_1(e^{j(\pi-\omega)})|$. Another desirable property is symmetry, i.e., the wavelet basis is symmetric or antisymmetric. It requires all filters in the filter bank to possess exactly linear phase, whose impulse responses are symmetric or antisymmetric.

3. The proposed orthogonal wavelet filter banks using allpass filters

In this section, we describe several classes of orthogonal wavelet filter banks without or with symmetry. The proposed classes of orthogonal wavelet filter banks are composed of two real allpass filters or a complex allpass filter.

3.1 Orthogonal wavelet filter banks without symmetry

In some applications of signal processing, for example, speech and acoustic signal processing, wavelet filters are required to have minimal phase response rather than exactly linear phase. Therefore, wavelet basis is not necessarily symmetric or antisymmetric. In the following, we discuss two classes of orthogonal wavelet filter banks without symmetry [20].

3.1.1 Filter bank using real allpass filters

We firstly consider a pair of IIR filters $H_0(z)$ and $H_1(z)$ that are based on a parallel connection of two real allpass filters as shown in **Figure 2**, i.e.,

$$\begin{cases} H_0(z) = \frac{1}{2}\{A_{N_1}(z^2) + z^{-2K-1}A_{N_2}(z^2)\} \\ H_1(z) = \frac{1}{2}\{A_{N_1}(z^2) - z^{-2K-1}A_{N_2}(z^2)\} \end{cases}, \tag{3}$$

where K is integer, $A_{N_1}(z)$ and $A_{N_2}(z)$ are real allpass filters of order N_1 and N_2 respectively. Let the synthesis filters $G_0(z) = H_0(z^{-1})$ and $G_1(z) = H_1(z^{-1})$, then the PR condition in Eq.(2) is satisfied.

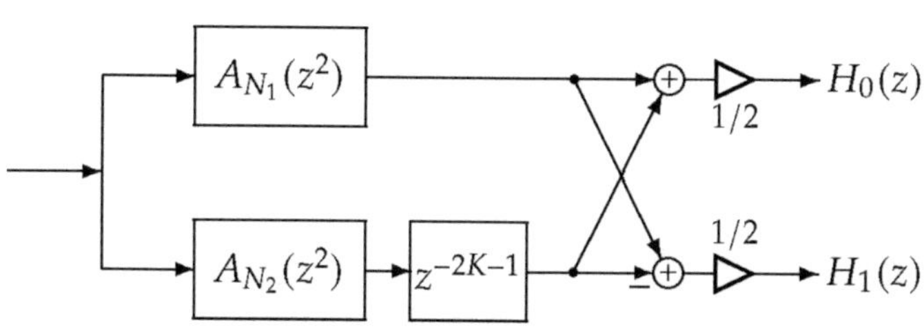

Figure 2.
Filter bank using real allpass filters. $A_{N_1}(z), A_{N_2}(z)$ are real allpass filters of order N_1 and N_2. $H_0(z), H_1(z)$ are lowpass and highpass filters.

From Eq.(3), we have

$$H_0(z) = \frac{1}{2} A_{N_2}(z^2)\{A_N(z^2) + z^{-2K-1}\}, \tag{4}$$

where $A_N(z)$ is a real allpass filter of order $N = N_1 + N_2$, and defined as

$$A_N(z) = \frac{A_{N_1}(z)}{A_{N_2}(z)} = z^{-N} \frac{\sum_{n=0}^{N} a_n z^n}{\sum_{n=0}^{N} a_n z^{-n}}, \tag{5}$$

where a_n is real coefficient, and $a_0 = 1$.

Let $\theta(\omega)$ be the phase response of $A_N(z)$, the magnitude responses of $H_0(z)$ and $H_1(z)$ are given by

$$\begin{cases} |H_0(e^{j\omega})| = |\cos\left(\frac{\theta(2\omega)}{2} + \left(K + \frac{1}{2}\right)\omega\right)| \\ |H_1(e^{j\omega})| = |\sin\left(\frac{\theta(2\omega)}{2} + \left(K + \frac{1}{2}\right)\omega\right)| \end{cases}. \tag{6}$$

It is clear that the magnitude responses satisfy $|H_0(e^{j\omega})| = |H_1(e^{j(\pi-\omega)})|$ and the following power-complementary relation;

$$|H_0(e^{j\omega})|^2 + |H_1(e^{j\omega})|^2 = 1, \tag{7}$$

which means that the filter bank is orthogonal.

For $H_0(z)$ and $H_1(z)$ to be a pair of lowpass and highpass filters, the desired phase response of $A_N(z)$ is given by

$$\theta_d(\omega) = -\left(K + \frac{1}{2}\right)\omega = -\tau\omega, \tag{8}$$

where $\tau = K + \frac{1}{2}$. From the regularity of wavelets, it is known that an additional flatness condition is required to impose on $H_0(z)$, i.e.,

$$\left.\frac{\partial^k |H_0(e^{j\omega})|}{\partial\omega^k}\right|_{\omega=\pi} = 0 \quad (k = 0, 1, \cdots, L-1), \tag{9}$$

where L is integer. Hence, the resulting wavelet function will have L consecutive vanishing moments. This flatness condition can be obtained if $H_0(z)$ contains L zeros located at $z = -1$.

For the maximally flat filters, the closed-form formula is given by

$$a_n = \binom{N}{n} \prod_{i=1}^{n} \frac{N - \tau - i + 1}{\tau + i}. \tag{10}$$

Once a set of filter coefficients a_n are obtained, we compute poles of $A_N(z)$ and then assign the poles inside the unit circle to $A_{N_1}(z)$ as its poles and the poles outside the unit circle to $A_{N_2}(z)$ as its zeros. Therefore, we can obtain causal stable analysis filters $H_0(z)$ and $H_1(z)$, then the synthesis filters $G_0(z)$ and $G_1(z)$ are anti-causal stable.

In many applications of signal processing, frequency selectivity is also thought of as a useful property from the viewpoint of signal band-splitting. However, regularity and frequency selectivity somewhat contradict each other. For this reason, design of $H_0(z)$ that has the best possible frequency selectivity for the given flatness condition has been also discussed in [20].

Example 1: We consider design of filter banks using two real allpass filters with $N_1 = N_2 = 2$ and $K = 0$. By setting $L = 9, 5, 1$, we have designed $H_0(z)$ by using the design method proposed in [20]. The magnitude responses are shown in **Figure 3**, and the scaling and wavelet functions are shown in **Figure 4**, respectively. When $L = 9$, it is seen that $H_0(z)$ is the maximally flat filter, and it is the elliptic filter if $L = 1$. It is clear in **Figure 3** that the magnitude error increases with an increasing L, and in **Figure 4** that the scaling and wavelet functions decline more rapidly.

3.1.2 Filter bank using complex allpass filter

We consider a pair of $H_0(z)$ and $H_1(z)$ using a single complex allpass filter as shown in **Figure 5**, i.e.,

$$\begin{cases} H_0(z) = \dfrac{1}{2}\left\{A_N(z) + \hat{A}_N(z)\right\} \\ H_1(z) = \dfrac{z^{-1}}{2j}\left\{A_N(z) - \hat{A}_N(z)\right\} \end{cases}, \tag{11}$$

where $A_N(z)$ and $\hat{A}_N(z)$ are complex allpass filters of order N, and their coefficients are mutually complex conjugate. Let $G_0(z) = H_0(z^{-1})$ and $G_1(z) = H_1(z^{-1})$ similarly. From the orthogonality, $A_N(z)$ and $\hat{A}_N(z)$ must satisfy [7]

$$A_N(z) = \pm j\hat{A}_N(-z), \tag{12}$$

which means that if α is a pole of $A_N(z)$, then $-\alpha^*$ is a pole of $A_N(z)$ also. Consequently, $A_N(z)$ has a pair of poles $(\alpha, -\alpha^*)$ or a single pole $j\beta$, where β is real, α is complex and α^* denotes the complex conjugate of α.

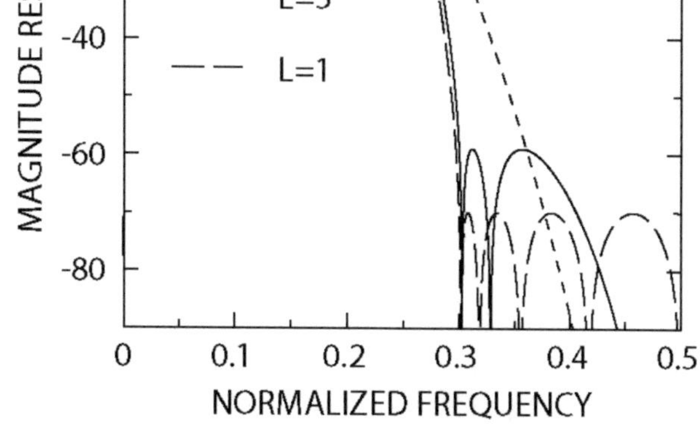

Figure 3.
Magnitude responses of $H_0(z)$ in example 1.

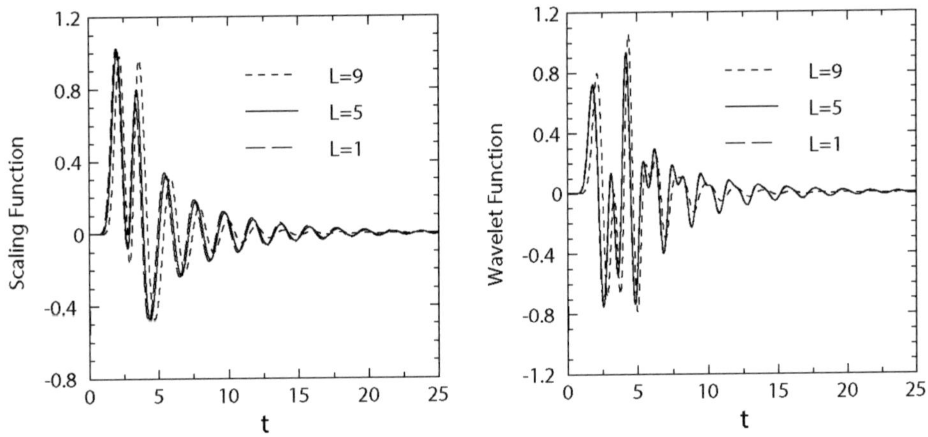

Figure 4.
Scaling and wavelet functions in example 1.

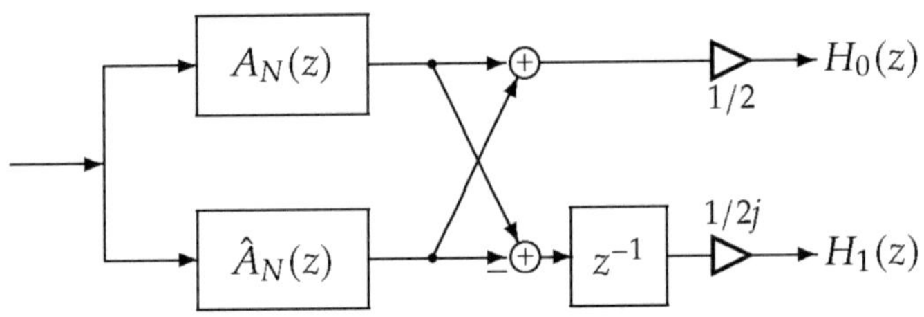

Figure 5.
Filter bank using complex allpass filter. $A_N(z), \hat{A}_N(z)$ *are complex allpass filters of order N.* $H_0(z), H_1(z)$ *are lowpass and highpass filters.*

From Eq.(11),

$$H_0(z) = \frac{1}{2}\hat{A}_N(z)\{A_{2N}(z) + 1\}, \tag{13}$$

where $A_{2N}(z)$ is a complex allpass filter of order $2N$, and defined by

$$A_{2N}(z) = \frac{A_N(z)}{\hat{A}_N(z)} = \frac{e^{j\eta}}{e^{-j\eta}} \frac{\sum_{n=0}^{N_1} a_{2n}\{z^{2n} + z^{-2n}\} + j\sum_{n=0}^{N_2} a_{2n+1}\{z^{2n+1} + z^{-2n-1}\}}{\sum_{n=0}^{N_1} a_{2n}\{z^{2n} + z^{-2n}\} - j\sum_{n=0}^{N_2} a_{2n+1}\{z^{2n+1} + z^{-2n-1}\}}, \tag{14}$$

where $\eta = \pm\pi/4$ or $\pm 3\pi/4$, a_n is real and $a_0 = 1/2$, $N_1 = N/2$ and $N_2 = N/2 - 1$ if N is even, and $N_1 = N_2 = (N-1)/2$ if N is odd.

Therefore, the phase response $\theta(\omega)$ of $A_{2N}(z)$ is given by

$$\theta(\omega) = 2\eta + 2\tan^{-1}\frac{\sum_{n=0}^{N_2} a_{2n+1}\cos((2n+1)\omega)}{\sum_{n=0}^{N_1} a_{2n}\cos(2n\omega)}, \tag{15}$$

and the magnitude responses of $H_0(z)$ and $H_1(z)$ are

$$\begin{cases} |H_0(e^{j\omega})| = |\cos\frac{\theta(\omega)}{2}| \\ |H_1(e^{j\omega})| = |\sin\frac{\theta(\omega)}{2}| \end{cases}, \tag{16}$$

which satisfies the power-complementary relation in Eq.(7).

The closed-form formula of the maximally flat filters can be given by

$$a_n = C_n \prod_{i=1}^{n} \frac{i-N-1}{i+N}, \tag{17}$$

where $C_{2n} = 1$ and $C_{2n+1} = \tan\eta$. Therefore, we compute poles of $A_{2N}(z)$ and assign the poles inside the unit circle to $A_N(z)$ as its poles to obtain causal stable analysis filters $H_0(z)$ and $H_1(z)$. Thus, the synthesis filters $G_0(z)$ and $G_1(z)$ are anti-causal stable. Design of $H_0(z)$ having the best possible frequency selectivity for the given degrees of flatness has been also discussed in [20].

Example 2: We consider design of filter banks using a complex allpass filter with $N = 4$ and $L = 8, 4, 0$. We have designed $H_0(z)$ by using the design method proposed in [20]. The magnitude responses are shown in **Figure 6**, and the scaling and wavelet functions are shown in **Figure 7**, respectively. It is seen in **Figure 6** that $H_0(z)$ is the maximally flat filter if $L = 8$, and the elliptic filter if $L = 0$ that does not have any zero located at $z = -1$ and is different from that in *Example 1*.

3.2 Orthogonal symmetric wavelet filter banks

In many applications of image processing, digital filters are required to have exactly linear phase. Therefore, the impulse responses of wavelet filters need to be symmetric or antisymmetric, and the generated wavelet bases are symmetric or antisymmetric also. In the following, we discuss two classes of orthogonal symmetric wavelet filter banks composed of allpass filters: HSS [21] and WSS [22] wavelet filter banks.

3.2.1 Filter bank using real allpass filters

To obtain exactly linear phase, we constitute a pair of $H_0(z)$ and $H_1(z)$ in **Figure 2** by using an allpass filter $A_N(z)$ as

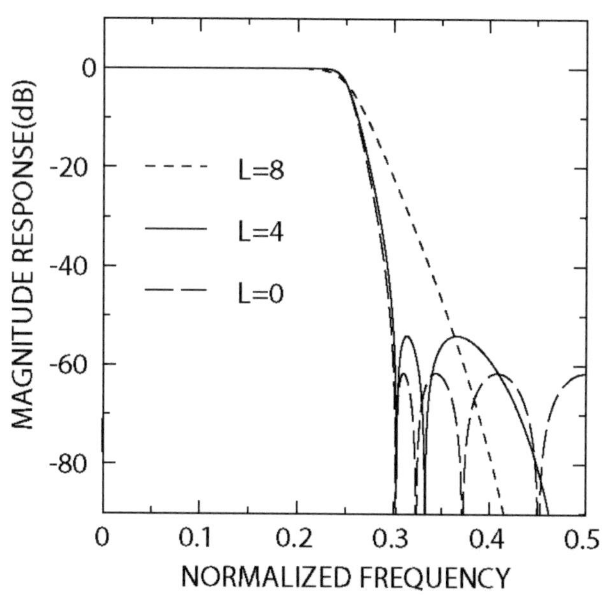

Figure 6.
Magnitude responses of $H_0(z)$ in example 2.

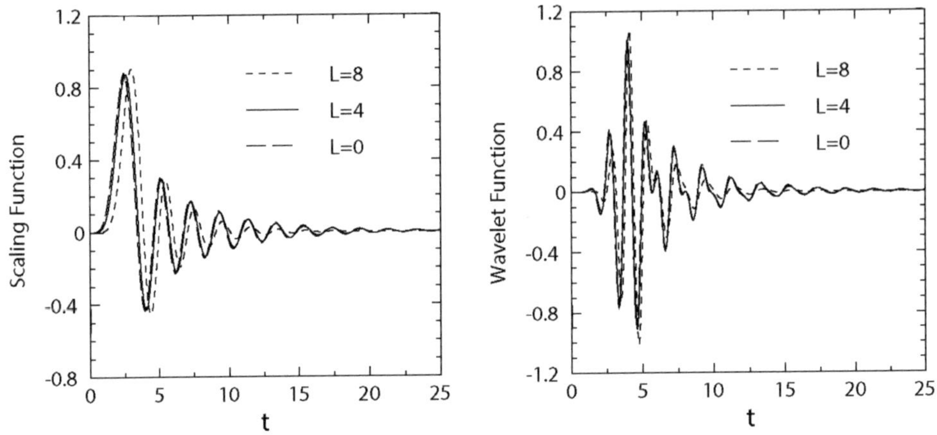

Figure 7.
Scaling and wavelet functions in example 2.

$$\begin{cases} H_0(z) = \frac{1}{2}\{A_N(z^2) + z^{-2K-1}A_N(z^{-2})\} \\ H_1(z) = \frac{1}{2}\{A_N(z^2) - z^{-2K-1}A_N(z^{-2})\} \end{cases}. \tag{18}$$

Let $\theta(\omega)$ be the phase response of $A_N(z)$, then the frequency responses of $H_0(z)$ and $H_1(z)$ are given by

$$\begin{cases} H_0(e^{j\omega}) = e^{-j\left(K+\frac{1}{2}\right)\omega}\cos\left\{\theta(2\omega) + \left(K+\frac{1}{2}\right)\omega\right\} \\ H_1(e^{j\omega}) = je^{-j\left(K+\frac{1}{2}\right)\omega}\sin\left\{\theta(2\omega) + \left(K+\frac{1}{2}\right)\omega\right\} \end{cases}. \tag{19}$$

It is clear in Eq.(19) that $H_0(z)$ and $H_1(z)$ have exact linear phase response and satisfy the power-complementary relation in Eq.(7). The filter has a group delay of $K+\frac{1}{2}$, and its impulse response is HSS. Therefore, the design problem of the wavelet filter banks becomes the phase approximation of allpass filter $A_N(z)$. For $H_0(z)$ and $H_1(z)$ to be a pair of lowpass and highpass filters, the desired phase response of $A_N(z)$ is

$$\theta_d(\omega) = -\left(\frac{K}{2} + \frac{1}{4}\right)\omega = -\tau\omega, \tag{20}$$

where $\tau = \frac{K}{2} + \frac{1}{4}$. The filter coefficients a_n of the maximally flat filters can be computed by Eq.(10). Design of wavelet filters having the best possible frequency selectivity for the given degrees of flatness has been also discussed in [21]. It has been pointed out in [21] that we must choose $K = \cdots, -7, -6, -3, -2, 1, 2, 5, 6, \cdots$ if N is odd and $K = \cdots, -5, -4, -1, 0, 3, 4, 7, \cdots$ if N is even, in order to obtain a pair of reasonable lowpass and highpass filters to avoid the undesired zero and bump.

Example 3: We consider design of the maximally flat wavelet filter banks with $N = 4$ and $L = 9$. We have designed $A_N(z)$ with $K = 0$ and $K = 1$. The magnitude responses of $H_0(z)$ are shown in **Figure 8**. It is seen in **Figure 8** that $H_0(z)$ with

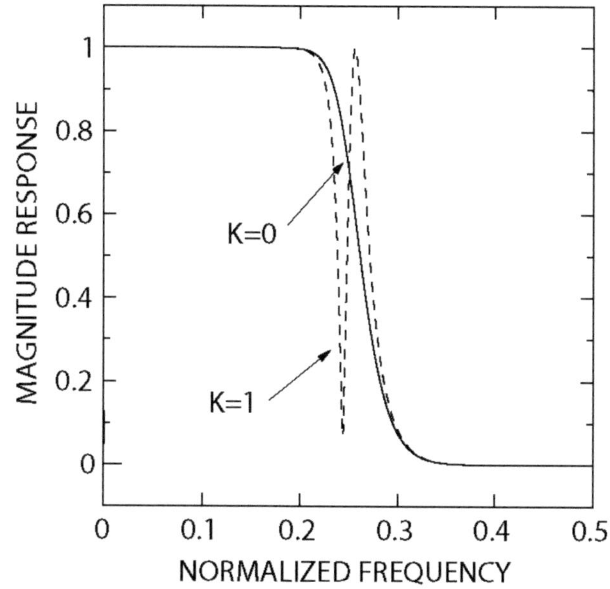

Figure 8.
Magnitude responses of $H_0(z)$ in example 3. $H_0(z)$ has an undesired zero and bump when $K = 1$.

$K = 1$ has the undesired zero and bump nearby $\omega = \pi/2$. The generated scaling and wavelet functions are shown in **Figure 9** respectively. It is seen in **Figure 9** that the scaling functions are symmetric, while the wavelet functions are antisymmetric. Although $H_0(z)$ with $K = 0$ and $K = 1$ have the same degrees of flatness, it is seen that the scaling and wavelet functions of $K = 1$ decline more slowly than that of $K = 0$, because of the undesired zero and bump. Therefore, we should not choose $K = 1$ in this case.

3.2.2 Filter bank using complex allpass filter

We consider again $H_0(z)$ and $H_1(z)$ using a complex allpass filter in **Figure 5**,

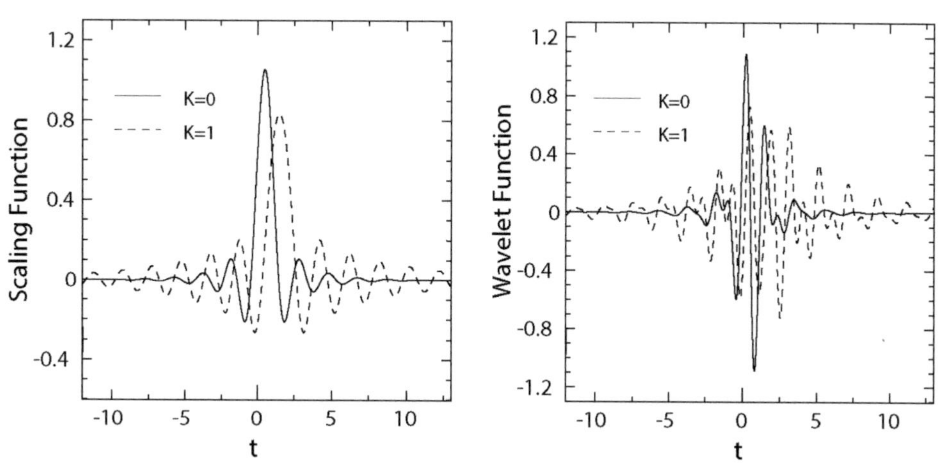

Figure 9.
Scaling and wavelet functions in example 3. The symmetric point is dependent on the group delay of $K + \frac{1}{2}$.

$$\begin{cases} H_0(z) = \frac{1}{2}\left\{A_N(z) + \hat{A}_N(z)\right\} \\ H_1(z) = \frac{z^{-1}}{2j}\left\{A_N(z) - \hat{A}_N(z)\right\} \end{cases}. \tag{21}$$

To obtain exactly linear phase, $A_N(z)$ and $\hat{A}_N(z)$ must satisfy

$$\hat{A}_N(z) = \frac{1}{A_N(z)}, \tag{22}$$

which means that if α is a pole of $A_N(z)$, then $1/\alpha$ is a pole of $A_N(z)$ too. In addition to orthogonality, $A_N(z)$ has a quadruplet of poles $(\alpha, -\alpha^*, 1/\alpha, -1/\alpha^*)$ or a pair of poles $(j\beta, 1/j\beta)$. Therefore, we have

$$A_N(z) = e^{j\eta} z^{-N} \frac{a_0 + ja_1 z + a_2 z^2 + \cdots + a_2 z^{N-2} + ja_1 z^{N-1} + a_0 z^N}{a_0 - ja_1 z^{-1} + a_2 z^{-2} + \cdots + a_2 z^{-N+2} - ja_1 z^{-N+1} + a_0 z^{-N}}, \tag{23}$$

where N is even, a_n is real and $a_0 = 1$. The phase response $\theta(\omega)$ of $A_N(z)$ is given by

$$\theta(\omega) = \eta + 2\varphi(\omega), \tag{24}$$

where if $M = N/2$ is even,

$$\varphi(\omega) = \tan^{-1} \frac{\sum_{n=0}^{M/2-1} a_{2n+1} \cos(M - 2n - 1)\omega}{\frac{a_M}{2} + \sum_{n=0}^{M/2-1} a_{2n} \cos(M - 2n)\omega}, \tag{25}$$

and if $M = N/2$ is odd,

$$\varphi(\omega) = \tan^{-1} \frac{\frac{a_M}{2} + \sum_{n=0}^{(M-3)/2} a_{2n+1} \cos(M - 2n - 1)\omega}{\sum_{n=0}^{(M-1)/2} a_{2n} \cos(M - 2n)\omega}. \tag{26}$$

Thus, we have

$$\begin{cases} H_0(e^{j\omega}) = \cos\theta(\omega) \\ H_1(e^{j\omega}) = e^{-j\omega} \sin\theta(\omega) \end{cases}. \tag{27}$$

It is clear that $H_0(z)$ and $H_1(z)$ have exactly linear phase responses and satisfy the power-complementary relation in Eq.(7). Its impulse response is WSS. Therefore, the design problem of wavelet filter banks becomes the phase approximation of $A_N(z)$ in Eq.(23).

For the maximally flat filters, the closed-form formula is given in [22] by

$$a_n = C_n \binom{N}{n}, \tag{28}$$

where $C_{2n} = 1$ and $C_{2n+1} = -\tan\frac{\eta}{2}$. Design of wavelet filters having the best possible frequency selectivity for the given degrees of flatness has been also discussed in [22]. It has been pointed out in [22] that we must choose $\eta = \pm\pi/4$ if M is even and $\eta = \pm 3\pi/4$ if M is odd.

Example 4: We consider design of the filter banks with $N = 6$ and $\eta = -3\pi/4$. We have designed $A_N(z)$ with $L = 0, 2, 4, 6$ by using the design method proposed in [22]. The magnitude responses of $H_0(z)$ are shown in **Figure 10**, and the scaling and

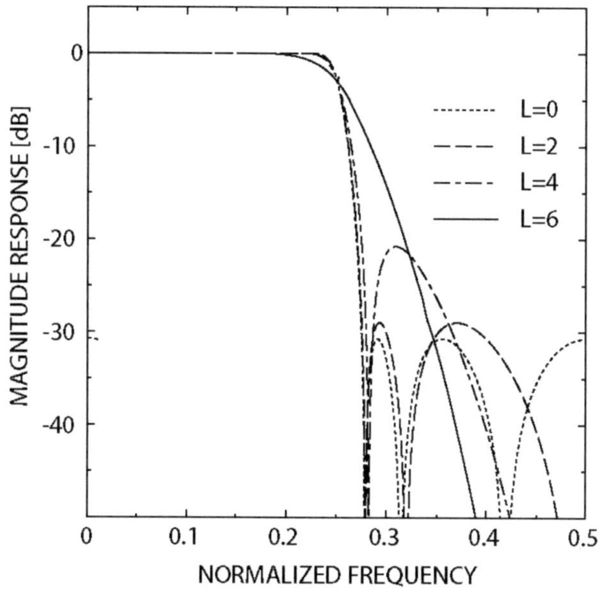

Figure 10.
Magnitude responses of $H_0(z)$ in example 4.

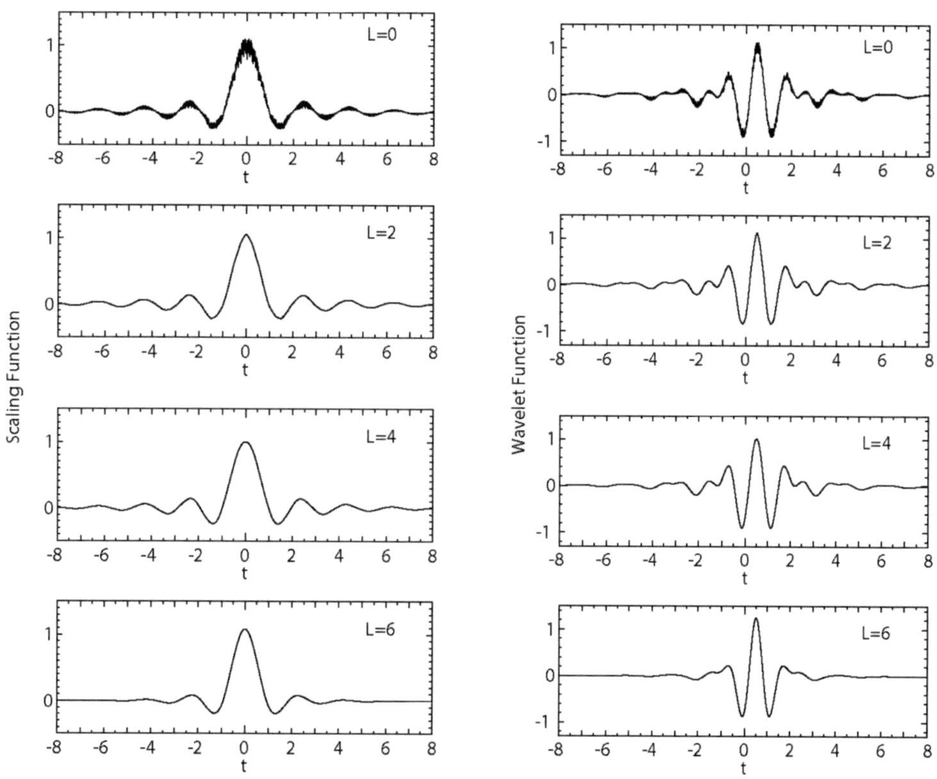

Figure 11.
Scaling and wavelet functions in example 4.

wavelet functions are shown in **Figure 11**, respectively. It is seen in **Figure 10** that $H_0(z)$ with $L = 6$ corresponds to the maximally flat filter, and $H_0(z)$ with $L = 0$ is the minimax filter that has no zero located at $z = -1$. In **Figure 11**, the scaling and wavelet functions are not continuous because the regularity condition is not satisfied when $L = 0$, and become more smooth with an increasing L. Both the scaling and wavelet functions are symmetric.

4. Lifting-based wavelet filter banks using allpass filters

The lifting scheme proposed in [16] and [17] is an efficient tool for constructing second generation wavelets, and has advantages such as faster implementation, fully in-place calculation, reversible integer-to-integer transforms, and so on. It has been proved in [18] and [19] that every FIR wavelet filter bank can be decomposed into a finite number of lifting steps, thus this allows the construction of an integer version of the wavelet transform. Such integer wavelet transforms are invertible, and then are attractive in lossless coding applications. Conventionally, the lifting scheme is often used to construct a class of biorthogonal wavelet filter banks. It has been shown in [18] that the orthogonal wavelet filter banks can also be realized by the lifting scheme. However, it is not always possible for IIR wavelet filter banks to be decomposed into a finite number of lifting steps. For example, it is difficult to realize the IIR orthogonal wavelet filter banks discussed in Section 3 by using a finite number of lifting steps.

Now, we restrict ourselves to the lifting scheme with two lifting steps [10] as shown in **Figure 12**. Let $H_0(z)$ and $H_1(z)$ be a pair of lowpass and highpass filters,

$$\begin{cases} H_0(z) = \frac{1}{2}\{z^{-2K_1-1} + P(z^2)\} \\ H_1(z) = z^{-2K_2} - Q(z^2)H_0(z) \end{cases}, \tag{29}$$

then $G_0(z) = H_1(-z)$ and $G_1(z) = -H_0(-z)$. It is clear in **Figure 12** that the PR condition is structurally satisfied. Therefore, the design of $H_0(z)$ and $H_1(z)$ becomes how to determine $P(z)$ and $Q(z)$. In the following, we describe two classes of near symmetric wavelet filter banks by using real allpass filters in the lifting scheme: causal stable biorthogonal wavelet filter bank [23] and orthogonal wavelet filter bank [24].

4.1 Causal stable wavelet filter banks

We use two real allpass filters in lifting steps, i.e., $P(z) = A_{N_1}(z)$ and $Q(z) = A_{N_2}(z)$, and thus,

$$\begin{cases} H_0(z) = \frac{1}{2}\{z^{-2K_1-1} + A_{N_1}(z^2)\} \\ H_1(z) = z^{-2K_2} - A_{N_2}(z^2)H_0(z) \end{cases}. \tag{30}$$

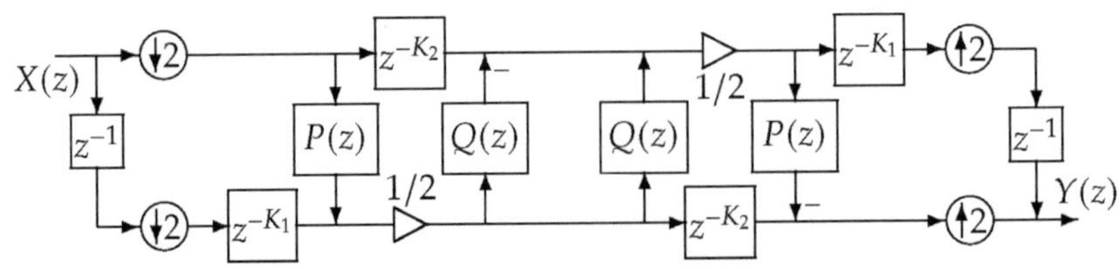

Figure 12.
Lifting scheme with two lifting steps [10]. X(z) is input and Y(z) is output. P(z), Q(z) are filters.

Let $\theta_1(\omega)$ be the phase response of $A_{N_1}(z)$, the frequency response of $H_0(z)$ is given by

$$H_0\left(e^{j\omega}\right) = e^{j\left(\frac{\theta_1(2\omega)}{2}-\left(K_1+\frac{1}{2}\right)\omega\right)} \cos\left(\frac{\theta_1(2\omega)}{2} + \left(K_1 + \frac{1}{2}\right)\omega\right). \tag{31}$$

For $H_0(z)$ to be lowpass filter, the desired phase response of $A_{N_1}(z)$ is

$$\theta_{1d}(\omega) = -\left(K_1 + \frac{1}{2}\right)\omega = -\tau_1\omega, \tag{32}$$

where $\tau_1 = K_1 + \frac{1}{2}$. According to Appendix, the order of $A_{N_1}(z)$ is required to be $N_1 = K_1$ or $N_1 = K_1 + 1$ to obtain causal stable allpass filter.

Ideally, $H_0\left(e^{j\omega}\right) = 0$ in the stopband of $H_0(z)$, then $H_1\left(e^{j\omega}\right) = e^{-j2K_2\omega}$, having linear phase response from Eq.(30). In the passband of $H_0(z)$, $H_0\left(e^{j\omega}\right) = e^{-j(2K_1+1)\omega}$ ideally, thus,

$$\begin{aligned} H_1\left(e^{j\omega}\right) &= e^{-j2K_2\omega} - e^{j\theta_2(2\omega)}e^{-j(2K_1+1)\omega} \\ &= -2je^{j\left(\frac{\theta_2(2\omega)}{2}-\left(K_1+K_2+\frac{1}{2}\right)\omega\right)} \sin\left(\frac{\theta_2(2\omega)}{2} - \left(K_1 - K_2 + \frac{1}{2}\right)\omega\right), \end{aligned} \tag{33}$$

where $\theta_2(\omega)$ is the phase response of $A_{N_2}(z)$. Therefore, in the stopband of $H_1(z)$, the desired phase response of $A_{N_2}(z)$ is

$$\theta_{2d}(\omega) = \left(K_1 - K_2 + \frac{1}{2}\right)\omega = -\tau_2\omega, \tag{34}$$

where $\tau_2 = K_2 - K_1 - \frac{1}{2}$. Similarly, the order of $A_{N_2}(z)$ is required to be $N_2 = K_2 - K_1$ or $N_2 = K_2 - K_1 - 1$ to obtain causal stable allpass filter. Therefore, once N_1 and N_2 are given, we can obtain causal stable wavelet filter banks by appropriately choosing K_1 and K_2. The maximally flat filters can be designed by using Eq.(10). $H_0(z)$ and $H_1(z)$ have approximately linear phase response.

Example 5: We consider design of the maximally flat wavelet filter banks with $N_1 = N_2 = 6$. We have designed $A_{N_1}(z)$ with $K_1 = 5$, and the magnitude response of $H_0(z)$ is shown in **Figure 13**. We then designed $A_{N_2}(z)$ with $K_2 = 11$ and $K_2 = 12$, and the magnitude responses of $H_1(z)$ are shown also in **Figure 13**. It is seen in **Figure 13** that $H_1(z)$ with $K_2 = 11$ has a large overshoot nearby $\omega = \pi/2$. To avoid this overshoot, we should choose $K_2 = N_2 + K_1 + 1$ if $K_1 = N_1 - 1$ and $K_2 = N_2 + K_1$ if $K_1 = N_1$. The scaling and wavelet functions generated by analysis and synthesis filters with $K_1 = 5$ and $K_2 = 12$ are shown in **Figure 14** respectively.

4.2 Orthogonal wavelet filter banks

The above-mentioned causal stable wavelet filter banks are biorthogonal (not orthogonal). Here we discuss a class of orthogonal wavelet filter banks using the lefting scheme. We use $P(z) = A_N(z)$ and $Q(z) = A_N(z^{-1})$, then,

$$\begin{cases} H_0(z) = \frac{1}{2}\left\{z^{-2K_1-1} + A_N\left(z^2\right)\right\} \\ H_1(z) = z^{-2K_2} - A_N\left(z^{-2}\right)H_0(z) \end{cases}. \tag{35}$$

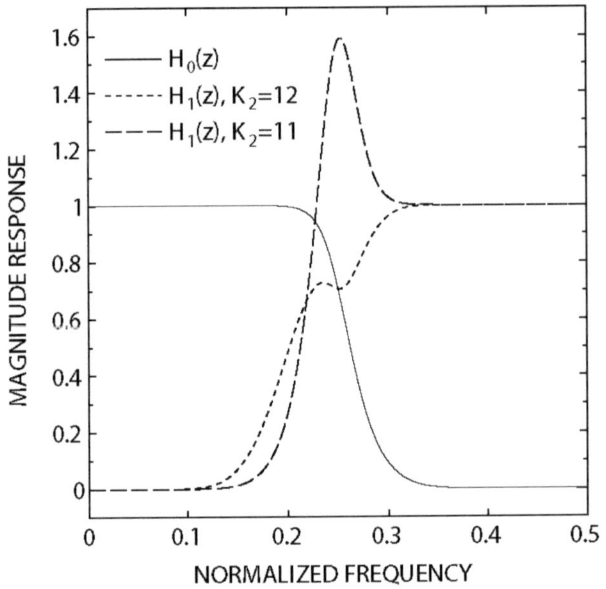

Figure 13.
Magnitude responses of $H_0(z)$ and $H_1(z)$ in example 5.

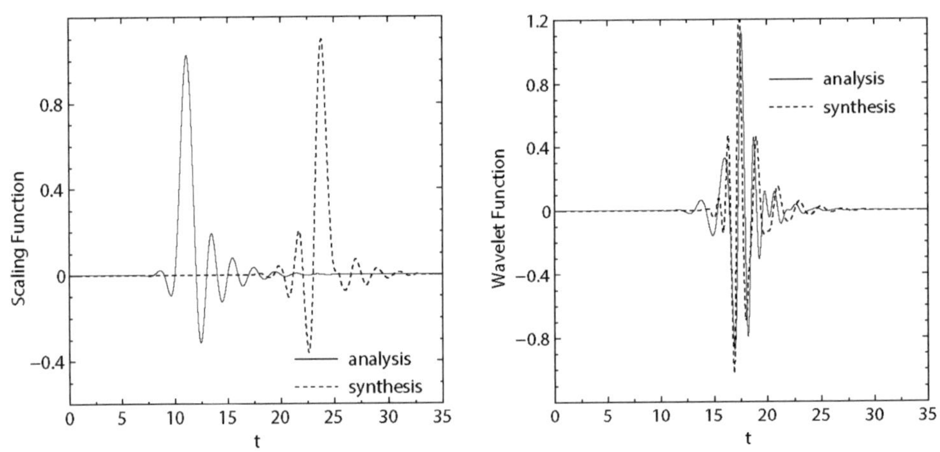

Figure 14.
Scaling and wavelet functions generated by analysis and synthesis filters in example 5. It is because the wavelet filter bank is biorthogonal, but not orthogonal.

Let $\theta(\omega)$ be the phase response of $A_N(z)$, the frequency response of $H_0(z)$ is the same as in Eq.(31), thus the desired phase response of $A_N(z)$ is

$$\theta_d(\omega) = -\left(K_1 + \frac{1}{2}\right)\omega = -\tau\omega, \tag{36}$$

where $\tau = K_1 + \frac{1}{2}$. To be orthogonal, we set $K_2 = 0$ and have

$$H_1(z) = 1 - A_N(z^{-2})\frac{1}{2}\{z^{-2K_1-1} + A_N(z^2)\} = \frac{1}{2}\{1 - z^{-2K_1-1}A_N(z^{-2})\}, \tag{37}$$

whose frequency response is

$$H_1\left(e^{j\omega}\right) = je^{-j\left(\frac{\theta(2\omega)}{2}+\left(K_1+\frac{1}{2}\right)\omega\right)}\sin\left(\frac{\theta(2\omega)}{2}+\left(K_1+\frac{1}{2}\right)\omega\right). \tag{38}$$

It is clear that the magnitude responses satisfy $|H_0(e^{j\omega})| = |H_1(e^{j(\pi-\omega)})|$ and the power-complementary relation in Eq.(7). Therefore, this class of wavelet filter banks is orthogonal and both $H_0(z)$ and $H_1(z)$ have approximately linear phase response. Design of this class of orthogonal wavelet filter banks has been discussed and applied to lossy to lossless image coding in [24].

Example 6: We consider design of the maximally flat orthogonal wavelet filter banks with $N = 2, 4, 6$. We have designed $A_N(z)$ with $K_1 = N - 1$. The magnitude

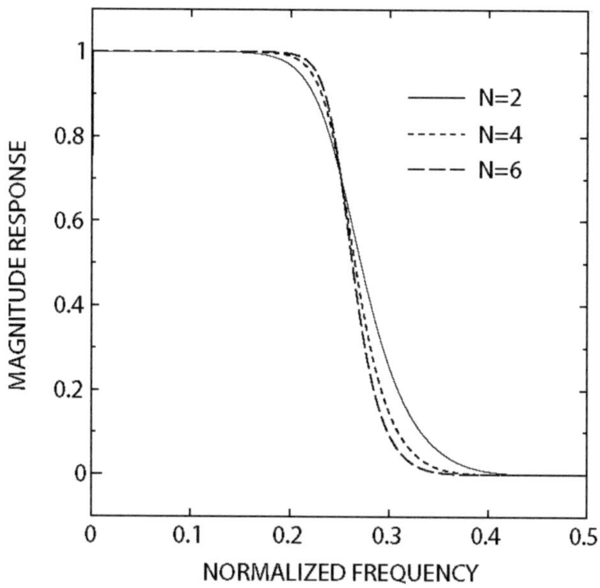

Figure 15.
Magnitude responses of $H_0(z)$ in example 6.

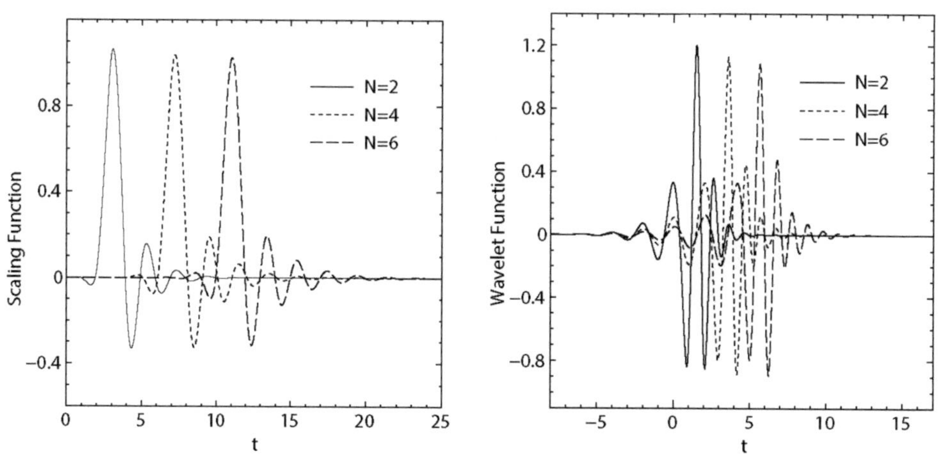

Figure 16.
Scaling and wavelet functions in example 6.

responses of $H_0(z)$ are shown in **Figure 15**. The generated scaling and wavelet functions are shown in **Figure 16**, respectively. It is seen in **Figure 16** that the wavelet functions are near symmetric.

5. Conclusions

In this chapter, we have proposed several new classes of wavelet filter banks with some properties of orthogonality, symmetry and causal stablity by using allpass filters, which are potential options for readers to choose wavelet basis in practical applications. As shown in **Table 1**, first class of wavelet filter banks in Section 3.1 is orthogonal, but asymmetric, its analysis filters is causal stable. Second class of wavelet filter banks in Section 3.2 is orthogonal and symmetric, but not causal. Third and fourth classes of wavelet filter banks are based on the lifting scheme. Third class in Section 4.1 is biorthogonal, causal stable and near symmetric, while fourth class in Section 4.2 is orthogonal and near symmetric, but not causal. There is no solution to all of orthogonality, symmetry and causal stablity. The wavlet filter banks using allpass filters have been extended to Hilbert transform pair of wavelets [25], 2D wavelet filter banks [26], and applied to lossy to lossless image coding [27–30] and scalable video compression [31]. It is possible also to extend them to higher dimension and irregural signal processing and to apply them to wavelet denoising, image fusion and so on.

Filter Bank Class	**Sec.3.1**	**Sec.3.2**	**Sec.4.1**	**Sec.4.2**	**D-8/8**	**D-9/7**
Filter Type	IIR	IIR	IIR	IIR	FIR	FIR
Orthogonality	○	○	×	○	○	×
Symmetry	×	○	Δ	Δ	×	○
Causal stablity	Δ	×	○	×	○	○

Table 1.
Comparison of the proposed classes of wavelet filter banks with the conventional wavelets D-8/8, D-9/7 in [1].

Acknowledgements

This work was supported by JSPS KAKENHI Grant Number 18 K11260.

Abbreviations

DWT	discrete wavelet transform
PR	perfect reconstruction
FIR	finite impulse response
IIR	infinite impulse response
HSS	half sample symmetric
WSS	whole sample symmetric

Appendix

Digital allpass filter is a computationally efficient signal processing building block and quite useful in many signal processing applications. One of the most widely used applications is phase or delay equalizer. Allpass filter possesses unit magnitude at all frequencies and is a basic scalar lossless building block. The interconnection of allpass filters has found numerous applications in practical filtering problems, such as low sensitivity filter structures, multirate filtering, filter banks and so on [13].

The transfer function of an Nth-order allpass filter is defined as

$$A_N(z) = z^{-N} \frac{\sum_{n=0}^{N} a_n z^n}{\sum_{n=0}^{N} a_n^* z^{-n}}, \tag{39}$$

where $a_n = a_{nr} + ja_{ni}$ is a complex coefficient in general, and a_n^* denotes the complex conjugate of a_n. When $a_{ni} = 0$, a_n is a real coefficient and $A_N(z)$ is a real allpass filter. Thus the real allpass filter is a special case of complex allpass filter. All poles and zeros of $A_N(z)$ occur in mirror-image pairs with respect to the unit circle, and then the frequency response $A_N(e^{j\omega})$ exhibits unit magnitude at all frequencies, i.e., $|A_N(e^{j\omega})| \equiv 1$ for all ω. The phase response of $A_N(z)$ is given by

$$\theta(\omega) = -N\omega + 2\tan^{-1} \frac{\sum_{n=0}^{N} \{a_{nr} \sin n\omega + a_{ni} \cos n\omega\}}{\sum_{n=0}^{N} \{a_{nr} \cos n\omega - a_{ni} \sin n\omega\}}. \tag{40}$$

If all poles locate inside the unit circle, then $A_N(z)$ is causal stable. The phase response decreases monotonically with an increasing frequency and $\theta(\pi) = \theta(-\pi) - 2N\pi$. If $A_N(z)$ is real allpass filter, $\theta(0) = 0$ and $\theta(\pi) = -N\pi$. When one pole locates at the origin, it is seen that $A_N(z) = z^{-1}A_{N-1}(z)$ due to $a_N = 0$. Then z^{-N} is a special case of $A_N(z)$ if all poles locate at the origin. When k poles locate outside the unit circle, we can divide $A_N(z)$ into two causal stable allpass filters $A_{N-k}(z)$ and $A_k(z)$, i.e.,

$$A_N(z) = \frac{A_{N-k}(z)}{A_k(z)}. \tag{41}$$

The phase response $\theta(\omega)$ of $A_N(z)$ is the phase difference between $A_{N-k}(z)$ and $A_k(z)$, and $\theta(\pi) = \theta(-\pi) - 2(N - 2k)\pi$. The design problem of allpass filters to approximate the specified phase response in the Chebyshev sense has been discussed in [14] and [15].

Author details

Xi Zhang
Department of Computer and Network Eng., University of Electro-Communications, Chofu-shi, Tokyo, Japan

*Address all correspondence to: zhangxi@uec.ac.jp

IntechOpen

© 2020 The Author(s). Licensee IntechOpen. This chapter is distributed under the terms of the Creative Commons Attribution License (http://creativecommons.org/licenses/by/3.0), which permits unrestricted use, distribution, and reproduction in any medium, provided the original work is properly cited.

References

[1] Daubechies I. Ten Lectures on Wavelets. SIAM. Philadelphia, PA; 1992

[2] Vaidyanathan PP. Multirate Systems and Filter Banks. Englewood Cliffs, NJ: Prentice Hall; 1993

[3] Vetterli M, Kovacevic J. Wavelets and Subband Coding. Prentice Hall PRT: Upper Saddle River: New Jersey; 1995

[4] Strang G, Nguyen T. Wavelets and Filter Banks. Wellesley Cambridge Press: Wellesley; 1996

[5] Taubman DS, Marcellin MW. JPEG2000: Image Compression Fundamentals, Standards and Practice. Kluwer Academic Publishers: Boston; 2002

[6] Antonini M, Barlaud M, Mathieu P, Daubechies I. Image coding using wavelet transform. IEEE Trans Image Processing. 1992; 1(2); 205–220

[7] Vaidyanathan PP, Regalia PA, Mitra SK. Design of doubly complementary IIR digital filters using a single complex allpass filter with multirate applications. IEEE Trans Circuits & Syst. 1987; 34(4); 378–389

[8] Smith MJT, Eddins SL. Analysis/ synthesis techniques for subband image coding. IEEE Trans Acoust Speech & Signal Processing. 1990; 38(8); 1446–1456

[9] Herley C, Vetterli M. Wavelets and recursive filter banks. IEEE Trans Signal Processing. 1993; 41(8); 2536–2556

[10] Phoong SM, Kim CW, Vaidyanathan PP, Ansari R. A new class of two-channel biorthogonal filter banks and wavelet bases. IEEE Trans Signal Process. 1995; 43(3); 649–665

[11] Creusere CD, Mitra SK. Image coding using wavelets based on perfect reconstruction IIR filter banks. IEEE Trans. Circuits & Systems for Video Technology. 1996; 6(5); 447–458

[12] Selesnick IW. Formulas for orthogonal IIR wavelet filters. IEEE Trans Signal Processing. 1998; 46(4); 1138–1141

[13] Regalia PA, Mitra SK, Vaidyanathan PP. The digital allpass filter: a versatile signal processing building block. Proceeding of IEEE. January 1988; 76(1); 19–37

[14] Zhang X, Iwakura H. Novel method for designing digital allpass filters based on eigenvalue problem. IEE Electronics Letters. 1993; 29(14); 1279–1281

[15] Zhang X, Iwakura H. Design of IIR digital allpass filters based on eigenvalue problem. IEEE Trans Signal Processing. 1999; 47(2); 554–559

[16] Sweldens W. The lifting scheme: A custom-design construction of biorthogonal wavelets. Appl Comput Harmon Anal. 1996; 3(2); 186–200

[17] Sweldens W. The lifting scheme: A construction of second generation wavelets. SIAM J Math Anal. 1997; 29 (2); 511–546

[18] Daubechies I, Sweldens W. Factoring wavelet transforms into lifting steps. J Fourier Anal Appl. 1998; 4; 247–269

[19] Calderbank AR, Daubechies I, Sweldens W, Yeo BL. Wavelet transforms that map integers to integers. Appl Comput Harmon Anal. 1998; 5(3); 332–369

[20] Zhang X, Yoshikawa T. Design of orthonormal IIR wavelet filter banks using allpass filters. ELSEVIER Signal Processing. 1999; 78(1); 91–100.

[21] Zhang X, Muguruma T, Yoshikawa T. Design of orthonormal symmetric wavelet filters using real allpass filters. ELSEVIER Signal Processing. 2000; .80(8); 1551–1559

[22] Zhang X, Kato A, Yoshikawa T. A new class of orthonormal symmetric wavelet bases using a complex allpass filter. IEEE Trans Signal Processing. 2001; 49(11); 2640–2647

[23] Zhang X, Yoshikawa T. Design of stable IIR perfect reconstruction filter banks using allpass filters. Electronics and Communications in Japan. Part 2. 1998; 81(5); 24–32

[24] Zhang X, Wang W, Yoshikawa T, Takei Y. Design of IIR orthogonal wavelet filter banks using lifting scheme. IEEE Trans Signal Processing. 2006; 54(7); 2616–2624

[25] Zhang X, Ge DF. Hilbert transform pairs of orthonormal symmetric wavelet bases using allpass filters. In: Proceedings of ICIP'07; September 2007; San Antonio Texas USA

[26] Zhang X. 2D orthogonal symmetric wavelet filters using allpass filters. In: Proceedings of ICASSP'13. May 2013; Vancouver Canada

[27] Kamimura S, Zhang X, Yoshikawa T. Wavelet-based image coding using allpass filters. Electronics and Communications in Japan. Part 3. 2002; 85(2); 13–21

[28] Zhang X, Kawai K, Yoshikawa T, Takei Y. Lossy to lossless image compression using allpass filters. In: Proceedings of ICIP'05; September 2005; Genova Italy

[29] Zhang X, Ohno K. Lossless image compression using 2D allpass filters. In: Proceedings of ICIP'08; October 2008; San Diego California USA

[30] Zhang X, Fukuda N. Lossy to lossless image coding based on wavelets using a complex allpass filter. International Journal of Wavelets, Multiresolution and Information Processing. 2014; 12(4); DOI: 10.1142/S0219691314600029

[31] Zhang X, Suzuki T. Scalable video coding using allpass-based wavelet filters. In: Proceedings of APCCAS'14; November 2014; Ishigaki Island Okinawa Japan 2014–11

Section 6

Recent Applications of Wavelet Theory

Chapter 14

A Wavelet Threshold Function for Treatment of Partial Discharge Measurements

Caio F.F.C. Cunha, Mariane R. Petraglia, André T. Carvalho and Antonio C.S. Lima

Abstract

Based on the wavelet transform filtering theory, the chapter will describe the elaboration of a wavelet threshold function intended for the denoising of the partial discharge phenomenon measurements. This new function, conveniently named Fleming threshold, is based on the logistic function, which is well known for its utility in several important areas. In the development is shown some variations in the application of the Fleming function, in an attempt to identify the decomposition levels where the thresholding process must be more stringent and those where it can be more lenient, which increases its effectiveness in the removal of noisy coefficients. The proposed function and its variants demonstrate excellent results compared to other wavelet thresholding methods already described in the literature, including the famous Hard and Soft functions.

Keywords: wavelet transform, threshold function, partial discharge, signal denoising

1. Introduction

The analysis of the Partial Discharges (PD) phenomenon, which manifests itself in the existing imperfections into the insulation of high voltage equipment have received global acceptance as an important tool for the predictive diagnosis of the operational conditions of these, allowing taking measures that can safeguard both the material and the power supply quality of the electrical system.

PD are short duration impulsive signals and, consequently, these can be detected in a wide frequency range, from a few kHz to GHz. Normally, there is a direct relationship between the frequency range where there is a higher incidence of PD pulses and the type of high-voltage equipment evaluated, e.g. transformers and generators usually emit pulses from a few tens of kHz up to about 30 MHz [1], whereas Gas Insulated Substations (GIS) are affected by very fast pulses ranging from 300 MHz to 3 GHz and for cables the spectrum covers frequencies from 300 MHz to 1 GHz.

Figure 1 shows two examples of measured PD pulses in two different HV equipment, a GIS and a hydro generator. Note the marked white noise presence. The pulses normally have an exponentially damped oscillatory shape or only an exponentially damped shape [2].

IntechOpen

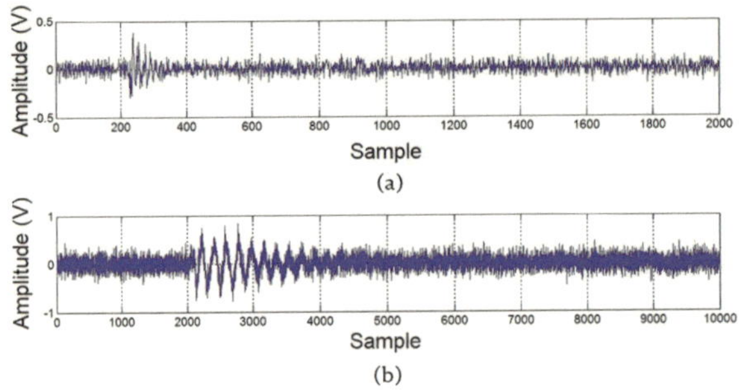

Figure 1.
PD pulses measured in: (a) GIS; (b) Hydro generator.

The proper diagnosis of equipment is closely related to the peak amplitude and shape of the pulses detected. Therefore, it is important to preserve the amplitude characteristics of the signal (especially the peaks), providing higher Signal to Noise Ratio (SNR) and lower Amplitude Error (EA).

The application of FFT and STFT filtering is not as effective in the treatment of non-stationary, transient, and stochastic signals as the PD [3], since these transforms do not allow a location in the time and frequency domain in the same way as the wavelet transform does [4, 5] (with better resolution in frequency and worse resolution in time for the low frequency components of the signal; and worse resolution in frequency and better resolution in time for the high frequency components of the signal). Therefore, the performance of these methods becomes limited in comparison with the wavelet denoising, which presents a capacity of self-adaptation to the signal.

Partial discharges, almost entirely, are electrically detected and quantified, exposing them to the extensive noise interferences that may compromise the PD signals measurement, limiting the diagnosis accuracy. Different signal processing tools have been used to extract the PD signals from these noise sources; among them, it is possible to highlight the Wavelet Transform (WT). The filtering by wavelet processing is recommended in the extraction of PD signals immersed in Gaussian noise [1, 6].

An efficient application of wavelet processing depends on the careful selection of the parameters that will concentrate the coefficients on the most suitable decomposition levels to minimize the PD signal information loss. Among these, we have the applied WT, the number of decomposition levels, the wavelet functions used in each of these levels, the method of estimating the threshold value of the obtained coefficients and the threshold function.

The choice of most of these parameters has already been widely explored in several works meant to PD processing [2, 6–13]. However, with respect to threshold functions, most studies do not focus on PD signals denoising but on audio signals [14–18], Electrocardiogram (ECG) [19–21] or images [22–28]. Therefore, there is a lack of a dedicated threshold function to improve the PD pulses denoising, in order to increase the precision in the diagnosis of High Voltage (HV) equipment.

Based on the WT filtering theory, this chapter will be described the development of a wavelet threshold function aiming to improve the noise reduction in PD measurements. The logistic function serves as an inspiration to this new function [29], which is well known for its usefulness in numerous areas. Since it is customary to associate functions of this type with something that refers them to the name of their developers (e.g., [25]), it was designated as Fleming threshold function.

The denoising performance of the proposed threshold function was compared with the traditional Hard and Soft functions and with twelve other thresholding functions. For a fair analysis of the filtering results, were used 2064 simulated and measured PD pulses contaminated with uniform white noise, Gaussian white noise, and Amplitude Modulated (AM) noise. The results showed that our proposal is able to overcome, qualitatively, and quantitatively, all the confronted functions.

2. Wavelet domain detection

Noise degrades the accuracy and precision of analysis, in addition to reducing the detection limit of the instrument applied in the PD measurements. Often the WT is a tool designed to attenuate continuous random noise (white noise), because after the decomposition of a signal in the wavelet domain can be noted that the average density of the coefficients is inversely proportional to the dyadic scale $1/2^j$ (j indicates the level of decomposition), i.e., half of the number of extreme local coefficients do not spread from a $1/2^j$ scale to the next $1/2^{j+1}$ scale, distributing it uniformly across the scales. As the wavelet coefficients distribution pattern of the PD signal (which tends to have its energy concentrated in few decomposition levels) differs from the noise pattern, it becomes easy to identify and separate the PD signals from the noise [30–32]. However, in wavelet denoising, the noise attenuation occurs not only to the white noise but also to the noise with frequency components that do not match the frequency components of the PD pulses.

Basically, the wavelet shrinkage denoising process involves three steps [13, 31]:

1. Determine the WT decomposition tree (discrete WT, wavelet packet transform, stationary WT or dual-tree complex WT) to be applied, the number of decomposition levels J and the wavelet function that will be employed on each of the j levels (where $j = 1, 2, \cdots, J$), and then perform the decomposition of the analyzed signal into its wavelet coefficients;

2. Calculate the threshold values using one of the threshold selection rules, which depend on statistical estimation of the noise level present in the signal. Apply the calculated value in a threshold function to thus reduce the coefficients of the noise figure and preserve the signal coefficients of interest, in our case the PD pulse;

3. Reconstruct the signal by applying the Inverse Wavelet Transform (corresponding to the decomposition tree selected in the first step) in the threshold coefficients, to obtain the filtered signal in the time domain.

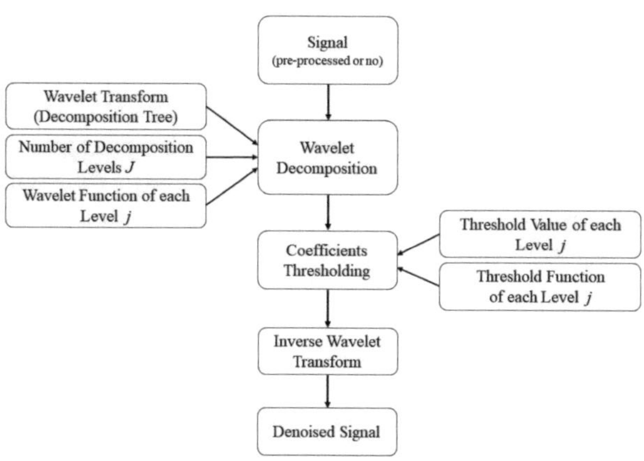

Figure 2.
Signal denoising steps by wavelet transform.

Figure 2 illustrates each of these steps involved in the wavelet denoising processing of a digitalized signal.

As several parameters are involved, they should be carefully selected according to the signal characteristics, in order to maximize their wavelet coefficients above the noise level. Thus, filtering performance is closely related to each of these parameters and some of these will have a greater influence on the quality of the result. The determination of these parameters shows to be an optimization challenge [33]. In this chapter, we will focus our attention on the improvement of the threshold function applied in the second step.

3. Fleming threshold function

In most of the wavelet denoising literature, especially those focused in the treatment of PD signals, the choice of the threshold function normally falls between the Hard and the Soft functions. Moreover, it is well known that for PD pulse filtering the Hard function tends to preserve more of the signal information, providing a higher SNR and a lower Amplitude Error (AE). However, the Hard estimate has discontinuities, being not differentiable, which ends up causing instability problems and sensitivity to small changes in the data pseudo-Gibbs effect. The Soft function is weakly differentiable and produces a high attenuation of the coefficients and, therefore, the reduction of the amplitude in the resulting signal.

In an attempt to get around these problems, many alternatives are being proposed. The main idea is to generate a high-derivative order thresholding function, which contributes to its use in optimization algorithms that look for the optimal parameters to be applied in the thresholding of each signal [34]. Therefore, the function becomes adaptable to the signal to be processed, improving the quality of the denoised signal.

When analyzing the threshold functions applied, whether in the area of PD, audio, ECG, or image processing, it is remarkable that those seek improvements by combining both, the preservation properties of the coefficients and magnitudes provided by the Hard function, as well as the differentiation and smoothness provided by the Soft function. In image processing, the smoothness property is interesting so that the resulting image shows more pleasant contours. In signal processing, such as audio, ECG, and PD pulse processing, it is important to achieve better preservation of signal magnitude (peak) and signal noise ratio.

For this reason, many authors have explored functions that correspond to an interpolation of the Soft and Hard alternatives. As an example, it is possible to mention functions such as: the Garrote described by Nasiri et al. in [19]; the Non Negative Garrote described in [12]; the Adaptive Shirinkage showed by Partha Ray in [35]; the Liu developed by Shan Liu in [16]; the Hui presented in [36]; Stein and Semi-Soft shown in [12]; and the functions described by Zhang et al. in [37, 38]. However, the majority of the functions cannot adapt to the different signals due to the fixed transition curve on the threshold value. In these functions, there is still a greater tendency to smooth the coefficients than to preserve them, not realizing that for PD signals it is appropriate for the function to be closer to the Hard them to the Soft threshold function, but still preserving some of the smoothness (differentiability) in the transition of the threshold value, which will allow an improvement in the EQM and CC.

Following this line of reasoning, we propose a new threshold function similar to the Hard but being differentiable for higher orders and being able to adjust to each signal. This proposal is based on the well-known logistic function, shown in **Figure 2**, widely used in artificial neural networks, demography, economics, probability, statistics, chemistry, etc. Eq. (1) represents the logistic function, where H is the maximum value of the curve, α controls the slope of this curve and x corresponds to the value of x at the midpoint of the sigmoid curve dictated by the numerator value (**Figure 3**). When the x value tends to $+\infty$ the curve approaches H and when it tends to $-\infty$ it approaches to zero.

$$f(x) = \frac{H}{1 + e^{-\alpha(x - x_0)}} \quad (1)$$

Eq. (1) enabled us to develop the threshold function for filtering signals in such a way that it circumvents the problems previously described. As the objective in the thresholding process is to preserve the coefficients above the threshold value, it is easy to see that the maximum value H will be the decomposed wavelet coefficient $w_{j,k}$ (which corresponds to the variable x). Thus, the function will maintain symmetry when we vary the inclination constant c of the curve. Finally, it is necessary to move the function along the abscissa axis so that the graph shown in **Figure 2** leaves the ordinate axis and stays over the threshold value, this is done by subtracting the variable x from the value where we want to move the function (x_0), i.e., the value of the coefficients $w_{j,k}$ must be subtracted from the threshold value λ. By making these adaptations, we obtain the following threshold function:

$$\eta_f\left(\omega_{j,k}, \lambda, c\right) = \frac{\omega_{j,k}}{1 + e^{-c\left(\omega_{j,k} - \lambda\right)}} = \frac{\omega_{j,k}}{1 + e^{c\left(-\omega_{j,k} + \lambda\right)}} \quad (2)$$

For a more efficient implementation, in which it is not necessary to worry about the fact that the coefficient $w_{j,k}$ is positive or negative, Eq. (2) can be rewritten using the signum (*sign*) function, which returns $+1$ if the value is positive and -1 if the value is negative. Thus, we have:

$$\eta_f\left(\omega_{j,k}, \lambda, c\right) = \frac{\omega_{j,k}}{1 + e^{c\left(\left(-sign\left(\omega_{j,k}\right) \times \omega_{j,k}\right) + \lambda\right)}} \quad (3)$$

With high c values, the curve inclination on the threshold point is such that it approaches the Hard function, but with a smoother (differentiable) transition. For low c values, the inclination of the function will act with less intensity on the coefficients below the threshold value and with greater intensity on the coefficients above this value, i.e., a large part of noisy coefficients may pass and there will be information losses on those coefficients that represent the signal of interest, in our case the PD pulse. With the appropriate choice of the c value for each processed signal, it is possible to obtain a significant improvement in the result of the PD wavelet denoising in relation to the Hard and the Soft functions.

Considering a threshold value $\lambda = 1$, **Figure 4** shows the behavior of the proposed threshold function (called the Fleming function) for different c values (3, 10, 20, 30, 50, 80, 100 e 200). With very low values of c there is the possibility of

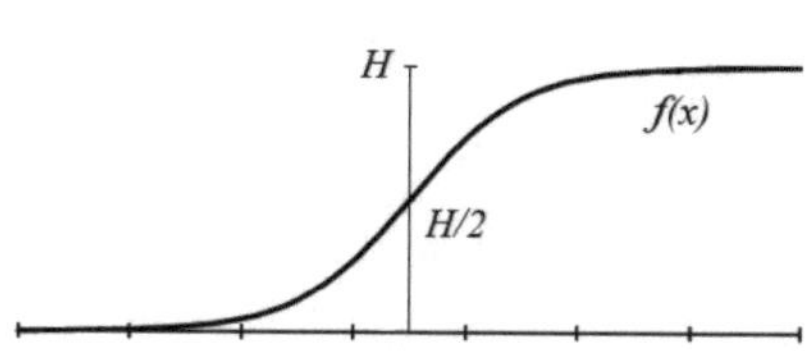

Figure 3.
Logistic function.

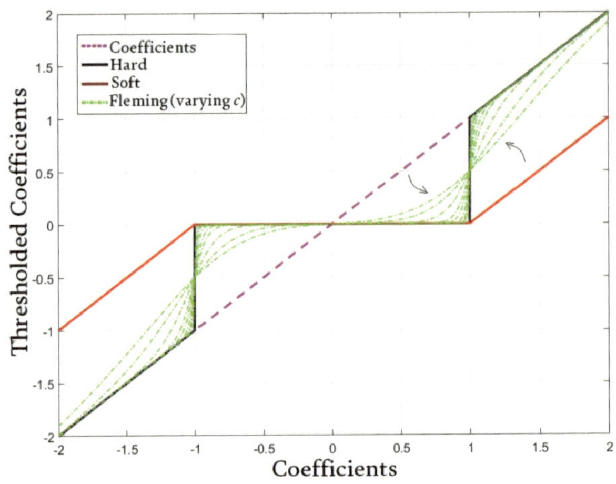

Figure 4.
Behavior of the Fleming threshold function to $\lambda = 1$.

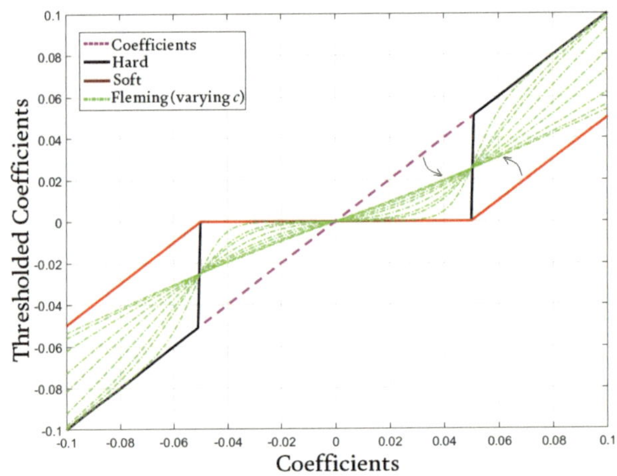

Figure 5.
Behavior of the Fleming threshold function to $\lambda = 0, 05$.

passing a large number of noisy coefficients, so it is indicated that the value of the constant be greater than or equal to 5.

In a PD evaluation, most measurements provide signals with amplitude around mV. Thus, if the WT technique is applied to filter the signals, its decomposed coefficients will also be in the mV range and by using a threshold rule (in our case *scaledep*) the threshold value λ will be small and usually smaller than 1, mainly for coefficients that contain more noise than the PD components. When we evaluate the threshold function for a small threshold value (e.g., $\lambda = 0, 05$), the accuracy with which the coefficients are attenuated becomes lower, as illustrated in **Figure 5**. Note that even for $c = 200$, most of the noisy coefficients can pass, different than what was seen for the threshold value $\lambda = 1$.

One solution to overcome this problem was to adapt Eq. (3) according to the threshold value when it is considered small (understand as small as $\lambda < 0, 5$), by simply changing the c constant that controls the inclination proportionally to the λ threshold values. With this, we can rewrite Eq. (3) as follows:

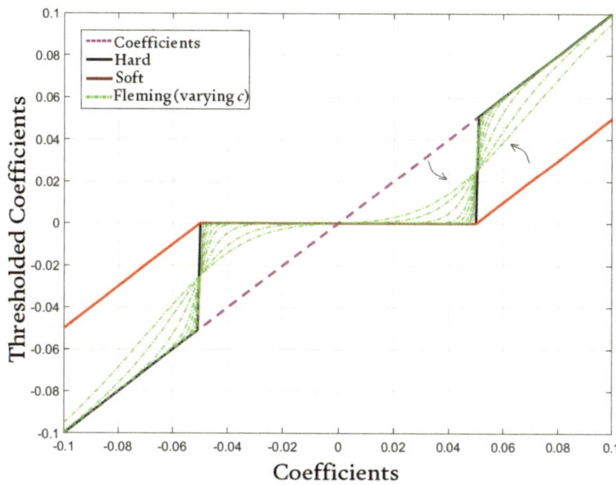

Figure 6.
Behavior of the modified Fleming threshold function to $\lambda = 0,05$.

$$\eta_f(\omega_{j,k}, \lambda, c) = \begin{cases} \dfrac{\omega_{j,k}}{1 + e^{c((-sign(\omega_{j,k}) \times \omega_{j,k}) + \lambda)}} & if\ \lambda \geq 0,5 \\ \dfrac{\omega_{j,k}}{1 + e^{\langle c/\lambda \rangle((-sign(\omega_{j,k}) \times \omega_{j,k}) + \lambda)}} & if\ \lambda < 0,5 \end{cases} \tag{4}$$

thus, when $\lambda < 0,5$ the lower the λ threshold value, the greater the rigor in discarding the coefficients (closer to the Hard function), with a significant improvement in the function's behavior, as shown in **Figure 6**.

Therefore, in Eq. (4) we have a function capable of adapting to different types of wavelet coefficients, varying between the Soft and the Hard threshold functions according to the c inclination value defined. Thus, there is a need to define how (and which) the inclination value should be applied to the coefficients.

3.1 Relevant wavelet coefficient identification

From the idea of identifying the most important coefficients to form the PD signal, used for the SNRBWS method, we were able to perform a variant on the threshold function. In this case, we chose to use kurtosis (K_u) as a statistical measure of the probability distribution's flatness [39] of the $\omega_{j,k}$ wavelet coefficient, because the tapered this curve, the farther from the Normal probability distribution (Gaussian), which is characteristic of the white noise presence. Therefore, kurtosis will serve as an indicator to know if we have noisy coefficients (kurtosis close to 3) or PD components (high kurtosis ≥3).

Figure 7 shows the detail coefficients at level $j = 1$ and the detail coefficients at level $j = 6$ with their respective histograms. Notice that in **Figure 7(a)**, formed almost exclusively by noise components, the histogram is very close to the Normal probability distribution, a fact that is confirmed by the kurtosis value equal to 2.9687; in the **Figure 7(b)** the coefficients have significant information about the PD pulse and the histogram is more tapered (leptokurtic), moving away from the Normal distribution, as indicated by the kurtosis value of 9.9612.

Then, to fulfill the task of identifying the most relevant coefficients to form the PD signal, it is enough to assume the following the condition regarding the kurtosis value: if the kurtosis of the coefficient is greater than 4, it must be considered important and the threshold function will make use of a lower c inclination

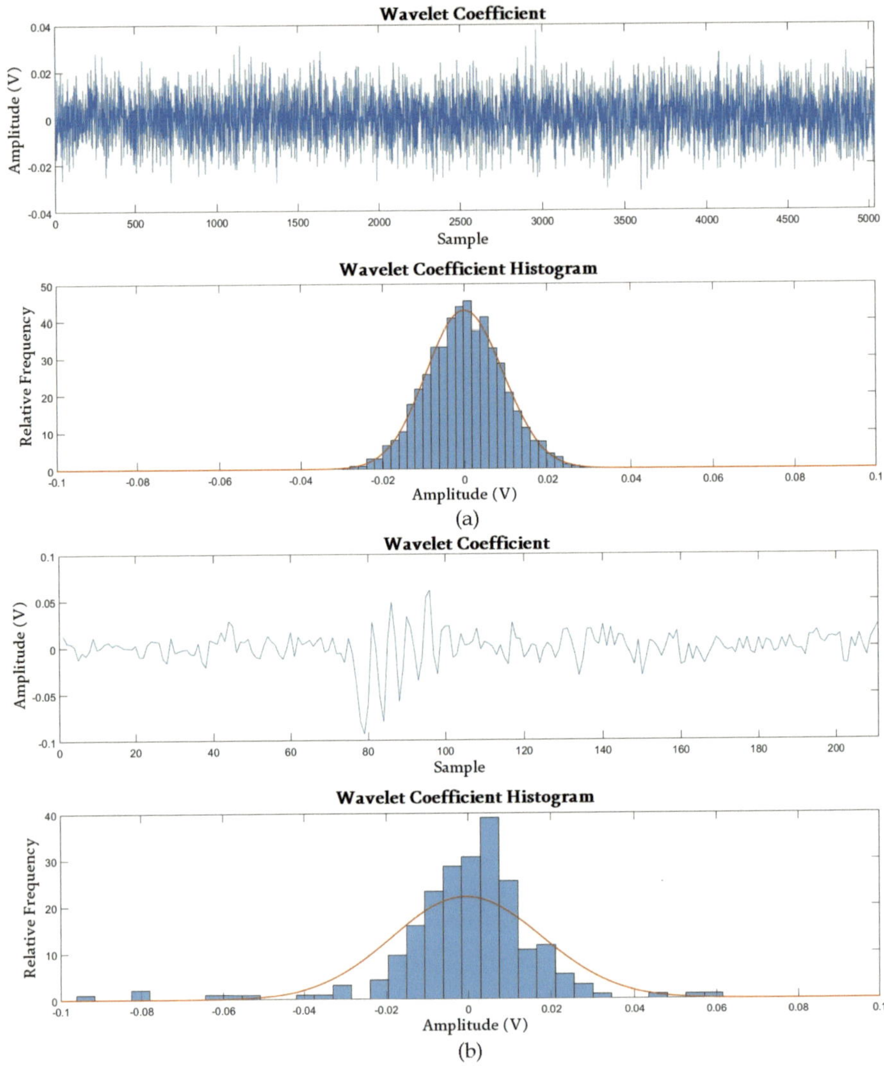

Figure 7.
Wavelet coefficients and histograms of a real PD pulse: (a) first detail coefficient (kurtosis = 2.9687); (b) sixth detail coefficient (kurtosis = 9.9612).

constant, allowing the passage of more coefficients, otherwise it will be considered as noisy coefficients and a much higher inclination constant must be assigned (in case $c = 10^{20}$), eliminating a greater amount of noise, which approximates our function of the Hard. In equational terms, we have the Eq. (5):

$$\eta_{f_2}(\omega_{j,k},\lambda,c) = \begin{cases} \dfrac{\omega_{j,k}}{1+e^{10^{20}\left(\left(-sign(\omega_{j,k})\times\omega_{j,k}\right)+\lambda\right)}} & if\ K_u(\omega_{j,k}) \leq 4 \\ \begin{cases} \dfrac{\omega_{j,k}}{1+e^{c\left(\left(-sign(\omega_{j,k})\times\omega_{j,k}\right)+\lambda\right)}} & if\ \lambda \geq 0,5 \\ \dfrac{\omega_{j,k}}{1+e^{\langle c/\lambda\rangle\left(\left(-sign(\omega_{j,k})\times\omega_{j,k}\right)+\lambda\right)}} & if\ \lambda < 0,5 \end{cases} & if\ K_u > 4 \end{cases} \tag{5}$$

4. Fleming threshold function

In order to perform the evaluation of the Fleming thresholding functions, we took 2064 signals and submitted to the wavelet denoising processes. Among these signals, we included real PD measurements from HV equipment and PD simulated with different levels of uniform white, Gaussian white and AM noise (created of the same way described in [40]). For each data, we compare the performance of our proposal against the classical Hard and Soft thresholding, along with 12 other thresholding functions mentioned in the Section 4.

In addition to thresholding, the wavelet shrinkage process also requires the choice of the decomposition tree, the mother wavelet, the decomposition levels number and the threshold value (λ) estimation method. As our goal is to evaluate only the thresholding functions performance, we change only these and keep fixed the other wavelet parameters necessary to the signal filtering. We chose to use the FWT structure, due to the ease of its implementation and because it is widely applied in the treatment of PD signals. We use the SNRBWS method to select the mother wavelet and the NWDLS method to find the decomposition levels number. In the threshold value estimative, we chose the *scaledep* method [3, 40, 41].

Since the Fleming function depends on an c inclination constant, which controls how the decomposed wavelet coefficients are eliminated or attenuated, we also compare the results for different values of this constant.

The comparisons were done using statistical parameters as Absolute Mean Error (AME), Mean Square Error (MSE), Root Mean Square Error ($RMSE$), Correlation Coefficient (CC), Normalized Correlation Coefficient (NCC), Energy Difference (EnD), Signal to Noise Ratio (SNR), Signal to Noise Ratio Difference ($DSNR$), Noise Level Reduction (NLR), kurtosis difference (Δk); and local similarity criteria that involve maximum Magnitude Error ($MEmax$), minimum Magnitude Error ($MEmin$), maximum Peak Time Variation ($PTVmax$), minimum Peak Time Variation ($PTVmin$) and Rise Time Variation (RTV). Some of these parameters are used to form a fitness function (J_{Apt}), composed by global similarity criteria (cs_g) and local similarity criteria (cs_l), that can determine the best filtering result. All these criteria were described in [42].

4.1 Investigating the better inclination value c

In a first analysis, it was investigated, through the J_{Apt} fitness criterion, what is the best c inclination value to be used in each alternative of the Fleming thresholding. **Table 1** evinced that $c = 5$ produces the highest amount of best results per threshold function. In **Table 2** both methods produce best results with a lower constant, in case $c = 10$ to Fleming and $c = 5$ to Fleming 2. In this way, it is possible to recommend not to use inclination values higher than 10.

Function	Best J_{Apt} results percentage								
	Inclination constant c								
	5	10	30	50	100	200	300	500	1000
Fleming	**47,09**	25,00	8,96	4,12	3,63	1,50	1,07	1,16	7,46
Fleming 2	**65,16**	11,82	5,52	2,37	1,79	1,55	0,78	1,16	9,84

Table 1.
Best results percentage (by J_{Apt}) comparison between Fleming functions to various inclination constants.

Function	Mean J_{Apt} results								
	Inclination constant c								
	5	**10**	**30**	**50**	**100**	**200**	**300**	**500**	**1000**
Fleming	2,71	**3,12**	1,02	0,09	0,73	0,49	0,07	0,93	0,96
Fleming 2	**1,34**	1,06	0,92	0,89	0,88	0,88	0,88	0,89	0,88

Table 2.
Mean value results (by J_{Apt}) comparison between Fleming functions to various inclination constants.

4.2 Comparison between Fleming, Hard and Soft threshold functions

The main objective of building a dedicated threshold function is to make it able to produce results superior to those of conventional functions. As seen, the most applied functions in wavelet coefficient filtering are Hard and Soft, not only for PD signals, but also for image processing, audio signals, etc.

First, we show in **Figure 8** the results of the comparison between the first proposed alternative using $c = 5$ against Hard and Soft functions. According to the J_{Apt}, we find that the proposed function achieves a higher percentage of better results than the Hard and Soft. As expected, due to its simplicity, the Soft thresholding is the fastest in runtime.

We then compare in **Figure 9** the second alternative proposed with the Hard and Soft functions. Note that there is a significant improvement in the number of better results, achieving superior performance in the *EMA* and *Δk* criteria, which did not occur with the first alternative of our function.

Therefore, is evidenced by the superiority of the proposed alternatives in relation to the amount of better results obtained compared to the usual Hard and Soft methodologies. The only drawback is that our second proposal needs a little more time to be processed, but it is a relatively low price to be paid to achieve better results in reducing noisy components of PD signals. Also, note that, compared with the Soft function, the Hard thresholding tends to provide a better preservation of the PD pulses amplitudes and of the SNR, which confirms the statements made in the literature [3, 22].

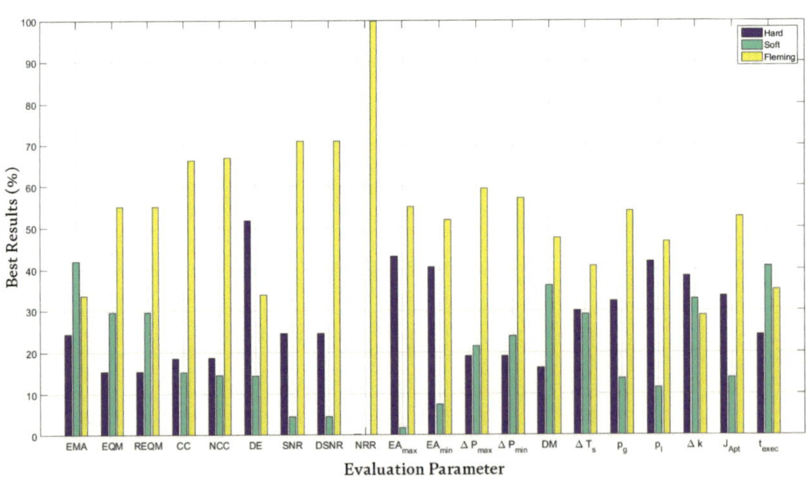

Figure 8.
Comparison of the better denoising results obtained between the Hard, Soft and Fleming threshold functions.

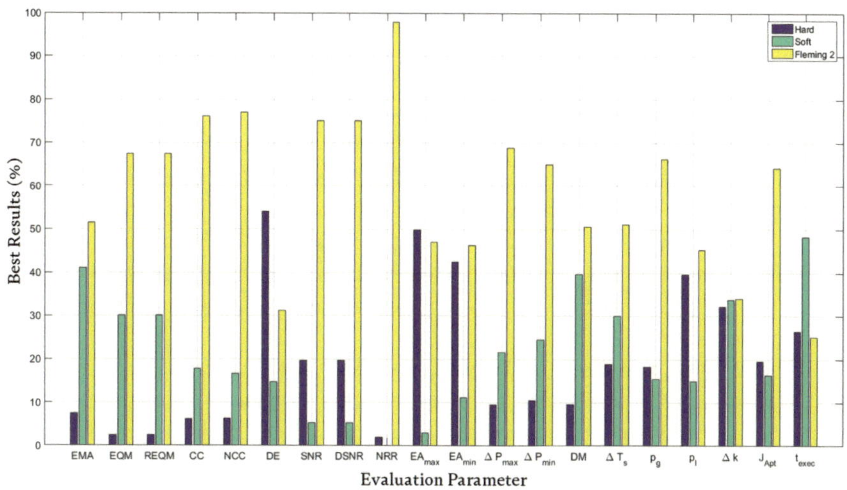

Figure 9.
Comparison of the better denoising results obtained between the Hard, Soft and Fleming 2 threshold functions.

In **Figure 10** is shown a signal consisting of 3 simulated PD pulses wrapped in white noise and in AM noise, which was created as performed in [3]. In addition, note the filtering results for the Hard, Soft, Fleming and Fleming 2 thresholding. The Soft function tends to considerably attenuate the pulses peak amplitudes; the Hard function shows greater preservation of these amplitudes; the Fleming function allows the passage of a little more noise with negligible amplitudes, but it achieves better preservation of the amplitudes than the Hard and Soft, while the Fleming 2 function is able to solve Fleming's problem by identifying the coefficients of greater importance. In this way, the Fleming 2 method presents better amplitudes preservation than the other functions and still manages to eliminate the low amplitude noise seen with the use of the Fleming. The filtering improvement is also indicated by the fitness function, with higher value (J_{Apt} = 16.4607) for the filtering result using the Fleming 2 thresholding.

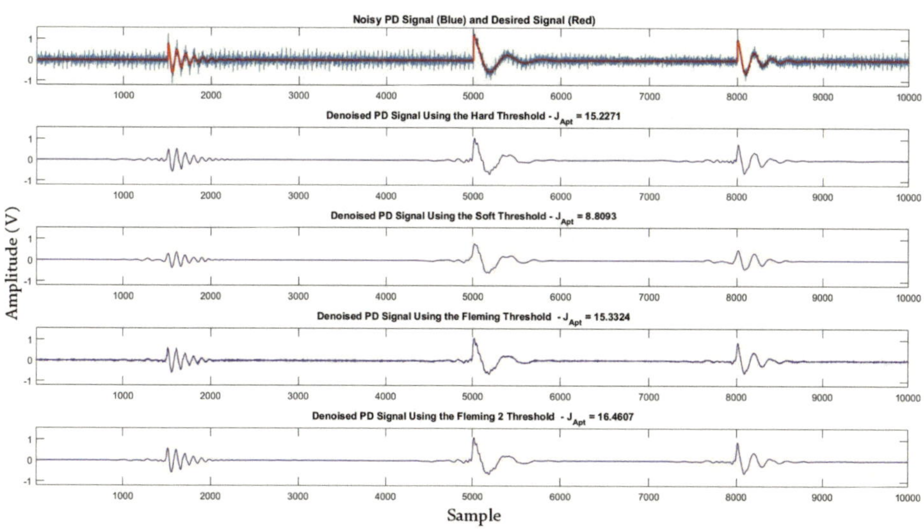

Figure 10.
Comparison of the better denoising results obtained between the Hard, Soft, Fleming and Fleming 2 threshold functions to a simulated PD signal.

4.3 Comparison between all threshold functions

With the results described in the previous subsection, we have a quantitative idea of the used method's capacity, but only with the average results is possible to have a real sense of the quality of each one. Taking advantage of the opportunity, we implement the various wavelet thresholding methods (mentioned in Section 4), including the: Adapt Shrink, Garrote, Hui, Liu, Non Negative Garrote (NNG), Semi Soft (SS), Stein, Zhang 1 (Z1), Zhang 2 (Z2), Zhang 3 (Z3), Zhang 4 (Z4), and Zhang 5 (Z5). The required variables for each of these alternatives were designated according to the specifications provided by the respective authors in the works that describe them.

Similarly to what was done in **Table 1**, we made a percentage evaluation of the amount of best results considering all threshold functions and the proposed Fleming functions (compared for a constant c = 5). From **Tables 3** and **4**, the bold values evidence that the Fleming threshold had a superior performance when compared to the other alternatives. In terms of fitness, the one with the highest amount of better filtering results was Fleming. The Stein function outperforms the others in execution time. The Soft function ends up losing space in practically all the evaluated criteria, confirming that it is not suitable to treat PD signals, due to the high attenuation generated in the wavelet coefficients processing.

Parameter	Threshold function							
	Hard	Soft	Fleming	Fleming 2	Adapt shrink	Garrote	Hui	Liu
AME	1,07	0,39	9,01	**31,49**	5,23	4,80	13,08	6,73
MSE	0,19	0,00	10,22	**33,38**	7,99	9,25	8,43	5,33
RMSE	0,19	0,00	10,22	**33,38**	7,99	9,25	8,43	5,33
CC	0,87	0,00	16,57	**39,87**	3,83	11,97	7,27	2,76
NCC	0,97	0,00	17,59	**40,26**	3,15	11,92	6,98	2,91
EnD	18,51	0,48	18,12	14,73	1,79	8,91	1,89	9,93
SNR	2,37	0,00	21,17	**42,49**	0,15	3,39	1,55	2,23
DSNR	2,37	0,00	21,17	**42,49**	0,15	3,39	1,55	2,23
NLR	0,00	0,00	0,00	0,00	0,00	0,00	0,00	0,00
MEmax	9,40	0,19	**20,78**	17,78	1,07	8,48	0,24	6,10
MEmin	8,19	0,10	18,70	**19,04**	2,28	10,17	1,31	8,28
PTVmax	0,73	0,00	6,15	3,20	2,96	5,43	0,48	2,57
PTVmin	1,16	0,00	7,27	4,36	2,86	5,09	0,58	2,96
DM	0,48	0,00	4,36	4,02	1,79	3,39	2,18	0,78
RTV	2,28	0,05	9,01	8,19	1,79	3,05	3,34	2,71
cs_g	3,25	0,10	20,01	**35,71**	0,78	5,62	4,89	5,96
cs_l	11,09	0,10	19,04	**20,83**	0,68	9,64	1,41	7,66
Δk	9,06	0,68	13,86	**18,75**	4,46	4,89	11,39	6,10
J_{Apt}	3,68	0,00	19,23	**35,03**	0,73	5,18	4,41	7,12
t_{exec}	1,60	3,59	4,22	0,00	18,27	0,15	0,10	4,51

Table 3.
Percentage of best results by evaluation parameters for all threshold functions.

Parameter	Threshold function							
	NNG	**SS**	**Stein**	**Z1**	**Z2**	**Z3**	**Z4**	**Z5**
AME	0,97	1,02	6,35	1,31	4,36	4,12	7,03	3,05
MSE	2,28	1,11	6,83	1,11	5,14	3,05	2,37	3,29
RMSE	1,70	1,11	7,41	1,11	5,14	2,96	2,37	3,39
CC	1,70	0,39	5,14	1,60	1,55	2,03	0,00	4,46
NCC	1,31	0,39	5,28	1,55	1,07	1,89	0,00	4,75
EnD	2,03	0,48	5,43	2,86	3,59	4,12	1,55	5,57
SNR	1,31	0,78	5,33	0,87	0,05	0,97	0,00	17,34
DSNR	1,21	0,78	5,43	0,87	0,05	0,97	0,00	17,34
NLR	0,00	0,00	0,00	26,26	0,00	0,00	0,00	**73,74**
MEmax	1,70	0,15	5,81	9,06	1,36	4,36	0,24	13,28
MEmin	1,11	0,05	11,05	5,18	2,71	5,28	0,15	6,40
PTVmax	0,00	1,16	1,79	6,15	5,33	20,16	2,52	**41,38**
PTVmin	0,00	1,16	1,26	6,06	4,99	18,90	1,31	**42,05**
DM	0,00	1,79	0,68	7,36	2,42	12,74	37,26	**20,74**
RTV	0,00	2,66	3,34	4,65	4,65	14,24	18,94	**21,08**
cs_g	4,02	2,57	7,95	0,24	1,07	3,59	0,00	4,26
cs_l	1,55	1,02	8,62	4,55	1,41	4,89	0,78	6,73
Δk	3,49	3,44	6,59	2,81	2,96	6,06	0,29	5,18
J_{Apt}	4,41	2,33	8,19	0,44	0,87	4,36	0,00	4,02
t_{exec}	0,24	2,37	9,93	18,02	0,00	3,05	0,58	**33,38**

Table 4.
Percentage of best results by evaluation parameters for all threshold functions.

Parameter	Threshold function							
	Hard	**Soft**	**Fleming**	**Fleming 2**	**Adapt shrink**	**Garrote**	**Hui**	**Liu**
AME	0,021	0,020	0,021	0,021	0,021	0,023	0,023	0,019
MSE	0,020	0,017	0,020	0,020	0,017	0,021	0,025	0,016
RMSE	0,029	0,031	0,029	0,029	0,030	0,030	0,038	0,028
CC	0,770	0,736	0,790	0,787	0,761	0,773	0,713	0,775
NCC	0,769	0,733	0,789	0,786	0,759	0,773	0,706	0,773
EnD	0,302	0,525	0,267	0,283	0,472	0,364	0,571	0,294
SNR	6,47	1,22	6,82	6,83	2,57	6,51	0,00	5,40
DSNR	3,05	−2,20	3,40	3,41	−0,85	3,09	−3,42	1,98
NLR	−37,33	−36,57	−37,55	−37,52	−36,86	−38,26	−36,40	−37,16
MEmax	14,13	39,75	12,97	14,77	35,08	13,78	43,78	20,53
MEmin	16,11	35,57	13,20	16,24	30,09	14,03	40,20	17,37
PTVmax	15,71	11,44	11,66	16,28	13,97	13,29	12,45	9,08
PTVmin	23,06	20,14	19,93	23,93	23,45	22,07	23,58	16,49

Parameter	Threshold function							
	Hard	Soft	Fleming	Fleming 2	Adapt shrink	Garrote	Hui	Liu
DM	4536,233	4536,207	4536,231	4536,188	4536,217	4536,216	4536,227	4536,236
RTV	11,31	9,87	12,64	11,64	13,94	15,11	10,78	9,53
cs_g	3,81	−5,27	5,99	4,31	3,88	3,19	3,86	2,69
cs_l	2,92	3,59	2,87	2,97	24,71	22,80	31,21	25,11
Δk	23,67	30,79	20,46	23,78	−4,51	1,08	−10,83	−0,55
J_{Apt}	0,88	−8,86	**3,12**	1,33	0,456	0,456	0,458	0,456
t_{exec}	0,459	0,455	0,48	0,461	−0,63	4,27	−6,97	2,15

Table 5.
Average results by evaluation parameters for all threshold functions.

Parameter	Threshold function							
	NNG	SS	Stein	Z1	Z2	Z3	Z4	Z5
AME	0,019	0,020	0,019	0,022	0,020	0,011	0,036	0,072
MSE	0,016	0,017	0,016	0,017	0,016	0,002	0,172	0,247
RMSE	0,028	0,031	0,028	0,032	0,029	0,017	0,058	0,092
CC	0,762	0,736	0,762	0,624	0,767	0,753	0,122	0,575
NCC	0,759	0,733	0,759	0,623	0,764	0,751	0,117	0,575
EnD	0,357	0,525	0,357	3269	0,477	0,359	0,978	10,527
SNR	4,18	1,23	4,18	2,84	2,52	−78,15	−16,80	4,09
DSNR	0,76	−2,19	0,76	−0,58	−0,90	−81,57	−20,22	0,67
NLR	−36,97	−36,57	−36,97	−43,48	−36,80	45,31	−34,20	−47,92
MEmax	26,41	39,72	26,41	27,26	35,11	98,39	89,02	39,01
MEmin	22,50	35,53	22,50	30,88	30,66	95,69	87,53	61,48
PTVmax	9,92	11,44	9,92	18,99	11,24	10,64	132,89	20,63
PTVmin	17,34	19,80	17,34	23,76	18,87	18,12	174,24	28,26
DM	4536,237	4536,233	4536,237	4536,195	4536,231	4536,246	4536,136	4536,172
RTV	9,57	9,87	9,57	15,23	11,57	9,07	14,38	20,10
cs_g	−0,58	−5,26	−0,58	−6,30	−1,44	−84,39	−35,96	−8,89
cs_l	2,90	3,58	2,90	4,26	3,49	5,86	12,31	5,33
Δk	27,85	30,79	27,85	31,62	26,35	30,61	43,79	36,17
J_{Apt}	−3,48	−8,84	−3,48	−10,56	−4,93	−90,25	−48,27	−14,22
t_{exec}	0,457	0,455	0,455	0,457	0,464	0,456	0,456	0,456

Table 6.
Average results by evaluation parameters for all threshold functions.

Also was evaluated the average results of the evaluation parameters, according to the **Tables 5** and **6**. Note that the fitness J_{Apt} = 3,20 of the Fleming function (using c = 10) is the highest among all the others, being followed by the Garrote and the Hard functions. Thus, the proposed alternatives achieve the objective of

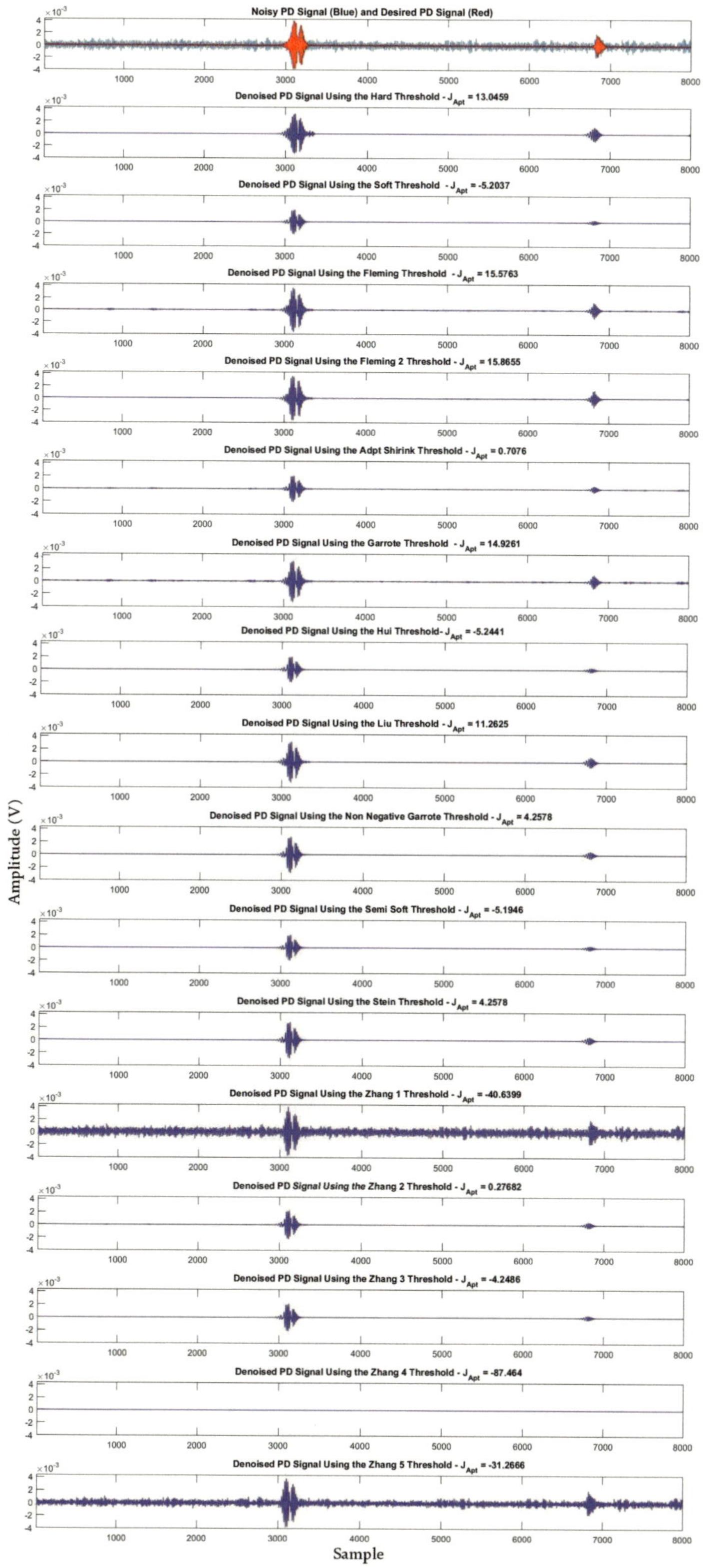

Figure 11.
Comparison of the better denoising results obtained between the all evaluated threshold functions to a measured PD signal from a hydro generator.

overcoming other methods, also providing a better qualitative result in the treatment of PD pulses.

Figure 11 exemplifies the wavelet shrinkage process using each of the thresholding functions discussed above for a PD signal measured from a hydro generator. In this case, note that the functions we have created are superior in preserving the amplitudes of the signals and eliminating the present noise, especially the filtering using the Fleming 2 function, which obtained the highest level of J_{Apt} compared to the other functions, followed by Fleming, Garrote and Hard functions. The Soft and the other thresholding alternatives end up causing deformations of the pulses waveforms and greater attenuation of these peak amplitudes. The Zhang 1 and Zhang 5 functions allow most of the noise to pass through the denoising process and the Zhang 4 function ends up eliminating the PD signal that we are interested in obtaining.

5. Conclusions

Was presented a new threshold function called Fleming, which combines the quality of a strongly differentiable function and a more flexible alternative, enabling its optimization to provide better results in the PD signals treatment, in order to preserve its important characteristics for the diagnosis of the HV equipment subjected to the partial discharge analysis. The proposal inspired by the well-known logistic function [29], which depends on a parameter that controls the inclination of the curve in the threshold value (calculated a priori). Also was created a variant of this same function, using a simple idea, but little investigated in the literature: identifying the decomposed coefficients with the greatest contribution in the desired signal recovering [3].

With the results described in Section 5, in which hundreds of signals (measured and simulated) were evaluated, the ability of the Fleming function and its Fleming 2 variant to overcome the most common functions such as Hard and Soft, as well as twelve other alternatives presented in some publications [15, 19, 22, 35–38]. The Fleming function can be applied with different inclination values, but for PD signals, the ideal is that these values are limited between 5 and 10 to provide the best results.

The Fleming 2 alternative showed the highest percentage of the best results and the Fleming alternative showed the highest average value in terms of amplitude. Thus, if the goal is to achieve a higher number of better results, the indicated is to threshold the wavelet coefficients using the Fleming 2 function, but if the idea is to achieve better average results, consider using the Fleming function. As for the average processing time, these functions are relatively fast when compared to the other evaluated functions, not falling far behind the classic ones Hard and Soft.

The application of the developed thresholding functions is extensible to other types of signals, such as acoustic emissions, electrocardiogram signals, image processing, among others. However, in each case it would be necessary to investigate the appropriate values of the inclination parameter $\boldsymbol{c}$.

Author details

Caio F.F.C. Cunha[1,2]*, Mariane R. Petraglia[1], André T. Carvalho[2]
and Antonio C.S. Lima[1]

1 COPPE/UFRJ, Federal University of Rio de Janeiro, Rio de Janeiro, Brazil

2 CEPEL/ELETROBRAS, Electrical Energy Research Center, Rio de Janeiro, Brazil

*Address all correspondence to: caioflemin@gmail.com

IntechOpen

© 2020 The Author(s). Licensee IntechOpen. This chapter is distributed under the terms of the Creative Commons Attribution License (http://creativecommons.org/licenses/by/3.0), which permits unrestricted use, distribution, and reproduction in any medium, provided the original work is properly cited.

References

[1] Carvalho A T, Lima A C, Cunha C F, Mariane R P. Identification of partial discharges immersed in noise in large hydro-generators based on improved wavelet selection methods. Measurement 2015; 75: 122–133. DOI: 10.1016/j.measurement.2015.07.050

[2] Cunha C F, Carvalho A T, Petraglia M R, Lima A C. A new wavelet selection method for partial discharge denoising. Electric Power Systems Research 2015; 125: 184–195. DOI: 10.1016/j.epsr.2015.04.005.

[3] Gaouda A, El-Hag A, Abdel-Galil T, Salama MM, Bartinikas R. On-line detection and measurement of partial discharge signals in a noisy environment. IEEE Transactions on Dielectrics and Electrical Insulation. 2008;**15**:1162-1173. DOI: 10.1109/TDEI.2008. 4591239

[4] Cunha CF, Carvalho AT, Petraglia MR, Lima AC. An improved scale dependent wavelet selection for data denoising of partial discharge measurement. IEEE International Conference on Solid Dielectrics (ICSD). 30 June-4 July 2013. Bologna. IEEE. 2013:100-104

[5] Hongxia L. Xuefeng Z. A method of second wavelet transform automated threshold for partial discharge signal extraction. In: Second International Conference on Digital Manufacturing and Automation (ICDMA). 5–7 August 2011. Zhangjiajie. IEEE. 2011. pp. 42-45. DOI: 10.1109/ICDMA.2011.19.

[6] Shin I, Soraghan J, Siew W. Detection of PD utilizing digital signal processing methods. Part 3: open-loop noise reduction. Electrical Insulation Magazine. IEEE. 2001;**17**:6-13. DOI: 10.1109/57.901611

[7] Li J, Sun C, Yang J. Adaptive de-noising for PD online monitoring based on wavelet transform. In: Proceedings of the SoutheastCon; 31 March 2006; Memphis: IEEE. 2006. pp. 71-74

[8] Li J, Jiang T, Grzybowski S, et al. Scale dependent wavelet selection for de-noising of partial discharge detection. Dielectrics and Electrical Insulation. IEEE Transactions; 2010; 17, 1705–1714. DOU: 10.1109/TDEI.2010.5658220.

[9] Evagorou D, Kyprianou A, Lewin P, et al. Feature extraction of partial discharge signals using the wavelet packet transform and classification with a probabilistic neural network. IET Science Measurement & Technology. 2010;**4**:177-192. DOI: 10.1049/iet-smt.2009.0023

[10] Zhaoheng D, Shanghe L, Lei W. Selection of the optimal wavelet bases for wavelet de-noising of partial discharge signal. In: 2nd International Conference on Signal Processing Systems (ICSPS); 5–7 July 2010; Dalian. IEEE. 2010. pp. 400-404

[11] Hao L, Lewin P, Hunter J, et al. Discrimination of multiple PD sources using wavelet decomposition and principal component analysis. IEEE Transactions on Dielectrics and Electrical Insulation. 2011; 18: 1702–1711. DOI: 10.1109/TDEI.2011. 6032842.

[12] Xiaorong W, Zongjun G, Yong S, et al. Extraction of partial discharge pulse via wavelet shrinkage. In: Proceedings of the 6th International Conference on Properties and Applications of Dielectric Materials; 21–26 June 2000; Xi'an. IEEE. 2002. pp. 685-688. DOI: 10.1109/ICPADM.2000.876113.

[13] Zhou X, Zhou C, Kemp IJ. An improved methodology for application of wavelet transform to partial discharge measurement denoising. IEEE Transactions on Dielectrics and

Electrical Insulation. 2005;**12**:586-594. DOI: 10.1109/TDEI.2005.1453464

[14] Vishwakarma D K, Kapoor R, Dhiman A, Goyal A, Jamil D. De-noising of audio signal using heavy tailed distribution and comparison of wavelets and thresholding techniques. In: International Conference on Computing for Sustainable Global Development (INDIACom); 11–13 March 2015; New Delhi. IEEE; 2015. p. 755–760.

[15] Tulsani H, Gupta R. 1-D signal denoising using wavelets based optimization of polynomial threshold function. In: 3rd International Conference on Reliability, Infocom Technologies and Optimization (ICRITO) (Trends and Future Directions). 8–10 October 2014; Noida. IEEE. 2014. pp. 1-5

[16] Zhang WQ. Song G X. A translation-invariant wavelet de-noising method based on a new thresholding function. In: International Conference on Machine Learning and Cybernetics. 5 November 2003; Xi'an. IEEE. 2003. pp. 2341-2345

[17] Dong X, Yue Y, Qin X, Wang X, Tao Z. Signal denoising based on improved wavelet packet thresholding function. In: 2010 International Conference on Computer, Mechatronics, Control and Electronic Engineering. 24-26 August 2014; Changchun. IEEE; 2014. p. 382–385.

[18] Guo Q, Yu S, Leng M, Liu W. Signal denoising using simplified SURE based linear expansion of thresholds in the wavelet domain. In: Congress on Image and Signal Processing. 27–30 May 2008; Sanya. IEEE. 2008. pp. 592-596

[19] Liu S, Li Y, Hu X, Liu L, Hao D. A novel thresholding method in removing noises of electrocardiogram based on wavelet transform. Journal of Information & Computational Science. 2013;**10**:5031-5041. DOI: 10.12733/jics20102845

[20] He H, Wang Z, Tan Y. Noise reduction of ECG signals through genetic optimized wavelet threshold filtering. In: IEEE International Conference on Computational Intelligence and Virtual Environments for Measurement Systems and Applications (CIVEMSA). 12–14 June 2015. Shenzhen. IEEE. 2015. pp. 1-6

[21] Tulsani H. Gupta R. Wavelet based optimized polynomial threshold function for ECG signal denoising. In: International Conference on Computing for Sustainable Global Development (INDIACom). 11-13 March 2015; New Delhi. IEEE;**2015**:1563-1566

[22] Nasri M, Nezamabadi-pour H. Image denoising in the wavelet domain using a new adaptive thresholding function. Neurocomputing. 2009;**72**: 1012-1025. DOI: 10.1016/j.neucom.2008.04.016

[23] Jangra S. Kumar S. A new threshold function for image denoising based on wavelet transform. International Journal of Engineering and Mathematical Sciences. 2012;**1**:60-65

[24] Mohideen S, Perumal S, Krishnan N. Selvakumar R. A novel approach for image denoising using dynamic tracking with new threshold technique. In: IEEE International Conference on Computational Intelligence and Computing Research (ICCIC). 28–29 December 2010. Coimbatore. IEEE. 2011. pp. 1-4

[25] Norouzzadeh Y, Rashidi M. Image denoising in wavelet domain using a new thresholding function. In: International Conference on Information Science and Technology. 26–28 March 2011. Nanjing. IEEE. 2011. pp. 721-724

[26] Chang SG, Yu B, Vetterli M. Adaptive wavelet thresholding for image denoising and compression. IEEE Transactions on Image Processing. 2000;**9**:1532-1546. DOI: 10.1109/83.862633

[27] Biswas M, Om H. An adaptive wavelet thresholding image denoising method. In: National Conference on Communications (NCC). 15-17 February 2013. New Delhi. IEEE; 2013. p. 1–5.

[28] Begum AS, Poornachandra S. Curvelet based image de-noising using beta-trim shrinkage for magnetic resonance images. 2014 International Conference on Science Engineering and Management Research (ICSEMR). 27-29 November 2014. Chennai. IEEE. 2014: 1-8

[29] Lumen. Build a logistic model from data [Internet]. Available from: https://courses.lumenlearning.com/ivytech-collegealgebra/chapter/build-a-logistic-model-from-data/.

[30] Ma X, Zhou C, Kemp I. Interpretation of wavelet analysis and its application in partial discharge detections. IEEE Transactions on Dielectrics and Electrical Insulation. 2002;**9**:446-457. DOI: 10.1109/TDEI.2002.1007709.

[31] Donoho DL, Johnstone IM. Adapting to unknown smoothness via wavelet shrinkage. Journal of the American Statistical Association. 1995; **90**:1200-1224. DOI: 10.1080/01621459.1995.10476626

[32] Li J, Sun C, Yang J. Adaptive de-noising for PD online monitoring based on wavelet transform. IEEE Southeast Conference. Memphis. 31 March – 2 April 2006. IEEE; 2006. p. 71–74.

[33] Satish L, Zaengl WS. Artificial neural networks for recognition of 3-d partial discharge patterns. IEEE Transactions on Dielectrics and Electrical Insulation. 1994;**1**:265-275. DOI: 10.1109/94.300259

[34] Li J, Cheng C, Jiang T, Grzybowski S. Wavelet de-noising of partial discharge signals based on genetic adaptive threshold estimation. IEEE Transactions on Dielectrics and Electrical Insulation. 2012;**19**:543-549. DOI: 10.1109/TDEI.2012.6180248

[35] Ray P, Maitra AK, Basuray A. Extract Partial Discharge signal using wavelet for on-line measurement. In: International Conference on Communications and Signal Processing (ICCSP). 3–5 April 2013. Melmaruvathur. IEEE. 2013. pp. 888-892. DOI: 10.1109/iccsp.2013.6577184.

[36] Huimin C, Ruimei Z, Yanli H. Improved threshold denoising method based on wavelet transform. Physics Procedia. 2012;**33**:1354-1359. DOI: 10.1016/j.phpro.2012. 05.222

[37] Zang H, Wang Z, Zheng Y. Analysis of signal de-noising method based on an improved wavelet thresholding. In: 9th International Conference on Electronic Measurement Instruments (ICEMI). 16–19 August 2009. Beijing. IEEE: 2009; p. 1–987–1–990.

[38] Zhang X. Thresholding neural network for adaptive noise reduction. IEEE Transactions on Neural Networks. 2001;**12**:567-584. DOI: 10.1109/72.925559

[39] Celso G, Steven WM, Edward O, James AY. Final state sensitivity: An obstruction to predictability. Physics Letters A. 1983;**99**:415-418. DOI: 10.1016/0375-9601(83)90945-3

[40] Cunha CF. Aplicação da Transformada Wavelet na redução de ruídos em medições de descargas parciais [dissertation]. Rio de Janeiro: Federal University of Rio de Janeiro (UFRJ; 2013

[41] Omari T, Bereksi-Reguig F. An automatic wavelet denoising scheme for heart sounds. International Journal of Wavelets, Multiresolution and Information Processing. 2015; 13:

1550016. DOI: 10.1142/S0219691315500162.

[42] Cunha CF, Carvalho AT, Mariane RP, Hélio PA, Lima AC. Proposal of a novel fitness function for evaluation of wavelet shrinkage parameters on partial discharge denoising. IET Science, Measurement & Technology. 2018;**12**:283-289. DOI: 10.1049/iet-smt.2017.0201

Chapter 15

Use of Daubechies Wavelets in the Representation of Analytical Functions

Paulo César Linhares da Silva

Abstract

This chapter aims to use Daubechies' wavelets as basis functions to generate analytical functions, thus being able to rewrite the Taylor series using these wavelets. This makes it possible to analyze functions with a high degree of complexity, in problems that require a high degree of precision in their solution. Wavelet analysis can be applied to practical problems that require a high degree of precision, for example, in the study and analysis of electromagnetic propagation in optical fibers, solutions of differential equations involving engineering problems, in the transmission of WiFi signals, in the treatment and analysis of biomedical images, detection of oil sources through the study of seismic signals.

Keywords: wavelets, Daubechies, analytical functions, basis functions, Taylor series

1. Introduction

Wavelets [1] were born from the need to generate functions, especially those that present singularities, high gradients, discontinuities both in the time domain and in the frequency domain. Wavelets enable the high-resolution analysis of functions with these characteristics. An example of a problem that occurs when generating functions with a Fourier base is the Gibbs phenomenon. Such a phenomenon occurs because there is no way to represent functions that present discontinuities, even adding more elements in the base that will generate the function. A characteristic of wavelets is that they do not produce such an effect.

Wavelets are widely used in the solution of numerical problems in several areas of knowledge such as image compression, Numerical Harmonic Analysis [2], financial analysis, oil detection, differential Equations [3, 4], biomedical signals, analysis of electromagnetic integral Equations [5], optical fibers [6], among others. Many of these applications use the specific properties of wavelets, such as coefficients that are determined numerically, multi-resolution analysis to decompose a signal, integrals, and derivatives obtained numerically, energy concentrated in its compact and base with orthogonal elements.

2. Short introduction to wavelet theory

For the development of topics presented in this chapter, the reader must have as a prerequisite knowledge of functional analysis, linear algebra, measure theory

IntechOpen

and integration, differential and integral calculus. It is important to note that the wavelet basis is for the wavelet transform as well as the trigonometric basis is for the Fourier transform. Generally, the term wavelet is also used as a wavelet transform. The following subsections present these initial prerequisites to the reader.

2.1 Preliminaries on Hilbert spaces

In this subsection, some mathematical concepts necessary for a better formal understanding of the wavelet tool are defined. The definitions, contained in this section, are due to the author [2].

Definition 2.1 The space $\mathcal{H}$ is said to be a Hilbert space, if an inner product $<,>$, associated with a standard $\| = \sqrt{<,>}$ has been defined in it. And a set of vectors $\{v_i\}$, for $i \in \mathbb{N}$ an orthonormal system is said if the internal product $<v_n, v_m> = \delta_{mn}$, for $m, n \in \mathbb{N}$.

Definition 2.2 A set of vectors $\{v_n\}$ is orthonormal, if and only if, for every finite set of complex numbers x_n, there is $\|\sum_n a_n x_n\|^2 = \sum |a_n|^2$, for $n \in \mathbb{N}$.

Definition 2.3 In Hilbert's $\mathcal{H}$ space, a set of vectors $\{v_n\}$ is said to be a Riez system, if there are constants $0 \leq c \leq C < \infty$ such that for any finite set of complex numbers x_n if you have:

$$c \sum |a_n|^2 \leq \|\sum_n a_n x_x\|^2 \leq C \sum_n |a_n|^2 \tag{1}$$

Definition 2.4 The space $L^2(\mathbb{R})$ is said to be an integrable square function space, that is,

$$L^2(\mathbb{R}) = \left\{ f : \mathbb{R} \mapsto \mathbb{C} : \int_{\mathbb{R}} |f|(x)|^2 dx < \infty \right\} \tag{2}$$

For $f, g \in L^2(\mathbb{R})$, define the inner product $<f, g> = \int_{\mathbb{R}} f(x)\overline{g(x)}dx$. On what, $\overline{g(x)}$ is the complex conjugate of the function $g(x)$.

In particular $\|f\| = \|f\|_2 = \left(\int_{\mathbb{R}} |f(x)|^2 dx\right)^{\frac{1}{2}}$, and f is said to be an integrable square.

Definition 2.5 Let $f : \mathbb{R} \mapsto \mathbb{C}$ be a function. The support of f, denoted by *suppf*, is the closing of the set $\{x \in \mathbb{R} : f(x) \neq 0\}$. A function f is said to have compact support if the *suppf* set is compact.[1]

Definition 2.6 We say that a function f is generated by the basis functions $\{f_1, \ldots f_n\}$, if coefficients exist $\{c_1, \ldots, c_n\}$ such that:

$$f = \sum_{i=1}^{n} f_i c_i \tag{3}$$

The concepts presented here about orthogonality and support of a f function, are fundamental to formalize the definition of wavelet. The following subsection presents the formal mathematical concept of wavelet.

[1] A set is said to be compact if it is limited and closed.

2.2 Definition of wavelet

This subsection aims to define wavelet [2], the main mathematical tool used in the development of this chapter. However, it is necessary to define the expansion and translation of mathematical operations beforehand.

Definition 2.7 *Given $a > 0$, the expansion operator, D_a, defined over a $f(x)$ function in L^1 or L^2 over $\mathbb{R}$, is given by, $D_a f(x) = a^{\frac{1}{2}} f(x)$.*

Definition 2.8 *Given $b \in R$, the translation operator, T_b, defined over a function $f(x)$, in L^1 or L^2 over $\mathbb{R}$, is given by, $T_b f(x) = f(xb)$.*

Thus, using the expansion and translation operations defined above, a family of functions $\psi_{j,k}(x)$ was built: $L^2 \to \mathbb{R}$, base orthogonal to $L^2(\mathbb{R})$.

$$\left\{\psi_{j,k}(x)\right\}_{j,k\in\mathbb{Z}} = \left\{2^{\frac{j}{2}}\psi\left(2^j x - k\right)\right\}_{j,k\in\mathbb{Z}} = \left\{D_{2^j} T_k \psi(x)\right\}_{j,k\in\mathbb{Z}} \tag{4}$$

The Definition 2.9, uses the family of functions $\left\{\psi_{j,k}(x)\right\}_{j,k\in\mathbb{Z}}$, to define the term mathematically wavelet.

Definition 2.9 A function $\psi(x)$ is called wavelet if the collection $\left\{\psi_{j,k}(x)\right\}_{j,k\in\mathbb{Z}}$ is an orthogonal basis on $L^2(\mathbb{R})$. Where j and k are the resolution and translation of wavelet respectively.

By varying the values of j and/or k, it is possible to analyze with greater precision, for example, the behavior of functions that present abrupt changes in values and discontinuity. This type of analysis makes the wavelet a tool as or more efficient than the basic Fourier functions.

The definition 2.10 is another way used to define a wavelet.

Definition 2.10 A wavelet[2] is a short duration wave, which has an average value equal to zero.

Due to the definition 2.10, wavelets resemble Fourier sine and cosine basis functions. Analogously to what is done in the Fourier transform, which has sine and cosine functions as base functions, in wavelet analysis, a function is decomposed into a base of wavelet functions.

The Fourier transform $F(\omega)$ expression of a $f(t)$ function is given by (5):

$$F(\omega) = \int_{-\infty}^{+\infty} f(t)e^{-i\omega t}dt \tag{5}$$

The expression (5) means that the Fourier transform is the sum of every $f(t)$ sign multiplied by a complex exponential, which can be separated into cosine and sinusoidal components in the real and complex parts, respectively.

Similarly, the expression of the wavelet transform $W_{j,k}(f)$ of a function $f(t)$, is given by (6):

$$W_{j,k}(f) = \int_{-\infty}^{+\infty} f(t)\psi_{j,k}(t)dt \tag{6}$$

Similarly, the expression of the wavelet transform (6) is the internal product of the signal to be transformed by a wavelet function.

---

[2] Anglophone term to designate a small wave, in the sense of having a fast duration.

In the following subsection, among the most varied types of wavelets, the Daubechies wavelets are highlighted, which are the basis for the development of this chapter.

2.3 Daubechies wavelet properties

At 1988, a family of compact support wavelets [7] is built by Ingrid Daubechies. This family of wavelets has highly well-located elements. Each member wavelet is governed by a set of N integer coefficients and $k = \{0.1, \dots, N-1\}$ coefficients through scale relations (7) and (8). The a_k and a_{1-k} coefficients, which appear in the (7) and (8), are called filter coefficients and verify the following relations:

$$\phi(x) = \sum_{k=0}^{N-1} a_k \phi(2x-k) \tag{7}$$

$$\psi(x) = \sum_{k=2-N}^{1} (-1)^k a_{1-k} \phi(2x-k) \tag{8}$$

In the **Figures 1** and **2** below, we have the graphical representation of the Daubechies wavelet functions ϕ and ψ of kind 4.

The functions ϕ in (7) and ψ in (8) are called the scale function ϕ and wavelet function ψ, respectively. The fundamental support of the scale function[3] is the interval $[0, N-1]$ as the fundamental support of wavelet function $\psi(x)$ is the interval $\left[1-\frac{N}{2}, \frac{N}{2}\right]$. In the case of $N = 4$, we have the graphs of the **Figures 1** and **2**.

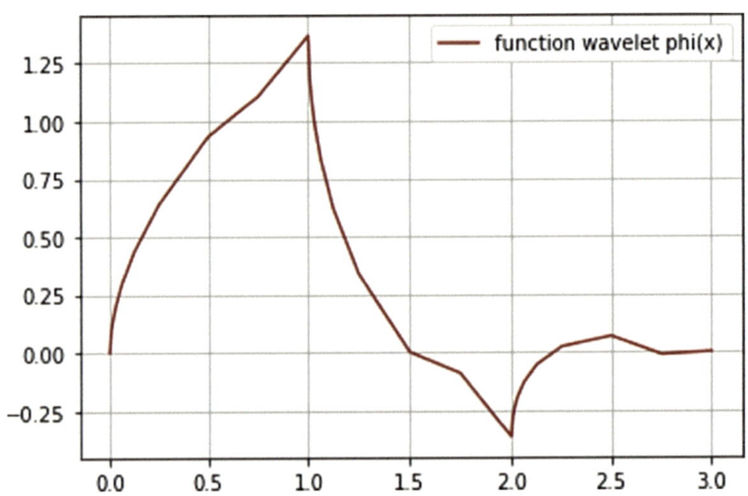

Figure 1.
Daubechies wavelets ϕ. Source: This figure was generated by the author using the python programming language.

[3] We emphasize that the scale function has energy concentrated in its support that is determined by the genus of the wavelet, that is, $supp(\phi) = [0, N-1]$, and that the total energy of the scale function is unitary, that is, $\int_{-\infty}^{+\infty} \phi dx = 1$.

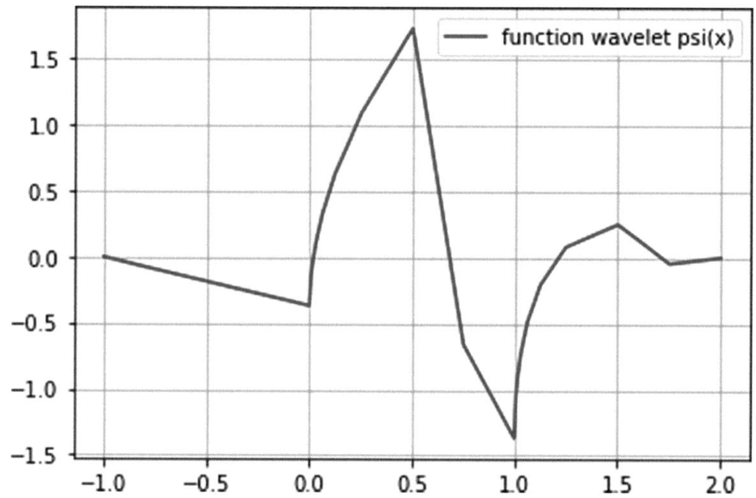

Figure 2.
Daubechies wavelet ψ. Source: This figure was generated by the author using the python programming language.

To determine the filter coefficients a_k and a_{1-k}, which appear in the (7) and (8), we use the relations (9)–(12) below.

$$\sum_{k=0}^{N-1} a_k = 2 \tag{9}$$

$$\sum_{k=0}^{N-1} a_k a_{k-m} = \delta_{0,m} \tag{10}$$

$$\sum_{k=0}^{N-1} (-1)^k a_{1-k} a_{k-2m} = 0 \tag{11}$$

$$\sum_{k=0}^{N-1} (-1)^k k^m a_k = 0, \quad m = 0.1, \ldots, \frac{N}{2} - 1, \tag{12}$$

where $\delta_{k,m}$ is the Kronecker Delta function.

3. Generating an analytical function of the type x^k using wavelets

Analytical functions are those that can be locally around a point x_0 expanded in a Taylor series, according to the following expression.

$$f(x) = \sum_{n=0}^{+\infty} \frac{f^{(n)}(x_0)}{n!} (x - x_0)^n \tag{13}$$

In general according to the author [8], any $f(x)$ function can be represented in terms of a wavelet base, as follows:

$$f(x) = \sum_{m=-\infty}^{+\infty} c_k \phi(x - m) = \sum_{m=-\infty}^{+\infty} c_m \phi_m(x). \tag{14}$$

The c_k coefficients are called moments of the scale functions. In particular, for $f(x) = x^k$, we have the expression (15), below:

$$x^k = \sum_{m=-\infty}^{+\infty} \frac{M_m^k}{2^{jk}} \phi(2^j x - m), \tag{15}$$

Since M_m^k the moment of the wavelet scales concerning the x^k monomial, where k is the degree of the polynomial, m and j are the translation and resolution of the ϕ wavelet. The justification for the construction of the equation is found in the work of [8–10], in which the author concludes that the c_m^j coefficients for approximating a monomial of the x^k form, using a Daubechies wavelet base ϕ, looks like this:

$$c_m^j = \frac{M_m^k}{2^{jk}} \tag{16}$$

The justification used in the approximation (15) of a polynomial function of type $f(x) = x^k$ derives from the number of null moments,

$$\int_{-\infty}^{+\infty} x^k \psi(x)dx = 0, \quad k = 0.1, \ldots, \frac{N}{2} - 1 \tag{17}$$

According to the Eq. (17), the N Daubechies Wavelet has $\frac{N}{2}$ vanish moments, being possible to represent a polynomial of degree at most $\frac{N}{2} - 1$, using the $\phi(x)$ scale function. The polynomial approximation using the scale function is formalized in the following definition.

Definition 3.1 A wavelet has p vanish moments (18), if and only if, the wavelet scale function ϕ can generate polynomials of degree up to $p - 1$ [Eq. (19)]. That is, the scale function alone can be used to represent these polynomials. The fact that it has more null moments means that the scale function can represent more complex functions.

$$\int_{-\infty}^{+\infty} x^m \psi(x)dx = 0; \quad m = 0, 1, \ldots, \frac{N}{2} - 1 \tag{18}$$

$$f(x) = p_1 + p_2 x + \ldots + p_{k-1} x^k, \quad k \leq \frac{N}{2} - 1 \tag{19}$$

In general, a Daubechies wavelet of kind N, properly translated and adjusted to the appropriate resolution level, generates a polynomial of degree k, with the relation between N and k given by $N = 2k + 2$. For example, to generate a polynomial of degree 1 a wavelet of Daubechies of kind 4 is necessary.

To generate a polynomial with $n + 1$ terms, in the function of Daubechies wavelets of genres $4, 6, 8, \ldots, N - 1$, we use the momentum equation and the polynomial expansion as a function of wavelets.

$$p(x) = \sum_{k=0}^{n} a_k x^k, \tag{20}$$

where x^k, takes the form

$$x^k = \sum_{m=-\infty}^{+\infty} \frac{M_m^k}{2^{jk}} \phi(2^j x - m) \tag{21}$$

Substituting the Eq. (21) in (20), we have:

$$p(x) = \sum_{k=0}^{n} \frac{a_k}{2^{jk}} \sum_{m=-\infty}^{+\infty} \phi(2^j x - m) M_m^k \tag{22}$$

where k is the degree of the polynomial j and m are the resolution and translation of the wavelet respectively.

In the next subsection, the calculation of the moment generating function, which appears in the expression 21 as a coefficient of x^k, is shown in detail.

4. Moment generating function

The calculation of the moment generating function according to the author [11] is of fundamental importance to approximate the functions by wavelets. The deduction of the moment-generating function now begins. For this, the mathematical expression is used

$$M_m^k = \int_{-\infty}^{+\infty} x^k \phi(x - m) dx \tag{23}$$

which refers to the moment of the wavelet scale ϕ in relation to the monomial x^k. For $m = k = 0$, in (23), we have:

$$M_0^0 = \int_{-\infty}^{+\infty} \phi(x) dx = 1. \tag{24}$$

Substituting $m = 0$ in the Eq. (23), we have:

$$M_0^k = \int_{-\infty}^{+\infty} x^k \phi(x) dx = \int_{-\infty}^{+\infty} x^k \sum_{s=0}^{N-1} a_s \phi(2x - s) dx \tag{25}$$

$$M_0^k = \sum_{s=0}^{N-1} a_s \int_{-\infty}^{+\infty} x^k \phi(2x - s) dx. \tag{26}$$

Note that the variable s, in the Eq. (26), also represents a translation.
Making the substitution $z = 2x$, $\frac{dz}{dx} = 2$, $dx = \frac{dz}{2}$, we have:

$$M_0^k = \sum_{s=0}^{N-1} a_s \int_{-\infty}^{+\infty} x^k \phi(2x - s) dx \tag{27}$$

$$M_0^k = \frac{1}{2^{k+1}} \sum_{s=0}^{N-1} a_s M_s^k \tag{28}$$

Using the substitution $x - m = t$, $\frac{dx}{dt} = 1$, $dx = dt$, in (23), we have:

$$M_m^k = \int_{-\infty}^{+\infty} x^k \phi(x - m) dx = \sum_{r=0}^{k} \binom{k}{r} m^{k-r} M_0^r \tag{29}$$

Now consider the equations:

$$M_0^k = \frac{1}{2^{k+1}} \sum_{s=0}^{N-1} a_s M_s^k \tag{30}$$

$$M_s^k = \sum_{r=0}^{k} \binom{k}{r} s^{k-r} M_0^r \tag{31}$$

Substituting M_s^k in M_0^k, we have: (note that $m = s$)

$$M_0^k = \frac{1}{2^{k+1}} \sum_{s=0}^{N-1} a_s \sum_{r=0}^{k} \binom{k}{r} s^{k-r} M_0^r \tag{32}$$

Now separate the last term of the sum (32), $(r = k)$, to place the term on the left side of the equation:

$$M_0^k = \frac{1}{2^{k+1}} \sum_{s=0}^{N-1} a_s \sum_{r=0}^{k-1} \binom{k}{r} s^{k-r} M_0^r + \frac{1}{2^{k+1}} \sum_{s=0}^{N-1} a_s \binom{k}{r} s^{k-k} M_0^k \tag{33}$$

Using the fact that $\sum_{s=0}^{N-1} a_s = 2$, we have:

$$M_0^k = \frac{1}{2(2^k - 1)} \sum_{r=0}^{k-1} \binom{k}{r} M_0^r \sum_{s=0}^{N-1} a_s s^{k-r} \tag{34}$$

Thus, the equations are obtained:

$$M_m^k = \sum_{r=0}^{k} \binom{k}{r} m^{k-r} M_0^r \tag{35}$$

$$M_0^k = \frac{1}{2(2^k - 1)} \sum_{r=0}^{k-1} \binom{k}{r} \sum_{s=0}^{N-1} a_s s^{k-r} M_0^r \tag{36}$$

From (35), (34), and (24), we get the moment generating function $M_m^k : \mathcal{W} \to \mathbb{R}$, where $\mathcal{W}$ is wavelet space, m is the translation of the scale function and k is the degree of the polynomial to be approximated.

$$M_m^k = \begin{cases} \dfrac{1}{2(2^k - 1)} \displaystyle\sum_{r=0}^{k-1} \binom{k}{r} \sum_{s=0}^{N-1} a_s s^{k-r} M_0^r, & \text{se} \quad m = 0; k \neq 0 \\ \displaystyle\sum_{r=0}^{k} \binom{k}{r} m^{k-r} M_0^r, & \text{se} \quad m \neq 0; k \neq 0 \\ 1, & \text{se} \quad m = k = 0, \end{cases} \tag{37}$$

The analytical expression for M_m^k was developed during the author's research [11] and to validate the results found, a comparative study was made with other numerical results [12, 13] of the scientific literature.

Similar to what was done with the calculation of the moments for the function ϕ, there is also the calculation of the moments for the function ψ. This is given by integral (38)

$$\int_{-\infty}^{+\infty} x^m \psi(x)dx = 0; \quad m = 0, 1, ..., \frac{N}{2} - 1. \tag{38}$$

The following is an example of the calculation of the moments for the case of Daubechies wavelets of a kind $N = 4$.

Example 4.1 In this example, the Daubechies wavelet of kind 4 is used to generate the analytical polynomial function $f(x) = x$. According to the definition 3.1, the scale function of Daubechies of genus $N = 4$, generates a line (polynomial of degree 1). To represent a 1 monomial with a 4 Daubechies wavelet in the $[0, 1]$ range, the translations $\phi(x), \phi(x+1), \phi(x+2)$, whose supports are $[0.3], [-1.2], [-2.1]$, that is:

$$x = \sum_{m=-2}^{0} \frac{M_m^1}{2^{jk}} \phi(2^j x - m) = \frac{M_{-2}^1}{2^j} \phi(x+2) + \frac{M_{-1}^1}{2^j} \phi(x+1) + \frac{M_0^1}{2^j} \phi(x) \tag{39}$$

The support of the linear combination (39), represented in **Figure 3**, is obtained by the intersection of the supports of the translations of the function $\phi(x)$. This intersection results in the interval $I = [0.1]$. This fact defines well the function to be integrated in the I range. In **Figure 3**, the number of translations of the function $\phi(x)$ to generate $f(x) = x$ is illustrated.

Figure 4 shows the graph of translated functions $\phi(x), \phi(x+1)$ and $\phi(x+2)$ respectively, that form a base to generate the function $f(x) = x$.

The calculation using the moment generating function depends on the Daubechies wavelet coefficients of kind 4. These coefficients are obtained by the Eqs. (9)–(12), which gives rise to the following non-linear system.

$$\begin{cases} a_0 + a_1 + a_2 + a_3 = 2 \\ a_0^2 + a_1^2 + a_2^2 + a_3^2 = 2 \\ a_0 a_2 + a_1 a_3 = 0 \\ -a_1 + 2a_2 - 3a_3 = 0 \end{cases} \tag{40}$$

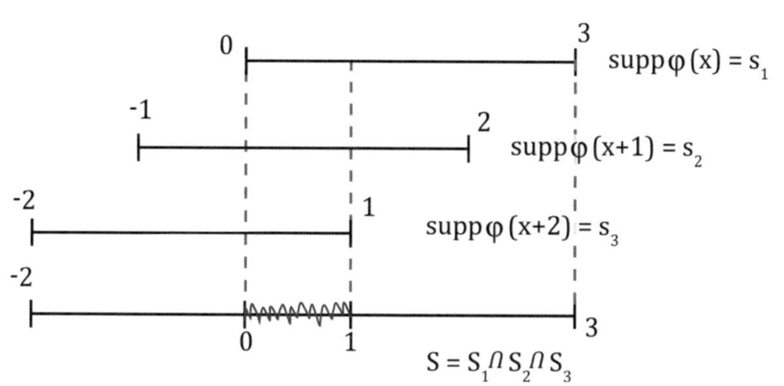

Figure 3.
Translations required to represent the analytical function $f(x) = x$ using Daubechies wavelets of kind 4. Source: Own authorship.

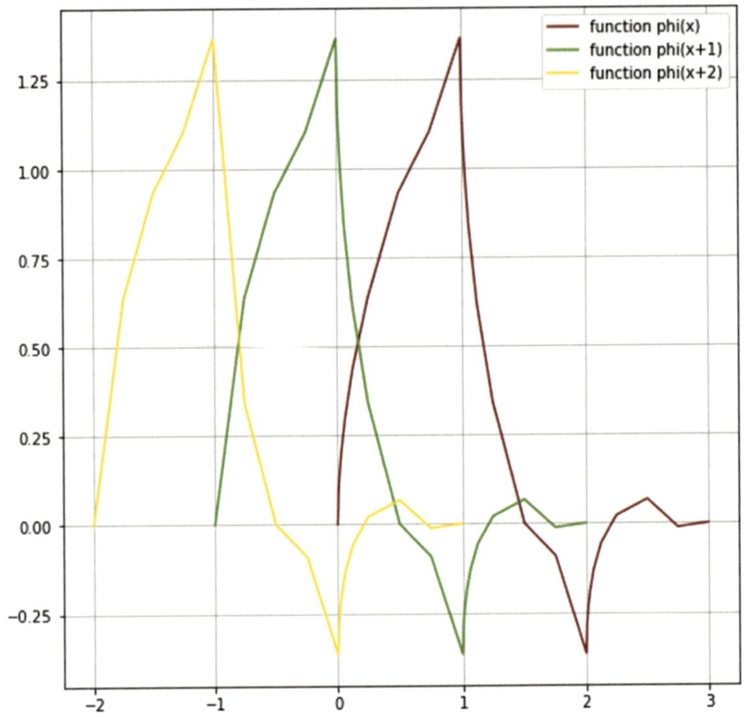

Figure 4.
Translations required to represent the analytical function $f(x) = x$ using Daubechies wavelets of kind 4. Source: This figure was generated by the author using the python programming language.

The solution of this system is the irrational numbers a_0, a_1, a_2, a_3, given by:

$$\begin{cases} a_0 = 0,683012701892219 \\ a_1 = 1.183012701892219 \\ a_2 = 0,316987298107781 \\ a_3 = -0,183012701892219 \end{cases} \tag{41}$$

Using the moment generating function for the case where $m = 0$, we have:

$$\begin{aligned} M_0^1 &= \frac{1}{2}\sum_{r=0}^{0}\binom{1}{r}M_0^r\sum_{s=0}^{3}a_s s^{1-r} = \frac{1}{2}M_0^0(a_1 + 2a_2 + 3a_3) = \frac{a_1 + 2a_2 + 3a_3}{2} \\ &= 0,633974600 \end{aligned} \tag{42}$$

Proceeding with the calculations, we obtain:

$$M_m^k = \sum_{r=0}^{1}\binom{1}{r}m^{1-r}M_0^r = m + M_0^1 = m + \frac{a_1 + 2a_2 + 3a_3}{2} = m + 0,633974600 \tag{43}$$

Replacing the value of m by $m = -1$, $m = -2$ and $k = 1$, we obtain:

$$M_{-1}^1 = -0,366025400 \tag{44}$$

$$M_{-2}^1 = -1,366025400 \tag{45}$$

So, the representation for the x polynomial (for a resolution $j = 0$) is:

$$x = 0,634\phi(x) - 0,366\phi(x+1) - 1.366\phi(x+2) \tag{46}$$

In **Figure 5**, we have the graphical representation of the function obtained of the expression (46). Here the function $f(x) = x$ is generated by linear combination of wavelets $\phi(x), \phi(x+1)$ and $\phi(x+2)$.

The representation for the expression (46) using the summation is given by,

$$x = \sum_{m_1=-2}^{0} \frac{M_{m_1}^1}{2^{jk}} \phi(2^j x - m_1) \tag{47}$$

The expression for writing polynomials of degrees $k = 2$ and $k = 3$ in terms of Daubechies wavelets is given by

$$x^2 = \sum_{m_2=-4}^{2} \frac{M_{m_2}^2}{2^{jk}} \phi(2^j x - m_2) \tag{48}$$

$$x^3 = \sum_{m_3=-6}^{0} \frac{M_{m_3}^3}{2^{jk}} \phi(2^j x - m_3) \tag{49}$$

See that to generate the polynomials (48), (49) is necessary to use Daubechies wavelets of kind 6 and 8, according with the definition 3.1.

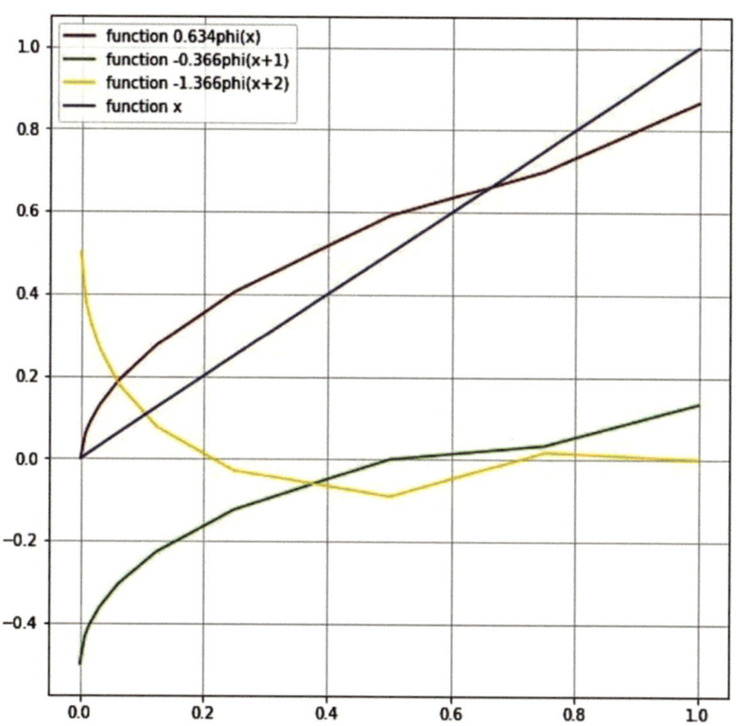

Figure 5.
Function $f(x) = x$ using Daubechies wavelets of kind 4. Source: This figure was generated by the author using the python programming language.

4.1 Taylor polynomial using Daubechies wavelets

The Taylor polynomial or Taylor series is an expression that allows the calculation of the local value of a function f using your derivatives. For this, the function f must be of class C infinite (represented by C^{∞}) which implies that the f is infinitely derivable in an interval containing a point x_0. The expression for the Taylor polynomial for the function f is as follows,

$$f(x) = \sum_{k=0}^{+\infty} \frac{f^{(k)}(x_0)}{k!}(x - x_0)^k \tag{50}$$

The expression (50) developed around $x_0 = 0$ is:

$$f(x) = \sum_{k=0}^{+\infty} \frac{f^{(k)}(0)}{k!}(x)^k \tag{51}$$

Making use of the expression (21), we have:

$$f(x) = \sum_{k=0}^{+\infty} \sum_{m=-\infty}^{+\infty} \frac{f^{(k)}(0)}{k!} \frac{M_m^k}{2^{jk}} \phi(2^j x - m) \tag{52}$$

The expression (52) is another way of writing Taylor's polynomial using Daubechies Wavelets.

Example 4.2 Consider the analytical function $f(x) = e^x$, using Daubechies wavelet of kind a $N = 4$ is possible to write this function f in terms of this wavelet. For this, Taylor's series development around the point $x_0 = 0$ of this function is given by:

$$e^x = \sum_{n=0}^{+\infty} \frac{x^n}{n!} \tag{53}$$

Using only two summation terms in the expression (53), we have:

$$e^x \approx \sum_{n=0}^{1} \frac{x^n}{n!} = 1 + x \tag{54}$$

Using the expression (46), we have:

$$e^x \approx 1 + x = 1 + 0,634\phi(x) - 0,366\phi(x+1) - 1.366\phi(x+2) \tag{55}$$

The expression (55) allows us to approximate the exponential function using a base of Daubechies wavelets. This type of approximation, although simple for this case, is very useful in the case of representation for functions other types.

In the following example, the expression (46) is used to approximate Taylor's series developments for the functions $s(x) = e^x, f(x) = \cosh(x), g(x) = \sinh(x)$ and $h(x) = \ln(1+x)$.

Example 4.3 For the functions $f(x) = \cosh(x), g(x) = \cos(x)$ and $h(x) = \sec(x)$. Taylor's series development of these functions around the point $x_0 = 0$ is:

$$\cosh(x) = \sum_{n=0}^{+\infty} \frac{x^{2n}}{(2n)!} \tag{56}$$

Function	Value in $x = 1$, Taylor Series[4]	Value in $x = 1$, Daubechies Wavelets[5]	Error %
$s(x) = e^x$	2.716666667	2.716735469443329	0.0025%
$f(x) = \cosh(x)$	1.543088161791753	1.543058311287478	0.0019%
$g(x) = \sinh(x)$	1.1750199840127897	1.1750591521108822	0.00376%
$h(x) = \ln(1+x)$	0.6456349203122008	0.6456349190214307	$1,9.10^{-7}$%

[4]*Calculation using Taylor Series.*
[5]*Calculation using Daubechies wavelets of kind 4.*

Table 1.
Comparison of the values obtained by the Taylor series and by Daubechies wavelets.

$$\sinh(x) = \sum_{n=0}^{+\infty} \frac{x^{2n+1}}{(2n+1)!} \tag{57}$$

$$\ln(1+x) = \sum_{n=1}^{+\infty} \frac{(-1)^{n+1} x^n}{n} \tag{58}$$

In order to verify the potentiality of the application of Daubechies wavelets we will calculate the value of the functions in (53), (56), (57) and (58) evaluated at point x = 1. Considering only 7 terms in each summation. For obtain the results using Daubechies wavelets we apply the expression (55) in each summation (53), (56), (57) and (58). In the **Table 1** we have a comparison between the calculation of the values of the functions $s(x) = e^x, f(x) = \cosh(x), g(x) = \sinh(x)$ and $h(x) = \ln(1+x)$ evaluated at point $x = 1$, using the Taylor series and the Daubechies wavelets of kind 4.

Table 1 appears here only as a way of showing the quality of the approximations using the Daubechies wavelets of kind 4. Obviously if we want more precise values, we must use Daubechies wavelets of the kind greater than 4. This will cause changes in the resolution and translation of each wavelet, but the result will be even better.

5. Conclusions

Daubechies wavelets are quite versatile mathematical tools. They can be used to analyze, generate, decompose a function, or even a signal that is represented by an analytical function. This type of application is widely used, for example, in electrical engineering in studies of magnetic fields and electric fields. The theory exposed in this chapter provides tools to carry out these studies. The use of the Taylor series as a way of approximating analytical functions is a very used technique in applied mathematics. Making use of the Taylor series with wavelets is another option to perform an approximation of analytical functions. In future work, we are researching other wavelets, for example Deslauries-Dubuc interpolets, that have an even better approach quality. As Deslauriers-Dubuc interpolets and others in research.

Acknowledgements

The author would like to thank UFERSA for support during my doctoral studies.

Author details

Paulo César Linhares da Silva
UFERSA, Mossoró-RN, Brazil

*Address all correspondence to: linhares@ufersa.edu.br

IntechOpen

© 2020 The Author(s). Licensee IntechOpen. This chapter is distributed under the terms of the Creative Commons Attribution License (http://creativecommons.org/licenses/by/3.0), which permits unrestricted use, distribution, and reproduction in any medium, provided the original work is properly cited.

References

[1] Daubechies I. Recent results in wavelets applications. Journal of Electronic Imaging, 1998, 7, 719–724.

[2] Walnut D. An Introduction to Wavelet Analysis. Applied and Numerical Harmonic Analysis. Birkha user Boston, 2002.

[3] Bertoluzza S. A wavelet collocation method for the numerical solution of partial differential equations. Applied and Computational Harmonic Analysis,3, 1–9, 1996.

[4] Choudhury A. Wavelet method for numerical solution of parabolic equations. Journal of Computational Engineering, 2014, 2014, 1–12, 2014. h ttps://doi.org/10.1155/2014/346731

[5] Robert L, Weng C. A study of wavelets for the solution of electromagnetic integral equations. IEEE Transactions on antennas and propagation, 1995, 43, 802–810.

[6] Silva P, Melo R, Silva J. Optical Fiber Coupler Analysis Using Daubechies Wavelets. Journal of Microwaves, Optoelectronics and Electromagnetic Applications (JMOe), 2020, 19(3), AoP 294-300. https://doi.org/10.1590/2179-10742020v19i3825

[7] Ingrid Daubechies. Ten lectures on wavelets. 1992, Society for Industrial and Applied Mathematics, USA.

[8] Burgos R. Análise de Estruturas Utilizando Wavelets de Daubechies e Interpolets de Deslauriers-Dubuc. PhD thesis, Pontifícia Universidade Católica, PUC, Setembro 2009.

[9] Burgos R. Finite elements based on deslauriers-dubuc wavelets for wave propagation problems, Applied Mathematics, 2016, 7, pp. 1490–1497.

[10] Burgos R. Solution of 1d and 2d poisson's equation by using wavelet scaling functions. Thermal Engineering, 2016, 15, pp. 68–75.

[11] Silva P, Silva J, Garcia A. Daubechies wavelets as basis functions for the vectorial beam propagation method, Journal of Electromagnetic Waves and Applications, 2019, 33:8, 1027-1041, DOI: 10.1080/09205071.2019.1587319

[12] Gopinath R and Burrus C. On the moments of the scaling function psi. Departament of Electrical and Computer Engineering-IEEE, 1992, 963–966,

[13] Butzer P, Fischer A, Ruckforth K. Scaling functions and wavelets with vanishing moments. Computers Math. Applic, 1994, 27, 33–39.

Chapter 16

Higher Order Haar Wavelet Method for Solving Differential Equations

Jüri Majak, Mart Ratas, Kristo Karjust and Boris Shvartsman

Abstract

The study is focused on the development, adaption and evaluation of the higher order Haar wavelet method (HOHWM) for solving differential equations. Accuracy and computational complexity are two measurable key characteristics of any numerical method. The HOHWM introduced recently by authors as an improvement of the widely used Haar wavelet method (HWM) has shown excellent accuracy and convergence results in the case of all model problems studied. The practical value of the proposed HOHWM approach is that it allows reduction of the computational cost by several magnitudes as compared to HWM, depending on the mesh and the method parameter values used.

Keywords: higher order Haar wavelet method, convergence analysis, accuracy estimates, improvement of widely used Haar wavelet method

1. Introduction

Wavelets are most commonly used in signal processing applications to denoise the real signal, to cut a signal into different frequency components, to analyze the components with a resolution matched to its scale, also in image compression, earthquake prediction and other algorithms.

However, the current study is focused on the area where the use of wavelet methods shows a growth trend, i.e., in the solution of differential equations. Many different wavelets based methods have been introduced for solving differential and integro-differential equations. The Legendre wavelets are utilized to solve fractional differential equations in [1–4] and integro-differential equations in [5, 6]. In [7, 8], the Daubechies wavelet based approximation algorithms are derived to solve ordinary and partial differential equations. In [9], the Lucas wavelets are combined with Legendre–Gauss quadrature for solving fractional Fredholm–Volterra integro-differential equations. The series solution of partial differential equations through separation of variables is developed by using the Fourier wavelets in [10]. The Riesz wavelets- based method for solving singular fractional integro-differential equations was developed in [11]. In the studies in [12], the Galerkin method was combined with the quadratic spline wavelets for solving Fredholm linear integral equations and second-order integro-differential equations. The Chebyshev wavelets method for partial differential equations with boundary conditions of the telegraph type is examined in [13].

The simplest of all wavelet-based approaches was introduced by Alfred Haar already in 1910 [14]. The Haar wavelet-based approach for solving differential and integro-differential equations was introduced in 1997 [15, 16]. Based on the Haar wavelet method (HWM), Chen and Hsiao in [15, 16] proposed an approach where the higher order derivative involved in the differential or integro-differential equations is expanded into the series of Haar wavelets. This approach is based on the nature of Haar functions. Due to the piece-wise constant nature of the Haar functions they are not differentiable but are integrable. In [15, 16], the problems of the lumped and distributed parameter system and those of linear time delayed systems were solved. The Chen and Hsiao approach-based HWM was adapted successfully for solving a wide class of differential, integro-differential and integral equations [17–45]. Pioneering work in the development of Haar wavelet-based techniques was conducted by Lepik [17–23], covering ordinary and partial differential equations [17, 19, 21], integro-differential equations [18], integral equations [20], and fractional integral equations [22]. The HWM approaches and their applications are summarized in a monograph [23]. The HWM is adapted for the analysis of nonlinear integral and integro-differential equations in [24–27], covering one- and multi-dimensional problems. Solid mechanics, particularly composite structures, are examined using the HWM in [28–33]. These studies cover free vibration analysis of orthotropic plates [28], functionally graded composite structures [30–32], delamination detection in composite beams [29], and other structures.

Some recent trends in the development and application of the HWM can be outlined as solutions of fractional differential and integro-diffrential equations [34–38] as well as the development of a non-uniform and adaptive grid. In the case of fractional differential or integro-differential equations, two principally different HWM approaches regarding to wavelet expansion are available in the literature. The aim of the first approach is to expand the highest order fractional derivative included in the differential equation directly into Haar wavelets, i.e., direct conversion of the Chen and Hsiao approach for fractional differential equations. In [34–38], the Haar wavelet operational matrix of fractional order integration is introduced and implemented for solving differential and integro- differential equations. The aim of the second approach is to utilize the definitions of fractional derivatives (Caputo derivative, etc.) and convert fractional differential terms into integrals, which contain integer derivatives only. Such an approach has been introduced by Lepik in [22] and utilized in a number of papers [39–41]. The two approaches considered are implemented and compared in [42]. It is pointed out in [42] that the two approaches have the same rate of convergence if the order of the fractional derivative exceeds one ($\alpha > 1$). However, if the order of the fractional derivative is less than one ($\alpha < 1$), the second approach has the rate of convergence equal to two, but the rate of convergence of the first approach is $1 + \alpha$, i.e., less than two. Thus, in the case of $\alpha < 1$, the second approach has a higher convergence rate and can be preferred.

HWM with a nonuniform grid was introduced in [43] using a proportionally changing grid size. The same approach was utilized for the free vibration analysis of non-uniform axially graded beams in [44] and for solving singularly perturbed differential difference equations of neuronal variability in [45].

In most of the studies [17–45], it was concluded that HWM is simple to implement. In the review paper [46], it was pointed out that the HWM is efficient and powerful in solving a wide class of linear and nonlinear reaction–diffusion equations. However, the convergence theorem and accuracy estimates derived for the HWM in [47, 48] state that the order of convergence of the Chen and Hsiao approach-based HWM is equal to two. The latter result is rather modest in the context of engineering. Comparison of the HWM with widely used numerical

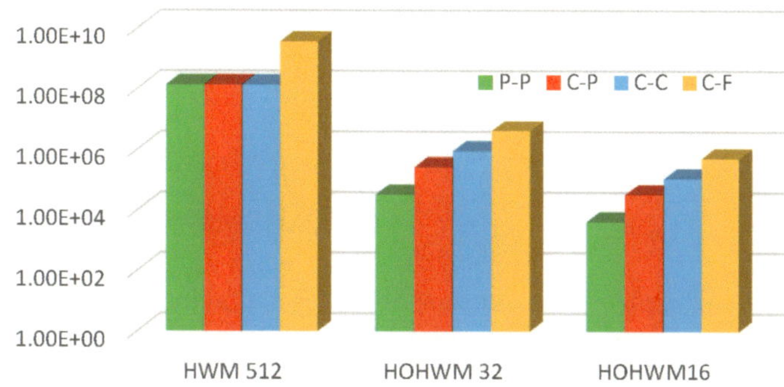

Figure 1.
Numerical complexity. Free vibration analysis of the nanobeam.

methods in engineering reveals that HWM needs principal improvement in order to compete with the differential quadrature method (DQM) [49].

The HOHWM as an improvement of HWM was recently introduced by Majak et al. in [50]. The convergence rate of the method was improved from 2 to 2 + 2 s, where s stands for the method parameter. This new method is currently underused, but the first results obtained have shown that a principal growth of the accuracy can be achieved with a minimum growth of complexity [51–53]. In [52], the free vibrations analysis of the Euler-Bernoulli nanobeam was performed. **Figure 1** shows the numerical complexity estimates of the HWM and HOHWM solutions yielding a similar absolute error. Here the numerical complexity is determined by number of main operations of the most complex subtask - solution of discrete algebraic system of equations [52].

The logarithmic scale is used in **Figure 1** since the complexity of the HWM appears several orders higher (10^8) than that of the HOHWM (10^3 ... 10^5). These results were obtained using the method parameter s = 1 (i.e., fourth order convergence). In practice, one of most important factors is the computational cost. In the case of the considered problem, the computational cost of the HOHWM solution is 10^3 ... 10^5 times lower than that of the HWM. The obtained results hold good in the case of all four boundary conditions considered: pinned-pinned (P–P), clamped-pinned (C-P), clamped-clamped (C-C), and clamped-free (C-F).

2. Theoretical basis of the HWM and the HOHWM

This section introduces the Haar functions and presents the theoretical basis of both, the HWM and the HOHWM, covering basic principles, algorithms, convergence and accuracy issues.

2.1 Haar functions

The HWM and the HOHWM use different approaches, but both use Haar function expansions for the approximation of derivatives. The Haar functions $h_i(x)$, are given as in [14].

$$h_i(x) = \begin{cases} 1 & for \quad x \in [\xi_1(i), \xi_2(i)) \\ -1 & for \quad x \in [\xi_2(i), \xi_3(i)) \\ 0 & elsewhere \end{cases} \tag{1}$$

where $i = m + k + 1$, $m = 2^j$ is a maximum number of square waves deployed in the interval $[A, B]$ and the parameter k indicates the location of the particular square wave,

$$\xi_1(i) = A + 2k\mu\Delta x, \xi_2(i) = A + (2k+1)\mu\Delta x, \xi_3(i) = A + 2(k+1)\mu\Delta x, \quad (2)$$

$$\mu = M/m, \Delta x = (B - A)/(2M), M = 2^J. \quad (3)$$

In Eq. (3) $M = 2^J$ stands for maximum resolution. Obviously, the Haar functions $h_i(x)$ form an orthonormal basis. The integrals of order n of the Haar functions (1) can be expressed as in [17]

$$p_{n,i}(x) = \begin{cases} 0 & for \quad x \in [A, \xi_1(i)) \\ \frac{(x - \xi_1(i))^n}{n!} & for \quad x \in [\xi_1(i), \xi_2(i)) \\ \frac{(x - \xi_1(i))^n - 2(x - \xi_2(i))^n}{n!} & for \quad x \in [\xi_2(i), \xi_3(i)) . \\ \frac{(x - \xi_1(i))^n - 2(x - \xi_2(i))^n + (x - \xi_3(i))^n}{n!} & for \quad x \in [\xi_3(i), B) \\ 0 & elsewhere \end{cases} \quad (4)$$

Formulas (4) hold for a general case where $i > 1$, in the case $i = 1$ holds $p_{n,1} = \frac{(x-A)^n}{n!}$. The interval $[A, B]$ in Eqs. (1)–(3) can be converted to the unit interval $[0, 1]$ by use of the exchange of variables $\tau = (x - A)/(B - A)$.

In the case of uniform mesh the collocation points can be introduced as.

$$x_l = \frac{2l - 1}{4M}, l = 1, \ldots, 2M. \quad (5)$$

The elements of the discrete $2M \bullet 2M$ Haar matrix can be expressed as values of Haar functions in collocation points given by Eq. (5)

$$H_{il}^{2M} = h_i(x_l). \quad (6)$$

The elements of the matrix of n–th order integrals of the Haar function can be evaluated as

$$(P_n)_{il}^{2M} = p_{n,i}(x_l), \quad (7)$$

where $p_{n,i}(x)$ is defined by formulas (4).

2.2 Haar wavelet method (HWM)

The Chen and Hsiao approach based HWM, utilized in [15–49]. can be considered as a commonly/widely used HWM.

2.2.1 Method description

Let us consider first the n-th order ordinary differential equation in the general form as.

$$G\left(x, u, u', u'', \dots u^{(n-1)}, u^{(n)}\right) = 0. \tag{8}$$

Let us assume that $f(x)$ is an integrable square and finite function. The Haar wavelet expansion for the function $f(x)$ is given as (here a_i stand for the unknown wavelet coefficients).

$$f(x) = \sum_{i=1}^{\infty} a_i h_i(x). \tag{9}$$

According to the HWM approach introduced by Chen and Hsiao in [15, 16] the highest order derivative involved in Eq. (8) is expanded into the series of Haar wavelets, i.e.,

$$\frac{d^n u(x)}{dx^n} = \sum_{i=1}^{\infty} a_i h_i(x), \tag{10}$$

Based on the definition of the Haar function (1)–(3), Eq. (10) can be expressed as.

$$\frac{d^n u(x)}{dx^n} = a_1 h_1 + \sum_{j=0}^{\infty} \sum_{k=0}^{2^j-1} a_{2^j+k+1} h_{2^j+k+1}(x). \tag{11}$$

The solution of the differential Eq. (8) can be obtained by integrating Eq. (11) n-times as.

$$u(x) = \frac{a_1 x^n}{n!} + \sum_{j=0}^{\infty} \sum_{k=0}^{2^j-1} a_{2^j+k+1} p_{n,2^j+k+1}(x) + B_T(x). \tag{12}$$

In Eq. (12) $p_{n,2^j+k+1}(x)$ stand for n -th order integrals of the Haar functions given by Eq. (4) and $B_T(x)$is a boundary term. Obviously, in the numerical analysis the finite number of the terms corresponding to the fixed maximum resolution $(N = 2M)$ can be considered as.

$$u(x) = \frac{a_1 x^n}{n!} + \sum_{j=0}^{2M} \sum_{k=0}^{2^j-1} a_{2^j+k+1} p_{n,2^j+k+1}(x) + B_T(x). \tag{13}$$

The integration constants included in the boundary term $B_T(x)$ can be determined from the boundary conditions. Substituting the solution (13) and its derivatives in the differential Eq. (8) and employing discrete collocation points (5), we obtain $2M \cdot 2M$ algebraic system of equations for determining unknown wavelet coefficients a_i . Finally, when the wavelet coefficients a_i are known, the solution of the differential Eq. (8) can be evaluated using expression (13). Note that the collocation points defined by Eq. (5) correspond to uniform mesh. Obviously, various non-uniform meshes can be utilized instead of Eq. (5).

Let us consider next a partial differential equation in the general form as

$$R\left(x, y, \frac{\partial u}{\partial x}, \frac{\partial u}{\partial y}, \frac{\partial^2 u}{\partial x^2}, \frac{\partial^2 u}{\partial y^2}, \frac{\partial^2 u}{\partial x \partial y}, \dots, \frac{\partial^{n+q} u}{\partial x^n \partial y^q}\right) = 0, \tag{14}$$

where n and p stand for the highest order derivatives with respect to x and y, respectively. The solution domain is considered rectangle $[0, L_1]x[0, L_2]$. According to the Chen and Hsiao approach the highest order derivative involved in the differential equation is expanded into Haar wavelets (in this case of 2D expansion)

$$\frac{\partial^{n+q} u}{\partial x^n \partial y^q} = \sum_{i=1}^{2M_1} \sum_{l=1}^{2M_2} a_{il} h_i(x) h_l(y). \tag{15}$$

In Eq. (15) $2M_1$ and $2M_2$stand for the number of grid points with respect to x and y coordinates, respectively. Such a 2D wavelet expansion was introduced by Lepik in [19] and is most commonly used. Similar to the 1D case, integrating the relation (15) n-times with respect to x and q times with respect to y, we obtain the solution of the differential Eq. (14) as [19].

$$u(x,y) = \sum_{i=1}^{2M_1} \sum_{l=1}^{2M_2} a_{il} p_{n,i}(x) p_{q,l}(y) + B_T(x,y). \tag{16}$$

The boundary term $B_T(x,y)$ includes $n+q$ integration constants which can be determined from the boundary conditions. Substituting Eq. (16) in the differential Eq. (14) and satisfying the obtained equation at the collocation points (e.g., in uniform grid points), we obtain an algebraic system of rank $(2M_1)^2 \bullet (2M_2)^2$ with respect to the wavelet coefficients. By substituting the wavelet coefficients in Eq. (16), the solution of the differential Eq. (14) can be evaluated at any point in the given domain.

2.2.2 Convergence theorem and accuracy estimates

The convergence theorem for the Chen and Hsiao based HWM was proved by Majak et al. in [47].

Theorem. Let us assume that$f(x) = \frac{d^n u(x)}{dx^n} \in L^2(R)$ is a continuous function on$[0,1]$ and its first derivative is bounded.

$$\forall x \in [0,1] \ \exists \eta : \ \left|\frac{df(x)}{dx}\right| \le \eta, n \ge 2 \text{ (boundary value problems)}. \tag{17}$$

Then, the Haar wavelet method based on the approach proposed by Chen and Hsiao in [15, 16] will be convergent, i.e., the L^2 -norm of the error function $|E_M|$ vanishes as J goes to infinity. The order of convergence is equal to two.

$$\|E_M\|_2 = O\left[\left(\frac{1}{2^{J+1}}\right)^2\right]. \tag{18}$$

The proof is given in [47]. The error bound is derived as.

$$\|E_M\|_2 \le \frac{\eta C_n}{6}\left(\frac{1}{2^{J+1}}\right)^2 = \frac{4}{9}\frac{\eta}{(floor(n/2)!)^2}\left(\frac{1}{2^{J+1}}\right)^2. \tag{19}$$

In the particular case where $n = 1$, the error bound can be derived as.

$$\|E_M\|_2 \le \frac{\eta}{\sqrt{3}}\left(\frac{1}{2^{J+1}}\right)^2. \tag{20}$$

The error bounds (19) and (20) are main/biggest error terms determining the rate of convergence. A detailed accuracy analysis of the HWM for the fourth order ordinary differential equations is performed in [48], where two error terms are pointed out as.

$$\|E_M\|_2 \leq \frac{\eta}{12}\left[\left(\frac{1}{N}\right)^2 + \frac{1}{28}\left(\frac{1}{N}\right)^4\right]. \tag{21}$$

It appears that the second error term is the fourth order term, which does not play any role in the standard HWM application. However, this information is important in cases where extrapolation is employed for obtained solutions. For example, by applying the Richardson extrapolation, the first error term is canceled and the order of convergence increases from two to four (the value three is omitted since the third order term in error estimate is missing). Furthermore, it has been shown in [48] that the error estimate includes even order terms only. This aspect can be considered in further improvement of the HWM.

Obviously, at the same assumptions, the multi-dimensional Haar wavelet method is also convergent and the rate of convergence is equal to two.

The obtained results will be validated by a number of case studies by computing the numerical rates of convergence and comparing the obtained and theoretical results. These results are confirmed in [47–53] and in other papers.

2.3 Higher order Haar wavelet method

As mentioned above, the HOHWM was introduced in [50] as an improvement of the widely used Chen and Hsiao approach based HWM. The HOHWM is based on:

- higher order wavelet expansion,
- algorithm for determining complementary integration constants.

It can be pointed out that utilizing the higher order wavelet expansion itself does not provide substantial increase of the rate of convergence and accuracy. The algorithm used for determining complementary integration constants plays key role.

2.3.1 Method description

Let us consider first the n-th order ordinary differential Eq. (8).

According to the the Haar wavelet expansion is expressed as.

$$\frac{d^{n+2s}u(x)}{dx^{n+2s}} = \sum_{i=1}^{\infty} a_i h_i(x), s = 1, 2, \ldots \tag{22}$$

In the simplest case, where$s = 1$, the $n + 2$ order derivative is expanded into Haar wavelets. The even values 2 s are used, based on the analysis of error estimates of the HWM given in the previous section. Integrating the expression (22) $n + 2$ times with respect to x we obtain the solution of the differential Eq. (8) as.

$$u(x) = \frac{a_1 x^{n+2s}}{(n+2s)!} + \sum_{j=0}^{\infty}\sum_{k=0}^{2^j-1} a_{2^j+k+1} p_{n+2s,2^j+k+1}(x) + S_{BT}(x) + H_{BT}(x). \tag{23}$$

The boundary terms $S_{BT}(x)$ and $H_{BT}(x)$ include $n + 2s$ integration constants c_r.

$$S_{BT}(x) = \sum_{r=0}^{n-1} c_r \frac{x^r}{r!}, H_{BT}(x) = \sum_{r=n}^{n+2s-1} c_r \frac{x^r}{r!}. \tag{24}$$

The integration constants $c_0, c_1, \dots, c_{n-1}$ can be determined from the boundary conditions. To determine the remaining 2 s integration constants, the following two algorithms are proposed by authors in [50].

Using selected uniform grid points (nearest to the boundary from both sides).

$$x_i = \frac{i}{N}, x_i = 1 - \frac{i}{N}, i = 0, \dots, s - 1. \tag{25}$$

Using selected Chebyshev–Gauss–Lobatto grid points (nearest to the boundary from both sides).

$$x_i = \frac{1}{2}\left[1 - \cos\frac{(i-1)\pi}{(N-1)}\right], i = 1, \dots, s, i = N - s + 1, \dots, N. \tag{26}$$

In the particular case $s = 1$ the differential Eq. (8) is satisfied in the boundary points.

$$G\left(0, u(0), u'(0), u''(0), \dots u^{(n-1)}(0), u^{(n)}(0)\right) = 0, G\left(1, u(1), u'(1), u''(1), \dots u^{(n-1)}(1), u^{(n)}(1)\right) = 0. \tag{27}$$

Obviously in the latter case the two algorithms considered above, coincide.

3. Numerical convergence analysis and Richardson extrapolation

The derivations of the numerical estimates of the order of convergence, as well as extrapolation formulas can be found in [54] and are omitted herein for the sake of conciseness. Let us denote the numerical solutions on a sequence of nested grids by F_{i-2}, F_{i-1}, F_i, corresponding to grid sizes $h_{i-2}/h_{i-1} = h_{i-1}/h_i = 2$. Then the order of convergence of the numerical method can be estimated by the formula.

$$k_i^E = \log\left(\frac{F_{i-1} - F_{exact}}{F_i - F_{exact}}\right) / \log(2), \tag{28}$$

if the exact solution F_{exact} is known. If the exact solution is unknown, the following formula can be employed [54].

$$k_i = \log\left(\frac{F_{i-2} - F_{i-1}}{F_{i-1} - F_i}\right) / \log(2). \tag{29}$$

The accuracy of the results can be improved by employing the Richardson extrapolation formula as [54].

$$R_i = F_i + \frac{F_i - F_{i-1}}{2^k - 1}. \tag{30}$$

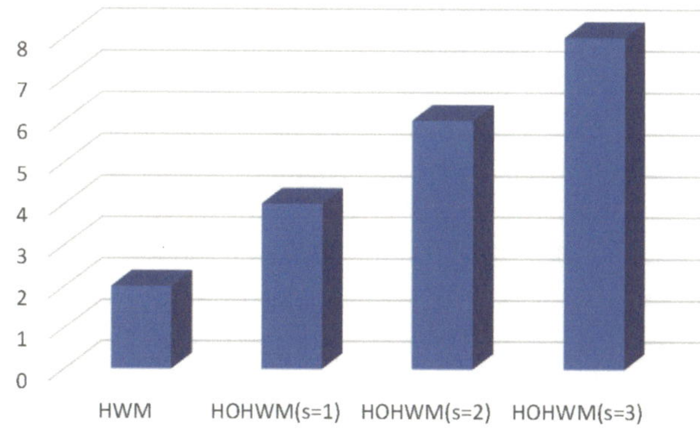

Figure 2.
Numerical rates of convergence for the HWM and the HOHWM.

The accuracy of the extrapolated results R_i can be estimated by applying formulas (28) or (29) to these results. The numerical rates of convergence computed for case study, described in Section 5.1, are depicted in **Figure 2**.

The numerical rates of convergence determined are in agreement with convergence theorem for the HWM (Section 2.2.2) and relation given for the HOHWM in Section 2.3 (the rate of convergence of the HOHWM is equal to 2 + 2 s).

4. Complexity analysis

As pointed out above, the accuracy and numerical/time complexity are two key characteristics for any numerical method, algorithm. The computing time is often used as a measure of complexity of algorithms using particular software. More general approach applied commonly in algorithm theory is to estimate the number of basic operations required by each algorithm. The latter approach is independent of software used and does not even require execution of the algorithms. For this reason, in the current study the numerical complexity of the algorithms is estimated based on the number of basic operations.

According to the HWM and the HOHWM algorithms the solution of the differential equation is obtained from the solution of the discrete algebraic system of equations and certain additional operations for composing the linear system and evaluation of the solution in given points. The mentioned additional operations are similar for both methods and have lower asymptotic complexity than the solution of the algebraic system of equations. Thus, the numerical complexity of the HWM and the HOHWM can be compared based on number of basic operations needed for solving algebraic system of equations determined by the rank of the algebraic system of equations (systems are similar by structure).

In the case of the same number of collocations points N, the ranks of the algebraic systems corresponding to the HWM and the HOHWM are equal to N and N + 2 s (here s = 1, 2 or 3), respectively. Furthermore, in the cases where the 2 s complementary integrations constants are determined analytically, the rank of the algebraic system of equations of the HOHWM reduces to N. Thus, in the case of the same mesh used the numerical complexity of the HWM and HOHWM is similar (or equal depending on implementation). However, these solutions have principally different accuracy (see **Tables 1-5**) and such comparison is rather theoretical.

In practice, it is important to compare methods, providing the same accuracy. In the following the given accuracy is fixed by absolute error less than 2.0e-10 and the complexities of the HWM and the HOHWM are compared in **Figure 3**. The

	HWM			Extrapolated results		
N	Solution at point t = 0.5	Absolute error	Converg. rate	Solution at point t = 0.5	Absolute error	Converg. rate
4	0.60256316864	1.72E-03				
8	0.60386098486	4.27E-04	2.0150	0.60429211947	4.49E-06	
16	0.60418124220	1.06E-04	2.0037	0.60428798114	3.56E-07	3.6598
32	0.60426104700	2.66E-05	2.0009	0.60428765000	2.44E-08	3.8645
64	0.60428098202	6.64E-06	2.0002	0.60428762718	1.59E-09	3.9385
128	0.60428596477	1.66E-06	2.0001	0.60428762569	1.02E-10	3.9680
256	0.60428721039	4.15E-07	2.0000	0.60428762560	6.40E-12	3.9901

Table 1.
HWM results and extrapolated results (Richardon extrapolation).

	HOHWM(s = 1)			HOHWM (s = 2, VPA)		
N	Solution at point t = 0.5	Absolute error	Converg. rate	Solution at point t = 0.5	Absolute error	Converg. rate
4	0.60426829567	1.93E-05		0.60428745306474	1.73E-07	
8	0.60428616352	1.46E-06	3.7247	0.60428762225323	3.34E-09	5.6915
16	0.60428752673	9.89E-08	3.8865	0.60428762553096	6.06E-11	5.7827
32	0.60428761918	6.41E-09	3.9477	0.60428762559057	1.03E-12	5.8780
64	0.60428762518	4.08E-10	3.9748	0.60428762559158	1.68E-14	5.9367
128	0.60428762557	2.56E-11	3.9876	0.60428762559160	2.69E-16	5.9679
256	0.60428762559	1.60E-12	4.0076	0.60428762559160	4.25E-18	5.9853

Table 2.
HOHWM ($s = 1$) and HOHWM ($s = 2$, VPA).

	HOHWM (s = 3,VPA)		
N	Solution at point t = 0.5	Absolute error	Converg. rate
4	0.604287625766393	1.75E-10	
8	0.604287625565219	2.64E-11	2.7282
16	0.604287625591526	7.19E-14	8.5195
32	0.604287625591597	2.18E-16	8.3631
64	0.604287625591598	7.32E-19	8.2201
128	0.604287625591598	2.69E-21	8.0890
256	0.604287625591598	9.17E-23	4.8738

Table 3.
HOHWM ($s = 3$, VPA).

logarithmic scale is used in **Figure 3**, since the complexities of the HWM and the HOHWM differ by several magnitudes.

In the case of the HWM the absolute error 2.0e-10 was reached by use of 16,384 collocation points (corresponding algebraic system has 16,384 equations). In the case of the HOHWM the same accuracy was achieved by use just 64, 16 or 4

	HWM			HOHWM (s = 1)		
N	Solution at point x = 0.5	Absolute error	Converg. rate	Solution at point x = 0.5	Absolute error	Converg. rate
4	5.5271847185	9.63E-02		5.4468387805	1.59E-02	
8	5.4504966936	1.96E-02	2.2961	5.4315095153	6.20E-04	4.6851
16	5.4355789218	4.69E-03	2.0639	5.4309280380	3.85E-05	4.0087
32	5.4320493805	1.16E-03	2.0155	5.4308919203	2.40E-06	4.0053
64	5.4311787173	2.89E-04	2.0038	5.4308896716	1.50E-07	4.0014
128	5.4309617728	7.23E-05	2.0010	5.4308895312	9.36E-09	4.0003
256	5.4309075816	1.81E-05	2.0002	5.4308895225	5.85E-10	4.0001
512	5.4308940366	4.51E-06	2.0001	5.4308895219	3.66E-11	3.9999
1024	5.4308906505	1.13E-06	2.0000	5.4308895219	2.29E-12	3.9968

Table 4.
Comparison of the HWM and the HOHWM (s = 1).

	HWM			HOHWM (s = 1)		
N	Solution at point x = 0.5	Absolute error	Converg. rate	Solution at point x = 0.5	Absolute error	Converg. rate
4	7.9429919221	2.31E-01		7.7081052185	4.12E-03	
8	7.7555236804	4.33E-02	2.4139	7.7125990532	3.78E-04	3.4430
16	7.7224700620	1.02E-02	2.0789	7.7122525729	3.19E-05	3.5669
32	7.7147496939	2.53E-03	2.0189	7.7122227612	2.12E-06	3.9125
64	7.7128508612	6.30E-04	2.0047	7.7122207750	1.34E-07	3.9795
128	7.7123780687	1.57E-04	2.0012	7.7122206490	8.43E-09	3.9949
256	7.7122599897	3.93E-05	2.0003	7.7122206411	5.27E-10	3.9986
512	7.7122304774	9.84E-06	2.0001	7.7122206406	3.30E-11	3.9974
1024	7.7122230998	2.46E-06	2.0000	7.7122206406	2.13E-12	3.9573

Table 5.
Comparison of the HWM and the HOHWM (s = 1).

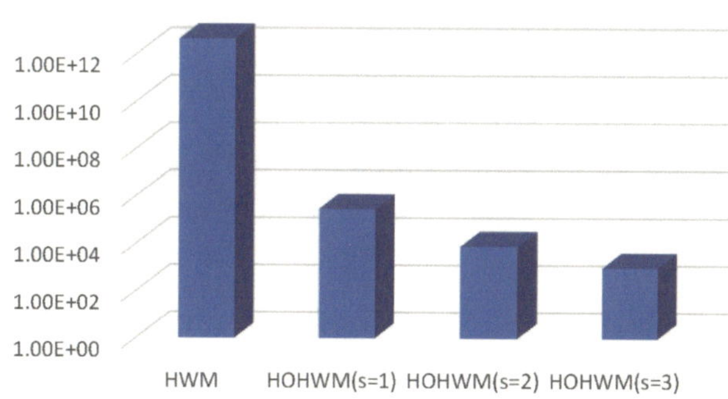

Figure 3.
Numerical complexities of the HWM and the HOHWM.

collocations points corresponding to the s = 1, s = 2 or s = 3, respectively (i.e. the algebraic system needed to solve is reduced to 64, 16 or 4 equations). Thus, it can be concluded, that making use of the HOHWM instead of the HWM will lead to principal reduction of numerical complexity of the solution.

The practical value of the developed HOHWM approach is reduction of computational cost of the solution by several magnitudes (directly determined by numerical complexity). It should be noted that making use of the HOHWM instead of the HWM, especially in the cases $s > 1$ will increase implementation complexity, but not substantially.

5. Case studies

In the following, the two case studies are performed in order to validate the accuracy and convergence of the recently introduced HOHWM and compare results with HWM.

5.1 Linear ordinary differential equations

As a rule, the new methods are validated on the samples where the exact solution is known. Herein, the linear ordinary differential equations are considered as the first sample problem. Let us consider a sample problem solved in [17] by applying the HWM

$$G\left(t, u(t), \frac{du(t)}{dt}, \frac{d^2u(t)}{dt^2}\right) = \frac{d^2u(t)}{dt^2} + \alpha\frac{du(t)}{dt} + \beta u(t) - \gamma f(t) = 0, \tag{31}$$

where $f(t)$ is a given function $(f(t) = \cos(2t))$, p, q and $r$are constant parameters $(\alpha = 0.05, \beta = 0.15, \gamma = 1)$. The initial conditions $u(0) = 0$, $\frac{du}{dt}(0) = 1$ are utilized.

In the case of the HWM, the second order derivative is expanded into Haar wavelets as

$$\frac{d^2u(t)}{dt^2} = aH. \tag{32}$$

In Eq. (32), a and H stand for the coefficient vector (row vector) and the discrete Haar matrix given by formulas (6), respectively. The solution of the differential Eq. (31) is obtained by integrating relation (32) twice with respect to t and satisfying initial conditions

$$u(t) = aP_2 + t, \tag{33}$$

where the elements of the matrix P_2 are defined by (7) and the coefficient vector a is determined by substituting the solution (33) and its derivatives in Eq. (31) as

$$a = (\gamma f(t) - \alpha - \beta t)[H + \alpha P_1 + \beta P_2]^{-1}. \tag{34}$$

In the case of the HOHWM and s = 1, the fourth order derivative is expanded into Haar wavelets as

$$\frac{d^4u(t)}{dt^4} = aH. \tag{35}$$

The solution of the differential Eq. (31) is obtained by integrating relation (35) four times with respect to t and satisfying initial conditions

$$u(t) = aP_4 + c_3\frac{t^3}{6} + c_2\frac{t^2}{2} + t, \tag{36}$$

The remaining two integration constants in Eq. (36) can be determined by satisfying Eq. (31) at the boundary points $t = 0$ and $t = 1$. The latter two algebraic equations can be added to the algebraic system obtained by substituting the solution (36) and its derivatives in Eq. (31). In the latter case, the algebraic system includes $2M + 2$ equations. An alternate approach is to determine the remaining two integration constants analytically from the same conditions and replace to algebraic system.

The numerical results obtained by utilizing the HWM and the HOHWM are compared in **Tables 1-3**.

It can be observed from **Tables 1-3** that in the case of the HWM, the order of convergence tends to two and in the case of the HOHWM, it tends to 2 + 2 s, i.e., to four if s = 1, to six if s = 2 and to eight if s = 3. Use of HOHWM provides a principal increase of accuracy. The maximum accuracy obtained by the use of the HWM at 256 collocation points (N = 256) has been achieved by using the HOHWM at 16 collocation points if s = 1, and at 4 collocation points if s = 2. In the case of the HOHWM and s = 3, the accuracy achieved at 4 collocation points was significantly higher than that of the HWM with 256 collocation points.

In **Figure 4** are shown the error ratios for different mesh (N = 4,16,64 and 256). The absolute error of the HWM is divided by error of the HOHWM, where blue, green and gray colors correspond to the HOHWM parameter s values 1,2 and 3, respectively. Thus, in the case of mesh N = 4, making use of the HOHWM instead of the HWM reduced the absolute error 8.91E+01 (s = 1) to 9.83E+06 (s = 3) times. In the case of mesh N = 256, the use of the HOHWM reduced the absolute error 2.59E +05 (s = 1) to 4.53E+15 (s = 3) times. Since the error ratio depends strongly on the mesh used, the logarithmic scale was used in **Figure 4**.

The numerical analysis is performed using MATLAB software. Since the accuracy achieved by the use of the HOHWM in the case of s = 2 and s = 3 exceeds the limits of the double precision computing, the variable precision computing (VPA) was used.

Note that this is needed only in the case of particular problems and large mesh where the accuracy exceeds the limits of double precision computing.

5.2 Nonlinear Lienard equations

The nonlinear differential equation given as

$$\frac{d^2u(x)}{dx^2} + f(u)\frac{du(x)}{dx} + g(u) = 0 \tag{37}$$

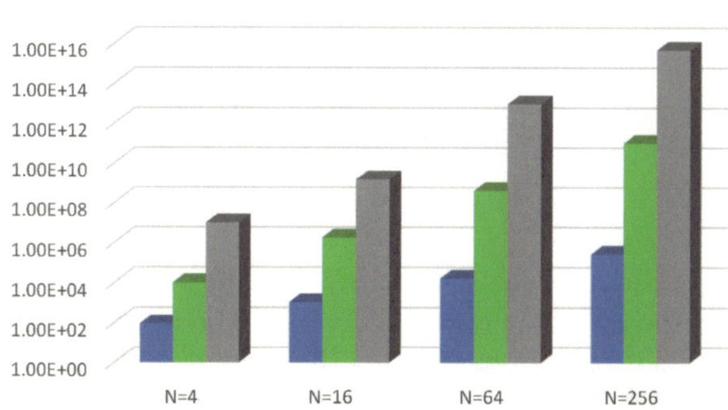

Figure 4.
Ratios of the absolute error of the HWM and the HOHWM.

is known as the Lienard equation. In the following, it is assumed that $f(u) = 0$ and $g(u) = 0$. Let us consider solution in the interval $[0, 1]$ and assume the boundary conditions in the form.

$$u(0) = u_0, u(1) = u_1. \tag{38}$$

In the current study the nonlinear differential Eq. (37) is linearized by applying the quasi-linearization technique as [55].

$$\frac{d^2u_{r+1}(x)}{dx^2} + \frac{du_{r+1}(x)}{dx}u_r(x) + \frac{du_r(x)}{dx}u_{r+1}(x) - \frac{du_r(x)}{dx}u_r(x) = 0. \tag{39}$$

Obviously, Eq. (39) can be solved iteratively with respect to r.

In the case of the HWM, the second order derivative is expanded into Haar wavelets

$$\frac{d^2u_{r+1}(x)}{dx^2} = aH \tag{40}$$

and the solution of the Lienard Eq. (37) can be derived as

$$u(x) = a_{r+1}P_2 - a_{r+1}xP_2(1) + 2xb_r. \tag{41}$$

In the case of the HOHWM and $s = 1$ the fourth order derivative is expanded into Haar wavelets

$$\frac{d^4u_{r+1}(x)}{dx^4} = aH \tag{42}$$

and the solution of the Lienard Eq. (37) can be derived as

$$u(x) = a_{r+1}P_4 - a_{r+1}((b_ry - z)P_4(1) - b_ryP_3(1) - yP_2(1)) + b_rz - {b_r}^2y,$$
$$y = \frac{x^3 - 3x}{6 + 2b_r}, z = \frac{3x}{3 + b_r}. \tag{43}$$

In Eqs. (41) and (43), the value of the parameter b_r depends on the particular boundary conditions applied. Let us consider first the following boundary conditions.

$$u(0) = 0, u(1) = 6\tanh(3) \tag{44}$$

then the exact solution is $u(x) = 6\tanh(3x)$ and $b_r = 6\tanh(3)$. The results obtained by the use of the HWM and the HOHWM are compared in **Table 4** (point x = 0.5 is used).

Next let us consider the following boundary conditions.

$$u(0) = 0, u(1) = 8\tanh(4) \tag{45}$$

In the latter case, the exact solution is $u(x) = 8\tanh(4x)$ and $b_r = 8\tanh(4)$. The results obtained by the use of the HWM and the HOHWM are compared in **Table 5**.

It can be observed from **Tables 4, 5** that the rates of convergence of the HWM and the HOHWM (with s = 1) tend to two and four, respectively. The accuracy obtained using the HWM with maximum resolution 2 M = 1024 is achieved in the case of the HOHWM with only 32 collocation points.

6. Conclusions

The HOHWM introduced recently by authors as an improvement of the HWM in order to compete with the numerical methods widely used in engineering. It was shown that using the HOHWM instead of the HWM will improve principally the accuracy of the solution and increase the rate of convergence in the case of all problems studied. It was found that the rate of convergence of the HOHWM depends on the model parameter s and is equal to 2 + 2 s.

From a practical point of view, it is important that the HOHWM can achieve the same accuracy as the HWM with significantly lower mesh and reduced computational cost.

In the simplest case of the HOHWM where s = 1, the order of the convergence of the HOHWM is equal to four. The user can select suitable s value depending on the accuracy requirements of a particular problem considered.

In future study, the new method proposed can be extended/adapted for solving a wide class of differential and integro-differential equations, including fractional differential equations, multidimensional problems, nonlinear boundary value problems arising in engineering design.

Acknowledgements

The study was supported by Estonian Centre of Excellence in Zero Energy and Resource Efficient Smart Buildings and Districts, ZEBE, TK146 funded by the European Regional Development Fund (grant 2014–2020.4.01.15–0016).

Author details

Jüri Majak[1*], Mart Ratas[1], Kristo Karjust[1] and Boris Shvartsman[2]

1 Tallinn University of Technology, Tallinn, Estonia

2 Estonian Entrepreneurship University of Applied Sciences, Tallinn, Estonia

*Address all correspondence to: juri.majak@taltech.ee

IntechOpen

© 2020 The Author(s). Licensee IntechOpen. This chapter is distributed under the terms of the Creative Commons Attribution License (http://creativecommons.org/licenses/by/3.0), which permits unrestricted use, distribution, and reproduction in any medium, provided the original work is properly cited.

References

[1] Mohammadi F, Cattani C, A generalized fractional order Legendre wavelet Tau method for solving fractional differential equations. Journal of Computational and Applied Mathematics, 2018;339:306–316.

[2] Yuttanan B, Razzaghi M, Legendre wavelets approach for numerical solutions of distributed order fractional differential equations. Applied Mathematical Modelling, 2019;70:350–364.

[3] Rayal A, Verma SR, Numerical analysis of pantograph differential equation of the stretched type associated with fractal-fractional derivatives via fractional order Legendre wavelets. Chaos, Solitons and Fractals, 2020;139: 110076.

[4] Jafari H, Yousefi SA, Firoozjaee MA, Momani S, Khalique CM, Application of Legendre wavelets for solving fractional differential equations. Computers & Mathematics with Applications, 2011; 62:1038–1045.

[5] Venkatesh SG, Ayyaswamy SK, Raja Balachandar S, Legendre approximation solution for a class of higher-order Volterra integro-differential equations. Ain Shams Engineering Journal, 2012;3:417–422.

[6] Setia A, Prakash B, Vatsala A, Numerical solution of fourth order fractional integro-differential equation by using Legendre wavelets. Neural, Parallel and Scientific Computations, 2015;23(2):377–386.

[7] Mishra V, and Jindal S, Wavelet Galerkin Solutions of Ordinary Differential Equations. Int. Journal of Mathematical Analysis, 2011;5(9): 407–424.

[8] Shiralashettia SC, Angadib LM, Desh AB, Daubechies Wavelet Based Multigrid and Full Approximation Scheme for the Numerical Solution of Parabolic Partial Differential Equations. International Journal of Modern Mathematical Sciences, 2018;16(1):58–75.

[9] Dehestani H, Ordokhani Y, Razzaghi M, Combination of Lucas wavelets with Legendre–Gauss quadrature for fractional Fredholm–Volterra integro-differential equations. Journal of Computational and Applied Mathematics, 2021; 382:113070.

[10] Sokhal S, Verma SR, A Fourier wavelet series solution of partial differential equation through the separation of variables method. Applied Mathematics and Computation, 2020; 388:125480.

[11] Mohammad M, Trounev A, Implicit Riesz wavelets based-method for solving singular fractional integro-differential equations with applications to hematopoietic stem cell modeling. Chaos, Solitons & Fractals, 2020;138: 109991.

[12] Černá D, Finěk V, Galerkin method with new quadratic spline wavelets for integraal and integro-differential equations. Journal of Computational and Applied Mathematics, 2020;363: 426–443.

[13] Heydari MH, Hooshmandasl MR, Ghaini FMM, A new approach of the Chebyshev wavelets method for partial differential equations with boundary conditions of the telegraph type. Applied Mathematical Modelling, 2014; 38(5–6): 1597–1606.

[14] Haar A, Zur Theorie der orthogonalen Funktionen systeme. Mathematische Annalen 1910;69:331–371.

[15] Chen CF, Hsiao CH, Haar wavelet method for solving lumped and distributed-parameter systems. IEE Proc Contr Theor Appl 1997;144(1):87–94.

[16] Hsiao CH, State analysis of the linear time delayed systems via Haar wavelets. Math Comput Simulat 1997;44 (5):457–70.

[17] Lepik Ü. Numerical solution of differential equations using Haar wavelets. Math Comput Simulat 2005; 68:127–43.

[18] Lepik Ü, Haar wavelet method for nonlinear integro-differential equations. Appl Math Comput 2006;176:324–33.

[19] Lepik Ü, Solving PDEs with the aid of two dimensional Haar wavelets. Comput Math Appl 2011;61:1873–1879.

[20] Lepik Ü, Application of the Haar wavelet transform to solving integral and differential Equations. Proc Estonian Acad Sci Phys Math 2007;56 (1):28–46.

[21] Lepik Ü, Numerical solution of evolution equations by the Haar wavelet method. Appl Math Comput 2007;185: 695–704.

[22] Lepik Ü, Solving fractional integral equations by the Haar wavelet method. Appl Math Comput, 2009;214(2):468–78.

[23] Lepik Ü, Hein H, Haar wavelets: with applications. New York: Springer; 2014. 207 p.

[24] Islam SU, Aziz I, Al-Fhaid AS, An improved method based on Haar wavelets for numerical solution of nonlinear integral and integro-differential equations of first and higher orders. J Comput Appl Math 2014;260: 449–69.

[25] Aziz I, Islam SU, Khana F, A new method based on Haar wavelet for the numerical solution of two-dimensional non-linear integral equations. J Comput Appl Math 2014;272:70–80.

[26] Aziz I, Islam SU, New algorithms for the numerical solution of nonlinear Fredholm and Volterra integral equations using Haar wavelets. J Comput Appl Math 2013;239(1):333–45.

[27] Erfanian M, Mansoori A, Solving the nonlinear integro-differential equation in complex plane with rationalized Haar wavelet. Mathematics and Computers in Simulation, 2019;165:223–237.

[28] Majak J, Pohlak M, Eerme M, Application of the Haar Wavelet-based discretization technique to problems of orthotropic plates and shells. Mech Compos Mater 2009;45(6):631–42.

[29] Hein H, Feklistova L, Computationally efficient delamination detection in composite beams using Haar wavelets. Mech Syst Signal Pr 2011; 25(6): 2257–70.

[30] Xie X, Jin G, Ye T, Liu Z. Free vibration analysis of functionally graded conical shells and annular plates using the Haar wavelet method. Appl Acoust 2014;85:130–42.

[31] Xie X, Jin G, Li W, Liu Z, A numerical solution for vibration analysis of composite laminated conical, cylindrical shell and annular plate structures. Compos Struct 2014;111: 20–30.

[32] Xie X, Jin G, Yan Y, Shi SX, Liu Z, Free vibration analysis of composite laminated cylindrical shells using the Haar wavelet method, Compos Struct 2014;109:169–77.

[33] Fan J, Huang J, Haar wavelet method for nonlinear vibration of functionally graded CNT-reinforced composite beams resting on nonlinear elastic foundations in thermal environment. Shock and Vibration, 2018; 1–16. https://doi.org/10.1155/2018/9597541.

[34] Li Y, Zhao W, Haar wavelet operational matrix of fractional order integration and its applications in

solving the fractional order differential equations, Applied Mathematics and Computation 2010;216:2276–2285.

[35] Yi M, Huang J, Wavelet operational matrix method for solving fractional differential equations with variable coefficients. Applied Mathematics and Computation 2014;230:383–394.

[36] Chen Y, Yi M, Yu C, Error analysis for numerical solution of fractional differential equation by Haar wavelets method. Journal of Computational Science 2012;3:367–373.

[37] Wang L, Ma Y, Meng Z, Haar wavelet method for solving fractional partial differentialequations numerically. Mathematics and Computation 2014;227:66–76.

[38] Saeed S Rehman M, Haar wavelet Picard method for fractional nonlinear partialdifferential equations. Applied MathematicsandComputation, 2015; 264: 310–322.

[39] Amin R, Shah K, Asif M, Khan I, Ullah F, An efficient algorithm for numerical solution of fractional integro-differential equations via Haar wavelet. Journal of Computational and Applied Mathematics, 2021;381:113028.

[40] Abdeljawad T, Amin R, Shah K, Al-Mdallal Q, Jarad F, Efficient sustainable algorithm for numerical solutions of systems of fractional order differential equations by Haar wavelet collocation method. Alexandria Engineering Journal, 2020;59(4):2391–2400.

[41] Xie J, Wang T, Ren Z, Zhang J, Quan L, Haar wavelet method for approximating the solution of a coupled system of fractional-order integral–differential equations. Mathematics and Computers in Simulation, 2019;163:80–89.

[42] Majak J, Shvartsman B, Pohlak M, Karjust K, Eerme M, Tungel E. Solution of fractional order differential equation by the Haar Wavelet method. Numerical convergence analysis for most commonly used approach. AIP Conference Proceedings,2016;1738: 480110. DOI: 10.1063/1.4952346.

[43] Lepik Ü, Solving integral and differential equations by the aid of non-uniform Haar wavelets. Applied Mathematics and Computation, 20088; 198(1): 326–332.

[44] Hein H, Feklistova L, Free vibrations of non-uniform and axially functionally graded beams using Haar wavelets. Engineering Structures, 2011; 33 (12):3696–3701.

[45] Raza A, and Khan A, Non-uniform Haar Wavelet Method for Solving Singularly Perturbed Differential Difference Equations of Neuronal Variability. Applications and Applied Mathematics, 2020;6:56–70.

[46] Hariharan G, Kannan K, Review of wavelet methods for the solution of reaction-diffusion problems in science and. Engineering. Appl Math Model 2014;38(3):799–813.

[47] Majak J, Shvartsman BS, Kirs M, Pohlak M, Herranen H, Convergence theorem for the Haar wavelet based discretization method. Composite Structures, 2015;126:227–232.

[48] Majak J, Shvartsman BS, Karjust K, Mikola M, Haavajõe A, Pohlak M, On the accuracy of the Haar wavelet discretization method. Composites Part B: Engineering, 2015;80:321–327.

[49] Kirs M, Karjust K, Aziz I, Ounapuu E, Tungel E, Free vibration analysis of a functionally graded material beam: evaluation of the Haar wavelet method. Proceedings of the Estonian Academy of Sciences, 2018;67 (1):1–9.

[50] Majak J, Pohlak M, Karjust K, Eerme M, Kurnitski J, Shvartsman BS,

New higher order Haar wavelet method: Application to FGM structures. Composite Structures, 2018; 201:72–78.

[51] Majak J, Pohlak M, Eerme M, Shvartsman B, Solving ordinary differential equations with higher order Haar wavelet method. AIP Conference Proceedings, 2019;2116:330002.

[52] Majak J, Shvartsman B, Ratas M, Bassir D, Pohlak M, Karjust K, Eerme M, Higher- order Haar wavelet method for vibration analysis of nanobeams, Materials Today Communications, 2020:25:101290.

[53] Ratas M, Salupere A, Application of Higher Order Haar Wavelet Method for Solving Nonlinear Evolution Equations. Mathematical Modelling and Analysis, 2020;25 (2):271–288. DOI: 10.3846/mma.2020.11112.

[54] Shvartsman B, Majak J, Numerical method for stability analysis of functionally graded beams on elastic foundation. Applied Mathematical Modelling, 2016;40 (4–5):3713–3719.

[55] Ratas M, Application of Haar Wavelet Method for Solving Nonlinear Evolution Equations. AIP Conference Proceedings, 2019;2116:330004. DOI: 10.1063/1.5114342.

Chapter 17

COVID-19 Outbreak and Co-Movement of Global Markets: Insight from Dynamic Wavelet Correlation Analysis

Maurice Omane-Adjepong, Imhotep Paul Alagidede and John Bosco Dramani

Abstract

The COVID-19 pandemic has in its short existence caused economic downturn and affected global markets. As would be expected, the occurrences of global crises or shocks often heighten uncertainties in international markets and increase correlations among them. Yet, not much is known of the actual impacts of COVID-19 on the behavior of global markets. This piece attempts to investigate whether the COVID-19 crisis has had any impact on the interrelationship structure of international markets using the cross-wavelet squared coherence and a dynamic wavelet correlation technique. It emerges that co-movements of the pairwise series become stronger (0.70–0.89) during the heightened periods labeled as epidemic and pandemic phases of COVID-19, than that of the periods that mark the pre-COVID-19 era (−0.49–0.36), hence announcing the influence of the crisis and eroding prospect of benefiting from a hedge instrument and/or a diversifier. Again, we observe that stock market-Global REITs have been the most influenced pair, showing significantly peaked co-movements (0.63–0.87) during the distinct phases of COVID-19. We attribute these developments to the loose monetary and financial measures implemented by central banks of the world. The findings hold important implications for economic and financial actors regarding diversification, hedging, and investment risk management.

Keywords: global markets, COVID-19 outbreak, co-movement, RWWC, portfolio diversification

JEL classification: C22, G15

1. Introduction

The COVID-19 pandemic has triggered untold uncertainties in most global financial and commodity markets. In March 2020, stock price fell intensely, mortgage-backed securities and yield spread of corporate bonds surged significantly, and U.S. Treasury bonds which usually serve as a safe-haven, plunged [1, 2]. Specifically, the behavior of global stock market appears illogical in presence of the pandemic to many investors. For instance, S&P500, one of the mega-size stock markets experienced three phases of changes as the rate of COVID-19 infection worsens. In phase one, it recorded a high value on 19th February 2020, prior to the declaration of the outbreak as a pandemic by the World Health Organization (WHO). However, a surge in the spread of the virus, exacerbated by soaring death rates, caused panic and created a colossal urgency to accumulate cash balances, sparking a concurrent selloff in stocks [3]. Thus, in phase two, the S&P500 plummeted by 34% reaching its low on March 23, 2020. In the third phase, S&P500 rose by 30% on April 30, 2020, despite the lockdown orders initiated by many countries to curb the spread of the virus. In hindsight, stock markets have performed well, generally, because on the eve of the COVID-19 outbreak been pronounced a pandemic, the ratio of market capitalization to GDP was higher compared to its level in 2007, and a little higher than the maximum value during the dot-com bubble [4]. Analysts attribute the rebound of the stock market partly to various loose monetary policy and other interventions pursued by central banks [4], which instilled confidence into shareholders, lightening fears of the health crisis. Besides, country-specific characteristics such as structural economic fragility and "at-risk" population, also seem to have had little effect on stock market reactions to the pandemic [5].

As the pandemic intensifies, bond trading has also encountered challenges, regardless of the asset's significance and essential role in the financial market system. In the face of the pandemic, the observed behavior of the bond market is similar to that of the global stock market. For instance, the U.S. Treasury bond and Canadian government bond markets, in phase one, witnessed an increase in demand for liquidity as investors embarked on a significant selloff [2]. In phase two, dealers curtailed the supply of liquidity, which deteriorated trading conditions. In the third phase, demand gradually reduced, due to some interventions introduced by the respective central banks of these giant economies.

In another development, commodity markets such as crude oil and the real estate markets equally experienced volatilities as the COVID-19 pandemic unfolds. On 20th April 2020, a barrel of West Texas Intermediate crude oil to be delivered in May recorded a negative price, implying sellers had to pay buyers [6]. Though the price for June also fell over a quarter on 27th April, it however remained a little above $12 a barrel. Crude oil market analysts attribute these fluctuations to a price-war between Russia and Saudi Arabia, which they claimed flooded the international market with crude oil and a slump in demand due to traveling and aviation restrictions imposed by countries following the pandemic [7].

A convergence of the uncertainties in the above global markets triggered an immediate deterioration of business environment with unintended negative consequences on commercial real estate markets. Demand for lease space slumped and continue to deteriorate as the pandemic unfolds due to the effects of social distancing and business closures across the globe. However, the impact appears to vary extensively across the real estate sectors. Whiles some sectors are severely and directly affected by the pandemic, others are less and indirectly affected. The performance of the stock market for Real Estate and Investment Trusts (REITs) reflects the differences in the degree of the uncertainties across different types of properties [8].

Though the individual global markets' responses to the COVID-19 pandemic may seem somehow similar, albeit yet to be determined actual worldwide impact quantification, there is a strong likelihood of a potential lead–lag co-movement among the markets, which may be induced or heightened by major news of the health crisis. There is an astronomical increasing number of empirical studies towards the reactions of individual global markets to the COVID-19 crisis. In a short epistle, Krugman [4] disclosed that the "*stock market is not the economy*", and that "*connection between stock growth and the expansion of the real sector of the economy lies within loose and nonexistence*". Capelle-Blancard and Desroziers [5] confirmed various interventions by governments, as well as the believe that loose monetary policy and lockdown initiatives stimulated the rebound of stocks, they further agreed with Krugman [4] that there exists a loose relationship between market fundamentals and stock market uncertainties. Jefferson [7] on the other hand established that projections of future crude oil prices are uncertain, however, in the absence of supply-side shocks, oil prices are likely to rebound by the end of the third quarter of 2020. Goodell and Goutte [9] investigate co-movement of Bitcoin with levels of COVID-19 fatalities and show that the levels of COVID-19 deaths cause a rise in Bitcoin prices. However, the analyses from the previous studies have failed to examine coherences and lead–lag behavior among conventional global markets as they react to the COVID-19 crisis. Again, to the best of our knowledge, the existing studies have not analyzed the interrelationship structure and reactions of global markets to distinctive stages of the COVID-19 outbreak.

In response to the identified gaps, this study attempts to offer fresh insights as to whether major news items of COVID-19 influence the interdependence structure of international markets. We contribute to the existing literature in two-fold. First, we explore the degree of co-movement and lead–lag relationship among aggregate global stock index, commodities, and the REITs market using the cross-wavelet squared coherence and a rolling-window wavelet correlation (RWWC) technique. In addition to its ability to address issues of nonlinearity, (non-) economic shocks, regime shifts, and non-stationarities, the RWWC approach possesses time-varying attributes that makes it possible to measure the temporal variations of cross-market correlations over time and frequency domains [10], with implications for heterogeneous market actors. Second, we examine the influence of the COVID-19's epidemic and pandemic stages on global market interrelatedness and determine whether international investors can hold positions in the markets to offset short-run investment losses during the crisis.

The remaining structure of the study is set as follows. Section 2 provides a review of the extant literature. Section 3 discusses the econometric techniques employed. Section 4 presents the results, while Section 5 provides conclusion and policy recommendations.

2. Literature review

Ramelli and Wagner [11] found a significant effect of world trade and global value chain on the value of corporations. Corporate bodies appeared profitable depending on the location of the epicenter of the pandemic. For instance, stocks of corporate bodies in China initially appeared risky as the pandemic unfolded while those in Europe were considered profitable. However, as the epicenter moved to Europe and America, investors perceived stocks of these regions to be unfavorable, causing the markets to behave feverishly. Investors were equally alarmed about the possibility of corporate bodies incurring high debts as well as the survival potentials of businesses with insufficient cash balances. Though the opportunity cost of holding cash balances appeared high, there was an increasing need to hold precautionary cash to soar the value of firms. Capelle-Blancard and Desroziers [5] revealed that investors in the stock markets were quick to respond to soaring cases of COVID-19, with advanced economies being highly affected. The authors also unearth that loose monetary and fiscal policies introduced by central banks and governments caused interest rates to fall, which moderated the fall in stock prices, making the market less responsive to the crisis. Baker et al. [12] argued that the stock market responded to the pandemic in a way that has never happened in history. The authors attributed this behavior to a host of factors such as governments' restraints on economic activity and the introduction of social distancing measures in the presence of a service dominated economy. Ehrmann and Jansen [13] revealed the presence of significant co-movement between stock returns and national stock markets. Investors placed a high value on global news and the effects were moderated by large oscillations in global stock markets. However, investors turned to place less value on firm-specific news, which caused stock returns and national stocks market to move together. The authors observed that, this relationship was significant for stocks that are characterized by low co-movement with national market, resulting in a convergence of beta across stocks.

Haddad et al. [14] in studying the disruptions in debt markets as the pandemic unfolded made interesting revelations. The authors revealed that bonds that are usually regarded as safe havens recorded weighty losses which analysts found difficult to explain applying risk premium channel or standard default. Corporate bonds traded at a huge discount to their equivalent credit default swaps and this became worse for many safer bonds. Similarly, liquid bond exchange-traded funds witnessed a huge discount to their corresponding net asset value. These findings imply traders attempted to sell safer and high liquid securities to increase cash balances. However, these disruptions did not see the light of day as the market recovered in a matter of weeks. The authors attributed the fast recovery of the bonds market to the unparalleled measures the Fed introduced by purchasing corporate bonds instead of extending credit. Fontaine et al. [2] and Kargar et al. [15] found that the market for bonds evolved in three phases as the pandemic worsens, using two-year benchmark bonds for Canada. The first phase witnessed a sharp rise in the demand for cash balances, which traders did well to accommodate but at a higher cost. The second phase experienced a massive decrease in the supply of cash balances by dealers, leading to a huge deficit as demand for cash kept soaring. The third and final phase saw trading activity and price of cash balance stabilizing due to the interventions by central banks to assist the financial sector.

Regarding the crude oil market, the Arezki et al. [16] pointed out that net oil-exporting economies face a dual shock emanating from the health crisis and a fall in prices of crude oil. However, the shock from the pandemic turned to lead and influence the collapse in oil prices. This manifested itself through the traveling restrictions placed on the aviation industry, self-isolation, and social distancing and complete lockdowns measures introduced by governments around the globe [7]. Elsewhere, Barbosa et al. [17] revealed that the dual shock on net exporting countries negatively affected the financial and structural health of the oil sector in an unprecedented manner. Since the intensity and length of the health crisis are uncertain, the authors suggested net exporting countries should introduce fundamental intervention to reverse the trend to make the industry profitable again.

In the Real Estate and Investment Trusts market, Schnure [8] observed that the social distancing and lockdown measures caused almost all businesses in the global economy to shut down. There is a high probability that most of these businesses may find it very difficult to honor their rent in the near future, which will affect negatively on cash flow of property owners. Again, the authors explained that hikes in unemployment would cause unspeakably high rent default by households. Coibion et al. [18] found persistence in low inflation, heightened uncertainty, and lower mortgage rates as the pandemic worsened. The authors attributed this low consumer spending and collapse in demand for office space as employees work from home.

It is important to mention that the existing literature is yet to explain whether COVID-19 has influenced the interdependence structure of global markets. Besides, the influence of the crisis at the onset stage and its transition into a pandemic has rarely been explored. Thus, applying time-frequency estimation methods, this paper augments the literature by assessing the extent of co-movement and the direction of linkages (lead–lag relationship) among selected major global market, amid COVID-19 major news. Our study provides international investors with further insight as they seek to diversify their investment portfolios by purchasing securities that do not or less co-move to minimize losses under the heightened market periods of COVID-19.

3. Data and methods

The data for the empirical analysis and the econometric approach for the multidimensional dynamic correlation measure is explained in this section. As preliminary to the main analysis, static descriptive measures, and a correlation matrix based on a global measure, computed under different time samples are provided.

3.1 Data

Daily price levels of MSCI All Country World Index (MSCI ACWI), S&P GSCI Energy Index, S&P GSCI Non-Energy Index, and S&P Global Real Estate and Investment Trusts (SREITGUP), which spans from January 01, 2016, to August 17, 2020 (giving us 1165 realizations after cleaning and synchronization the timestamps) are considered and used as proxy for global markets[1]. The US dollar denominated price level datasets are sourced from the Bloomberg database terminal. The daily prices are converted to percentage log changes: $r_{i,t} = In(P_{i,t}/P_{i,t-1}) \times 100$, for $i = 1, \ldots, 4$ and $t = 2, \ldots, 1165$, where $P_{i,t}$ and $P_{i,t-1}$ denote the close of day prices for global market i at day t and $t-1$ respectively. The time sample is foremost divided into pre-COVID-19 crisis (January 01, 2016, to December 07, 2020) and periods marking the onset and duration of COVID-19 (December 08, 2019, to August 17, 2020).

To establish the dynamic impact of the COVID-19 outbreak on the interrelationship structure of international markets and its implication thereon, the crisis period is further separated into epidemic (December 08, 2020, to March 10, 2020) and pandemic (March 11, 2020, to August 17, 2020) stages. The dating of these distinct periods is based on the announcements or timelines of the first patient who was reported to have developed symptoms of the Wuhan coronavirus (on December 08, 2019) and the subsequent declaration of the outbreak by WHO as a pandemic (on March 11, 2020)[2].

Table 1 shows summary moment measures for the global markets during the crisis and non-crisis periods. Compared to the non-crisis, the average daily returns fairly decrease under the crisis period, with all markets recording low negative skewness values, indicative of a high tendency of reaping non-positive investment returns. Noticeably, the markets appear to be more volatile as we transition from

---

[1] International bonds, bills, and the currency markets are beyond our scope, perhaps, they could be considered in future studies.

[2] For the applicable COVID-19 timelines, refer to: https://www.weforum.org/agenda/2020/04/coronavirus-spread-covid19-pandemic-timeline-milestones/ (Accessed on July 29, 2020).

	Period I: Pre-COVID-19 crisis				Period II: COVID-19 crisis			
	Mean	**Std Dev**	**Skew**	**Kurt**	**Mean**	**Std Dev**	**Skew**	**Kurt**
MSCI ACWI	0.0359	0.6733	−0.7606	4.6379	0.0131	2.0655	−1.1728	7.2169
GSCI Energy	0.0486	1.8979	0.1638	3.2746	−0.2060	5.1673	−1.3398	9.8393
GSCI Non-Enr	0.0123	0.5844	−0.1489	0.7892	0.0019	0.7723	−0.7365	2.2106
SREITGUP	0.0137	0.6934	−0.4641	1.8341	−0.1105	2.7274	−1.6971	9.5034

Table 1.
Descriptive measures.

	Period I: Pre-COVID-19 crisis				Period II: COVID-19 crisis			
	(a)	**(b)**	**(c)**	**(d)**	**(a)**	**(b)**	**(c)**	**(d)**
(a) MSCI ACWI	1				1			
(b) GSCI Energy	0.3668	1			0.4671	1		
(c) GSCI Non-Enr	0.2921	0.2511	1		0.5813	0.4340	1	
(d) SREITGUP	0.5598	0.1608	0.1450	1	0.8962	0.3764	0.5599	1

Table 2.
Pearson correlation matrix.

the pre-crisis to the crisis period. For instance, the volatility measures for MSCI ACWI, S&P GSCI Energy, and the Global REITs (SREITGUP) increased almost three-fold between the two periods. Meaning, the COVID-19 crisis has ushered in periods of heightened uncertainty and created a high financial risk environment. Besides, the return distributions of the markets exhibit relatively more leptokurtic features under the crisis, hence giving rise to extreme return realizations.

Results of the Pearson's unconditional correlations for the market pairs are presented in **Table 2**. Except for MSCI ACWI-Global REITs (with correlation measure of 0.5598), low positive correlations (ranging from 0.1450 to 0.3668) characterizes the pairs before the coming into being of the crisis. However, all the pairwise correlations increased in magnitude from low positive to moderately low positive values (0.3764 to 0.5813), with a peaked measure of 0.8962 for MSCI ACWI-Global REITs. These static-based measures signal the influence of the COVID-19 crisis on cross-market relationship, which seems to be in line with the literature that suggests that financial markets tend to move closely together (i.e., increased correlation or co-movement intensity) during turmoil or crisis. Yet, estimations that are more robust are undertaken to validate these early detections.

3.2 RWWC methodology

To examine the interrelationship structure between the daily percentage log changes of MSCI ACWI, S&P GSCI Energy Index, S&P GSCI Non-Energy Index, and S&P Global REITs Index across time and frequency, we employed a dynamic correlation version of the wavelet correlation approach by Gençay et al. [19]. The rolling-window wavelet correlation (RWWC) method, which was introduced by Ranta [20, 21] has recently gained traction in the economics and financial literature, perhaps due to its ability to unearth the temporal variations of the wavelet correlation for distinct time series, by incorporating a dynamic measure under a multidimensional setting. Using the Maximal Overlap Discrete Wavelet

Transformation (simply, MODWT) methodology (see, [22–24]; etc.) and following Gençay et al. [19], we express the MODWT-based unbiased estimator of the wavelet correlation for pairs of the market series, X_t and Y_t for scale λ_j as:

$$\tilde{\rho}_{XY} = \frac{\operatorname{cov}\left(\tilde{W}_{X,j,t}, \tilde{W}_{Y,j,t}\right)}{\left(\operatorname{var}\left(\tilde{W}_{X,j,t}\right)\operatorname{var}\left(\tilde{W}_{Y,j,t}\right)\right)^{1/2}} = \frac{\tilde{\gamma}_{XY}\left(\lambda_j\right)}{\tilde{\sigma}_X\left(\lambda_j\right)\tilde{\sigma}_Y\left(\lambda_j\right)} \tag{1}$$

where, $\tilde{\gamma}_{XY}\left(\lambda_j\right)$ represent the unbiased estimator of the wavelet covariance for the wavelet constituents $\tilde{W}_{X,j,t}$ and $\tilde{W}_{Y,j,t}$ involving the pair of distinct series, and $\tilde{\sigma_X^2}\left(\lambda_j\right)$ and $\tilde{\sigma_Y^2}\left(\lambda_j\right)$ denote unbiased measures of the wavelet variances for X and Y at scale λ_j.

We specify the MODWT-based unbiased estimator of the wavelet variance as:

$$\tilde{\sigma}_X^2\left(\lambda_j\right) = \frac{1}{\tilde{N}_j}\sum_{t=L_j-1}^{N-1} \tilde{W}_{j,t}^2 \tag{2}$$

where, $\tilde{w}_{j,t}$ represent the jth-level MODWT wavelet constituents for market variable X, $L_j = (2^j - 1)(L-1) + 1$ give the length of the scale λ_j wavelet filter, and $\tilde{N}_j = N - L_j + 1$ present the number of wavelet constituents unaffected by the boundary. Next, an expression for the computation of a random interval that captures non-spurious wavelet correlations [21] and offers an approximate $100(1-2p)\%$ confidence interval is deduced from the extended work of Whitcher et al. [25]:

$$\left[\tanh\left\{h\left[\tilde{\rho}_{XY}\left(\lambda_j\right)\right] - \frac{\Phi^{-1}(1-p)}{\left(\tilde{N}_j-3\right)^{1/2}}\right\}, \tanh\left\{h\left[\tilde{\rho}_{XY}\left(\lambda_j\right)\right] + \frac{\Phi^{-1}(1-p)}{\left(\tilde{N}_j-3\right)^{1/2}}\right\}\right] \tag{3}$$

where, $h(\tilde{\rho}_{XY}) = \tanh^{-1}(\tilde{\rho}_{XY})$ describes Fisher's z-transformation and $\tilde{N}_j$ remains the number of wavelet constituents that correspond to a particular scale.

To provide short-term investment solutions for traders during the ongoing COVID-19 crisis, we considered to perform our analysis at the lower frequency bands or investment horizons, thus, the $2 \sim 4$ day band (D1: intraweek), $4 \sim 8$ day band (D2: weekly), and the $8 \sim 16$ day band (D3: fortnight), which are associated respectively with scales λ_j of the MODWT time-scale decomposition, where $j = 1, \ldots, 3$. We computed the RWWC using a 100-day window size, which we rolled forward one day (or datapoint) at a time, and centred around time t. With a sample size of 1165, we obtained a total of (1165 minus 100 = 1065) windows. We later varied the window size to 120 (or half-year) to verify the sensitivity of the results to the choice of window length. The two window sizes, $w = 100$ and $w = 120$ truncated on June 04, 2020, and May 20, 2020, respectively, giving us relatively less information losses compared to using higher window sizes. On the other hand, selecting very low window lengths plays down on the power of the test, therefore our choices are not misplaced.

4. Empirical results

We begin the main analysis with results from the cross-wavelet squared coherence method of Grinsted et al. [26]. Under this well-known technique, the estimator of interdependence is rooted on a continuous wavelet transform (see, [9, 21, 27]; etc., for detailed explanation) rather than the discrete wavelet transform. Besides, unlike the RWWC, the wavelet coherence does not incorporate a dynamic measure in its computation. Therefore, taken as a precursor to the RWWC analysis, we initially gleaned the direction and nature of the interdependence structure that emerges for the market pairs using the wavelet coherence plots shown in *Appendix B*. A general observation of *Appendix B* reveals that the arrows are mostly pointing to the right, implying the series are positively correlated, with patches of varying significant coherences predating the year 2020. Observably, the post-2020 period recorded a high degree of coherences for pairs involving stock market-Global REITs and stock market-energy commodities, which witnessed long stretches of white contours over the frequency bands. These detected peaks in coherences conform to the contagion effect literature that projects high co-movements (or increased correlation intensity) for financial markets during and/or after the occurrences of turmoils or major crises. This finding is in line with the findings of Polanco-Martínez et al. [24] who reported strong correlation (0.56–0.87) among global financial markets during episodes of heightened economic crises, particularly during the 2008 financial meltdown.

Specifically, in *Appendix B*, we first focus on the behavior of the global stock market and energy commodity pair. As noted, a significantly strong positive correlation could be gleaned between the 16 and 128 trading day bands, with purloins of co-movement within the intraweek to fortnight trading frequencies. The strongest level of coherences falls within March and May 2020, where the equity market clearly leads energy commodity. This post-2020 co-movement pattern could perhaps be attributed to the lockdown measures, which brought global transportation to a temporary halt, hence the lagging of global energy commodity in its interrelationship with stock market. This finding confirms those of Nguyen et al. [28] who established strong evidence of co-movement (ranging from 0.62–0.89) between stock and energy markets at the peak of the global financial crisis 2007–2009. Similar visibly strong coherences, which intensifies at the beginning of 2020 and beyond, conspicuously grows from the intraweek to half-year trading day bands for the stock market and Global REITs pair. It is also important to stress that the coherences between stock market-Global REITs stretch over longer periods than what could be witnessed for the other global market pairs. These zones of strong correlations are suggestive of contagion impact resulting from the imposition of measures, implemented by world governments and their central banks to avert meltdown of the global financial markets.

With respect to stock market and non-energy commodity in *Appendix B*, we notice isles of significantly segmented zones of coherences, which mark periods, before, during, and after the outbreak of COVID-19. Coherences between energy and non-energy commodities appear generally weak, with few moderately low correlations concentrated between frequency bands of 8–32, and fairly distributed across time. Similarly, besides the 32–128 trading day bands of the opening months of 2020, coherences between energy commodity and global REITs are equally weak. Finally, we observe patches of moderately low significant co-movements between non-energy commodity and Global REITs, which appears mounded within the medium-to-long-run frequencies (16–128) with weak coherences below the fortnight band, coupled with a nonhomogeneous lead–lag relationship.

Our RWWC analysis in **Figure 1**, drawn from a dynamic version of the discrete wavelet transform is initially estimated using a 100-day window length using Eq. (3). The estimations from the dynamic approach reveal thought-provoking findings that may be hardly discernible with static or global measures. The

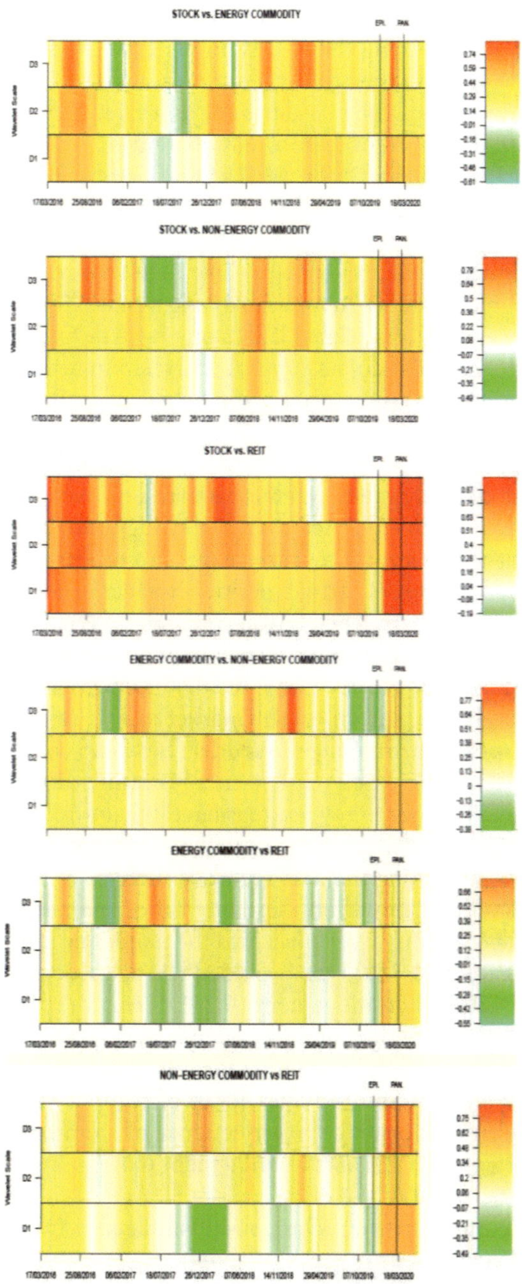

Figure 1.
A 100-day RWWC for pairs of selected global markets. Note: the strength of the dynamic wavelet correlation for the pairs is displayed by the heat map colors, which ranges from weak to high (thus, from blue, cyan, green, white, yellow, orange, to red respectively), where red (blue) denotes highest (lowest) wavelet correlation coefficients within a 95% confidence interval (refer to the web version of the article for color representation).

horizontal axis of **Figure 1** depicts timelines (or time intervals) and the vertical axis represents frequency bands or investment horizons categorized into D1 (intraweek), D2 (weekly), and D3 (fortnight)[3]. A glance from the RWWC results in **Figure 1** shows that the market pairs are predominantly characterized by weak to

[3] For recent applications, readers may refer the works by Polanco-Martíneza et al., (2018); Omane-Adjepong and Alagidede [10]; Nguyen et al., (2020); etc.

moderately low positive correlations, with few abrupt zones of inverse correlations that completely disappear during the uncertain periods ushered in by the global outbreak of COVID-19. This latter finding signals a negligible or unlikely opportunity of benefiting from any of the market assets as a safety net tool or instrument to hedge against short-term losses of an international investment portfolio. Generally, except for stock market and the Global REITs pair, we observe fairly low but steadily increasing correlations for the markets over the frequency bands before the onset of COVID-19.

Particularly in **Figure 1**, we notice vast yellow to less warm orange regions with scores of white and green patches before the dating of the COVID-19 crisis for markets pairs of non-energy and Global REITs on one side, and energy versus Global REITs at the other side. Similar colors from the correlation heatmap could also be somehow advance to describe interrelationship behavior for the pairs involving energy and non-energy commodities, as well as stock market and (non-) energy commodity. These pre-COVID-19 co-movement patterns mark an era that is generally dominated by moderately fewer interactive markets, except for the stock market and Global REITs pair, which exhibited moderately high interactions. Besides, the latter market pair witnessed strong co-movement across the trading frequencies in the second to the third quarter of 2016, a period which coincides with the UK's referendum on June 23, 2016, to leave European Union. Perhaps, the uncertainties induced by the referendum accounted for such high cross-market interactions[4].

As viewed from **Figure 1**, the onset and the distinct phases of COVID-19 has had cause to alter the correlation patterns of the market pairs. For instance, the RWWC measure becomes strong (warm orange to reddish heatmap colors) during the epidemic phase of COVID-19 for stock market-energy commodity and energy commodity-Global REITs, only to reverse to moderately low correlations, as observed under the pre-COVID-19 period, thereby signaling a temporal effect of the crisis on the market pairs. The remaining markets recorded increasing co-movements in the early period of the epidemic, however, these intensifies and peaks in the latter part of the epidemic, and subsequently overflows into the pandemic period. Our finding confirms that of Samadi et al. [29] who provided strong evidence to the effect that energy market exhibited low co-movement (0.36) during the pre-Ccovid-19 episode, which later heightened (0.88) during the pandemic era. The results presuppose that COVID-19 has exerted varying influences on the relatedness of global markets, and as a result decreased, to a large extent, the tendency of reaping diversification gains. These strong co-movements could be a consequence of looming heightened financial instabilities, compelling central banks to implement loose monetary and financial measures to curtail the effect of the crisis.

From the above context, it would be non-advantageous to hold a position in pairs of these global markets during the ongoing crisis, more importantly, for the stock market and Global REITs. As the evidence suggests, the latter pair react more to global shocks or crises, perhaps, owing to the reason that both markets are subjected to similar circumstances, hence, in periods where stock prices plummet, REITs are not overly immune to the perils of falling stock prices.

To ensure the robustness of our results we conduct a sensitivity test by increasing the rolling-window size to 120-days and report the results in *Appendix A*, estimated using Eq. (3). Conspicuously, we observe similar trends for the market pairs, and therefore conclude that our results are invariant with the size of the rolling-window.

---

[4] It is imperative knowing that the weight composition of MSCI ACWI and SREITGUP captures several developed markets in Europe.

5. Conclusions

This study investigates the behavior of global markets amid the onset and the different stages of COVID-19. We represent the first period as pre-COVID-19 (January 1, 2016, to December 07, 2019) – this period is characterized by high financial market growth and stability; the second period as an epidemic (December 07, 2020, to March 10, 2020), which is considered as relatively high volatility in a specific country (China) or region; and the last as a pandemic stage (March 11, 2020, to August 17, 2020), branded as a period of heightened global markets instability. Four selected international markets, namely MSCI AWCI, S&P GSCI Energy, S&P GSCI Non-energy, and S&P Global REITs are used for the analysis. We accomplished the goal of the study by applying both static and dynamic measures to ascertain the extent to which COVID-19 has influence the interrelationship structure of global markets.

Overall, we detect that COVID-19, through its different stages has generally affected the relatedness patterns of global markets. Thus, co-movements of the markets become stronger during the heightened periods of COVID-19's epidemic and pandemic, and as a result, erodes, to a greater extent, the likelihood of diversification benefits. These increases in co-movement are attributed to the loose monetary and financial measures as well as stringent interventions imposed central banks and government, worldwide, as panic remediations to curtail global economic meltdown.

In conclusion, hitherto the general observation that global markets comove during episodes of heightened crisis, our study provides a strong evidence that these correlations are not just strong during the entire period but stronger at the peak of the crisis or pandemic stage (March 11, 2020, to August 17, 2020, in the case of COVID-19). This situation is likely to vary from country to country due to the degree of aggressive interventions and restrictions introduced by respective central banks and governments. The findings of this study indicate that health crisis (COVID-19) can have important implications for global markets through some transmission channels. It is thus important that policy-makers, through research, begin to identify these important channels and fashion both institutional and regulatory policies to address them. Future research can widen the scope to cover relevant aspects of this study by asking the following questions; what accounts for the differences in global markets or countries with dissimilar reactions to COVID-19 pandemic? Can these differences in reaction across global markets and countries be attributed to different approaches to the conduct of monetary policy or institutional characteristics? Can the differences in responses of global markets be due to approaches adopted by various countries in handling the pandemic? Even though COVID-19 health crisis has been pronounced a global pandemic, its negative impact has not been equally distributed, leading to dissimilar responses across countries. By asking and addressing the above questions using different techniques, researchers will produce findings with a strong heterogeneity.

Acknowledgements

The authors acknowledge the efforts of Mia Vulovic. We are also grateful to the editor and all prospective reviewers. As the usual caveat goes, all views and errors/omissions are that of authors, and do not in any way represent the positions of our institutions.

Conflict of interest

The authors declare no conflict of interest.

A. Appendix A

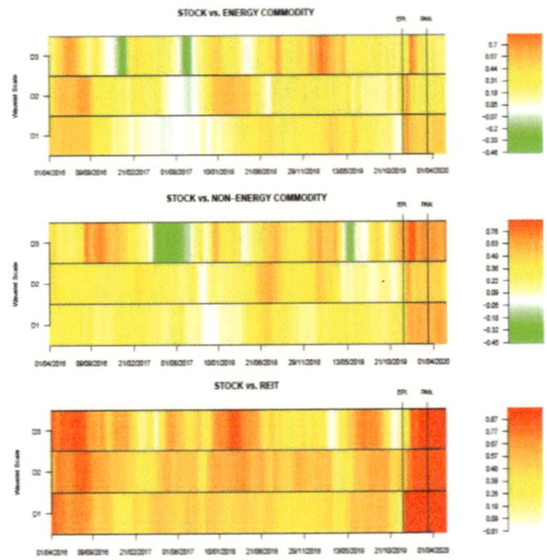

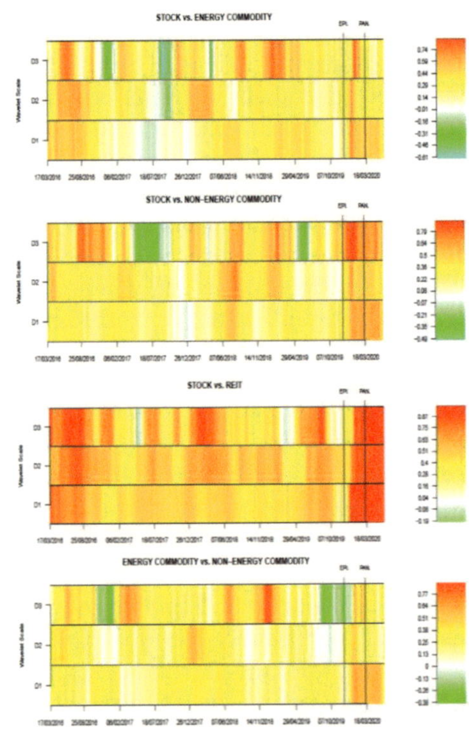

Appendix A: A 120-day RWWC for pairs of selected global markets.

Note: the strength of the dynamic wavelet correlation for the pairs is displayed by the heat map colors, which ranges from weak to high (thus, from blue, cyan, green, white, yellow, orange, to red respectively), where red (blue) denotes highest (lowest) wavelet correlation coefficients within a 95% confidence interval (refer to the web version of the article for color representation).

B. Appendix B

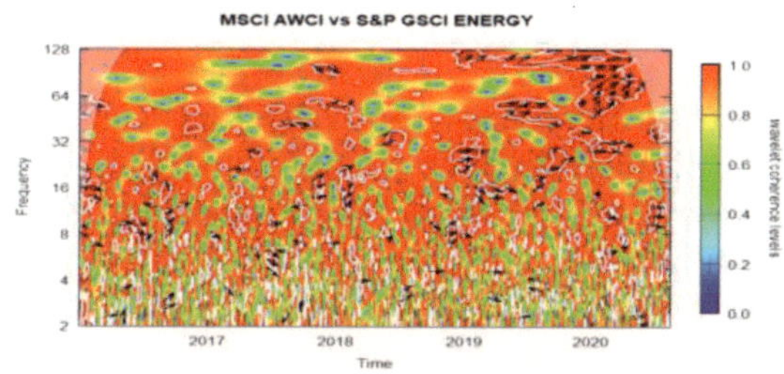
MSCI AWCI vs S&P GSCI ENERGY
Frequency
128
64
32
16
8
4
2
2017
2018
2019
2020
Time
wavelet coherence levels
1.0
0.8
0.6
0.4
0.2
0.0

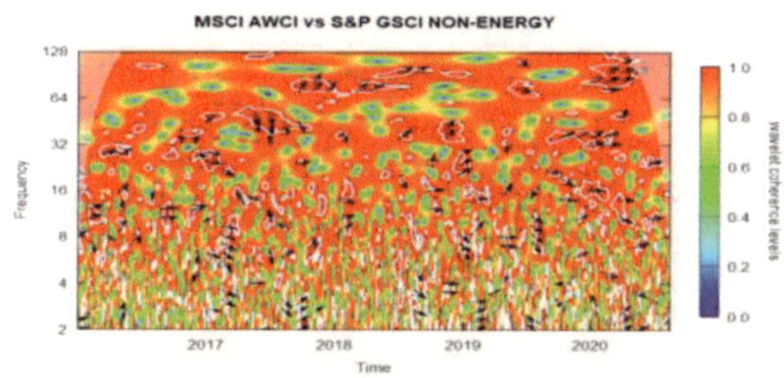
MSCI AWCI vs S&P GSCI NON-ENERGY
Frequency
128
64
32
16
8
4
2
2017
2018
2019
2020
Time
wavelet coherence levels
1.0
0.8
0.6
0.4
0.2
0.0

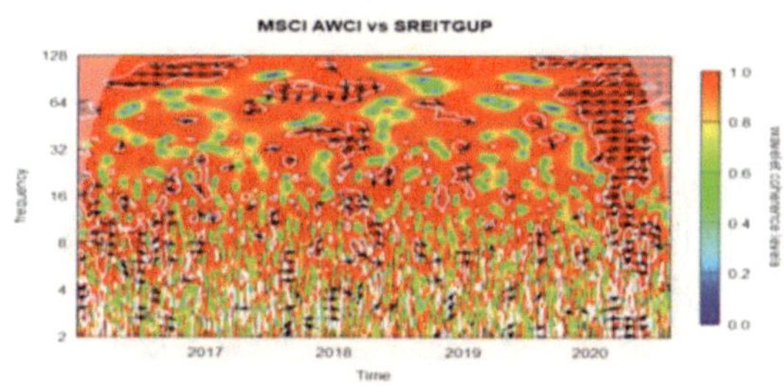
MSCI AWCI vs SREITGUP
frequency
128
64
32
16
8
4
2
2017
2018
2019
2020
Time
wavelet coherence levels
1.0
0.8
0.6
0.4
0.2
0.0

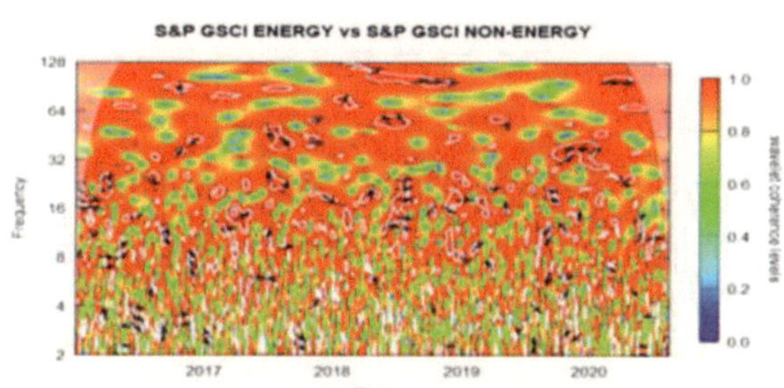
S&P GSCI ENERGY vs S&P GSCI NON-ENERGY
Frequency
128
64
32
16
8
4
2
2017
2018
2019
2020
Time
wavelet coherence levels
1.0
0.8
0.6
0.4
0.2
0.0

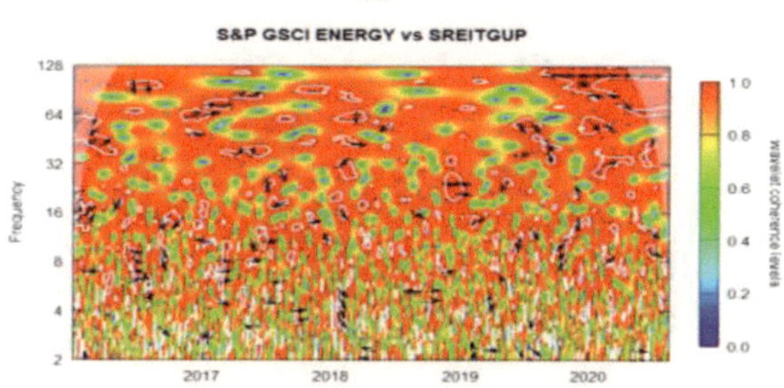
S&P GSCI ENERGY vs SREITGUP
Frequency
128
64
32
16
8
4
2
2017
2018
2019
2020
Time
wavelet coherence levels
1.0
0.8
0.6
0.4
0.2
0.0

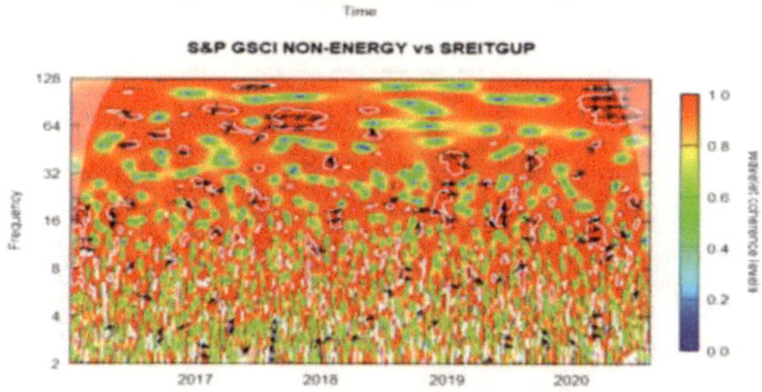
S&P GSCI NON-ENERGY vs SREITGUP
Frequency
128
64
32
16
8
4
2
2017
2018
2019
2020
Time
wavelet coherence levels
1.0
0.8
0.6
0.4
0.2
0.0

Appendix B: Cross-wavelet squared coherence with phase difference for selected pairs of global markets.

Note: In the wavelet coherence plots, the time interval (01.04.2016–2017.08.2020) is represented by the horizontal axis, while the vertical axis gives the frequencies (ranging from a 2-day to 128-day or half-year band). The degree of coherence is described by the heatmap, where warmer greenish to reddish colors denote medium-to-high interrelationship, and the light to deep blue indicates weak to uncorrelated markets. The 5% statistically significant coherence is displayed within the zones bounded by the white contours, and also confined to the "cone of influence" (the bell-shaped region), beyond that, coherence estimates become spurious. The direction of coherence is detected through the phase arrows, where the left and right black arrows denote that the two market series are out-of-phase (negative correlation or opposite movement) and in-phase (positive correlation or same direction of movement) respectively. Down pointing arrows put the first series as a leader; upward arrows mean the second is leading; right and down means the first series is leading; right and up suggest the first is lagging; the first series lags when the arrows point left and down, and leads the second series when the directional arrows point left and up. In all our plots, the first and second series correspond to the positions of the figure caption.

Author details

Maurice Omane-Adjepong[1*], Imhotep Paul Alagidede[1] and John Bosco Dramani[2]

1 Wits Business School, University of the Witwatersrand, 2 St. David's Place, Parktown, Johannesburg, 2193, South Africa

2 Economics Department, Kwame Nkrumah University of Science and Technology, Kumasi, Ghana

*Address all correspondence to: omane.maurice@gcuc.edu.gh

IntechOpen

© 2020 The Author(s). Licensee IntechOpen. This chapter is distributed under the terms of the Creative Commons Attribution License (http://creativecommons.org/licenses/by/3.0), which permits unrestricted use, distribution, and reproduction in any medium, provided the original work is properly cited.

References

[1] Aramonte, S., & Avalos, F., 2020. The recent distress in corporate bond markets: cues from ETFs (No. 6). Bank for International Settlements.

[2] Fontaine, J.S., Ford, H., & Walton, A., 2020. COVID-19 and bond market liquidity: Alert, isolation and recovery. Staff Analytical Note 2020–2014.

[3] Shiller, R., & Malkiel, B., 2020. *Does Covid-19 prove the stock market is inefficient?* Pairagraph. Available online at: https://www.pairagraph.com/dialog ue /c93c449006c344ce94e6e2e8fbe7aba 3. (Accessed August 05, 2020).

[4] Krugman, P., 2020. *Crashing Economy, Rising Stocks: What's Going On?* New York Times. Available online at: https://www.nytimes.com/2020/04/30/ opinion /economy-stock-market-corona virus.html. (Accessed August 30, 2020).

[5] Capelle-Blancard, G., & Desroziers, A., 2020. The stock market is not the economy? Insights from the COVID-19 crisis. Insights from the COVID-19 Crisis. CEPR Covid Economics.

[6] The Economist. 2020. *Oil and commodity prices are where they were 160 years ago*. The Economist Magazine. Available online at: https://www.ec onomist.com/ graphic-detail/2020/04/ 27/oil-and-commodity-prices-are-whe re-they-were-160-years-ago. (Accessed August 15, 2020).

[7] Jefferson, M., 2020. A crude future? COVID-19s challenges for oil demand, supply and prices. Energy Research & Social Science, 68, p.101669. https://doi. org/ 10.1016/j.erss.2020.101669.

[8] Schnure, C., 2020. *The outlook for REITs during the COVID-19 crisis*. REIT Newsletter for Advisors. Available online at: https://www.reit.com/news/b log/market-commentary/outlook-re its-during-covid-19-crisis. (Accessed August 10, 2020).

[9] Goodell, J.W., & Goutte, S., 2020. Co-movement of COVID-19 and bitcoin: Evidence from wavelet coherence analysis. Finance Research Letters, p.101625. https://doi.org/ 10.1016/j.frl.2020.101625.

[10] Omane-Adjepong, M., & Alagidede, I. P., 2020. Dynamic linkages and economic role of leading cryptocurrencies in an emerging market. Asia-Pacific Financial Markets https://doi.org/10.1007/s10690-020-09306-4.

[11] Ramelli, S., & Wagner, A.F., 2020. Feverish stock price reactions to COVID-19. Swiss Finance Institute Research Paper (20–12).

[12] Baker, S.R., Bloom, N., Davis, S.J., Kost, K.J., Sammon, M.C., & Viratyosin, T., 2020. The unprecedented stock market impact of COVID-19 (No. w26945). National Bureau of Economic Research.

[13] Ehrmann, M., & Jansen, D.J., 2020. Stock Return comovement when investors are distracted: More, and more homogeneous. CEPR Discussion Paper No. DP14713.

[14] Haddad, V., Moreira, A. & Muir, T., 2020. When selling becomes viral: Disruptions in debt markets in the covid-19 crisis and the fed's response (No. w27168). National Bureau of Economic Research.

[15] Kargar, M., Lester, B., Lindsay, D., Liu, S., Weill, P.O., & Zúñiga, D., 2020. Corporate bond liquidity during the COVID-19 Crisis. NBER Working Paper No. 27355.

[16] Arezki, R., Fan, R.Y., & Nguyen, H., 2020. Covid-19 and oil price collapse: Coping with a dual shock in the Gulf cooperation council. ERF Policy Brief No. 52.

[17] Barbosa, F., Bresciani, G., Graham, P., Nyquist, S., & Yanosek, K., 2020. Oil and Gas after COVID-19: The Day of Reckoning or a New Age of Opportunity? McKinsey and Company.

[18] Coibion, O., Gorodnichenko, Y., & Weber, M., 2020. The cost of the Covid-19 crisis: lockdowns, macroeconomic expectations, and consumer spending. NBER Working Paper No. 27141.

[19] Gençay, R., Selçuk, F., & Whitcher, B., 2002. An Introduction to Wavelets and Other Filtering Methods in Finance and Economics. Academic Press.

[20] Ranta, M., 2010. Wavelet multiresolution analysis of financial time series. Ph.D Thesis, University of Vaasa, Finland Acta Wasaensia.

[21] Ranta, M., 2013. Contagion among major world markets: A wavelet approach. International Journal of Managerial Finance, 9(2): 133–149. https://doi.org/ 10.1108/ 17439131311307556.

[22] Fernández-Macho, J., 2012. Wavelet multiple correlation and cross-correlation: A multiscale analysis of Eurozone stock markets. Physica A: Statistical Mechanics and its Applications, 391, 1097–1104. https://doi.org/10.1016/j.physa.2011.11.002.

[23] Percival, D.B., & Walden, A.T., 2006. Wavelet methods for Time Series Analysis. Cambridge University Press, 2006.

[24] Polanco-Martínez, J.M., Fernández-Macho, J., Neumann, M.B., & Faria, S. H., 2018. A pre-crisis vs. crisis analysis of peripheral EU stock markets by means of wavelet transform and a nonlinear causality test. Physica A: Statistical Mechanics and its Applications, 490, 1211–1227. http://dx.doi.org/10.1016/ j.physa.2017.08.065.

[25] Whitcher, B., Guttorp, P., & Percival, D.B., 2000. Wavelet analysis of covariance with application to atmospheric time series. Journal of Geophysical Research, 105(D11): 941–962. https://doi.org/10.1029/ 2000JD900110.

[26] Grinsted, A., Moore, J.C., & Jevrejeva, S., 2004. Application of the cross wavelet transform and wavelet coherence to geophysical time series. Nonlinear Processes in Geophysics, 11, 561–566. https://doi.org/10.5194/npg-11-561-2004.

[27] Omane-Adjepong, M., & Dramani, J.B., 2018. On the dynamic effects of global commodities on stock market blocks in Africa. Applied Economics Letters, 25(11): 800–805. https://doi.org/10.1080/13504851.2017.1368978.

[28] Nguyen, T.T.H., Naeem, M.A., Balli, F., Balli, H.O., & Vo, X.V., 2020. Time-frequency comovement among green bonds, stocks, commodities, clean energy, and conventional bonds. *Finance Research Letters*. (In-Press). https://doi.org/10.1016/j.frl.2020.101739.

[29] Samadi, A.H., Owjimehr, S., and Halaf, Z.N., 2020. The Cross-Impact between Financial Markets, Covid-19 Pandemic, and Economic Sanctions: The Case of Iran. *Journal of Policy Modeling*. Article in Press.